Common 1W Zener Diodes

	Voltage		Voltage
1N4733	5.1	1N4744	14
1N4734	5.6	1N4745	15
1N4735	6.2	1N4746	18
1N4736	6.8	1N4747	20
1N4738	8.2	1N4749	24
1N4739	9.1	1N4753	36
1N4740	10	1N4754	39
1N4742	12	1N4756	47
		1N4764	100

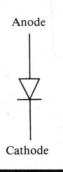

Anode — Cathode

Cathode

Cathode   Anode

Bipolar and JFET Transistors in TO92 Plastic Case

(Japanese manufacturers may use R203a package with B pin-out)

Bipolar		Pin-out	h_{fe}	f_T	I_C	Application
NPN	PNP	Code		MHz	mA	
2N3904	2N3906	A	100	300	200	general
2N4124	2N4126	A	120	300	300	general
2N4401	2N4403	A	100	250	600	general
	2N5086	A	150	40	50	low noise
2N5088		A	300	50	50	low noise
	2N5208	C	20	300	50	rf
2N5222		A	20	450	50	rf
2N5551		A	50	100	600	high volt, 140 V, 0.6 A
2N6427		A	20k	130	500	Darlington
2N6429A		A	500	100	10	low noise
2N6517	2N6520	A	40	40	500	high volt, 350 V, 0.5 A

JFET			g_{fs}	I_{DSS}	C_{iss}	
N-chan	P-chan		mmhos	mA	pF	
2N3819		DA	2–6	20	8	general
	2N3820	DA	1–5	15	32	general
2N5459	2N5462	DD	2–6	16	7	general
2N5486		DD	4–8	20	5	rf
2N5640		DD	–	2	–	switch $R_{on} = 100$

Packaging

Three-Terminal Device Packaging

Pin-Out patterns

	Pin number		
Code	1	2	3
A	E	B	C
B	E	C	B
C	B	E	C
DA	S	G	D
DB	S	D	G
DD	D	S	G

Voltage Regulators

78xx or
LM340-xx
positive
voltage

— Output
— Ground
— Input

Ground

xx = 5, 6, 8, 12, 15, 18, 20, 24 V

79xx or
LM320-xx
negative
voltage

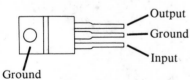

— Output
— Input
— Ground

Input

TO3

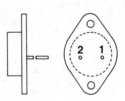

TO5

TO92

R203a

T220AB

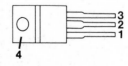

Fortney: Principles of Electronics   Harcourt Brace Jovanovich

Principles of Electronics:
Analog and Digital

Principles of Electronics: Analog and Digital

Lloyd R. Fortney

Duke University

Harcourt Brace Jovanovich, Publishers

and its subsidiary, Academic Press

San Diego   New York   Chicago   Austin   Washington, D.C.
London   Sydney   Tokyo   Toronto

Preface for the Instructor

Although electronics is a well established part of the modern science and engineering curriculum, the course content varies widely from place to place. The subject is a rich and varied one, but most modern textbooks provide poor support for any course that attempts to move beyond the elementary skills level. In particular, the commom failure to develop an adequate description of signals obscures the interrelationship of frequency response, phase shift, and impulse response and precludes the development of the mathematical tools necessary for effective circuit analysis or design. Without an introduction to these methods the student is poorly prepared even to seek answers in more advanced texts.

If derivations are omitted, sufficient attention cannot be given to the approximations that are an integral part of most electronic analysis, and the motivation for many designs remains hidden. The student often comes away with some understanding of a few circuits but no idea when and where to apply them. The presentation in *Principles of Electronics: Analog and Digital* is more mathematical than usual but is in many ways also more practical: It covers a broad range of material, concentrates on the general rather than the specific circuit, and makes every effort to explain the analysis techniques fully. One aim has been to produce a book that will serve both as a text and as a self-study reference. The supporting mathematics in the last three chapters is particularly important in this regard: It summarizes a number of loosely related mathematical techniques and presents them in a consistent electronic notation.

This book contains sufficient material for a two-semester course in electronics, but with suitable omissions it can be adapted to a variety of one-semester courses. The level of the material assumes previous courses only in introductory physics and calculus, but an earlier or concurrent course in differential equations will allow a student to recognize the generality of the linear systems ideas. Because of the increased attention to analysis, this text can support a wide range of analytical and investigative laboratory exercises.

A brief description of the chapter contents follows.

Chapter 1, "DC Circuits" is a traditional review of DC circuits but with emphasis on equivalent circuits.

In Chapter 2, "AC Circuits," simple AC circuits are described first with differential equations, then with complex impedances. The equivalent four-terminal network is introduced, with its transfer function **H,** and an example of a unity-gain, ideal amplifier.

Chapter 3, "Signals, Filters, and Amplifiers" provides applications of **H** to signal analysis with emphasis on frequency dependence and impulse response and includes an introduction to amplifier models and negative feedback.

Chapter 4, "Transformers," covers differential equation and complex impedance treatment of transformers leading to simplified schematic models.

Chapter 5, "The *PN* Junction and Diode Circuits" contains a descriptive treatment of *PN* junction physics and introduces linear models for biased two-terminal devices with circuit applications.

Chapter 6, "Single Transistor Circuits," describes bipolar and FET transistor physics and investigates single transistor circuits for both bipolar and FET devices using simplified models. Emphasis is placed on DC biasing and AC analysis for the calculation of input impedance, voltage gain, and output impedance.

Chapter 7, "Multiple Transistor Circuits," offers brief descriptions of the more important circuits, with particular attention to those used in integrated circuits.

Chapter 8, "Operational Amplifiers," is a three-level treatment of operational amplifiers with negative feedback under conditions of infinite open loop gains, constant open loop gains, and frequency-dependent open loop gains. A practical discussion of feedback-induced oscillations is included.

Chapter 9, "Digital Circuits," covers binary numbers, Boolean algebra, combinational logic, various flip-flops, and registers, with discussion of design techniques.

Chapter 10, "Data Acquisition and Process Control," describes transducers and data domain-changing circuits, particularly those used at the computer interface.

Chapter 11, "Computers and Device Interconnection," introduces computer internal architecture and interface standards including device interconnection and grounds.

Chapter 12, "Signal Analysis," develops Fourier and Laplace transforms with applications to continuous signals and formalizes the relationship between a circuit's frequency response and its impulse response.

Chapter 13, "Noise and Statistics," introduces statistical methods with applications to noise analysis in circuits.

Chapter 14, "Discrete Signal Analysis," includes techniques needed for computer-aided signal and system analysis and for the design of digital filters.

The material in the first nine chapters is derived from my course in introductory electronics taught in the physics department to a mixed group of science and engineering students, both undergraduate and graduate. Most of our physics majors take the course in the second semester of their sophomore year, concurrently with differential equations. The laboratory portion of the course is not limited to circuit fabrication but emphasizes measurement and design, providing many students with their first loosely structured experience at the theory-to-experiment interface.

Most of the material in Chapters 10 through 14 is covered in a second course, the subject of which is the application of computers to data acquisition, process control, and signal analysis. The inclusion of discrete signal analysis greatly expands the range of laboratory experiments that are available for this course.

Depending on the mathematical level of the students, the available lecture time, and the general goals of the course, at least three subsets of the analog material can be covered in a one-semester introduction:

1. To get the full benefit of the transfer function approach, it is necessary to establish the connection between circuits, transfer functions, $j\omega$ and **s**, poles and zeros, and the impulse response function $h(t)$. Adherence to this goal will require the expenditure of more time in the early passive circuit chapters. A guide to the minimum material required for this option is indicated in the Contents by the superscripts "R" (read only for background) and "O" (omit).

2. A more conventional electronics course would need to omit in addition all discussion of poles and zeros or impulse response. When the impulse response is omitted, most of the nonideal operational amplifier material in the last third of Chapter 8 should also be omitted. Although the complex frequency **s** will still appear in some retained material, it can simply be treated as a shorthand notation for $j\omega$.

3. Some instructors might wish to develop the linear systems mathematics more fully or to get to the applications in Chapters 10 and 11. The extra time may be found by omitting most of the transistor material in Chapters 6 and 7 and is justified by the overwhelming importance of operational amplifiers and other integrated circuits relative to discrete transistors. Chapter 6 may be reduced to a single-parameter (current gain) analysis of the bipolar transistor in the common collector configuration. Chapter 7 may be similarly shortened to just a general description of the differential amplifier's DC bias and input impedance and the push-pull amplifier's output impedance.

At the outset, the electronics course presents the instructor with three big advantages: The subject is relatively easy, its predictions are easily verified in the laboratory, and students perceive the material as important. It is hard to imagine a better arena for demonstrating the application of theoretical ideas to experimental realities, and it seems wasteful to present the subject as just a technical skill. As one who has worked extensively with analog and digital circuits, both before and after learning many of the circuit analysis methods presented in this text, I can attest to both their intuitive and their calculational value.

I am indebted to my students for their enthusiasm and patience, to family and friends for their encouragement, to colleagues and reviewers for their suggestions, and to an always helpful editorial and production staff.

Preface for the Student

The foundations of electronics were established by the fundamental observations of Michael Faraday (1791–1867) and others, and the discovery of the electron by J. J. Thompson in 1897, but its development proceeded relatively slowly until the importance of radar became apparent at the beginning of World War II. During that period and extending until about 1955, the expanding field of electronics depended heavily on the principle of electron emission from the hot cathode of a vacuum tube. By modern standards, vacuum-tube electronics was expensive, bulky, hot, unreliable, and even dangerous because of the high voltages present in the tubes. With the discovery of the transistor by J. Bardeen, W. H. Brittain, and W. B. Shockley in 1948, the stage was set for the electronics explosion of today.

Because of the availability of inexpensive, prefabricated microelectronic circuits on silicon chips, it is now relatively easy to produce a custom electronic solution to a specific measurement or display problem. But this very ease has led to a kind of trap, for a reasonably clever person can now produce operational electronic devices using only trial-and-error empirical methods. Although this situation can have short-term advantages, the lack of fundamental knowledge produces a ''cookbook'' electronics in which the experimenter is unable to tailor a circuit to his or her individual needs. Indeed, it often leads to a kind of design ''magic,'' whereby a poorly conceived initial design, with many ''corrective'' appendages, is painstakingly reproduced in successive applications. A scientist or engineer working on state-of-the-art problems will sooner or later need to design and not just reproduce a circuit.

In this book I have followed my natural inclinations as a physicist and presented the fundamental principles of the subject, applied wherever possible to the simplest circuit examples. The mathematical methods used in this development are chosen for their applicability to practical (and usually more complicated) circuits. The student who has mastered the material of this text should be qualified to attempt the design of both circuits and systems and will be prepared to make effective use of advanced engineering texts or application notes from the literature.

Contents

Chapter 2   AC Circuits

Chapter 3   Signals, Filters, and Amplifiers

Chapter 4-Transformers

Chapter 5   The PN Junction and Diode Circuits

Chapter 6   Single Transistor Circuits

Appendixes

1

DC Circuits

1.1 Introduction

The traditional development of electronic circuit analysis is divided into direct current (DC) circuits and alternating current (AC) circuits. Although early electrical circuits could often be identified as being AC or DC, most modern circuits do not fall strictly into either category. Consequently, the notation DC and AC has come to signify a division whereby DC circuit analysis deals only with constant currents and voltages, and AC analysis deals with time-varying voltage and current signals whose time average values are zero. For most circuits, a complete analysis must include both parts.

To introduce the basic techniques of circuit analysis with minimum complications, the discussion in this chapter is limited to circuits having only DC components: constant voltage sources, constant current sources, and resistors. Although much of the material is commonly taught in introductory physics courses and may constitute a review for some, the methods of two-terminal networks and equivalent circuits introduced here may not be so familiar. Equivalent circuits are an important aid to understanding circuit behavior and will be used extensively in later chapters.

1.2 Basic Concepts

Electronics deals with charges, electric and magnetic fields, and potentials, but does not provide the detailed description obtainable from the more exact techniques of classical and quantum physics. However, the approximate methods of electronics greatly simplify the treatment of energy sources, conductors, and insulators in complex geometries, and the results are adequate for most practical applications. It is important to recognize this inherently approximate nature of electronics and be willing to approximate further to simplify a given problem.

1.2.1 Current

Although most macroscopic objects are electrically neutral, the behavior of the positive protons bound in the nucleus is quite different from that of the much lighter negative electrons. Whereas the nucleus and inner atomic electrons of a solid are fixed in position, some of the outer electrons may break free of their parent atoms and move about in the material. If by some external means a macroscopic electric field is produced within the material, these "free" electrons will migrate back along the field lines, giving rise to a charge current.

This aggregate motion of charges can be identified with a current I, defined by

$$I = \frac{dq}{dt} \tag{1.1}$$

where dq is the quantity of positive charge crossing a specified surface in a time dt. In the MKS system of units, universally used in electronics, the current I is given in amperes and q in coulombs. An ampere is thus a coulomb of charge per second and is abbreviated by the symbol A.

Beginning students often worry that this current definition seems to be in conflict with the fact that the charges in motion are usually negative, and unnecessary confusion often results from trying to keep track of the motion of electrons. In fact, it is experimentally difficult to determine the sign of the moving charges, and only occasionally is it of any importance. The best approach is to assume the motion of positive charges (thereby eliminating one source of sign confusion) and worry about the details only when they are needed to explain specific effects such as the operation of diodes.

1.2.2 Potentials

Since the currents used in electronics are generally confined to wires, the generalized vector properties of the electric field are unimportant. It is therefore more convenient to take the electrostatic potential V rather than the electric field as the motivating influence for electric charge. The change in this potential across a distance $d\mathbf{r}$ in an electric field $\mathbf{E}$ is given by

$$dV = -\mathbf{E} \cdot d\mathbf{r} \tag{1.2}$$

The change in potential energy U of a charge q is related directly to the change in its electric potential V by

$$dU = q \, dV \tag{1.3}$$

and from this equation we see that the units of potential must be energy per unit of charge. The MKS unit of potential is the volt, which is abbreviated V and defined as a joule per coulomb. Thus a positive charge q moving along an electric field line is moving from a higher to a lower voltage (or potential) and losing potential energy.

The change in potential across some macroscopic object is usually expressed simply as V but *always* represents the difference between two potentials, as indicated by

$$V = V_{21} = V_2 - V_1 = \int_{V_1}^{V_2} dV \qquad (1.4)$$

A good analogy can be made between the flow of charge in a conductor and the downhill flow of water in a stream. This analogy is especially helpful if we remember that, by defining the positive direction for water flow (current), we are also defining the downhill direction (higher to lower potential). Although obvious at this point, this simple relationship is often confused or overlooked, producing equations with a wealth of incorrect signs.

1.2.3   The Electromotive Force or EMF

If we suddenly introduce a length of conducting material into an electric field, charge will flow in the material only until the additional field generated by the redistributed charge cancels the applied field within the material. In a short time the net macroscopic electric field and current *within* the material will again be zero. To maintain a potential drop and flow of charge within the material, some external device must provide a continuing source of external energy. Such a device is known as an electromotive force or EMF and is found in the electronics laboratory in the form of batteries, power supplies, and signal generators.

The two types of schematic EMF symbols applicable to DC circuits are shown in Figure 1.1a and represent the ideal voltage source and the ideal current source. An ideal voltage source has the ability to maintain a constant voltage across its terminals independent of the current it must deliver. At the opposite extreme, an ideal current source provides a constant current and will generate whatever voltage is needed to produce the current. An ideal voltage source is thus capable of producing an infinite current, whereas an ideal current source is capable of producing an infinite voltage.

1.2.4   Ohm's Law and Resistance

For most materials a linear equation known as Ohm's law,

$$V = RI \qquad (1.5)$$

adequately describes the relationship between the voltage V applied across an object ($V_2 - V_1$) to the current I through the object. The proportionality constant R is known as the resistance and is a function of both the material and the shape of the object. Resistance is measured in ohms, commonly indicated by the symbol omega, Ω. Ohm's law and its AC extension, which we will develop in the next chapter, are at the root of most electronic calculations.

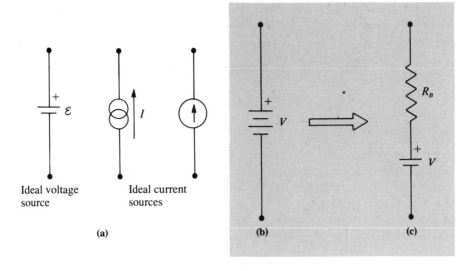

Figure 1.1   (a) Commonly used schematic symbols for the ideal DC voltage source and the ideal current source. (b) The symbol for a battery, a nearly ideal voltage source, and (c) its representation in terms of ideal elements.

   The resistivity (or its inverse, the conductivity) is a property of the material alone and specifies how much the material impedes the motion of charge. The resistivity parameter ρ has units of ohm-meters, and in terms of this parameter, the resistance R of a length L of material with constant cross-sectional area A is

$$R = \frac{\rho L}{A} \qquad (1.6)$$

The resistivity of a particular material shows a small variation with the temperature of the material, increasing with temperature for most metals but decreasing for carbon, the conducting material in the most common resistors. Except for those materials that under certain conditions exhibit the phenomenon known as superconductivity, all actual conductors have nonzero resistivity.
   The linear relationship of Ohm's law follows from the assumption that a material's resistivity is independent of the applied voltage or current. Although this assumption is found to be correct for most materials under normal operating conditions, a special class of materials known as semiconductors shows dramatic variations from linearity. This nonlinearity has important consequences, which are discussed in Chapter 5.

1.3   The Schematic Diagram

In electronics, a laboratory arrangement of circuit elements is represented by a schematic diagram composed of idealized elements each of which represents some property of the actual circuit. Since an actual element often has more than one property that must be represented, the schematic will often have more elements than the actual circuit. As an example, consider the circuit shown in Figure 1.2.

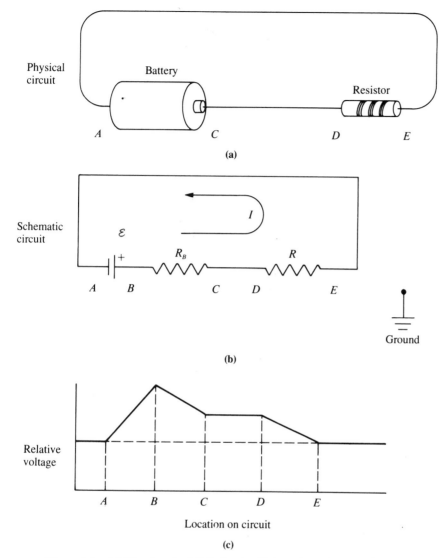

Figure 1.2 *(a) Simple actual DC circuit, (b) its schematic equivalent,
and (c) the relative voltage at various points.*

The actual circuit consists of a battery, some wire, and a resistor; but the battery,
which cannot supply an infinite current, is represented on the schematic by an
ideal voltage EMF and a series resistor R_B. Even the wire on a schematic differs
from real wire in that the schematic wire is assumed to have zero resistance. This
has the advantage that the voltage must be the same at all points along an unbro-
ken length of schematic wire, but it also means that any significant resistance in
the actual wire must be shown as one or more schematic resistor elements.

The potential or voltage at various points on the schematic of this simple circuit
is shown in Figure 1.2c. Note that the zero of the voltage scale on this figure is
completely arbitrary; only voltage differences are defined. It is common, but not
always necessary, to connect some part of the circuit to earth or ground, which is

taken, for convenience and by convention, to be at zero volts. The grounded point on the circuit becomes the zero of potential, and all other voltages in the circuit can then take on specific values. When neither a ground nor any other voltage reference is shown explicitly on a schematic, it is useful for purposes of discussion to adopt the convention that the bottom line on a circuit is at zero potential, as for example in Figure 1.6.

Independent of which point on a schematic is at zero voltage, the current direction is always "downhill" across a resistor. The current direction through an EMF may be either "uphill" or "downhill." Uphill indicates that the EMF is adding electrical energy to the circuit, whereas downhill indicates that the EMF is absorbing electrical energy, as when a battery is being recharged.

1.4 Power in DC Circuits

The energy flow in an electronic circuit is generally described in terms of power, but it will be helpful first to investigate the movement of charge in a material. A charge dq falling free across a potential drop V will gain kinetic energy dK in the amount

$$dK = V\,dq \qquad (1.7)$$

If the charge is truly free, as in a vacuum, this energy increment will stay with the charge, showing up as an acceleration. This effect is present in television and video display tubes (also known as cathode-ray tubes or CRTs), where electrons are accelerated to a high velocity before striking a phosphorescent screen. However, in a material such as a resistor, the "free" charge does not continue to gain kinetic energy while falling in a potential gradient. Instead, its acquired energy is transferred by collision into the random motion (heat) of atoms in the material. The net effect is that the "free" charge gains an insignificant amount of kinetic energy, moves with an essentially constant *drift velocity,* and effectively converts potential energy directly into heat.

Power is generally defined as the change in energy per unit time, expressed in units of watts, and given the symbol W. In electronic terms the power P is defined by the equation

$$P = \frac{V\,dq}{dt} = VI \qquad (1.8)$$

where V is the voltage across and I the current through any two points of a circuit. Most often this equation is applied to a single circuit element and can then be expressed as the voltage across the element times the current through it. The sign of the energy flow (out of the electrical domain or into it) is determined by the current direction relative to the potential change. If the charges flow downhill without accelerating as described above, energy is being converted to heat and leaving the electrical domain; if the charges flow uphill as in an EMF, some nonelectrical energy source (mechanical, chemical, nuclear) must be driving them and adding energy to the electrical domain.

Equation (1.8) combined with Ohm's law gives the well-known expression

$$P = RI^2 \tag{1.9}$$

which expresses the power dissipated by a current I in resistor R. Be aware that the application of this equation is limited to resistors where the current-voltage relationship is given by Ohm's law. For devices where the current-voltage relation is nonlinear (as for diodes and transistors), we must fall back on the fundamental expression of Eq. (1.8).

1.5   Series and Parallel Connections

Although the circuits formed by the interconnection of resistors and sources may be arbitrarily complex, two simple connections are encountered repeatedly: the series connection, where a common current passes through each circuit element; and the parallel connection, where a common voltage is applied across each circuit element. The simplest configurations involve only two resistors, but the expressions for the total or equivalent resistance of these combinations are among the most useful in electronics.

The series connection of two resistors is shown in Figure 1.3. Using Ohm's law twice, the voltage drop $V_A - V_B$ is found to be

$$V_{AB} = R_1 I + R_2 I = (R_1 + R_2)I \tag{1.10}$$

It is apparent that the current I will produce the same voltage drop V_{AB} whether it passes through R_1 and R_2 in sequence or through an equivalent resistor:

$$R_{eq} = R_1 + R_2 \tag{1.11}$$

Thus, we can schematically replace two resistors in series by the single equivalent resistance specified by this equation.

Figure 1.3   The series connection of two resistors showing the equivalent resistance.

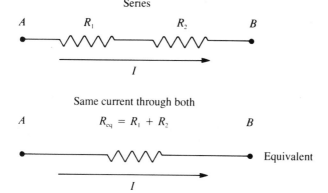

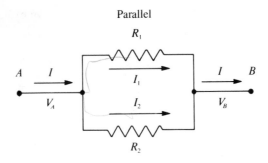

Parallel

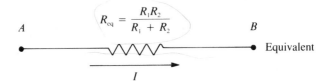

Figure 1.4   *The parallel connection of two resistors, again showing the equivalent resistance.*

The parallel configuration of two resistors is shown in Figure 1.4. Here a common voltage exists across each resistor, and the current I must split into two parts I_1 and I_2, pass separately through the resistors, and recombine to I again at point B. Writing the total current and then applying Ohm's law to each resistor gives

$$I = I_1 + I_2 = \left(\frac{1}{R_1} + \frac{1}{R_2}\right)V_{AB} = \frac{R_1 + R_2}{R_1R_2}V_{AB} \qquad (1.12)$$

The equivalent resistance for this parallel combination of two resistors is then

$$R_{eq} = \frac{R_1R_2}{R_1 + R_2} \qquad (1.13)$$

Many circuits can be reduced in complexity by the repeated application of these replacement rules for series and parallel elements; an example is shown in Figure 1.5. However, not all circuits can be reduced by these rules! The bridge circuit of Figure 1.8 is one such irreducible configuration: As long as the current through R_5 is significantly different from zero, this circuit has no resistors in either the series or the parallel configuration.

1.5.1   The Voltage Divider

Certain circuit arrangements appear frequently. One of these is the voltage divider shown in Figure 1.6a. If a voltage V is applied across the series resistors R_1 and R_2 as shown in the figure, then the common current through each is

$$I = \frac{V}{R_1 + R_2} \qquad (1.14)$$

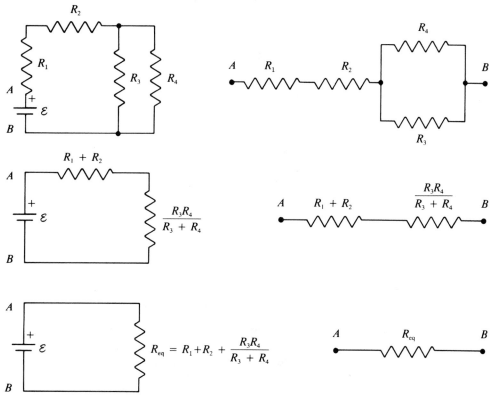

Figure 1.5 *An example of the reduction of a circuit or part of a circuit by the repeated use of the series and parallel rules.*

This current develops a voltage across R_2 given by

$$V_2 = IR_2$$

(1.15)

$$V_2 = \frac{R_2}{R_1 + R_2}V$$

which is seen to be a fraction of the applied voltage. Since one or both of the resistors, R_1 and R_2, could be equivalent resistances formed by parallel or series combinations of other resistors, this equation can be applied to more complicated circuits such as the one shown in Figure 1.6c.

Example 1.1 Find the voltage V_2 developed by the circuit shown in Figure 1.6c when $V = 15$ V, $R_1 = 400\Omega$, $R_2 = 2$ kΩ, and $R_3 = 8$ kΩ.

The parallel combination of resistor R_2 and R_3 can be replaced by the equivalent resistance

$$R_{23} = \frac{R_2 R_3}{R_2 + R_3} = 1.6 \text{ k}\Omega$$

Using this value as R_2 in Eq. (1.15) gives

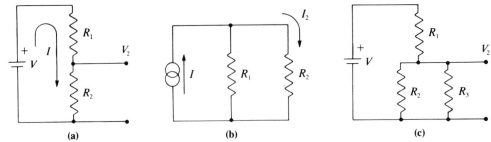

Figure 1.6 *Two of the most common circuits are (a) the voltage divider and (b) the current divider. (c) An example of a more complicated circuit that can still be treated as a voltage divider.*

$$V_2 = \frac{(1.6 \text{ k}\Omega)(15)}{(0.4 \text{ k}\Omega) + (1.6 \text{ k}\Omega)} = 12 \text{ V}$$

1.5.2   The Current Divider

Another common circuit configuration is the current divider shown in Figure 1.6b. The applied current I passing through the equivalent resistance formed by the parallel combination of resistors R_1 and R_2 generates a voltage given by

$$V = \frac{R_1 R_2}{R_1 + R_2} I \tag{1.16}$$

This voltage appears across R_2 generating a current I_2 given by

$$I_2 = \frac{R_1}{R_1 + R_2} I \tag{1.17}$$

This result will find more practical application in later chapters, after we have learned to make current sources from transistors and operational amplifiers.

1.6   Kirchhoff's Laws

The preceding discussion has been based implicitly on two fundamental principles of physics: the conservation of charge and the conservation of energy. When expressed in a form suitable for application to electronic circuits, these two principles are known as Kirchhoff's laws. Conservation of energy requires a zero algebraic sum for the voltage changes V_i encountered on an imaginary passage around any closed circuit loop:

$$\text{Closed loop} \qquad \sum_i V_i = 0 \tag{1.18}$$

Conservation of charge at any point is guaranteed by requiring a zero algebraic sum for the currents I_k into that point:

$$\text{Point} \quad \sum_k I_k = 0 \qquad (1.19)$$

This equation requires that the total charge flowing into a point of the circuit equal the total charge flowing out.

As formulated here, these laws apply only to time-independent voltage and current variables. However, the basic principles of charge and energy conservation are equally valid for time-dependent variables, and with the introduction of complex variables in the next chapter we will be able to reinterpret these expressions and apply them to the analysis of AC circuits.

Before proceeding to the application of these laws, we need to define some terminology.

Element: a resistance (more generally an impedance) or an EMF

Node: a point where three or more current-carrying elements are connected

Branch: one element or several in series connecting two adjacent nodes

Interior loop: a circuit loop that is not subdivided by a branch

For example, every element in Figure 1.8 is in a separate branch, with the four nodes labeled 1–4 and the three interior loops labeled *A, B,* and *C.*

By applying Kirchhoff's laws to the loops and nodes of a circuit, it is possible to write a set of linear algebraic equations whose variables are the various currents in the circuit. The analysis of a circuit of arbitrary complexity is thus reduced to a routine procedure where the most difficult task is to write a correct set of equations. The solution of the resulting linear algebraic equations follows a well-defined prescription.

Before any attempt is made to apply Eqs. (1.18) and (1.19) to a circuit, it is necessary to define the voltages and currents. The importance of a clear, correct, and properly labeled sketch cannot be emphasized too strongly. The currents can be defined either as branch currents or as loop currents, providing two different approaches to the problem.

1.6.1 Branch Current Method

The method generally taught in elementary physics requires the use of both Eqs. (1.18) and (1.19). To use this method it is necessary to label the current in each branch of the circuit as shown in Figure 1.7. Each arrow indicates the positive direction for a *presumed* current and need not correspond to the direction of actual current. When completed, this process will label the current and define its positive sense through each circuit element. If a labeled direction turns out to be wrong, the algebraic solution for that current will simply be negative.

There are six branches and thus six unknown currents on the schematic of Figure 1.7, and we therefore need six independent equations to obtain an algebraic solution. However, this circuit has five loops and four nodes as shown in Figure 1.8: We could write nine equations. An arbitrary choice will not always yield an

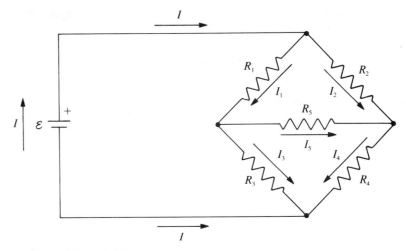

Figure 1.7 A bridge circuit showing the six independent currents I, I_1, I_2, I_3, I_4, and I_5, which must be labeled before proceeding with the branch current method.

independent set of equations. A selection that always works is to use only interior loops and all but one of the nodes.

To show the technique, we will obtain the six equations from Figure 1.7 using the interior loops *A*, *B*, and *C* and nodes 1, 2, and 3 from Figure 1.8. The branch current method produces an excessive number of variables and is generally not a good method for circuits with more than two loops.

Starting in the lower left corner of the circuit, we make an imaginary trip around loop *A*, writing the algebraic sum of the voltage changes encountered. The hardest part is getting the signs of the voltage changes correct. Just bear in mind

Figure 1.8 The same bridge circuit showing five loops and four nodes, which could be used with Kirchhoff's laws.

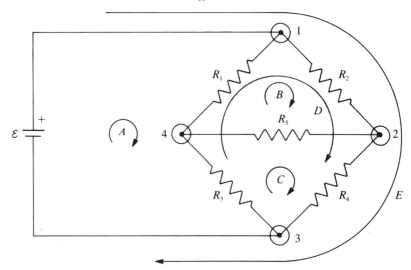

when crossing a resistor that the direction of the labeled current defines "downhill" and therefore determines the sign of the voltage change.

The equation for loop A reflects a voltage increase going across the battery and a voltage decrease across R_1 and R_3:

$$\text{Loop } A \qquad + \mathscr{E} - R_1I_1 - R_3I_3 = 0 \qquad (1.20)$$

Note that this expression contains only voltage differences; it does not matter where the actual zero of potential is located.

The loop B equation is obtained in a similar manner: Starting at point 4, we get a voltage increase going against the current through R_1, a decrease across R_2, and another increase across R_5:

$$\text{Loop } B \qquad + R_1I_1 - R_2I_2 + R_5I_5 = 0 \qquad (1.21)$$

Starting at point 3 for loop C yields

$$\text{Loop } C \qquad R_3I_3 - R_5I_5 - R_4I_4 = 0 \qquad (1.22)$$

The point equations are easily found by adding the currents directed into a node, subtracting the currents directed outward, and setting the result equal to zero:

$$
\begin{aligned}
\text{Node 1} \qquad & I - I_1 - I_2 = 0 \\
\text{Node 2} \qquad & I_2 + I_5 - I_4 = 0 \\
\text{Node 3} \qquad & I_3 + I_4 - I = 0
\end{aligned}
\qquad (1.23)
$$

There is little value in reviewing here the traditional pencil-and-paper methods for solving this set of linear equations. The numerical solution of any network with particular component values is easily accomplished from this point using standard computer programs for the solution of systems of linear equations, and we will not have need in the following chapters for the algebraic solution of any system of more than three variables.

1.6.2   Loop Current Method

In the loop current method, the independent current variables are taken to be the circulating currents in each of the interior loops; Figure 1.9 shows three such currents on the bridge circuit of the previous example. These circulating currents provide a complete description and automatically satisfy the point equations of the branch current method.

To analyze a circuit using the loop current method, it is necessary to label the interior loop currents on a diagram, use Eq. (1.18) to obtain an expression for the voltage changes around each interior loop, and solve these equations to find the loop currents. Depending on the problem, it may ultimately be necessary to algebraically sum two loop currents in order to obtain the needed interior branch currents.

Applied to the bridge circuit, the loop current method thus has only three variables and generates three loop equations:

$$
\begin{aligned}
\text{Loop } A \qquad & \mathscr{E} - R_1(I_a - I_b) - R_3(I_a - I_c) = 0 \\
\text{Loop } B \qquad & -R_1(I_b - I_a) - R_2I_b - R_5(I_b - I_c) = 0 \\
\text{Loop } C \qquad & -R_3(I_c - I_a) - R_5(I_c - I_b) - R_4I_c = 0
\end{aligned}
\qquad (1.24)
$$

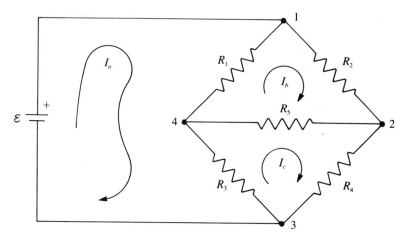

Figure 1.9 *The bridge circuit showing the currents used in the loop current method.*

For easier solution, each of these equations can be rearranged to give

$$
\begin{aligned}
(R_1 + R_3)I_a & & -R_1I_b & & -R_3I_c & = \mathcal{E} \\
R_1I_a & -(R_1 + R_2 + R_5)I_b & & +R_5I_c & = 0 & \quad (1.25) \\
R_3I_a & & +R_5I_b & -(R_3 + R_4 + R_5)I_c & = 0 &
\end{aligned}
$$

This set of three equations in three unknowns is now in standard form and can be solved for the loop currents I_a, I_b, and I_c using the method of determinants as outlined in Appendix A.

Although the loop current method yields fewer equations than the branch current method, it is usually necessary to find the branch current through a specific circuit element. By comparing Figures 1.7 and 1.9, we see that exterior branch currents such as I_2 can be related to a single loop current (I_b), whereas interior branch currents such as I_5 are given by the difference of two loop currents ($I_c - I_b$).

 Example 1.2 With R_L disconnected, determine the voltage between points A and B on the circuit of Figure 1.10.

The first step is to calculate the current through the various elements of the circuit, and the loop current method is easiest. Do not be misled by the wires connected to points A and B; since they carry no current, they are not branches, and the points where they connect are not nodes. With the loop currents drawn as shown, the two interior loop equations are

$$V_1 - 2RI_1 - 4R(I_1 - I_2) = 0$$

and

$$4R(I_1 - I_2) - 2RI_2 = 0$$

These equations are easily solved to give

$$I_1 = \frac{0.3V_1}{R}$$

and

$$I_2 = \frac{0.2V_1}{R}$$

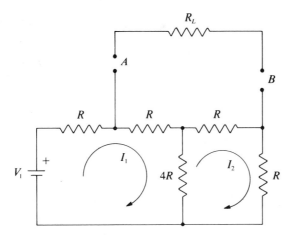

Figure 1.10 *A three-loop circuit is reduced to two loops by removing the load resistor R_L.*

Finally, starting with an assumed voltage V_A at point A, we can write

$$V_A - RI_1 - RI_2 = V_B$$

which can be rearranged to give

$$V_A - V_B = R(I_1 + I_2) = 0.5V_1$$

A useful added exercise is to reverse the directions of one or both of the currents on the figure, then rework the problem with attention to the sign differences.

1.7 The Equivalent Circuit

Early in this chapter we developed rules that allow us to replace any combination of series- and parallel-connected resistors with a single equivalent resistor. That simplifying replacement was possible because we had a linear relationship (Ohm's law) between voltage and current. Since the loop and node equations for general circuit arrangements retain this linear relationship, the concept of a simplifying replacement can be extended from a single element to the circuit itself.

The technique of replacing part of a circuit with a simpler equivalent circuit is useful because it permits us mentally to disconnect and isolate a part of the circuit, analyzing it without regard for the removed portion. After the isolated part has been reduced to a simpler equivalent, it can be used to redraw and simplify the original circuit.

Consider the unseen circuit in the box of Figure 1.11a. The details of this circuit are unimportant, but it could be the circuit of Figure 1.10, with the terminals A and B protruding from the box. The behavior of the hidden circuit can be studied by attaching a resistor R_L across these exposed terminals and measuring the resulting voltage V_{AB} and current I through the resistor. In the DC case, only these three experimental quantities R_L, I, and V_{AB} can be determined outside the box, and if the resistor obeys Ohm's law, only two are really independent.

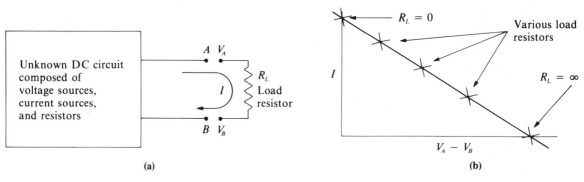

(a)   (b)

Figure 1.11   (a) Unknown two-terminal network and (b) its current versus voltage curve.

Each value of the resistance R_L will generally produce a different voltage and current, and a plot of I verses V_{AB} for different choices of R will yield a straight line (the load line) as shown in Figure 1.11b. There are three obvious choices for the two parameters needed to describe the load line: I_{short}, the intercept on the current axis when $R_L = 0$; V_{open}, the intercept on the voltage axis when $R_L = \infty$; and the slope of the line. The slope of the load line is related to an equivalent resistance associated with the circuit in the box and is obviously determined by the intercepts.

A circuit that has only two points of interest (points A and B in Figures 1.10 and 1.11) is known as a two-terminal network. Some additional examples of two-terminal networks are shown in Figure 1.12.

Figure 1.12   Two-terminal network examples.

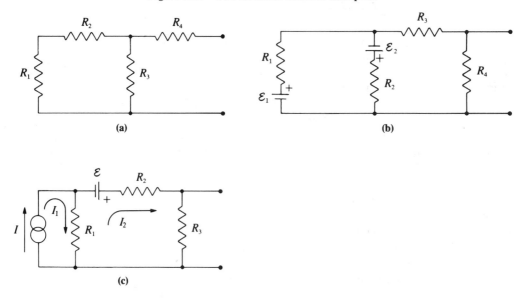

Although the actual circuit in the box of Figure 1.11a could be a complicated two-terminal network with multiple sources and resistors as shown in Figure 1.12, it could also be either of the simple single-source, single-resistor networks shown in Figure 1.13; all will produce a straight load line as shown in Figure 1.11b. The load line does not distinguish between these different circuit configurations, but if we assume that the box contains one of the elementary circuits of Figure 1.13, then the load line can be used to determine the component values of that circuit.

The circuit of Figure 1.13a, composed of an ideal voltage source and a series resistor, is known as a Thevenin equivalent circuit. Thevenin's theorem states that this circuit, with correctly chosen components, can be used to replace a two-terminal network composed of any combination of voltage sources, current sources, and resistors. The circuit of Figure 1.13b, composed of an ideal current source and a parallel resistor, has the same properties and is known as a Norton equivalent circuit. These equivalent circuits have direct extensions to AC circuit analysis, and the dramatic reduction in circuit complexity possible with these two-element replacements is extremely useful for both calculation and understanding.

When the Thevenin and Norton circuits are used to describe the same two-terminal network, their resistors are identical, making the transition from one equivalent to the other quite easy. The following derivation of this relationship provides some insight into the use of these equivalent circuits.

If both equivalent circuits are used separately to describe the same network (the one in the box of Figure 1.11a), then all three circuits must produce the same load line measured across terminals A and B. If the circuit elements of the Thevenin and Norton circuits are properly chosen to produce the response of the original network, then the two equivalent circuits must respond identically. With the terminals of the Thevenin and Norton equivalents left open as shown in Figure 1.13 ($R_L = \infty$), we have

$$V_{AB}(\text{open}) = V_{\text{Th}} \qquad (1.26)$$

or

$$V_{AB}(\text{open}) = I_N R_N \qquad (1.27)$$

Figure 1.13 *(a) Thevenin and (b) Norton equivalent circuits.*

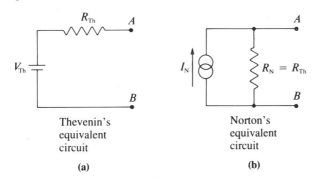

Thevenin's equivalent circuit

(a)

Norton's equivalent circuit

(b)

Note that $V_{AB}(\text{open}) = V_{Th}$ because there is no current in the series resistor R_{Th}. If we connect terminals A and B on the equivalent circuit with an ideal wire, we can write the short circuit current through that wire as

$$I_{AB}(\text{short}) = \frac{V_{Th}}{R_{Th}} \tag{1.28}$$

or as

$$I_{AB}(\text{short}) = I_N \tag{1.29}$$

Note that on the Norton equivalent, the shorting wire allows no voltage drop across R_N and therefore no current. The result is that the current in the shorting wire is I_N. Combining Eqs. (1.26) to (1.29) to eliminate I_N yields

$$V_{Th} = \frac{R_N}{R_{Th}} V_{Th} \tag{1.30}$$

which can be true only if

$$R_{Th} = R_N \tag{1.31}$$

1.7.2 Determination of Thevenin and Norton Circuit Elements

Given a general two-terminal network, the process of finding the elements of a Thevenin or Norton equivalent can be reduced to two steps. A helpful memory aid is to apply these steps to one of the two basic circuits of Figure 1.13.

Step 1
   Thevenin: Calculate the open-circuit voltage, V_{AB}. This voltage is V_{Th}.
   Norton: Calculate the short-circuit current through an imagined shorting wire between A and B. This current is I_N.
Step 2, Both: Determine both V_{Th} and I_N from step 1 and divide the first by the second to get the equivalent resistance.
Alternate Step 2: A generally quicker method for step 2 is to short (replace with a wire) all voltage sources, open (remove) all current sources, and then simply calculate the equivalent resistance remaining between A and B. However, some caution is necessary with this method, since it is incorrect to so manipulate the nonphysical ''algebraic'' voltage and current sources to be introduced later.

1.7.3 Circuit Simplification with a Thevenin Equivalent

The equivalent circuit can be a useful step toward the solution of a network problem such as the bridge circuit of Figure 1.7. To solve for the voltage drop across R_5, we first simplify the circuit by removing R_5 as shown in Figure 1.14. The remaining circuit is treated as a two-terminal network, with terminals A and B being the ends that were originally attached to R_5. Once we have the elements of

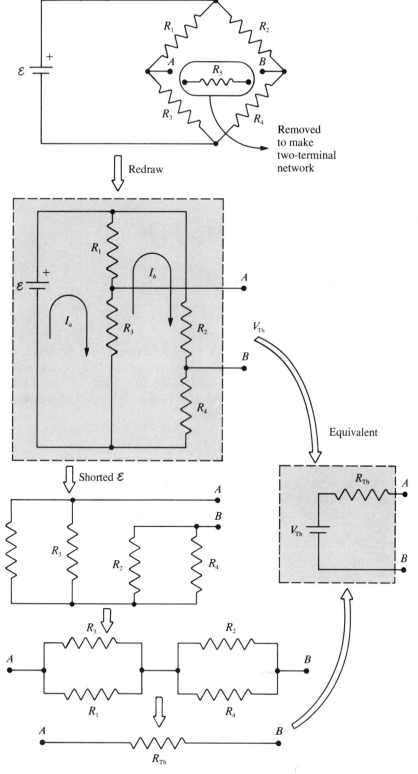

*Figure 1.14   The bridge circuit with R_5 removed and its conversion to
a Thevenin equivalent circuit problem.*

the equivalent two-terminal Thevenin circuit, we can reattach R_5 to the equivalent as in Figure 1.15 and calculate the current in the single-loop problem.

The diagrammatic steps for finding the Thevenin equivalent are shown in Figure 1.14. Removing R_5 changes the three-loop bridge circuit to a simpler two-loop circuit that could be solved by the loop current method described above. However, the analysis of this reduced circuit is further simplified by assuming an ideal voltage source that has no internal series resistance. As a consequence, we can use the voltage divider expression of Equation 1.14 to determine that

$$V_A = \mathscr{E}\left(\frac{R_3}{R_1 + R_3}\right)$$

and   (1.32)

$$V_A = \mathscr{E}\left(\frac{R_4}{R_2 + R_4}\right)$$

The difference of these two voltages is the open-circuit voltage of the two-terminal network and is thus the Thevenin equivalent voltage

$$V_{\text{Th}} = V_{AB}(\text{open}) = \mathscr{E}\left(\frac{R_3}{R_1 + R_3} - \frac{R_4}{R_2 + R_4}\right) \tag{1.33}$$

To complete the equivalent circuit, we need to determine R_{Th}. Since the voltage source is a true energy source, we can apply the simpler, second version of step 2. Shorting the voltage source, as required by the rule, leaves a circuit composed only of resistors. After redrawing as shown in Figure 1.14, application of the rules for parallel and series connections produces

$$R_{\text{Th}} = \frac{R_1 R_3}{R_1 + R_3} + \frac{R_2 R_4}{R_2 + R_4} \tag{1.34}$$

If the original problem was to find the current through R_5, we can reconnect the Thevenin equivalent as shown in Figure 1.15. The resulting single-loop circuit can be easily evaluated to give the current through R_5,

$$I_5 = \frac{V_{\text{Th}}}{R_{\text{Th}} + R_5} \tag{1.35}$$

Figure 1.15   Equivalent circuit for the complete bridge.

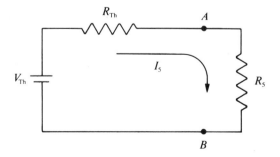

and the desired closed-circuit voltage across R_5,

$$V_{AB} = R_5 I_5 \qquad (1.36)$$

1.7.4 Reduction of a Two-Terminal Network

The simplified configuration of a Thevenin or Norton equivalent circuit is an important aid in understanding the operation of complex circuits. Consider the circuit in Figure 1.16, which shows an actual current source connected in parallel with an actual voltage source. Both actual sources have associated resistances. If the intention was to use this two-terminal network as an EMF to drive a further unspecified circuit, it would clearly be useful to first reduce this circuit to one of the basic equivalents. An interesting question is whether the combination behaves most like an ideal voltage source (a small R_{Th}) or most like an ideal current source (a large R_{Th}).

By shorting the voltage source and opening the current source, we can determine the equivalent resistance for either Thevenin's or Norton's circuit. The resulting circuit is shown on the figure and yields

$$R_N = R_{Th} = \frac{R_1 R_2}{R_1 + R_2} \qquad (1.37)$$

Norton's method is easier to apply to this circuit; the steps are shown in the figure. Note that the shorting wire fixes A and B at the same voltage and Ohm's law then requires that the current through R_1 be zero. Because of the series EMF, the current through R_2 is not zero and is given by

$$I_2 = \frac{\mathcal{E} - 0}{R_2} \qquad (1.38)$$

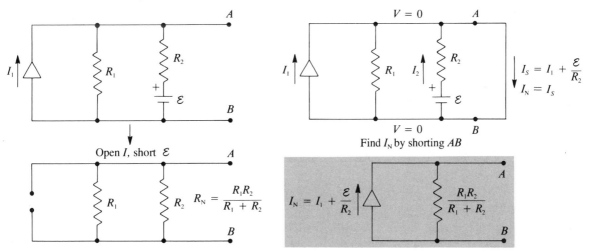

Figure 1.16 *Steps in converting a circuit to its Norton equivalent.*

The short circuit current I_S is therefore

$$I_S = I_1 + \frac{\mathcal{E}}{R_2} \tag{1.39}$$

These results are summarized on the Norton equivalent circuit at the bottom of Figure 1.16. If this Norton equivalent is to represent a nearly ideal current source, R_N should be large. Since the original circuit represented an actual current source and an actual voltage source, we would expect R_2 to be much smaller than R_1. Armed with this knowledge about the relative sizes of the resistors, Eq. (1.37) can be approximated by

$$R_N = \frac{R_1 R_2}{R_1 + R_2} \simeq \frac{R_1 R_2}{R_1} = R_2 \tag{1.40}$$

This expression shows that the equivalent resistance R_N is small, making the combined circuit behave like a nearly ideal voltage source. (Take note that as an aid to simplifying the algebra for a particular circuit, we will often use approximation methods similar to that indicated in this equation. The general procedure is to replace a sum by its largest term.)

> **Example 1.3**   Find the Norton equivalent circuit elements I_N and R_N for the two-terminal network of Figure 1.12c.
>
> Following step 1 above, we first short the output terminals with a wire; this wire is in parallel with R_3 and forces the voltage drop across R_3 and the current through R_3 to be zero. The Norton current I_N is just the current in this shorting wire. The reduced circuit can now be analyzed in terms of two clockwise interior loop currents I and I_N. We write the loop equation for the right-hand loop as
>
> $$R_1(I - I_N) + \mathcal{E} - R_2 I_N = 0$$
>
> where R_3 does not appear because the current through it is zero. Solving for I_N gives
>
> $$I_N = \frac{\mathcal{E} + R_1 I}{R_1 + R_2}$$
>
> We can find R_N most easily by shorting the voltage source, opening the current source, and calculating the resistance remaining between the two terminals. The result is just the series combination of R_1 and R_2 in parallel with R_3:
>
> $$R_N = \frac{(R_1 + R_2)R_3}{R_1 + R_2 + R_3}$$
>
> An alternative way to work this problem is first to replace the current source I and its shunt resistor R_1 with a Thevenin equivalent consisting of a voltage source IR_1 and a series resistor R_1. The above equation for I_N then follows immediately.

1.8   Measurement

The accurate experimental measurement of circuit operating characteristics is an acquired skill, and no amount of textbook discussion can substitute for laboratory experience. Although many elegant laboratory instruments exist for making measurements of circuit behavior, the most common instruments by far are the mul-timeter (volts, ohms, amperes), the oscilloscope, and the logic probe, which is used only on digital circuits. Of these, the oscilloscope is the most versatile, the most expensive, and the least likely to compromise or obscure a measurement. In addition to the DC measurements needed for the circuits of this chapter, the os-cilloscope can also display the time variation of a signal.

When using a laboratory instrument to measure some aspect of a circuit's be-havior, three points must be considered: the correct operation of the instrument, the effect of the instrument on the circuit, and the measurement error. Because of the diversity and variation among brands, no specific discussion of the correct operation of any measuring instrument will be given here.

However, the effect of a measuring instrument on the circuit can be covered with some degree of generality. Whenever we connect an instrument to a circuit, we effectively make the circuit into a two-terminal network with the two terminals being the points of connection to the instrument; see Figure 1.17. The instrument itself can also be considered to be a two-terminal network, generally passive (without voltage or current sources) except for the multimeter when used as an ohmmeter.

When used to measure DC signals, a voltmeter or oscilloscope can be thus characterized by a Thevenin equivalent resistance, known as the input impedance of the instrument. For an oscilloscope, the input resistance is a relatively large $10^6 \, \Omega$, and can be increased to $10^7 \, \Omega$ by using a $10\times$ probe. The increasingly common digital versions of the multimeter have a similarly large input impedance,

Figure 1.17   Test circuit with voltage being measured with an oscilloscope and the equivalent circuits for each.

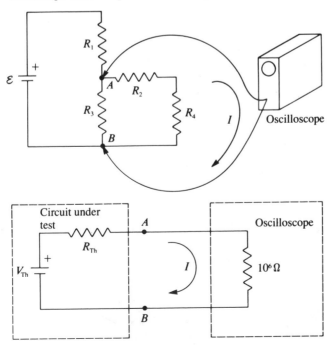

but the traditional dial readout volt-ohm meter has a much lower input impedance when used as a voltmeter. Its impedance is generally specified as a function of the full-scale voltage setting, typically 20,000 Ω/V.

If such an instrument is connected to a test circuit to measure a voltage, it will invariably draw some current from the circuit. This will, in turn, modify all voltages on the test circuit to some degree. The measurement is then not of the original circuit but rather of the circuit and instrument combination. If the effect on the circuit is significant, the best solution is to obtain a better instrument that has a higher input impedance. The only alternative is to correct for the effect of the instrument by additional calculation.

The third point to consider is the uncertainty in a particular measurement, usually called the estimated error or simply the measurement error. Be aware that measurement error is not defined as the difference between a measured result and the expected result. Rather it comes either directly or indirectly from the experimental ability to determine a value for a given quantity. In electronics, the measurement error usually stems either from an uncertainty in the actual values of components or from the limited accuracy of a particular instrument reading. If the difference between a measured value and its expected (or theoretical) value is much larger than the estimated measurement error, the fault may lie most anywhere, even in the expectation.

Frequently, the quantity of interest is not measured directly but is derived from one or more measurements. It is then necessary to propagate the measurement errors through the formulas that relate the quantity of interest to the measured quantities. A straightforward procedure exists for doing this propagation of errors for the case where the measured quantities are independent variables.

Assume that the following equation describes the relationship between the derived quantity Z and the directly measured or otherwise uncertain quantities X and Y:

$$Z = Z(X, Y) \tag{1.41}$$

The error in Z as a result of an error in X is reasonably related to the change in Z with respect to X (holding Y constant) times the error in X. The formal statement of this relation is

$$\Delta Z_X = \frac{\partial Z}{\partial X} \Delta X \tag{1.42}$$

and the similar one for the error in Y is

$$\Delta Z_Y = \frac{\partial Z}{\partial Y} \Delta Y \tag{1.43}$$

where the symbol $\partial Z/\partial X$ means take the derivative of Z with respect to X while treating Y as a constant, and $\partial Z/\partial Y$ means with respect to Y while treating X as a constant. Under the assumption that X and Y are independent variables, these two sources for an error in Z can be combined by

$$\Delta Z = \sqrt{(\Delta Z_X)^2 + (\Delta Z_Y)^2} \tag{1.44}$$

and expanded to

$$\Delta Z = \sqrt{\left(\frac{\partial Z}{\partial X}\Delta X\right)^2 + \left(\frac{\partial Z}{\partial Y}\Delta Y\right)^2} \tag{1.45}$$

The derivation of Eq. (1.44) is given in Section 13.3.3.

As an example of the application of this method for the propagation of errors, consider the very simple circuit of Figure 1.18 and assume that the intention is to determine the current I using a voltmeter or oscilloscope. For added simplicity we will assume that the current drawn by the measurement instrument can be neglected. The current through R_2 can be deduced directly from the measurement of V using Ohm's law:

$$I = \frac{V}{R_2} \tag{1.46}$$

Now consider the estimated error associated with this result if we have a 10% uncertainty in the voltage reading. At first glance, it might seem that the error in I would also be 10%, but in the laboratory, the resistor R_2 will not be exactly known either; the most common resistors have uncertainties of 20%. A more complete estimate of the uncertainty in I requires the use of Eq. (1.45) to combine the two sources of error. To use this equation we need the change in I resulting from a change in V,

$$\Delta I_V = \frac{\Delta V}{R_2} \tag{1.47}$$

and the change in I resulting from a change in R_2,

$$\Delta I_R = -\left(\frac{V}{R_2^2}\right)\Delta R_2 \tag{1.48}$$

Since it is possible to change R_2 while keeping V fixed (remember that since $\mathscr{E}$ and R_1 are not known, the current I can have any value), these are independent variables and the combined error in the dependent variable I can be expressed as

$$\Delta I = \sqrt{(\Delta I_V)^2 + (\Delta I_R)^2} \tag{1.49}$$

Figure 1.18 *Test circuit for determining the current I by measuring the voltage drop across a known resistor R_2.*

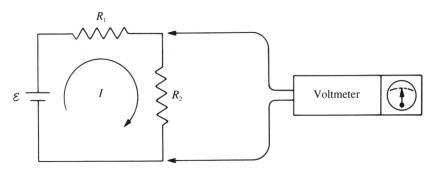

Expanding the currents in terms of voltage and resistance gives

$$\Delta I = \sqrt{\left(\frac{\Delta V}{R_2}\right)^2 + \left(\frac{V}{R_2^2}\Delta R_2\right)^2} \tag{1.50}$$

This expression will simplify if everything is expressed in fractional errors as in the following steps:

$$\Delta I = \sqrt{\left(\frac{V}{R_2}\right)^2\left(\frac{\Delta V}{V}\right)^2 + \left(\frac{V}{R_2}\right)^2\left(\frac{\Delta R_2}{R_2}\right)^2} \tag{1.51}$$

$$= \frac{V}{R_2}\sqrt{\left(\frac{\Delta V}{V}\right)^2 + \left(\frac{\Delta R_2}{R_2}\right)^2} \tag{1.52}$$

$$= I\sqrt{\left(\frac{\Delta V}{V}\right)^2 + \left(\frac{\Delta R_2}{R_2}\right)^2} \tag{1.53}$$

Putting in the error estimates given above, this equation yields

$$\frac{\Delta I}{I} = \sqrt{(0.1)^2 + (0.2)^2} = 0.22 \tag{1.54}$$

showing that the fractional error in the current is more than twice that in the voltage measurement. Obviously it would be a waste to spend time or money on a better voltage measurement when a more precisely known resistance is really needed.

Do not be misled by the simplicity of this example. A more complicated algebraic relationship between the measured and derived quantities can lead to multiplicative factors that make the percent error in the derived quantity either much larger or much smaller than the percent error in the measurement estimates.

Circuits used for measurement are often designed to minimize the propagated error. The bridge circuit of Figure 1.7 is a good example: When pairs of components are carefully matched ($R_1 = R_2$ and $R_3 = R_4$) so that the current I_5 through R_5 is nearly zero, the fractional change in I_5 becomes very sensitive to small changes in the values of the matched resistors; hence a relatively poor measurement of I_5 can still determine a resistance ratio quite well. As a result, this circuit is used extensively for precision measurements where the physical effect being measured produces a change in one of the resistor values R_1 to R_4.

Example 1.4   The temperature-sensitive LM334 two-terminal integrated circuit made by National Semiconductor has an output current that varies linearly with temperature. The device is shown as a variable current source in Figure 1.19. If the current is given by the equation

$$I = 0.0334T \ \mu A$$

where T is the temperature in Kelvin, choose R_1 so that V_{AB} is approximately zero when $T = 25°C$, then find dV/dT.

The voltage at terminal A is given by

$$V_A = 5.0 \text{ V} - R_1 I$$

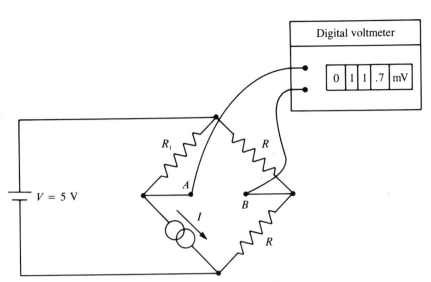

Figure 1.19 *A temperature-sensitive current source and a digital voltmeter in a bridge circuit produce a voltage output sensitive to temperature.*

and the voltage at terminal B is 2.5 V. The voltage difference V_{AB} is then

$$V_{AB} = 2.5 \text{ V} - R_1 I$$

This difference voltage can be read directly off a high-input-impedance digital voltmeter connected between these two terminals as shown.

At $T = 298$ K (25°C), the predicted current through the device is $I = 9.9532$ μA, and if R_1 is chosen to be 250,000 Ω, the voltage difference V_{AB} will be only 11.7 mV.

The change in this voltage with temperature is given by the derivative

$$\frac{dV_{AB}}{dT} = -R_1 \frac{dI}{dT}$$

where $dI/dT = 0.0334$ μA/°C. Using the R_1 just found, this expression evaluates to

$$\frac{dV_{AB}}{dt} = -(2.5 \times 10^5)(0.0334 \times 10^{-6})$$

$$= -8.35 \text{ mV/°C}$$

One worry with this design is the temperature variation of the resistors. Ordinary carbon resistors vary by about −0.5% for each degree increase in temperature. If all three resistors and the LM334 are thermally connected and running at a common temperature, how much error does this introduce into the instrument?

Problems

1. Number 22 gauge copper wire has a diameter of 0.71 mm. If the resistivity of copper is 1.7×10^{-8} Ω-m, what is the resistance of a 2-m length of the wire?

2. A common type of resistor is made from carbon granules compressed into a cylindrical shape. If the resistivity of a sample of this material is 0.5 Ω-m at room temperature, determine the cross-sectional area required to make a 1 MΩ resistor. Take the cylinder length to be 3 mm.

3. Precision resistors are often formed using a thin film of metal evaporated onto a ceramic material. If aluminum (2.6×10^{-8} Ω-m resistivity) is deposited to form a wire with a 10^{-12} m^2 cross section, what length would be required to produce a resistance of 1 kΩ?

4. *(a)* For a series circuit composed of a constant voltage source and a resistor R, sketch the current as a function of the resistance. *(b)* For a series circuit composed of a constant current source and a resistor R, sketch the voltage developed across the resistor as a function of R.

5. Assuming that the resistors in Problem 4 are variable, sketch the power dissipated in each resistor as a function of its resistance.

6. The 12-V power supply shown in Figure P1.6 is being used to charge a single-cell nickel-cadmium battery. *(a)* If the battery typically develops 1.3 V while being charged, what size resistor is needed to produce a 50-mA current? *(b)* Assuming that the battery develops 1.5 V when fully charged, calculate the current. *(c)* How much power is being dissipated as heat in the battery after it is fully charged? *(d)* How much power is being dissipated in the resistor R?

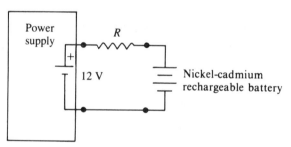

Figure P1.6

7. Starting with the basic equation $P = VI$ and Ohm's law, *(a)* derive Eq. (1.9), and *(b)* a similar expression in terms of the resistance R and the voltage developed across the resistor.

8. Show that the power input by the voltage source in Figure P1.8 is equal to the power dissipated by the two resistors.

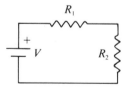

Figure P1.8

9. Make a sketch similar to Figure 1.2c for the circuit shown in Figure P1.9. If point C is grounded, what is the voltage (with respect to ground) at point A?

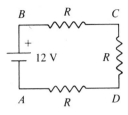

Figure P1.9

10. Determine the equivalent resistance of each of the circuits in Figure P1.10. [*Ans.*: $2R/3$, $3R/2$, R, $5R/3$]

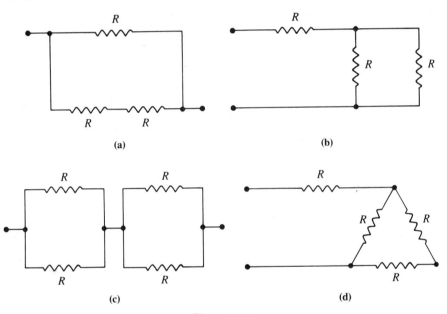

Figure P1.10

11. Determine an expression for the voltage V_2 on the voltage divider of Figure 1.6c.

12. Determine an expression for the current I_3 in resistor R_3 of the circuit shown in Figure P1.12.

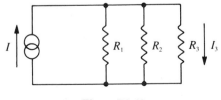

Figure P1.12

13. Rework Example 1.2 with both of the assumed current arrows reversed.

14. Use the loop current method to determine the voltage developed across the terminals *AB* in the circuit of Figure P1.14. [*Ans.*: $(V_1 + V_2)/5$]

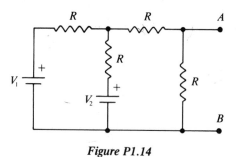

Figure P1.14

15. Use a Thevenin equivalent circuit to work the previous problem. (*Hint*: Temporarily remove the resistor connecting terminals *A* and *B*, then replace it after the Thevenin equivalent has been determined.)

16. If a battery develops 1.5 V across its terminals when unloaded but only 1.3 V when it is connected to a 100-Ω load, what is the internal impedance of the battery?

17. Find the Thevenin equivalent components V_{Th} and R_{Th} for the circuit shown in Figure P1.17.

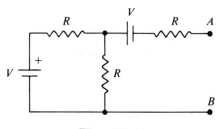

Figure P1.17

18. Find the Norton equivalent components I_N and R_N for the circuit shown in Figure P1.18. [*Ans.*: $(V/R + I)/2$, $2R$]

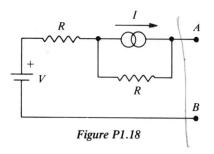

Figure P1.18

19. (*a*) For different settings of the variable resistor (potentiometer) in Figure P1.19, what range of voltage can be obtained for V_{AB}? (*b*) Find the elements of a Norton equivalent circuit when $V_{AB} = 5$ V.

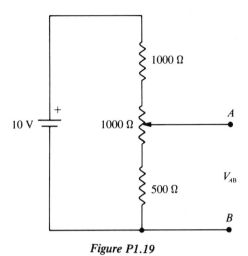

Figure P1.19

20. *(a)* Add one resistor to the circuit in Figure P1.20 in a position outside the box such that the Thevenin equivalent R_{Th} of the resulting circuit is less than 900 Ω. *(b)* If R_{Th} is to be 90 Ω, determine the value of the added resistor. *(c)* What will be the value of V_{Th} in this equivalent circuit? *(d)* Sketch the new two-element equivalent circuit and label the voltage and resistance values.

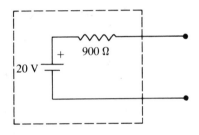

Figure P1.20

21. On the bridge circuit of Figure 1.14, take R_5 to be infinite, let $R_1 = R_3 = R_4 = R$, and assume that R_2 is a variable resistor parameterized in terms of its fractional change x by the expression $R_2 = R(1 + x)$. Show that the Thevenin equivalent voltage is $V_{Th} = [x/(4 + 2x)]\,\mathscr{E}$.

22. On the bridge circuit of Figure 1.14, let $R_1 = R_2 = R_4 = R$ and $R_3 = R + \Delta R$. *(a)* Find expressions for the Thevenin equivalent circuit elements and *(b)* solve for the current I_5 through a resistor R_5 that is much smaller than R. Express this current as a function of $x = \Delta R/R$. [*Ans.:* $I_5 = \mathscr{E}x/4R$]

23. An ideal voltmeter has an infinite input impedance and draws no current, whereas an ideal ammeter has a zero impedance and develops no voltage across its terminals. An ammeter displays the current I as indicated in Figure P1.23 on page 32. Show how to make a nearly ideal 0–10 mA ammeter using an ideal 0–100 μV voltmeter and a resistor.

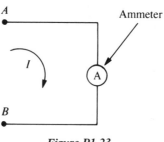

Figure P1.23

24. If each of the resistors in Figure 1.17 is 1 MΩ and the oscilloscope with a 1X probe shows a 2-V offset, what is the magnitude of the battery EMF?

25. Devise an experimental circuit to determine the value of an unknown resistor R_U. You may assume a known resistor R, a known voltage source V, and a voltmeter.

26. A circuit like that of Figure 1.18 with $R_1 = R_2 = 1$ kΩ is used to determine the value of the battery EMF. If the voltmeter is ideal and reads 10 V with a ±10% uncertainty, what is the voltage of the battery and what is the error estimate on this voltage? Assume that the resistors are known to ±20%. [*Ans.*: 20 V, 18%]

27. Assume that R_1 in the voltage divider shown in Figure P1.27 is a thermistor that makes small fractional changes in its resistance as the temperature changes.

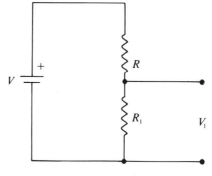

Figure P1.27

(*a*) Show that for small changes in R_1, the change in V_1 can be written

$$\Delta V_1 = \frac{R_1 R}{(R_1 + R)^2}\left(\frac{\Delta R_1}{R_1}\right)V$$

(*b*) Using this expression, show that for a small change in R_1, the output change in V_1 is maximized when $R = R_1$.

2

AC Circuits

2.1 Introduction

In this chapter, we introduce circuits where current and voltage signals may vary with time. These elementary circuits are loosely classified as alternating current (AC) circuits, even though in a specific case the current may not actually alternate in direction. Compared to DC circuits, the analysis of AC circuits is relatively complicated because the direct application of Kirchhoff's laws now produces a differential equation for each circuit loop.

Following a summary of capacitance and inductance, we will introduce time-dependent circuit analysis by writing and solving the differential equations for some simple AC circuits. In the remainder of the chapter, we will outline a more practical, algebraic approach to AC circuit analysis, which uses complex variables and sinusoidal signals.

2.2 AC Circuit Elements

In the discussion of DC circuits, we noted that EMFs put energy into the electronic system and resistors remove it by conversion to heat. With time-varying currents, we must also consider variations in the energy stored in the electric field associated with every potential difference, and in the magnetic field generated by every current. The ability of a circuit element to store energy in an electric or magnetic field is measured by the element's capacitance C and its inductance L, respectively. Even though present to some degree in all circuit elements (including actual resistors), these field energies are constant and unimportant when all voltages and currents are constant.

In some high-frequency applications, significant amounts of energy can be removed from the electronic domain by electromagnetic radiation resulting from the acceleration of charges; and even though this effect is obviously important to applications such as radio and television transmission, throughout this text we will assume that such radiation losses are small and can be ignored.

The resistance, capacitance, and inductance of a length of ordinary wire are sometimes significant by themselves, especially for long or closely spaced wires. Although these quantities are actually distributed over the length of the wire, it is convenient and usually adequate to localize them into what are known as "lumped impedances." In the DC case this approximation allows us to replace a real wire with an ideal wire and a series resistor equal to the wire's resistance. Distributed impedances are discussed further in Section 12.6.1.

Circuit elements designed specifically to introduce capacitance into a circuit at a particular point are known as capacitors (or sometimes condensers), and those designed for inductance are known as inductors. Collectively, these energy storage devices are known as reactive elements, to distinguish them from energy dissipative resistive elements.

2.2.1 Capacitance

A pair of parallel plates is the usual model for a capacitor in introductory physics, even though relatively few actual capacitors have this simple shape. An example of a parallel-plate capacitor is shown in Figure 2.1. Note that the opposite charges, of magnitude Q, on each of the plates are mutually attracted onto the inside of the conducting plates, where they exactly terminate the electric field lines. Usually some type of dielectric material is placed between the plates to increase the capacitance and to provide mechanical stability.

Whatever the configuration, the capacitance C between two appropriate surfaces is defined by

$$V = \frac{Q}{C} \tag{2.1}$$

where Q is the magnitude of the charge distributed on either surface, and V is the potential difference between the surfaces. The ideal capacitance is a constant, dependent only on the physical configuration of materials, and not a function of either V or Q. Differentiating this equation and using $I = dQ/dt$, we obtain

$$\frac{dV}{dt} = \frac{I}{C} \tag{2.2}$$

Strictly speaking, this charge current I exists only in the wires leading to the capacitor plates, whereas the displacement current I_D spans the gap between the plates and has the same magnitude. In electronics, the quantities I and I_D are merged into a single algebraic current, which passes through the capacitor like current through a resistor.

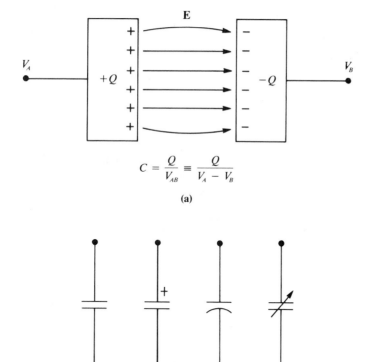

$$C = \frac{Q}{V_{AB}} \equiv \frac{Q}{V_A - V_B}$$

(a)

Electrolytic   Variable
capacitor   capacitor

(b)

Figure 2.1 (a) Parallel-plate capacitor showing charge and field. (b) Schematic symbols for capacitance.

A capacitor serves four principal functions in an electronic circuit. The first follows from Eq. (2.1), the others from Eq. (2.2).

1. Because it can store charge and energy, a capacitor can be used as a (nonideal) source of current and voltage.

2. Since it will pass AC but not DC current, a capacitor can be used to connect parts of a circuit that must operate at different DC voltages.

3. When used in combination with a resistor that limits the current, a capacitor will smooth the sharp edges from a voltage signal.

4. When charged or discharged by a constant current, a capacitor will develop a voltage signal with a constant slope.

The MKS unit of capacitance is the farad (coulomb/volt), but the typical capacitance is measured in microfarads (μF $= 10^{-6}$ farad) or even as small as picofarads (pF $= 10^{-12}$ farad). Commercial capacitors come in several different types, categorized mainly by the dielectric material. Comments on the various types are given in Table 2.1. Capacitors are further distinguished by their "work-

Table 2.1   Capacitor Types

Dielectric	Capacitance Range	Comments
Paper	0.001 μF–5 μF	Obsolete, bulky
Polyester/Mylar	1000 pF–10 μF	*Inexpensive
Polycarbonate	100 pF–10 μF	Stable
Polypropylene	2 pF–0.5 μF	Stable
Polystyrene	2 pF–0.02 μF	Stable
Ceramic	1 pF–5 μF	*Compact, inexpensive
Mica	1 pF–0.01 μF	Stable, bulky
Glass and quartz	1 pf–1000 pF	High precision, stable
Al oxide	1 μF–0.1 F	Polarized, bulky in large sizes
Ta oxide	0.1 μF–0.1 F	*Polarized, compact, low voltage

*Frequently used types.

ing voltage,'' above which the dielectric is likely to break down and become a conductor. The working voltage will vary with the dielectric, the capacitance, and the physical size of the capacitor; it ranges from a few volts to thousands of volts and must be taken into account when choosing a capacitor for an application.

Electrolytic capacitors provide the largest capacitance values but are also the least stable and the most failure-prone. This type uses a very thin layer of oxide to separate one capacitor plate from a conducting electrolytic medium that serves as the other plate. Most electrolytic capacitors are asymmetric devices with a definite polarity: One terminal is labeled positive, and in operation this terminal must always be at a higher potential than the other.

The schematic symbol for capacitance, shown in Figure 2.1b, represents an ideal capacitor and is used to describe any physical configuration that can store significant amounts of electric field energy. It may thus represent either an actual capacitor in the circuit or the ''stray'' capacitance between any two points on the circuit where a potential difference, and hence an electric field, can be developed. Most stray capacitances can be ignored because of their small size, but some may need to be included explicitly on the schematic in order to describe a circuit's behavior satisfactorily.

The stray capacitance associated with some part of a circuit can be estimated from the parallel-plate capacitor model whose equation is

$$C = \frac{\kappa \, \epsilon_0 A}{d} \tag{2.3}$$

where κ is the dielectric constant of the dielectric material, ϵ_0 is the permittivity of free space, A is the area of a plate, and d is the plate separation. Thus we see that a resistor made with a very thin layer of high-resistance material will have a larger stray capacitance than one made from a long cylinder of low-resistance material. Similarly, two long circuit wires that lie close together will have a larger capacitance than two that are widely separated.

An actual capacitor may exhibit other electrical properties besides capacitance. If the dielectric is imperfect (as with a reversed biased electrolytic), the capacitor may pass DC current; the path for this leakage current would be represented schematically by a resistor in parallel with the capacitor. At high frequencies, even the alternating magnetic field generated by the current through the capacitor can be significant, making it necessary to draw the schematic equivalent as a capacitor and an inductor (see below) connected in series.

2.2.2 Inductance

The usual model for an inductor is the solenoid shown in Figure 2.2a. By Faraday's law of self-inductance, a changing current in a circuit induces a back EMF in the circuit that opposes the change, and although there are contributions from all parts of the circuit, the multiple loops of the solenoidal inductor concentrate the effect so that most of the back EMF appears across its terminals. With the sign definitions in Figure 2.2, the inductance L is defined by

$$V = L \frac{dI}{dt} \tag{2.4}$$

Figure 2.2 (a) Solenoidal inductor showing the current and the resulting field. (b) Schematic symbols for inductors.

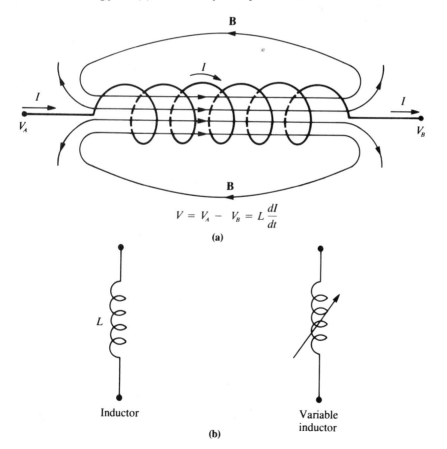

$$V = V_A - V_B = L \frac{dI}{dt}$$

(a)

L

Inductor

Variable inductor

(b)

where *V* is the potential drop across the inductor (the back EMF), and *dI/dt* is the derivative of the current through the indicator. Ideally, the inductance is determined by the physical configuration of materials and is not a function of either *V* or *I*. In the MKS system of units the inductance is measured in henrys (H).

In many ways the inductor is the electronic opposite of the capacitor: It shows no voltage drop if a DC current is applied; and for a given applied voltage the inductor acts to limit the rate of change of the current, a larger inductor producing a smaller *dI/dt*. The inductor thus behaves like a simple wire (the capacitor is like an open circuit) in a DC analysis, and tends to smooth or resist sudden changes in the current (the capacitor smooths sudden changes in the voltage).

Actual inductors are all formed from coils of wire but exhibit significantly different characteristics depending on construction details such as the spacing between the wires and the properties of the core material around which the wire is wrapped. A ferromagnetic core material of iron or ferrite alloy is often used to increase the inductance of a coil at the cost of some nonlinearity between current and field. Typical values for inductors used in signal applications range from less than 1 μH (10^{-6} henry) for a couple of wire loops on an air core to many henrys for a ferrite core inductor. For a given core material, a larger value of inductance requires more turns of wire, resulting in either a physically larger inductor or one with higher intrinsic resistance resulting from the use of smaller-diameter wire.

The schematic symbol for an inductor shown in Figure 2.2b represents an ideal inductor; it is used whenever a circuit configuration exists where energy storage in a magnetic field must be considered. On a schematic, this symbol usually represents the coil of a physical inductor, but it may also be used to describe the stray inductance of any arrangement of conducting wire.

There is no simple formula for calculating the stray inductance associated with a loop of wire in a circuit, but we can use the energy formulas from introductory physics to gain some insight. The energy *U* stored in a magnetic field can be written as

$$U = \frac{1}{2}LI^2 \tag{2.5}$$

where *I* is the current in the loop. Far from ferromagnetic materials, the energy can also be written as the volume integral

$$U = \frac{1}{2\mu_0} \int_{\text{all space}} \mathbf{B}^2 \, dV \tag{2.6}$$

where **B** is the magnetic field strength and μ_0 is the magnetic permeability of free space. Combining these two expressions we obtain

$$L = \frac{1}{\mu_0 I^2} \int_{\text{all space}} \mathbf{B}^2 \, dV \tag{2.7}$$

Note that the inductance is not a function of the current, because the magnetic field **B** is also proportional to *I*. For a given current, the integral of Eq. (2.7) can be reduced by shrinking the area of the current loop so that the field contributions generated by opposing current elements tend to cancel over a larger volume of space. This reduction can be accomplished by making a physically smaller circuit

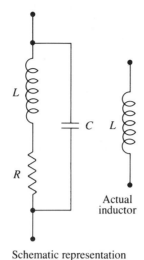

Schematic representation

Figure 2.3   It is usually necessary to represent a physical inductor with three ideal schematic elements.

or by specific wire arrangements such as twisted pairs or coaxial cables. Less dramatically, the stray inductance of a length of wire can also be decreased by increasing the wire diameter.

An actual inductor is far from the ideal inductance represented by the schematic symbol. The coil is often formed from a long length of fine wire with adjacent loops separated only by a thin layer of insulating varnish; the resulting inductor will generally have a significant series resistance and a troublesome parallel capacitance. These physical effects can be represented by ideal schematic circuit elements as shown in Figure 2.3.

2.3   The Circuit Equations

Kirchhoff's loop rule (Eq. 1.17) can still be used to obtain the voltage sum equation around each loop of an AC circuit if we use Eqs. (2.1) and (2.4) to specify the voltage drop across capacitors and inductors. These two equations replace Ohm's law when the Kirchhoff path crosses a reactive element. Application of the loop current method to a general AC network of ideal elements will yield a set of linear differential equations with constant coefficients, one for each inner loop of the circuit. When necessary, this set of equations can be solved by standard methods to yield time-dependent functions for charge, current, and voltage.

The general form of these time-dependent solutions is a "transient response" term describing the circuit's return to equilibrium after all EMFs have been suddenly set to zero, plus a "steady-state" term describing the circuit's long-term behavior when driven by a sinusoidal source. As we will see in the next chapter, the transient and steady-state terms are interrelated to the extent that a full knowledge of one implies the other.

In the next three sections, we will discuss the transient response terms for three simple circuits that do not have signal sources, but that are initially assumed to be in a nonequilibrium condition.

2.3.1   The RC Circuit

The circuit shown in Figure 2.4a has no signal source, but we take the capacitor to be initially charged with $+Q_0$ on the top plate and $-Q_0$ on the bottom. At $t = 0$, the voltage across the charged capacitor is

$$V_0 = \frac{Q_0}{C} \tag{2.8}$$

and produces an initial loop current

$$I_0 = \frac{V_0}{R} \tag{2.9}$$

For $t > 0$, charge flows off the capacitor, causing the voltage across the capacitor and the loop current to decrease with time.

The function describing this time variation is found by writing the Kirchhoff voltage sum equation for the circuit loop. The time derivative of Eq. (1.17) produces an alternative form of Kirchhoff's loop rule,

$$\sum_i \frac{dV_i}{dt} = 0 \tag{2.10}$$

which is better suited to the present circuit. When this sum rule is applied to the circuit in Figure 2.4a, the resistance term is obtained from the derivative of Ohm's

Figure 2.4   (a) RC circuit and (b) the voltage as a function of time. (c) RL circuit.

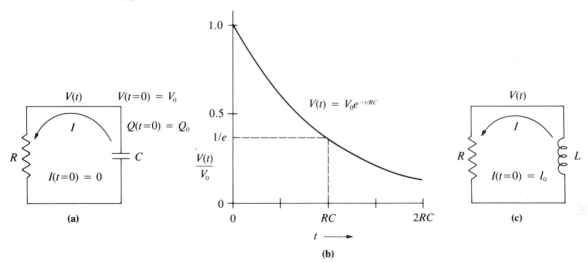

law, $dV/dt = R \, dI/dt$, and the capacitance term from Eq. (2.2). A clockwise traversal of the loop gives

$$\frac{R \, dI}{dt} + \frac{I}{C} = 0 \tag{2.11}$$

This differential equation is easily solved because it can be rewritten with the variables neatly separated on opposite sides of the equation:

$$\frac{dI}{I} = -\frac{1}{RC} \, dt \tag{2.12}$$

The integral from the initial current I_0 at $t = 0$ to a general I at a time t is

$$\int_{I_0}^{I} \frac{dI}{I} = -\frac{1}{RC} \int_0^t dt \tag{2.13}$$

and it integrates to

$$\ln\left(\frac{I}{I_0}\right) = \frac{-t}{RC} \tag{2.14}$$

The solution for $I(t)$ is obtained by taking the antilog of both sides:

$$I(t) = I_0 e^{-t/RC} \tag{2.15}$$

With the loop current known, Ohm's law gives the voltage across the resistor (and capacitor):

$$V(t) = RI(t) = V_0 e^{-t/RC} \tag{2.16}$$

The quantity RC is called the time constant τ of this circuit; it will have units of seconds if R is in ohms and C is in farads. When rewritten in terms of τ, Eq. (2.16) takes on a more general form

$$V(t) = V_0 e^{-t/\tau} \tag{2.17}$$

which will reappear in the discussion of many other circuits. As shown in Figure 2.4b, the time constant τ defines the time needed for $V(t)$ to fall to $1/e$ of its initial value, a result that is easily seen from Eq. (2.17). The RC time constant appears regularly in electronics, whenever a capacitor is charged or discharged through a series resistance.

The RL circuit of Figure 2.4c produces a current expression with the same exponential time dependence as Eq. (2.17) if we assume an initial loop current at $t = 0$. For this circuit, the time constant is found to be L/R; the derivation is left as a problem.

2.3.2   The LC Circuit

A series circuit composed of an inductor and a capacitor is shown in Figure 2.5. We will again assume that the capacitor holds its maximum charge at $t = 0$. Since this circuit has no resistor to remove energy from the electronic system, at

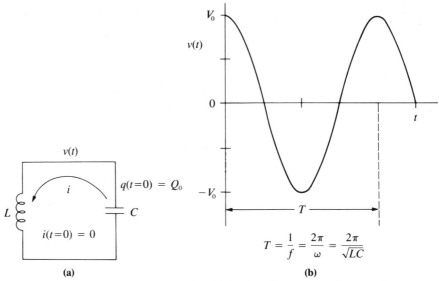

Figure 2.5 *(a) The LC circuit and (b) the voltage across the capacitor as a function of time.*

$t > 0$ we can reasonably expect to find an oscillatory exchange of energy between the fields of the capacitor and inductor. For the current direction shown in Figure 2.5, Kirchhoff's expression for the loop voltage sum is

$$L\frac{di}{dt} + \frac{q}{C} = 0 \tag{2.18}$$

Using $i = dq/dt$, this expression can be converted to a second-order differential equation in q:

$$L\frac{d^2q}{dt^2} + \frac{q}{C} = 0 \tag{2.19}$$

Unlike the differential equation of the *RC* example, this equation cannot be separated for integration. Instead, we must assume a general solution and adapt it to this specific problem. As a trial solution, we take a general sinusoidal expression

$$q(t) = Q_0 \cos(\omega t + \theta) \tag{2.20}$$

If $q(t)$ is to have its maximum value at $t = 0$ as assumed, then θ must be zero. To test the validity of this expression, we also need the time derivatives

$$\frac{dq}{dt} = -Q_0\omega \sin(\omega t) \tag{2.21}$$

and

$$\frac{d^2q}{dt^2} = -Q_0\omega^2 \cos(\omega t)$$

Substitution of this assumed solution for q and d^2/dt^2 into Eq. (2.19) gives

$$-LQ_0\omega^2 \cos(\omega t) + \frac{Q_0}{C} \cos(\omega t) = 0 \qquad (2.22)$$

and we see that the equation will be satisfied only when $\omega^2 = 1/LC$. This specific frequency,

$$\omega_r = \frac{1}{\sqrt{LC}} \qquad (2.23)$$

is called the natural or resonant frequency of the circuit.

The assumed solution of Eq. (2.20) has thus been restricted to an expression that applies specifically to the circuit of Figure 2.5:

$$q(t) = Q_0 \cos(\omega_r t) \qquad (2.24)$$

The derivative of $q(t)$ yields the current through the capacitor,

$$i(t) = \frac{dq}{dt} = -Q_0\omega_r \sin(\omega_r t) \qquad (2.25)$$

and this current can also be expressed as

$$i(t) = -I_0 \sin(\omega_r t) = I_0 \cos\left(\omega_r t + \frac{\pi}{2}\right) \qquad (2.26)$$

The voltage drop across the capacitor (and across the inductor) also has a sinusoidal form and is given by

$$v(t) = \frac{q(t)}{C} = \frac{Q_0}{C} \cos(\omega_r t) = V_0 \cos(\omega_r t) \qquad (2.27)$$

These results confirm the assumption of oscillatory behavior described by sinusoids. Note that the amplitude and phase ($\theta = 0$) are determined by the initial conditions, but that the frequency is determined entirely by the values of the circuit elements.

2.3.3 The LCR Circuit

The introduction of a series resistor into the previous circuit, as shown in Figure 2.6a, results in a more realistic circuit that will ultimately dissipate its initial field energy as resistive heating. The differential equation for this circuit is

$$L\frac{d^2Q}{dt^2} + R\frac{dQ}{dt} + \frac{Q}{C} = 0 \qquad (2.28)$$

and its solution takes three different functional forms (three different transient response functions) as shown in Figures 2.6b–c. The voltage solution is shown, although similar functions also describe the time variation of charge and current. The functional form favored by the circuit depends on the relative size of R with respect to the other components:

1. If $R^2 < 4L/C$, the circuit is said to be underdamped, yielding a transient response function that is the product of a sinusoidal function and an exponential

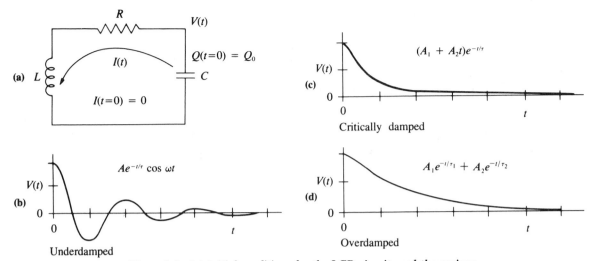

Figure 2.6 *(a) Initial conditions for the LCR circuit, and the various ways in which voltage developed across the capacitor can decay as a function of time: (b) underdamped, (c) critically damped, and (d) overdamped.*

function; the result is an oscillation at an angular frequency slightly less than $1/\sqrt{LC}$ with an amplitude that decays exponentially with time.

2. For $R^2 > 4L/C$, the circuit will be overdamped, and the transient response will be given by the sum of two decaying exponentials without an oscillatory part.

3. For circuit values such that $R^2 = 4L/C$, the circuit is critically damped, with a nonoscillatory decaying response of the form $(A_1 + A_2 t)e^{-t/\tau}$, where A_1, A_2, and τ are constants. Critical damping is particularly important because it separates the two broad classes of underdamped and overdamped solutions.

We will further discuss the underdamped response of this series *LCR* circuit in Section 2.8.1 (*Q* Factor) but will defer a complete analysis of its transient behavior until Chapter 3.

2.4 AC Signals

An AC signal is strictly defined to have a time-averaged value of zero. With this definition, we can think of a general time-varying signal as being composed of a constant (DC) component and an alternating (AC) component. Thus, in a circuit where the current is always in the same direction but fluctuates in magnitude, the total current has a DC part that describes the average current and an AC part that describes the fluctuation.

To help distinguish between the various signals (AC, DC, or combined), we will always use a lowercase letter to identify an AC signal whose time averaged value is zero. A signal designated by a capital letter may be either constant or a general time-varying quantity having both an AC and a DC part. Some examples of signals and their AC parts are shown in Figure 2.7.

All of the signals shown in Figure 2.7 have the additional property of being periodic, meaning that the signal exactly repeats itself after every time interval T, called the period. Most of the mathematical development in the following chapters will apply only to periodic signals, but this restriction has no serious consequences when applied to actual signals. Even if an observed signal is totally random and without periodic content, it can be treated as one that starts to repeat itself from the beginning as soon as we stop observing it. The "period" of such an inherently nonperiodic signal can thus be taken to be the same as the observation time.

Although periodic AC signals can assume any shape that has a time-averaged value of zero, the simplest oscillatory function is the sinusoid. The general sinusoid can be written in several ways, but the most appropriate for our purposes is

$$y(t) = A \cos(\omega t + \theta) \qquad (2.29)$$

Figure 2.7   Time-dependent signals and their AC parts.

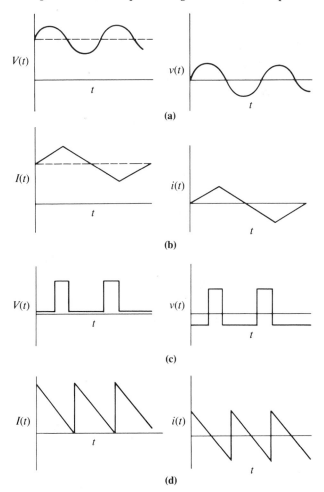

The angle θ is known as the initial phase angle and determines the starting value of the function at $t = 0$. With the proper choice of initial phase, this equation can be used to describe any sinusoidal oscillation of amplitude A and angular frequency ω. In particular, if we set $\theta = \pi/2$, this equation becomes a sine function with $y(0) = 0$ and $(dy/dt)(0) = A$. For mathematical convenience, most of the AC voltage and current signals discussed in this text will have a sinusoidal time dependence describable by the above equation. When a more complicated AC wave shape is present, it can be treated as a linear combination of sinusoidal signals as discussed in Chapter 3.

Figure 2.8 shows an example of two sinusoidal signals that can be described by Eq. (2.29). The signals have a common frequency but different amplitudes and phases. As can be seen from the figure, the signals v_1 and v_2 do not arrive at corresponding points on their cycles at the same time: v_1 is always the first to reach its peaks and zeros. In this situation, we say that the signal v_1 "leads" v_2 or that v_2 "lags" behind v_1. Although this lead or lag between the two signals can be expressed as a time difference ($t_2 - t_1$ in Figure 2.8), it is more common to speak of a phase difference between the two signals. In this figure, v_1 leads v_2 (or v_2 lags behind v_1) by $\pi/4$ radians.

Figure 2.8   Two sinusoidal signals of the same frequency but with different amplitude and phase.

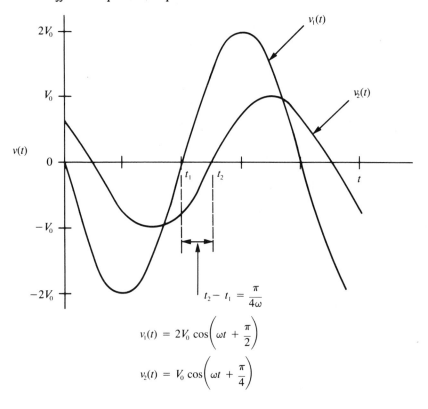

$$v_1(t) = 2V_0 \cos\left(\omega t + \frac{\pi}{2}\right)$$

$$v_2(t) = V_0 \cos\left(\omega t + \frac{\pi}{4}\right)$$

Before proceeding with the analysis of AC circuits, it is convenient to introduce a different mathematical form for oscillatory time dependence. Although the sinusoidal oscillation of charge, voltage, and current in an electronic circuit is properly described by Eq. (2.29), algebraic equations involving trigonometric functions soon become unwieldy when applied to anything but the most elementary circuits. Fortunately, there exists an extremely useful mathematical relation, known as Euler's equation, which connects these geometrically important trigonometric functions to the more easily manipulated algebra of complex exponents.

With the angles expressed in radians, Euler's equation is written

$$e^{j\phi} = \cos \phi + j \sin \phi \qquad (2.30)$$

(To avoid confusion with currents, we have adopted the engineering practice of representing $\sqrt{-1}$ with the symbol j rather than i.) This extremely powerful identity can be verified by replacing each of the three terms with its Taylor series expansion. Equation (2.30) describes a complex number of unit magnitude—the left side as a complex exponential, and the right side in the more common $A + jB$ form. The latter form defines a point on the complex plane at a location (A, B) much like an (X, Y) coordinate on the real plane.

Euler's equation makes it possible to visualize the complex exponential $e^{j\phi}$ as a unit phasor (vector) on the complex plane; the trigonometric expressions in Eq. (2.30) clearly require that the phasor make an angle ϕ with the real axis as shown in Figure 2.9a. Although phasors have some unusual properties, the normal rules of vector addition, subtraction, and multiplication by scalars still apply.

Since Euler's equation connects the algebra of complex exponents on the left with the geometry of the complex plane on the right, it can be used to produce many unexpected relationships. Consider the following unit magnitude examples: $e^{j0} = e^{j2\pi} = 1$, $e^{j\pi/2} = e^{j5\pi/2} = j$, $e^{j\pi} = -1$, $e^{j3\pi/2} = -j$, and so on. Study these examples carefully and use Euler's equation to locate them on the complex plane; it is important to be able to visualize immediately the connection between a complex exponential and the corresponding phasor on the complex plane.

A few definitions will simplify the manipulation of complex numbers and expressions. (Note that we use boldface type to identify a number that is, in general, complex.) If a complex number if given by

$$\mathbf{C} = Ce^{j\phi} = A + jB \qquad (2.31)$$

where A, B, and C are all real, then its complex conjugate $\mathbf{C}^*$ is found by replacing each j with $-j$:

$$\mathbf{C}^* = Ce^{-j\phi} = A - jB \qquad (2.32)$$

The magnitude of a complex number is always real and can be expressed as the square root of the product of the number with its complex conjugate:

$$C = |\mathbf{C}| = \sqrt{\mathbf{C}\mathbf{C}^*} = \sqrt{A^2 + B^2} \qquad (2.33)$$

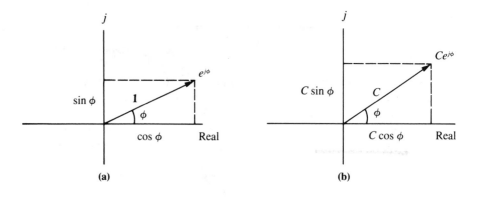

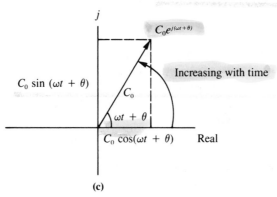

Figure 2.9 *(a) Constant phasor with unit amplitude at an arbitrary angle ϕ with respect to the real axis. (b) A time-dependent phasor of amplitude A that moves counterclockwise as time increases. (c) At t = 0 this phasor makes an angle θ with the real axis.*

From Figure 2.9b, which shows the phasor corresponding to **C**, it is clear that ϕ is related to A and B by the expression

$$\phi = \tan^{-1}\left(\frac{B}{A}\right) \tag{2.34}$$

Euler's equation makes it possible to use a complex exponential to describe a sinusoidal oscillation. If we replace the constant phase angle ϕ with $\omega t + \theta$ and make use of Euler's equation, the complex number of Eq. (2.32) can be written

$$\mathbf{c}(t) = C_0 e^{j(\omega t + \theta)} = C_0 \cos(\omega t + \theta) + jC_0\sin(\omega t + \theta) \tag{2.35}$$

The real part of $\mathbf{c}(t)$ is thus equal to Eq. (2.29). Using the principle of superposition, which applies to all linear differential equations, it can be shown that the full complex expression of $\mathbf{c}(t)$ is an alternative solution to any differential equation satisfied by the real or imaginary part alone.

This new solution does not imply that the physical quantities of charge, current, and voltage are complex. The complex phasor is simply a mathematical convenience, and whenever a comparison is to be made with a laboratory measurement, it is necessary to take the real (or imaginary) part of a phasor to be the physical solution. Nevertheless, it is substantially easier to delay the conversion to a real physical solution until after the conclusion of all algebraic steps. This deferral is correct and proper as long as we are considering only sums and differences of oscillating signals; it fails when signals must be multiplied, as in the calculation of power. (See Section 2.12.)

It is sometimes convenient to rewrite $\mathbf{c}(t)$ with the time dependence as a factor:

$$\mathbf{c}(t) = C_0 e^{j\theta} e^{j\omega t} \tag{2.36}$$

The factor $C_0 e^{j\theta}$ is just a complex constant $\mathbf{C}_0$ and is completely determined by the initial conditions of the problem. In terms of this constant, which carries both amplitude and initial phase information, the complex expression for a general oscillatory signal reduces to

$$\mathbf{c}(t) = \mathbf{C}_0 e^{j\omega t} \tag{2.37}$$

Note that any time-dependent phasor will be instantaneously real when it crosses the real axis at 0 and π radians and likewise imaginary at $\pi/2$ and $3\pi/2$ radians. The phasor $\mathbf{c}(t)$ is shown in Figure 2.9c at a particular time when $\omega t + \theta = \pi/4$. Because the time dependence of the complex exponent of $\mathbf{c}(t)$ causes the angle between the phasor and the positive real axis to increase continually, the phasor must spin around the origin in a counterclockwise direction as time increases.

Example 2.1 Show that $e^{j\theta} + e^{-j\theta} = 2 \cos \theta$.

With Euler's equation, we first expand both exponential terms back to their $a + jb$ form:

$$e^{j\theta} + e^{-j\theta} = \cos \theta + j \sin \theta + \cos(-\theta) + j \sin(-\theta)$$

Combining the real and imaginary terms separately and using the relations

$$\cos(-\theta) = \cos \theta$$

and

$$\sin(-\theta) = - \sin \theta$$

we get

$$e^{j\theta} + e^{-j\theta} = 2 \cos \theta + j(\sin \theta - \sin \theta) = 2 \cos \theta$$

This result is useful for converting complex voltage differences to observable quantities, as we will see in Example 2.2

2.6 *AC Signal Sources*

For the purposes of this chapter, an AC signal source is defined as any EMF capable of providing a time-dependent voltage or current signal of the form

$$v(t) = V_0 \cos(\omega t + \theta)$$

or

$$i(t) = I_0 \cos(\omega t + \phi)$$

(2.38)

The schematic representations of ideal AC EMFs are shown in Figure 2.10. As with direct current circuits, when a signal is produced by the output of an actual device, a signal generator or just another circuit, it must be described schematically by an ideal EMF plus additional circuit elements; this problem will be discussed further in Section 2.11, which introduces AC equivalent circuits.

For analysis purposes, we will replace the real functions of Eqs. 2.38 with the complex expressions

$$\mathbf{v}(t) = V_0 e^{j(\omega t + \theta)}$$

and

$$\mathbf{i}(t) = I_0 e^{j(\omega t + \theta)}$$

(2.39)

The real parts of these functions match Eqs. 2.38.

If the driving AC EMFs in a problem are described by these complex functions, then all steady-state AC signals will take on this same complex form, differing from the source only in amplitude and phase. To avoid carrying unnecessary symbols, one EMF in a circuit is generally defined to have a zero phase angle, equivalent to setting $t = 0$ at the peak of this signal. All other signals in the problem will then have their phase angles referred to this EMF. A sinusoidal voltage or current signal of a given frequency is thus described by two parameters, amplitude and phase, instead of the single parameter needed for a DC signal.

Example 2.2 An oscilloscope is used to observe the difference $\mathbf{v}_A - \mathbf{v}_B$ of two AC signals described by $\mathbf{v}_A = V_A e^{j(\omega t + \theta_A)}$ and $\mathbf{v}_B = V_B e^{j(\omega t + \theta_B)}$. What will be the amplitude of the displayed signal?

Figure 2.10 Schematic symbols for ideal AC voltage and current signal sources.

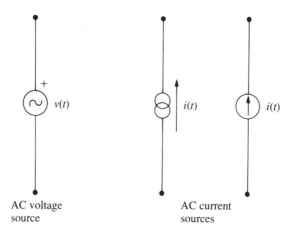

AC voltage
source

AC current
sources

The phasor representation of two signals and their difference at $t = 0$ is shown in Figure 2.11. As time progresses, all three phasors will rotate counterclockwise around the origin with angular frequency ω, but will maintain their relative positions, like a rigid body. Since the oscilloscope displays the real part of $\mathbf{v}_A - \mathbf{v}_B$, the magnitude of this difference phasor corresponds to the amplitude of the observed signal.

The two signals have the same angular frequency, allowing the $e^{j\omega t}$ term to be factored out of the difference signal:

$$\mathbf{v}_A - \mathbf{v}_B = (V_A e^{j\theta_A} - V_B e^{j\theta_B})e^{j\omega t}$$

Using Eq. 2.33, the amplitude of this difference can be written

$$|\mathbf{v}_A - \mathbf{v}_B| = \sqrt{(V_A e^{j\theta_A} - V_B e^{j\theta_B})(V_A e^{-j\theta_A} - V_B e^{-j\theta_B})}$$

Note that the time dependence has disappeared from this expression, since $e^{j\omega t}e^{-j\omega t} = 1$. Expanding the product and collecting similar terms yields

$$|\mathbf{v}_A - \mathbf{v}_B| = \sqrt{V_A^2 + V_B^2 - V_A V_B[e^{j(\theta_A - \theta_B)} + e^{-j(\theta_A - \theta_B)}]}$$

Using the result of Example 1.1, the magnitude can be written in the more obvious form

$$|\mathbf{v}_A - \mathbf{v}_B| = \sqrt{V_A^2 + V_B^2 - 2V_A V_B \cos(\theta_A - \theta_B)}$$

Note that the observed amplitude depends on the amplitude of $\mathbf{v}_A$ and $\mathbf{v}_B$ and also on their phase difference.

Figure 2.11 *Phasor representation of the signals of Example 2.2 at $t = 0$.*

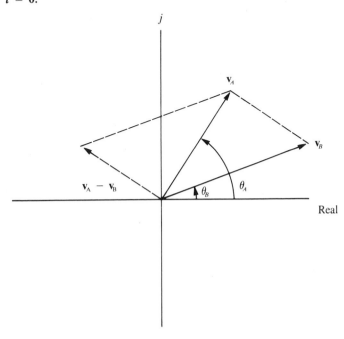

2.7   Circuits with Sinusoidal Sources and the Complex Impedance

Circuits loops containing sinusoidal voltage or current sources can be described by a differential equation that has a sinusoidal, time-dependent term for each source. Equations of this type are classified as inhomogeneous and always have solutions that contain a sinusoidal term at the frequency of each source. The solution for a particular problem thus reduces to finding the amplitude and phase of each of these sinusoidal terms.

A standard method for solving inhomogeneous differential equations is to assume a solution (in our case a loop current), differentiate it as needed, then substitute it back into the original equation. Invariably, the time dependence can be canceled, converting the differential equation to an algebraic equation in frequency and phase variables. By using the complex impedance **Z**, it is possible to skip the differential equation step and write the algebraic equations directly.

In the following discussion, we will show that the complex expression

$$\mathbf{v}(\omega, t) = \mathbf{Z}(\omega)\, \mathbf{i}(\omega, t) \tag{2.40}$$

can be used as an extension of Ohm's law to AC circuits. The voltage and current in this expression are rotating phasors with the functional form of Eq. (2.37). It is often convenient to cancel the common $e^{j\omega t}$ time dependence, producing a simpler expression connecting the complex amplitudes of these signals:

$$\mathbf{v}(\omega) = \mathbf{Z}(\omega)\, \mathbf{i}(\omega) \tag{2.41}$$

In the laboratory, we observe a real signal that we can associate with the real (or imaginary) part of these time-dependent phasors. Since the amplitude of a real signal is equal to the magnitude of its corresponding phasor, the magnitude of Eq. (2.41),

$$|\mathbf{v}(\omega)| = |\mathbf{Z}(\omega)|\, |\mathbf{i}(\omega)| \tag{2.42}$$

will describe the observed relationship between the amplitudes of the real, time-dependent signals.

2.7.1   Resistive Impedance

An expression for the resistive impedance $\mathbf{Z}_R$ can be derived using the simple AC circuit of Figure 2.12a. Using complex notation for voltages and currents, the application of Kirchhoff's voltage law around this loop yields

$$\mathbf{v}(t) = R\mathbf{i}(t) \tag{2.43}$$

Since this equation must be true at all times, $\mathbf{v}(t)$ and $\mathbf{i}(t)$ must have the same time dependence. If we assume a sinusoidal current signal of the form

$$\mathbf{i}(t) = \mathbf{i}e^{j\omega t} \tag{2.44}$$

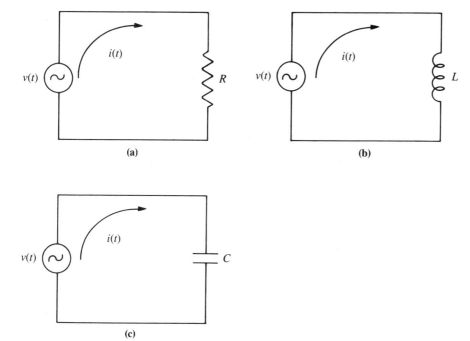

Figure 2.12 *An AC source driving (a) a resistor, (b) an inductor, and
(c) a capacitor.*

then the voltage signal must be given by

$$\mathbf{v}(t) = \mathbf{v}e^{j\omega t} \tag{2.45}$$

where

$$\mathbf{v} = R\mathbf{i} \tag{2.46}$$

Note that the complex amplitudes $\mathbf{i}$ and $\mathbf{v}$ define the initial conditions of both
phase and amplitude. Comparison with Eq. (2.41) shows that the impedance of a
resistor is given by

$$\mathbf{Z}_R = R \tag{2.47}$$

This impedance is frequency independent, so that the voltage developed across a
resistor depends only on the amplitude of the current, not on its frequency.

2.7.2 *Inductive Impedance*

The derivation of the inductive impedance $\mathbf{Z}_L$ follows the same technique. Using
the defining equation for inductance, Kirchhoff's voltage law applied to the loop
of Figure 2.12b yields

$$\mathbf{v}(t) - \frac{L\,d\mathbf{i}(t)}{dt} = 0 \tag{2.48}$$

The current $i(t)$ and voltage $v(t)$ are again given by Eqs. (2.44) and (2.45), and the derivative of the current is

$$\frac{d\mathbf{i}(t)}{dt} = j\omega \mathbf{i}e^{j\omega t} \tag{2.49}$$

Substituting for di/dt in Eq. (2.48), we obtain

$$\mathbf{v}(t) = j\omega L \mathbf{i}e^{j\omega t} \tag{2.50}$$

or

$$\mathbf{v}(t) = j\omega L \mathbf{i}(t) \tag{2.51}$$

Note that $\mathbf{v}(t)$ and $\mathbf{i}(t)$ must again have the same time dependence, and that it can be canceled to produce the time-independent equation

$$\mathbf{v} = j\omega L \mathbf{i} \tag{2.52}$$

Comparison with Eq. (2.41) shows that the impedance of an inductor is given by

$$\mathbf{Z}_L = j\omega L \tag{2.53}$$

Since an inductor produces a back EMF only in response to a changing current, it is not surprising that this impedance is zero for DC signals and increases with the frequency. Thus, a constant current produces no voltage drop across an inductor.

2.7.3 Capacitive Impedance

To determine the capacitive impedance $\mathbf{Z}_C$, we write the voltage equation for the loop of Figure 2.12c,

$$\mathbf{v}(t) - \frac{q}{C} = 0 \tag{2.54}$$

then differentiate it to get

$$\frac{d\mathbf{v}(t)}{dt} - \frac{\mathbf{i}(t)}{C} = 0 \tag{2.55}$$

Assuming that the driving voltage is given by Eq. (2.45), its derivative is

$$\begin{aligned}\frac{d\mathbf{v}(t)}{dt} &= j\omega \mathbf{v}_0 e^{j\omega t} \\ &= j\omega \mathbf{v}(t)\end{aligned} \tag{2.56}$$

Substituting for dv/dt in Eq. (2.55) and rearranging gives

$$\mathbf{v}(t) = \left(\frac{1}{j\omega C}\right)\mathbf{i}(t) \tag{2.57}$$

The time dependence again cancels, and comparison with Eq. (2.41) shows that the impedance of a capacitor is given by

$$\mathbf{Z}_C = \frac{1}{j\omega C} \tag{2.58}$$

A constant current sees the capacitor as a nonconductor—an open circuit—while a time-varying current can pass through the capacitor by means of the displacement current. This behavior is reflected in the variation of $\mathbf{Z}_C$ with frequency, infinite for DC signals and decreasing with increasing frequency.

2.7.4 Combined Impedances

Except for being complex, Eq. (2.41) can be used just as Ohm's law is used for DC circuits. If we use complex voltages and currents, and express the impedances with

$$\mathbf{Z}_R = R$$
$$\mathbf{Z}_L = j\omega L$$

and

$$\mathbf{Z}_C = \frac{1}{j\omega C} = \frac{-j}{\omega C}$$

(2.59)

then the techniques of DC circuit analysis can be applied to AC circuits. In particular, we can use Eq. (2.41) to define the voltage changes in a Kirchhoff loop equation, or we can combine complex impedances in series or parallel using the same rules developed earlier for resistors:

$$\text{Series} \quad \mathbf{Z}_{eq} = \mathbf{Z}_1 + \mathbf{Z}_2$$

$$\text{Parallel} \quad \mathbf{Z}_{eq} = \frac{\mathbf{Z}_1 \mathbf{Z}_2}{\mathbf{Z}_1 + \mathbf{Z}_2}$$

(2.60)

As an example, consider two capacitors C_1 and C_2 connected in parallel; since this connection increases the plate area available for charge storage, we know from elementary physics that it produces an equivalent capacitance of $C_1 + C_2$. From Eq. (2.60), the equivalent impedance is

$$\mathbf{Z}_{eq} = \frac{(1/j\omega C_1)(1/j\omega C_2)}{(1/j\omega C_1) + (1/j\omega C_2)}$$

(2.61)

and this expression reduces to

$$\mathbf{Z}_{eq} = \frac{1}{j\omega(C_1 + C_2)} = \frac{1}{j\omega C_{eq}}$$

(2.62)

as expected.

The general equivalent impedance $\mathbf{Z}_{eq}$ arising from a combination of fundamental elements will have both real (resistive) and imaginary (reactive) parts: the general form is

$$\mathbf{Z}_{eq} = R + jX$$

(2.63)

where the real quantity X is known as the reactance. The reactance of an inductor is ωL, and the reactance of a capacitor is $-1/\omega C$.

The impedance can also be expressed as an exponential,

$$\mathbf{Z}_{eq} = Z_0 e^{j\theta} \tag{2.64}$$

with the magnitude Z_0 given by

$$Z_0 = |\mathbf{Z}| = \sqrt{\mathbf{Z}^*\mathbf{Z}} = \sqrt{(R - jX)(R + jX)}$$

$$Z_0 = \sqrt{R^2 + X^2} \tag{2.65}$$

and the phase angle by

$$\theta = \arctan\left(\frac{X}{R}\right) \tag{2.66}$$

A phasor representing this general complex impedance is shown in Figure 2.13a. If X is positive, this phasor will lie in the first quadrant; if X is negative, the phasor will lie in the fourth quadrant. Since the reactance X is always a function of frequency, its presence will cause the magnitude and phase of a general impedance to vary with frequency.

*Figure 2.13   (a) A typical complex impedance phasor plotted on the complex plane. (b) The complex impedance for the series LCR circuit is the sum of one real and two imaginary terms, which sum like two-dimensional vectors to form the total **Z**.*

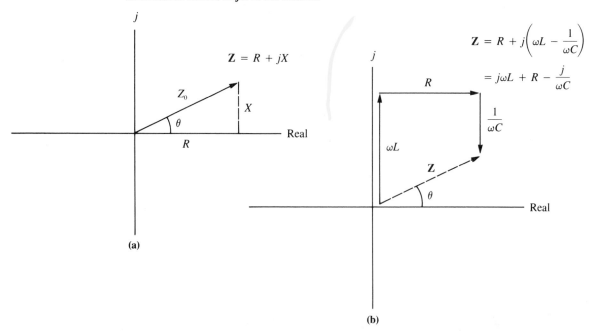

The equivalent impedance of a resistor, inductor, and capacitor connected in series is

$$\begin{aligned}
\mathbf{Z}_{eq} &= R + j\omega L + \frac{1}{j\omega C} \\
&= R + j\omega L - \frac{j}{\omega C} \\
&= R + j\left(\omega L - \frac{1}{\omega C}\right)
\end{aligned} \qquad (2.67)$$

Figure 2.13b shows this equivalent impedance as the vector sum of three phasors, R, $j\omega L$, and $-j/\omega C$. The change in $\mathbf{Z}_{eq}$ with frequency is evident from this figure: As the frequency increases, ωL gets larger and $1/\omega C$ gets smaller, causing the angle θ to increase toward its limiting value of $+90°$; as the frequency decreases, the angle will decrease, causing $\mathbf{Z}$ to be real at a frequency $\omega = 1/\sqrt{LC}$, and ultimately to approach its other limit at $-90°$.

Note that at the special frequency $\omega = 1/\sqrt{LC}$, the effects of the inductance and capacitance cancel, and the impedance becomes purely resistive.

Example 2.3 A signal generator with output impedance R_s is used to drive the circuit shown in Figure 2.14. Find the current $\mathbf{i}_s$.

It is convenient to work this problem by first reducing the inductor and three capacitors to an equivalent impedance. Using series and parallel impedance rules, the three capacitors reduce to an effective impedance of $2/(3j\omega C)$, which corresponds to a capacitance of $3C/2$. This combined capacitance is in series with the inductance

Figure 2.14 *The reduction of a two-loop circuit to a series circuit by use of an equivalent impedance.*

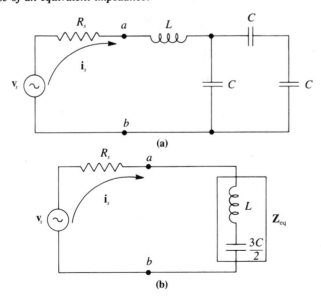

L, and we therefore find that the equivalent impedance of the inductor-capacitor combination is

$$\mathbf{Z}_{eq} = j\omega L + \frac{2}{j3\omega C}$$

With the circuit reduced to a single series loop, the expression for the current $\mathbf{i}_s$ is

$$\mathbf{i}_s = \frac{\mathbf{v}_s}{R_s + j\omega L + 2/(j3\omega C)}$$

2.7.5   Phase Angle Between Voltage and Current

The complex form of Ohm's law connects not only the amplitude of the voltage and current, but also their phases. This relationship is seen most clearly when the impedance in Ohm's law is expressed in the exponential form of Eq. (2.64):

$$\mathbf{v}(t) = Z_0 e^{j\theta}\mathbf{i}(t) \tag{2.68}$$

If we specify that the current has a phase angle of zero such that $\mathbf{i}(t) = I_0 e^{j\omega t}$, then the voltage developed across the impedance is given by

$$\mathbf{v}(t) = Z_0 I_0 e^{j(\omega t + \theta)} \tag{2.69}$$

The relationship between the voltage and current phasors just described is shown in Figure 2.15. The voltage and current phasors both spin counterclockwise as time increases, with the voltage leading (lagging if θ is negative) the current by an angle θ that is determined by the impedance.

If we suppress the $e^{j\omega t}$ time dependence and write Ohm's law as

$$\mathbf{v} = Z_0 e^{j\theta}\mathbf{i} \tag{2.70}$$

the meaning of θ is unchanged; no matter what angle we assign to the phasor $\mathbf{i}$ on the complex plane, the angle of phasor $\mathbf{v}$ will differ by θ. It is instructive to evaluate this relative phase angle for the elementary impedances, $\mathbf{Z}_R$, $\mathbf{Z}_L$, and $\mathbf{Z}_C$.

Figure 2.15   *A general phasor diagram showing the relationship between the voltage across an equivalent impedance and the current through it.*

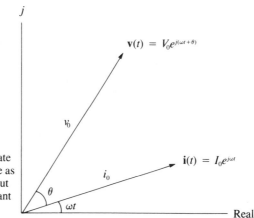

Both phasors rotate counterclockwise as time increases, but maintain a constant relative angle θ.

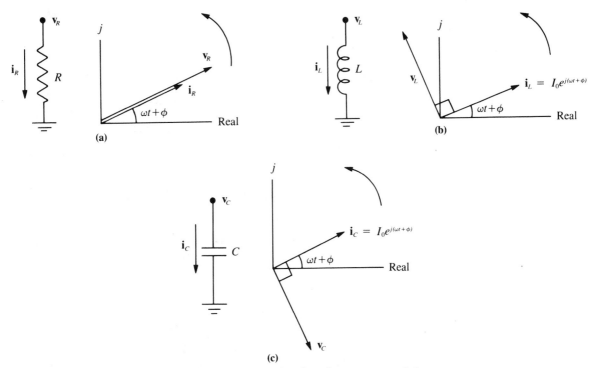

Figure 2.16 *Phasors representing the voltage across and the current through the elementary impedances. The current is assumed to have an arbitrary initial phase* ϕ.

When an impedance is composed of a simple resistor, we have

$$\mathbf{v}_R = R\mathbf{i}_R \tag{2.71}$$

Since R is a real number, its product with $\mathbf{i}_R$ changes only the length of the phasor and not its direction; the phasor $\mathbf{v}_R$ thus lies along the phasor $\mathbf{i}_R$, and the voltage across the resistor is in phase with the current through it as shown in Figure 2.16a.

When the impedance is an inductor, the voltage-current relationship becomes

$$\mathbf{v}_L = j\omega L\mathbf{i}_L \tag{2.72}$$

and since $j = e^{j\pi/2}$, we can write

$$\mathbf{v}_L = e^{j\pi/2}\omega L\mathbf{i}_L \tag{2.73}$$

Thus, the voltage across an inductor *leads* the current through it by 90° as shown in Figure 2.15b.

When the impedance is a capacitor, the voltage is related to the current by

$$
\begin{aligned}
\mathbf{v}_C &= \frac{1}{j\omega C}\mathbf{i}_C \\
&= -\frac{j}{\omega C}\mathbf{i}_C \\
&= \frac{1}{\omega C}e^{-j\pi/2}\mathbf{i}_C
\end{aligned}
\tag{2.74}
$$

This time the voltage phasor makes an angle of $-90°$ with the current phasor; it *lags* behind the current phasor by $90°$ as shown in Figure 12.14c.

The relative angle θ is typically a function of the frequency and for combinations of the elementary impedances can lie anywhere in the range $-90°$ to $+90°$. We will later introduce circuit elements capable of inverting a signal (multiplying it by -1), thereby producing phase differences that lie in the full range $-180°$ to $+180°$.

These methods can also be used to determine the relative phase between two voltages or two currents at different points on a circuit.

Example 2.4 When a 1-kHz, 1.0-V amplitude signal v_A is applied to the input of the circuit shown in Figure 2.17, the output signal v_B is found to have a 10-mV amplitude and to be lagging behind v_A by $30°$. Determine L and R_L for the actual inductor.

If we use v_A to define a phase angle of zero, then this 1-V amplitude, 6283 rad/s signal can be described by

$$v_A = 1.0e^{j6283t} \text{ V}$$

and the lagging, 10-mV amplitude v_B by

$$v_B = 0.01e^{(j6283t - \pi/6)} \text{ V}$$

Using the complex version of Ohm's law, the complex loop current $\mathbf{i}$ can be obtained from the voltage drop v_B across the 10-Ω resistor,

$$\mathbf{i} = \frac{v_B}{R}$$

giving

$$\mathbf{i} = 0.001e^{(j6283t - \pi/6)} \text{ mA}$$

Ohm's law applied to the unknown impedance $\mathbf{Z}$ would read

$$v_A - v_B = \mathbf{Zi}$$

but R is quite small, making the amplitude of v_B only about 0.01 V. This small term can be neglected with respect to the much larger v_A, and the equation can be rewritten as

$$\mathbf{Z} = \frac{v_A}{\mathbf{i}}$$

Figure 2.17   Circuit used to measure an unknown inductor.

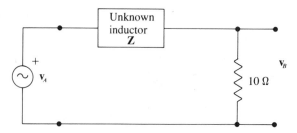

Substituting in the expressions for $\mathbf{v}_A$ and $\mathbf{i}$ yields

$$\mathbf{Z} = \frac{1.0e^{j6283t}}{0.001e^{j(6283t-\pi/6)}} = 1000e^{j\pi/6}\ \Omega$$

and Euler's equation gives

$$\mathbf{Z} = 1000\cos(30°) + j1000\sin(30°)\ \Omega$$
$$= 866 + j500\ \Omega$$

This result corresponds to an impedance of the form $R_L + j\omega L$, which for a 1-kHz signal frequency is

$$\mathbf{Z} = R_L + j6283L$$

Term-by-term comparison of these last two equations shows that the physical inductor has a resistance of 866 Ω and an inductance of 79.6 mH, but because of the approximations, the last digit in both of these answers is certainly suspect.

Although it is a lengthy task, the serious student will find it instructive to estimate the errors in these two results caused by 10% uncertainties in both the measurements (voltage and phase) and the value of the 10-Ω resistor, and then to compare these with the error introduced by neglecting $\mathbf{v}_B$ as was done above.

2.8   Resonance

We have already seen that circuits containing both capacitance and inductance, such as those shown in Figures 2.5 and 2.6, can exhibit an oscillatory transient response function. These electrical oscillators are very similar to mechanical systems such as a tuning fork or pendulum. Whereas the oscillation of a mechanical system results from an energy flow back and forth between kinetic and potential,

Figure 2.18   (a) A series LCR circuit and (b) its resonant behavior.

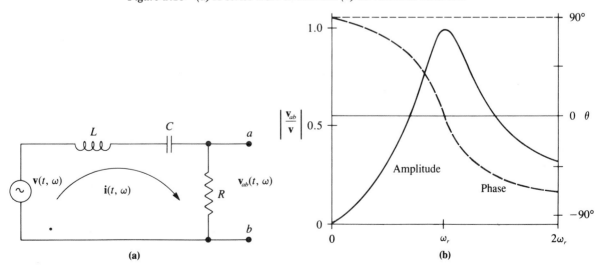

in an electrical system the flow is between energy stored in magnetic fields (kinetic) and energy stored in electric fields (potential).

As in a mechanical system that has been disturbed by a sharp blow, the natural frequency of an electrical circuit is most apparent when the circuit is disturbed (by an applied EMF) and then allowed to return freely to equilibrium. However, when such a circuit is driven by a sinusoidal EMF whose frequency is slowly varied, the natural frequency manifests itself in the phenomenon known as resonance.

The series *LCR* connection shown in Figure 2.18a is the simplest schematic circuit that exhibits the full range of resonant behavior. For this analysis, the frequency ω of the sinusoidal driving EMF will be treated as an independent variable, and this dependence will be emphasized by writing the EMF as $\mathbf{v}(j\omega)$. [The use of $j\omega$ as the independent variable rather than ω has no earth-shaking significance; j is a constant, so ω is still the variable. This notation is used for the sake of mathematical correctness, since we wish to retain various function symbols when making the transition to the **s**-plane notation of the next chapter.]

In complex notation, Ohm's law applied to the entire loop gives the loop current

$$\mathbf{i}(j\omega) = \frac{\mathbf{v}(j\omega)}{\mathbf{Z}} \tag{2.75}$$

where

$$\mathbf{Z} = R + j\left(\omega L - \frac{1}{\omega C}\right)$$

The voltage $\mathbf{v}_{ab}$ developed across the series resistor as a result of the current $\mathbf{i}$ is given by

$$\mathbf{v}_{ab}(j\omega) = R\mathbf{i}(j\omega) \tag{2.76}$$

Using Eq. (2.75) to eliminate $\mathbf{i}$, we obtain an expression for the voltage developed across the series resistor R in terms of the applied voltage $\mathbf{v}(j\omega)$:

$$\mathbf{v}_{ab}(j\omega) = \frac{R}{R + j(\omega L - 1/\omega C)}\, \mathbf{v}(j\omega) \tag{2.77}$$

This has the general form

$$\mathbf{v}_{ab}(j\omega) = \mathbf{H}(j\omega)\mathbf{v}(j\omega) \tag{2.78}$$

where $\mathbf{H}(j\omega)$ is known as the transfer function in the frequency domain. For this circuit, $\mathbf{H}$ can be expressed as an impedance ratio:

$$\mathbf{H}(j\omega) = \frac{\mathbf{v}_{ab}}{\mathbf{v}} = \frac{R}{R + j(\omega L - 1/\omega C)} \tag{2.79}$$

The transfer function $\mathbf{H}$ is useful because it describes many of a circuit's more important characteristics without reference to a specific input signal. In the present discussion, $\mathbf{H}$ contains all the information needed to characterize the resonance behavior of the *LCR* circuit.

The transfer function $\mathbf{H}$ can be put into the standard $a + jb$ form by multiplying numerator and denominator by the complex conjugate of the denominator:

$$\mathbf{H}(j\omega) = a(j\omega) + jb(j\omega) = \frac{R}{R + j(\omega L - 1/\omega C)} \cdot \frac{R - j(\omega L - 1/\omega C)}{R - j(\omega L - 1/\omega C)}$$

$$= \frac{R^2 - jR(\omega L - 1/\omega C)}{R^2 + (\omega L - 1/\omega C)^2} \tag{2.80}$$

We can also write $\mathbf{H}(j\omega)$ in exponential form,

$$\mathbf{H}(j\omega) = H(j\omega)e^{j\theta(\omega)} \tag{2.81}$$

where the real amplitude $H(j\omega)$ is

$$H(j\omega) = \frac{R}{\sqrt{R^2 + (\omega L - 1/\omega C)^2}} \tag{2.82}$$

and the phase $\theta(\omega)$ is

$$\theta(\omega) = \tan^{-1}\left[\frac{1/\omega C - \omega L}{R}\right] \tag{2.83}$$

As is typical of most problems, neither the $a + jb$ nor the polar form for $\mathbf{H}(j\omega)$ is as algebraically compact as the original impedance ratio form of Eq. (2.79).

If Eq. (2.78) is rewritten using the phasor form of Eq. (2.81), we get

$$\mathbf{v}_{ab} = He^{j\theta}\mathbf{v} \tag{2.84}$$

where all four variables are still functions of ω. Despite its complexity when expanded fully, the polar form for $\mathbf{H}$ used in this equation shows how $\mathbf{H}$ determines the phase and amplitude relationship between the two voltage phasors $\mathbf{v}_{ab}$ and $\mathbf{v}$. For example, if at some instant t_1, $\mathbf{v} = ve^{j\omega t_1}$, then at that instant $\mathbf{v}_{ab} = Hve^{j(\omega t_1 + \theta)}$.

Equation (2.82) shows that the magnitude of $\mathbf{H}(j\omega)$ is a maximum when the reactance term $\omega L - 1/\omega C$ is zero. This condition is known as resonance, and occurs when the frequency of the driving signal $\mathbf{v}$ equals the natural or resonant frequency of the circuit, given by

$$\omega_r = \frac{1}{\sqrt{LC}} \tag{2.85}$$

Equation (2.83) shows that for this circuit, the phase angle θ is zero at resonance. This phase angle, which connects the input and output signals as defined by Eq. (2.84), changes with the reactance $X = \omega L - 1/\omega C$. As specified by Eq. (2.83), θ is positive at low frequencies below resonance and negative at high frequencies, approaching limiting values of $+\pi/2$ and $-\pi/2$ radians at the respective extremes.

2.8.1 Q Factor

An important parameter used to describe a resonant circuit is its Q, or quality factor. This factor gives a quick measure of the strength or significance of a resonance, with a large Q identifying a highly resonant circuit. The Q is most simply defined by

$$Q = \frac{X_L}{R} = \frac{\omega L}{R} \qquad (2.86)$$

where R is the resistance in series with the inductor. This definition refers specifically to a circuit, but in a high-Q circuit where R is mainly the intrinsic resistance of the inductor, the inductor itself is often described as having a high or low Q. The capacitance, which is a necessary part of any resonant circuit, is not explicit in this definition but enters indirectly when ω is replaced by ω_r as discussed below.

The Q factor of a circuit is more fundamentally defined in terms of the energy flow in the circuit: The Q of a resonance is equal to 2π times the stored energy divided by the energy lost per cycle. The stored energy resides in the electric and magnetic fields of the capacitances and inductances, respectively, and we assume that the only energy loss from the circuit is through resistive heating.

These two definitions of Q are not exactly equivalent, but approach the same value as Q increases. Practically, if $Q > 10$ (meaning that less than a tenth of the stored energy is lost per cycle), the difference in the definitions is insignificant.

The energy loss definition of Q is most easily related to Eq. (2.86) by considering the underdamped voltage decay of the undriven series LCR circuit shown in Figure 2.6. Using the equation $V = Q/C$ (this Q is charge) together with its first and second derivatives, Eq. (2.28) for the charge flow in the LCR circuit can be converted to

$$LC \frac{d^2V}{dt} + RC \frac{dV}{dt} + V = 0 \qquad (2.87)$$

where V is the voltage across the capacitor. If we assume a complex decaying oscillatory solution for $V(t)$ of the form

$$\mathbf{V}(t) = V_0 e^{-\sigma t} e^{j\omega t} \qquad (2.88)$$

and substitute it and its derivatives into Eq. (2.87), we find that this $\mathbf{V}(t)$ is a solution only if

$$\sigma = \frac{R}{2L}$$

and

$$\omega = \sqrt{\frac{1}{LC} - \frac{R^2}{4L^2}} \qquad (2.89)$$

Although derived for the series LCR circuit, this oscillating voltage decay is typical of any highly resonant (large Q factor) circuit that has been excited and then left to return to equilibrium without further disturbance.

The energy stored in a capacitor at any instant is proportional to the square of the physical voltage across the capacitor. Since the physical voltage is just the real part of Eq. (2.88), squaring the whole complex expression will not give the correct result. However, Eq. (2.88) for $\mathbf{V}(t)$ describes a slowly shrinking phasor that is spinning around the origin on the complex plane, a phasor that is real whenever $e^{j\omega t} = 1$. This condition occurs at times

$$t_n = nT \qquad \text{for } n = 0, 1, 2, 3, \ldots \qquad (2.90)$$

where T is the period of the oscillation

$$T = \frac{1}{f} = \frac{2\pi}{\omega} \qquad (2.91)$$

For reasonably large values of Q, these times correspond closely to maxima in the real part of $\mathbf{V}(t)$ and thus identify those times when the voltage across the capacitor reaches a maximum. These are also the times when the current and its associated magnetic field are zero, moving the total stored energy to the electric field of the capacitance. At these particular times, the total energy in the circuit is proportional to the square of the voltage across the capacitor. We can therefore use Eqs. (2.88) and (2.90) to write

$$E(t_n) = E_0 e^{-2\sigma nT}$$
$$\qquad\qquad\qquad\qquad (2.92)$$
and
$$E(t_{n+1}) = E_0 e^{-2\sigma(n+1)T}$$

where the factor 2 comes from squaring the voltage. The energy definition of Q is then

$$Q = \frac{2\pi E_n}{E_n - E_{n+1}} \qquad (2.93)$$

or

$$Q = \frac{2\pi e^{-2\sigma nT}}{e^{-2\sigma nT} - e^{-2\sigma(n+1)T}} \qquad (2.94)$$

$$= \frac{2\pi}{1 - e^{-2\sigma T}}$$

For large Q, σT will be small and the exponential can be approximated by the first two terms of its Taylor series ($e^x = 1 + x + x^2/2! + \ldots$), yielding

$$Q = \frac{2\pi}{2\sigma T} \qquad (2.95)$$

This expression reduces to Eq. (2.86) if we replace σ with $R/2L$ and convert the period back to an angular frequency with $T = 2\pi/\omega$.

The Q factor is most often of interest at or near the resonant frequency of a circuit where it is directly related to the shape of the $|\mathbf{H}(j\omega)|$ curve plotted against frequency. Near the resonant frequency, ω can be replaced by the constant ω_r, thereby making Q a constant rather than a function of ω. Explicit reference to the circuit elements R, L, and C in Eq. (2.79) can be eliminated by substituting expressions derived from those for ω_r and Q:

$$\omega_r = \frac{1}{\sqrt{LC}} \qquad (2.96)$$

and

$$Q = \frac{\omega_r L}{R} \qquad (2.97)$$

With this substitution, Eq. (2.79) becomes

$$\mathbf{H}(j\omega) = \frac{1}{1 + jQ(\omega/\omega_r)[1 - (\omega_r/\omega)^2]}$$ (2.98)

The magnitude of this expression is plotted in Figure 2.19 for various values of Q. Note that larger Q values correspond to narrower resonance curves. In fact, Q can also be expressed as

$$Q = \frac{\omega_r}{\omega_2 - \omega_1}$$ (2.99)

where ω_1 and ω_2 are the frequencies in Figure 2.19 that satisfy the condition

$$|\mathbf{H}(j\omega)| = \frac{1}{\sqrt{2}}|\mathbf{H}(j\omega_r)|$$ (2.100)

Since $|\mathbf{H}|$ determines the shape of the output voltage and power is proportional to voltage squared, the frequencies ω_1 and ω_2 are known as the half-power points.

Figure 2.19   A log-log plot of $|\mathbf{H}(j\omega)|$ as a function of ω shows the resonant behavior of the series LCR circuit for various Q values.

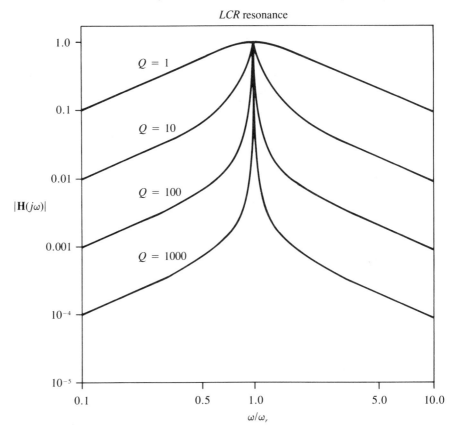

2.9 Four-Terminal Networks

The term network applies to any connected arrangement of idealized circuit elements (sources and impedances) and is often used to describe some mathematically convenient part of a complete circuit. For example, if we remove the driving EMF from the series *LCR* circuit of Figure 2.18, we are left with the four-terminal network shown in Figure 2.20a. The transfer function $\mathbf{H}(j\omega)$, defined for this circuit by Eq. (2.79), is also independent of the driving EMF and is thus determined by and associated with the four-terminal network.

Several *LCR* arrangements together with their transfer functions are shown in Figure 2.20. With the assumption that the current out of terminals *cd* is zero, each of these networks is simply a voltage divider, and $\mathbf{H}$ is always the impedance of the vertical branch divided by the series impedance. Thus, these transfer functions are all of the form

$$\mathbf{H}(j\omega) = \frac{\mathbf{Z}_3}{\mathbf{Z}_1 + \mathbf{Z}_3} \qquad (2.101)$$

where $\mathbf{Z}_3$ is the effective impedance of the vertical branch.

Figure 2.20 *Series LCR circuits drawn as four-terminal networks.*

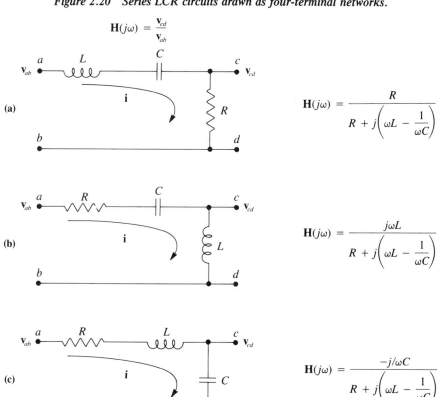

$$\mathbf{H}(j\omega) = \frac{\mathbf{V}_{cd}}{\mathbf{V}_{ab}}$$

(a)
$$\mathbf{H}(j\omega) = \frac{R}{R + j\left(\omega L - \dfrac{1}{\omega C}\right)}$$

(b)
$$\mathbf{H}(j\omega) = \frac{j\omega L}{R + j\left(\omega L - \dfrac{1}{\omega C}\right)}$$

(c)
$$\mathbf{H}(j\omega) = \frac{-j/\omega C}{R + j\left(\omega L - \dfrac{1}{\omega C}\right)}$$

2.10   Bode Plots

The complex transfer function $\mathbf{H}(j\omega)$ associated with a four-terminal network describes two simultaneously varying and interrelated variables, amplitude and phase. These variables are commonly displayed by plotting both the magnitude, $|\mathbf{H}(j\omega)|$, and the phase angle, $\theta(\omega)$, against the independent variable ω.

For example, Eqs. (2.81)–(2.83) give the transfer function for the network of Figure 2.20a in a form that explicitly displays the amplitude and phase variables, but even this relatively simple transfer function is algebraically unwieldy when so expressed. Although this algebraic complexity is of no particular consequence if a computer is employed to calculate the exact expressions for a particular case, it does tend to obscure the general characteristics of a circuit. Hence, we need to consider some approximate forms for the transfer function.

Still using the series *LCR* circuit as an example, Figure 2.19 reveals that for frequencies sufficiently far from resonance, $|\mathbf{H}(j\omega)|$ on a log-log plot can be well represented by straight lines. The following approximate procedures follow from this observation. They reduce the algebra significantly while retaining and emphasizing the essential features of most circuits.

2.10.1   Single-Term Approximations of $\mathbf{H}(j\omega)$

A great simplification results if we replace the exact expression for $\mathbf{H}(j\omega)$ with a number of single-term approximations, each varying with a different power of the frequency. The most general $\mathbf{H}(j\omega)$ is of the form

$$\mathbf{H}(j\omega) = \frac{\mathbf{P}(j\omega)}{\mathbf{D}(j\omega)} \tag{2.102}$$

where $\mathbf{P}(j\omega)$ and $\mathbf{D}(j\omega)$ are polynomials in $j\omega$ of the general form

$$\mathbf{P} = P_0 + P_1(j\omega)^1 + \cdots + P_n(j\omega)^n + \cdots + P_N(j\omega)^N$$

and
$$\mathbf{D} = D_0 + D_1(j\omega)^1 + \cdots + D_m(j\omega)^m + \cdots + D_M(j\omega)^M \tag{2.103}$$

Note that P_n and D_m are constant coefficients multiplying the integral powers of $j\omega$.

If, in some frequency region, we reduce $\mathbf{H}$ to a simple fraction by keeping only a single term (the largest at this frequency) in the numerator and a single term (again the largest) in the denominator, we obtain

$$\mathbf{H}(j\omega) = \frac{P_n(j\omega)^n}{D_m(j\omega)^m} \tag{2.104}$$

which can be written

$$\mathbf{H}(j\omega) = \frac{P_n}{D_m}(j\omega)^{(n-m)} \tag{2.105}$$

If $j^{(n-m)}$ is replaced by the equivalent form $(e^{j\pi/2})^{n-m}$, the amplitude and phase factors can be separated:

$$\mathbf{H}(j\omega) = \frac{P_n}{D_m} \omega^{(n-m)} e^{j(n-m)\pi/2} \tag{2.106}$$

The magnitude of $\mathbf{H}(j\omega)$ is thus proportional to an integral power of ω, and its phase angle is just $(n - m)\pi/2$.

The magnitude of $\mathbf{H}$ in a particular frequency region is then

$$|\mathbf{H}(j\omega)| = \omega^{(n-m)} \left|\frac{P_n}{D_m}\right| \tag{2.107}$$

and taking the base 10 log of both sides gives

$$\log_{10}|\mathbf{H}(j\omega)| = \log_{10}\left|\frac{P_n}{D_m}\right| + (n - m) \log_{10}\omega \tag{2.108}$$

Thus, the magnitude of this approximate form of the transfer function can be displayed on the log-log plot as a straight line of slope $n - m$.

At the extremes of low and high frequency, the approximation will always approach the actual transfer function, producing lines on the log-log plot that are the asymptotes of $|\mathbf{H}(j\omega)|$. At intermediate frequencies, the single-term approximation comes with no guarantees. However, it is often convenient to use these straight lines to sketch and describe $|\mathbf{H}(j\omega)|$, even when the approximation is poor.

Selecting the largest term in $\mathbf{P}$ and in $\mathbf{D}$ is generally straightforward. The lowest power of $j\omega$ (may be a constant term) will always dominate an expansion at the lowest frequencies, and the highest power will always dominate at the highest frequencies. At intermediate frequencies, the relative sizes of the various coefficients can have an overriding effect, but for most circuits we will consider, the frequency at which a term dominates will increase monotonically with the power of $j\omega$.

Following this procedure, the voltage ratio $\mathbf{H}(j\omega)$ for the series LCR circuit given in Eq. (2.98) can be approximated at low frequency by

$$\mathbf{H}_{\text{low}}(j\omega) = \frac{1}{-jQ} (\omega/\omega_h)(\omega_r/\omega)^2$$

$$= \frac{j\omega}{Q\omega_r} \tag{2.109}$$

This expression also gives us the phase shift between the output and input signals at sufficiently low frequencies: The $+j$ multiplier corresponds to $e^{j\pi/2}$, indicating that at low frequency the output voltage leads the input voltage by $\pi/2$ radians. The magnitude needed for plotting is now simply

$$|\mathbf{H}_{\text{low}}(j\omega)| = \frac{\omega}{Q\omega_r} \tag{2.110}$$

At high frequency the approximate form is

$$\mathbf{H}_{\text{high}}(j\omega) = \frac{1}{jQ(\omega/\omega_r)}$$

$$= -\frac{j\omega_r}{Q\omega} \qquad (2.111)$$

The $-j$ factor here shows that at high frequency the output voltage lags behind the input voltage by $\pi/2$ radians. Again the magnitude is easily found to be

$$\left|\mathbf{H}_{\text{high}}(j\omega)\right| = \frac{\omega_r}{Q\omega} \qquad (2.112)$$

These magnitudes are plotted in Figure 2.21 and are tangent at the frequency extremes of the curves plotted in Figure 2.19.

Figure 2.21 Single-term representation of the transfer functions plotted in Figure 2.19.

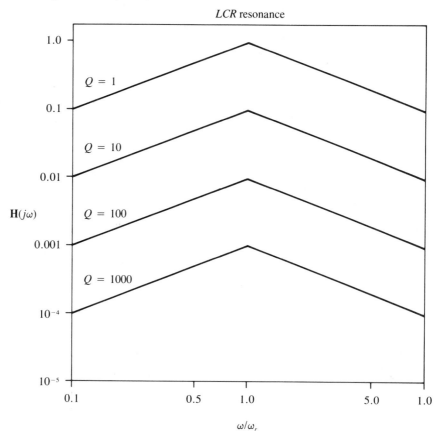

2.11   AC Equivalent Circuits

Extensions of Thevenin's and Norton's two-terminal equivalent circuit theorems play a fundamental role in the development of all equivalent circuits. These theorems, as stated in Chapter 1, are still valid when used with complex impedances and AC signal sources. The circuit reduction methods are similar to those used previously, but because of the frequency dependence of AC EMFs and impedances, an AC two-terminal equivalent circuit may contain more than two elements.

When all EMFs in the original circuit operate at the same frequency, the equivalent circuit will reduce to the standard forms shown in Figure 2.22a. However, note that the $\mathbf{Z}_{eq}$ in these circuits will in general be some combination of fundamental impedances and that the amplitude of the equivalent EMFs may be a function of frequency.

When there are several independent EMFs in the original circuit, they will all be represented in the equivalent circuit, either as voltage EMFs in series or current EMFs in parallel as shown in Figure 2.22b. The signals from these distinct sources always form algebraic sums. In this regard, note that a zero-amplitude voltage source behaves like an ideal wire, whereas a zero-amplitude current source behaves like an open circuit.

2.11.1   The Four-Terminal AC Equivalent Circuit

Any four-terminal network of linear elements can be replaced with an equivalent circuit with only four (generally complex) parameters. Although exact only for

Figure 2.22   (a) The impedance $\mathbf{Z}_{eq}$ needed for a two-terminal equivalent circuit will in general be a combination of fundamental impedances. (b) Independent AC EMFs operate at different frequencies and are represented in an equivalent circuit by separate sources.

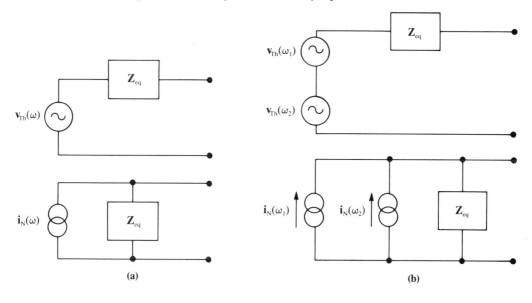

(a)   (b)

linear systems, the most important application of this four-terminal equivalent circuit is as a model for the AC operation of nonlinear systems of one or more transistors.

As shown in Figure 2.23, the form of the four-terminal equivalent circuit follows directly from Thevenin's two-terminal theorem. If a general four-terminal network (terminals *abcd*) in Figure 2.23a is driven by an ideal voltage source and nothing is connected to terminals *cd*, the output signal will be

$$\mathbf{v}_{cd}(\text{open}) = \mathbf{H}\mathbf{v}_{ab} \tag{2.113}$$

where **H** is the transfer function of the network. The notation (open) identifies this equation as valid only when nothing is connected to the output terminals *cd*.

But the entire circuit, source plus network, can be represented by a Thevenin two-terminal equivalent, and since Eq. (2.113) defines the open-circuit output voltage, it also defines the Thevenin equivalent voltage source as shown in the

Figure 2.23   Steps in the development of a general four-terminal network model: (a) the output loop, (b) the input loop, and (c) the complete four-terminal equivalent.

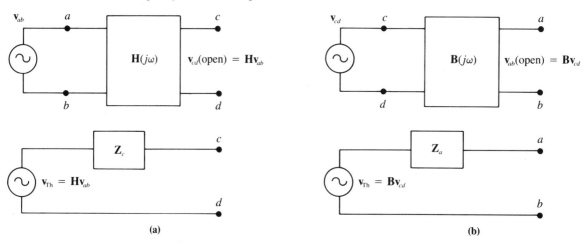

(a)   (b)

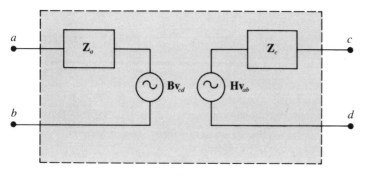

(c)

figure. The equivalent impedance Z_c needed to complete this Thevenin equivalent is found in the normal way by shorting terminals ab and calculating the impedance between terminals c and d.

If we reverse the original network and apply an input signal at cd as shown in Figure 2.23b, the network will usually have a different transfer function, labeled here as **B** for backward. In terms of this reverse transfer function, the open-circuit output voltage across ab is

$$\mathbf{v}_{ab}(\text{open}) = \mathbf{B}\mathbf{v}_{cd} \qquad (2.114)$$

again corresponding to a Thevenin equivalent voltage source as shown. The Thevenin equivalent impedance can be calculated for this configuration and will yield, in general a new impedance Z_a.

To form a complete model of the original four-terminal network, these two equivalents are schematically associated as shown in Figure 2.23c. The input and output of the four-terminal network now appears physically disconnected but remains algebraically connected by the **B** and **H** transfer functions. All four parameters, **H**, **B**, Z_a, and Z_c, are in general functions of frequency and are entirely determined by the four-terminal network. Since each of the two halves of this final circuit are Thevenin equivalents, it is clearly possible to replace one or both with the corresponding Norton equivalent.

2.11.2 Input and Output Impedances of the Four-Terminal Network

The four-terminal network is typically situated between a signal generator and a load impedance as shown in Figure 2.24a. In this figure, the network is represented by its four-terminal equivalent circuit. When the source $\mathbf{v}_s$ is fixed, there are four independent (complex for AC) variables outside the network, $\mathbf{i}_s$, $\mathbf{v}_{ab}$, $\mathbf{v}_{cd}$, and $\mathbf{i}_L$.

Because of the simple form of the model, the circuit has only two loops, producing two Kirchhoff loop equations,

$$\mathbf{v}_s - (\mathbf{Z}_s + \mathbf{Z}_a)\mathbf{i}_s - \mathbf{B}\mathbf{v}_d = 0 \qquad (2.115)$$

and
$$\mathbf{H}\mathbf{v}_{ab} - (\mathbf{Z}_c + \mathbf{Z}_L)\mathbf{i}_L = 0$$

But if we wish to solve these for the currents $\mathbf{i}_s$ and $\mathbf{i}_L$, we must eliminate $\mathbf{v}_{ab}$ and $\mathbf{v}_{cd}$ using

$$\mathbf{v}_{ab} = \mathbf{v}_s - \mathbf{Z}_S\mathbf{i}_s \qquad (2.116)$$

and
$$\mathbf{v}_{cd} = \mathbf{Z}_L\mathbf{i}_L$$

Substitution back into Eqs. (2.115) produces two equations that can be solved for $\mathbf{i}_s$ and $\mathbf{i}_L$. These solutions can then be combined with Eqs. (2.116) to give $\mathbf{v}_{ab}$ and $\mathbf{v}_{cd}$, completing the description.

Unfortunately, application of these equations to specific circuits of the type we have been studying is algebraically cumbersome, and would obscure the intended purpose. Instead we will make some general observations, then move on to an important special case.

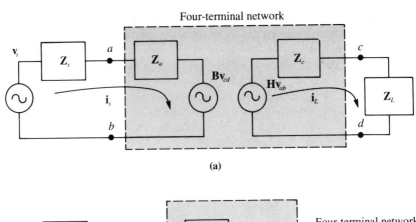

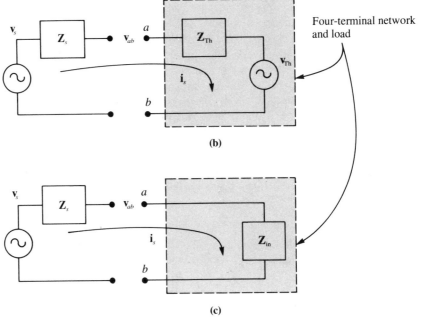

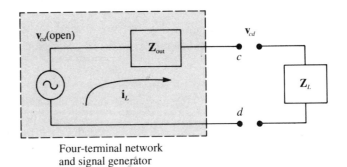

Figure 2.24 (a) A general circuit consisting of a signal generator, a four-terminal model, and a load impedance Z_L. The circuit to the right of ab can be replaced by (b) a Thevenin equivalent, or by (c) an impedance Z_{in}. (d) The circuit to the left of cd can be replaced by another Thevenin equivalent to define the output impedance Z_{out}.

If we disconnect the signal generator at terminals *ab* in Figure 2.24a but leave the load impedance $\mathbf{Z}_L$ connected, the left-opening two-terminal network can certainly be represented by a Thevenin equivalent as shown in Figure 2.24b. But if the actual circuit to the right of terminals *ab* has no sources, we would be most surprised to find a signal at terminals *ab* when the signal generator was disconnected. Thus, $\mathbf{v}_{Th}$ must be zero, and the two-terminal network will reduce to just an input impedance $\mathbf{Z}_{in}$ as shown in Figure 2.24c. This argument shows that $\mathbf{B}\mathbf{v}_{cd}$ and $\mathbf{H}\mathbf{v}_{ab}$ within the four-terminal equivalent model are not true sources; they function only as algebraic links between the input and output loops and are ultimately derived from sources outside the network (in this case by $\mathbf{v}_s$).

If, in Figure 2.24a, we disconnect the load $\mathbf{Z}_L$, the remaining right-opening two-terminal network can also be represented by the Thevenin equivalent shown in Figure 2.24d. Since the actual circuit now includes a true signal source $\mathbf{v}_s$, the Thevenin source labeled $\mathbf{v}_{cd}(\text{open})$ must be included.

As long as the reverse transfer function $\mathbf{B}$ is nonzero, the input impedance $\mathbf{Z}_{in}$ in Figure 2.24c will be a function of the load impedance $\mathbf{Z}_L$. In any case it is operationally defined by the equation

$$\mathbf{Z}_{in} = \frac{\mathbf{v}_{ab}}{\mathbf{i}_s} \tag{2.117}$$

where $\mathbf{i}_s$ is the current drawn from the signal generator.

As with any Thevenin equivalent circuit, the output impedance of Figure 2.24d is defined as the open-circuit output voltage divided by the short-circuit output current:

$$\mathbf{Z}_{out} = \frac{\mathbf{v}_{cd}(\text{open})}{\mathbf{i}_L(\text{short})} \tag{2.118}$$

The short circuit current $\mathbf{i}_L(\text{short})$ is just $\mathbf{i}_L$ for the special case when $\mathbf{Z}_L = 0$. If the reverse transfer function is nonzero, this output impedance will be a function $\mathbf{Z}_s$ and all four parameters of the four-terminal equivalent.

For a specific circuit, evaluation of $\mathbf{Z}_{out}$ from this definition requires the analysis of two distinct circuits: one with the output open to find $\mathbf{v}_{cd}(\text{open})$, and the other with the output shorted to find $\mathbf{i}_L(\text{short})$. Care must be taken with variables

Figure 2.25   The four-terminal model of a buffered amplifier.

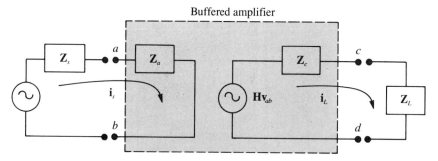

such as $\mathbf{i}_s$, $\mathbf{v}_{ab}$, and $\mathbf{Z}_{in}$, since they will be different in the two parts of the problem. Note that the ideal voltage source $\mathbf{v}_s$ can supply any current and is thus unaffected by changes in the load on the circuit.

2.11.3 Four-Terminal Equivalent Circuit for a Buffered Amplifier

An important special case of the four-terminal network of Figure 2.24a occurs when the reverse transfer function **B** is zero (or small enough that it can be neglected). The resulting circuit, shown in Figure 2.25, is a good model for the more complicated amplifier circuits that will be discussed in later chapters.

For this simplified circuit, the input impedance is just $\mathbf{Z}_a$, independent of the load impedance $\mathbf{Z}_L$. Consequently, $\mathbf{i}_s$ and $\mathbf{v}_{ab}$ are also independent of the load, and the two loops of the circuit are effectively isolated. The input loop forms a voltage divider, making

$$\mathbf{v}_{ab} = \frac{\mathbf{Z}_a}{\mathbf{Z}_s + \mathbf{Z}_a} \mathbf{v}_s \tag{2.119}$$

The open-circuit output voltage in Figure 2.24d is now just $\mathbf{Hv}_{ab}$ and is given by

$$\mathbf{v}_{cd}(\text{open}) = \frac{\mathbf{Z}_a}{\mathbf{Z}_s + \mathbf{Z}_a} \mathbf{Hv}_s \tag{2.120}$$

Figure 2.26 (a) An ideal amplifier used as a buffer between a signal generator and a load resistor. (b) The same signal generator driving the load resistor directly.

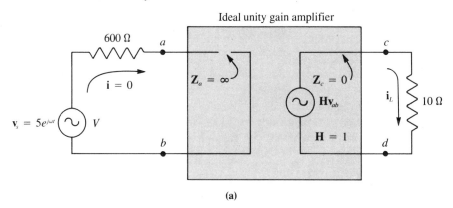

(a)

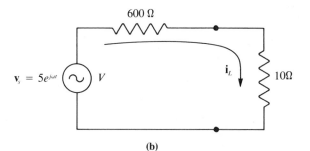

(b)

where $\mathbf{Z}_a$ is also $\mathbf{Z}_{in}$.

Example 2.5 The four-terminal model for an ideal "unity gain" amplifier would have $\mathbf{Z}_a = \infty$, $\mathbf{B} = 0$, $\mathbf{H} = 1$ (this is the unity gain), and $\mathbf{Z}_c = 0$. Connect this four-terminal device as a buffer between a signal generator with an output impedance of 600 Ω and a load impedance of 10 Ω. If the signal generator is set to have an open-circuit amplitude of 5 V, what is the amplitude of the current in the load resistor? How large would it be if the signal generator were hooked directly to the load resistor?

The buffered connection is shown in Figure 2.26a. Since the amplifier has an infinite input impedance, there is no current in the signal generator's output resistor; this condition makes $\mathbf{v}_{ab} = \mathbf{v}_s$.

The output impedance of the amplifier is zero, making $\mathbf{v}_{cd} = \mathbf{H}\mathbf{v}_{ab}$; and with $\mathbf{H} = 1$ these equations combine to give $\mathbf{v}_{cd} = \mathbf{v}_s$. The current in the load resistor is then given by

$$\mathbf{i}_L = \frac{\mathbf{v}_s}{R_L} = 0.5e^{j\omega t} \text{ A}$$

The observed current would thus have an amplitude of 500 mA. Without the amplifier the signal generator would drive a series combination of 610 Ω as shown in Figure 2.26b, and the current in the load resistor would be only 8.2 mA.

2.12 Power in AC Circuits

For the DC circuits of the last chapter, where voltages and currents were time-independent constants, we determined that the power was given by $P = VI$. For AC circuits, when voltages and currents are functions of time, this equation is correct only as an instantaneous description for the power. The instantaneous power is thus a time-dependent quantity given by

$$P(t) = V(t) I(t) \tag{2.121}$$

If the signals $V(t)$ and $I(t)$ are periodic with period T, we can define an average power loss as

$$\begin{aligned} P_{avg} &= \frac{1}{T} \int_0^T P(t) \, dt \\ &= \frac{1}{T} \int_0^T V(t) I(t) \, dt \end{aligned} \tag{2.122}$$

Since we must deal here with the product of two signals, the complex representation cannot be used. In terms of the real sinusoidal signals

$$v(t) = V_0 \cos(\omega t) \tag{2.123}$$

and

$$i(t) = I_0 \cos(\omega t + \phi)$$

the general expression for average power loss becomes

$$P_{avg} = \frac{V_0 I_0}{T} \int_0^T \cos(\omega t) \cos(\omega t + \phi) \, dt \tag{2.124}$$

where ϕ is the phase angle between the voltage and current signals. We evaluate this integral by first expanding the $\cos(\omega t + \phi)$ term with the trigonometric identity for the sum of two angles. After moving the constant $\cos \phi$ and $\sin \phi$ terms outside the integral, this expanded expression becomes

$$P_{avg} = \frac{V_0 I_0}{T} \left[\cos \phi \int_0^T \cos^2(\omega t)\, dt - \sin \phi \int_0^T \cos(\omega t)\sin(\omega t)\, dt \right] \quad (2.125)$$

Using the relationship between period and angular frequency $T = 2\pi/\omega$, and defining a new variable $\theta = \omega t$, these integrals can now be written

$$P_{avg} = \frac{V_0 I_0}{2\pi} \cos \phi \int_0^{2\pi} \cos^2 \theta\, d\theta - \sin \phi \int_0^{2\pi} \cos \theta \sin \theta\, d\theta \right] \quad (2.126)$$

The second integral is zero, since the integrand is equally positive and negative over the 360° range of integration, leaving only the first term to be integrated. Using either integral tables or the trigonometric identity $2\cos^2 \phi = 1 - \cos(2\phi)$, the first term reduces to

$$P_{avg} = \frac{1}{2} V_0 I_0 \cos \phi \quad (2.127)$$

Now that the average power expression has been derived using real functions, it can be expressed in terms of the complex phasors

$$\mathbf{v} = V_0 e^{j\omega t}$$

and

$$\mathbf{i} = I_0 e^{j(\omega t + \phi)} \quad (2.128)$$

Since

$$\mathbf{v}^*\mathbf{i} = V_0 I_0 e^{j\phi} \quad (2.129)$$

and $\mathbf{v}\mathbf{i}^*$ differs only slightly, Eq. (2.217) can be written

$$P_{avg} = \frac{1}{2} \text{Real}(\mathbf{v}^*\mathbf{i}) = \frac{1}{2} \text{Real}(\mathbf{v}\mathbf{i}^*) \quad (2.130)$$

Measurements of sinusoidal signals made with AC signal meters, as opposed to direct observation of the waveform display on an oscilloscope, will generally be expressed in terms of the "root-mean-square" or RMS signal rather than the more basic amplitude. This RMS voltage is defined as

$$V_{RMS} = \left[\frac{1}{T} \int_0^T v^2(t)\, dt \right]^{1/2}$$

$$= \left[\frac{V_0^2}{T} \int_0^T \cos^2(\omega t)\, dt \right]^{1/2} \quad (2.131)$$

and integrates to

$$V_{RMS} = \frac{V_0}{\sqrt{2}} \quad (2.132)$$

A similar definition of the RMS current results in

$$I_{RMS} = \frac{I_0}{\sqrt{2}} \qquad (2.133)$$

The relative phase angle ϕ between V to I is lost during these last integrations, but still appears in Eq. (2.127) for the average power. Using that equation, the average power can be written in terms of RMS signal values as

$$P_{avg} = V_{RMS}I_{RMS} \cos \phi \qquad (2.134)$$

The use of RMS values is common, partly because this last equation is similar to the DC result. The standard 110-V, 60-cycle EMF commonly available from the local power company is a good example. The 110 V is a nominal RMS value; the amplitude is closer to 160 V.

Problems

1. Given the two signals $v_1(t) = A \cos(\omega t + \theta_1)$ and $v_2(t) = B \cos(\omega t + \theta_2)$, show that the time $t_2 - t_1$ between zero crossings as shown in Figure 2.8 is $(\theta_1 - \theta_2)/\omega$.

2. Use Euler's equation to evaluate $e^{j\pi/2}$, $e^{j\pi}$, and $e^{j3\pi/2}$. Plot these three phasors on the complex plane.

3. Show that $e^{j(\theta + \pi/2)}$ is 90° ahead of $e^{j\theta}$ by using Euler's equation to plot the phasors when θ is 0, $\pi/2$, π, and $3\pi/2$ radians.

4. Write the phasor $10e^{j0.75\pi}$ in $a + jb$ form.

5. Find the phasor that equals $e^{j0.5\pi} + 3e^{j\pi}$.

6. If a complex voltage and current are related by the expression

$$\mathbf{v}(t) = (-1 + j\sqrt{3})\mathbf{i}(t)$$

what is the phase angle in degrees between the voltage and the current?

7. Express the complex number $1/(1 + j\sqrt{3})$ in the form $Ae^{j\theta}$.

8. Use the phasor notation and Euler's equation to derive the trigonometric identities $\sin 2\theta = 2 \sin \theta \cos \theta$ and $\cos 2\theta = \cos^2 \theta - \sin^2 \theta$.

9. On the complex plane, sketch the locus of points of the end of a phasor whose real part would describe the damped oscillation curve of Figure 2.6b. This phasor is described by Eq. (2.88).

10. Graph the real part of Eq. (2.88) when $V_0 = 1$ and *(a)* $(\sigma = \omega = 1 \text{ s}^{-1}$, *(b)* $\sigma = 10\omega = 1 \text{ s}^{-1}$, *(c)* $10\sigma = \omega = 1 \text{ s}^{-1}$.

11. Following the procedure used to derive the time constant for the RC circuit shown in Figure 2.5a, determine the time constant of the RL circuit shown in Figure 2.5c.

12. *(a)* Sketch the signal $v(t) = 10 - 5 \sin(2\pi 100t)$, and indicate the voltage at times $t = 0$, $t = 1/800$, and $t = 1/400$ s. *(b)* What is the amplitude of the AC part of this signal? *(c)* What is the period of this signal?

13. Sketch the signal of the previous problem as a phasor on the complex plane at the three times indicated; show the two terms separately as well as summed.

14. Ideal DC and AC EMFs are connected in series as shown in Figure P2.14. **(a)** Write a real and then a complex expression for the output signal at terminals *AB*. **(b)** Sketch the real signal as a function of time.

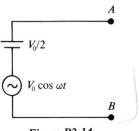

Figure P2.14

15. Use Eq. (2.41) and the methods of Chapter 1 to derive Eqs. (2.60) for two impedances connected in series and parallel.

16. An inductor and capacitor in parallel as shown in Figure P2.16 form a tank circuit. **(a)** Determine an expression for the impedance of this circuit. **(b)** What is the impedance when $\omega = 1/\sqrt{LC}$?

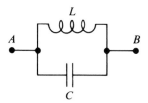

Figure P2.16

17. The tank circuit schematic shown in Figure P2.17 results from the use of a real inductor. **(a)** Find an expression for the impedance of this circuit. **(b)** If $L = 1$ H, $R = 100$ Ω, and $C = 0.01$ μF, what is the impedance when $\omega = 1/\sqrt{LC}$? **(c)** What is the impedance when ω is very small? **(d)** What is the phase angle between the voltage $\mathbf{v}_{AB}$ and $\mathbf{i}$ at resonance and at $\omega = 10^5$ rad/s?

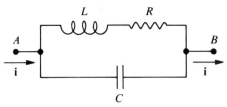

Figure P2.17

18. Sketch simplified versions of the circuit shown in Figure P2.18 that would be valid at: **(a)** $\omega = 0$; **(b)** very low frequencies but not $\omega = 0$; **(c)** very high frequencies but not $\omega = \infty$; **(d)** $\omega = \infty$.

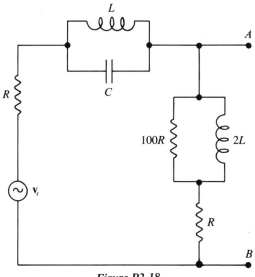

Figure P2.18

19. Consider the series *LCR* circuit of Figure 2.18 driven by a voltage phasor $\mathbf{v}(t) = v_0 e^{j\omega t}$. *(a)* At an angular frequency such that $\omega L = 2R$ and $1/\omega C = R$, write the current phasor in terms of $\mathbf{v}(t)$ and R. *(b)* At the instant when $\mathbf{v}(t)$ is exactly real, calculate the three phasors representing the voltage developed across the R, C, and L circuit elements. *(c)* Algebraically and with a sketch on the complex plane, show that the complex voltage sum around the closed loop is zero.

20. For the circuit shown in Figure P2.20, plot $|\mathbf{Z}_{eq}|$ as a function of frequency over the range $\omega = 1$ rad/s to $\omega = 10^6$ rad/s.

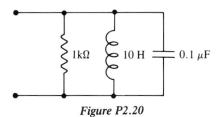

Figure P2.20

21. *(a)* Evaluate and sketch the phasor $\mathbf{A}(j\omega) = 1 + 10j\omega + \omega^2$ when $\omega = 0.01, 0.1, 1, 10,$ and 100 rad/s. *(b)* At which of these values can the phasor be reasonably represented by a single term of the polynomial? *(c)* What is the amplitude error introduced by the single-term approximation at 100 rad/s?

22. Consider a series circuit like that of Figure 2.18 with $L = 1$ H, $C = 1$ μF, and $R = 1$ kΩ. *(a)* At the resonant frequency, determine the complex relationship that connects $\mathbf{v}_{cb}$ and $\mathbf{v}_{ab}$. ($\mathbf{v}_c$ is the voltage at a point between L and C and $\mathbf{v}_{cb} = \mathbf{v}_c - \mathbf{v}_b$.) *(b)* sketch $\mathbf{v}_{cb}$ and $\mathbf{v}_{ab}$ on the complex plane at the instant when $\mathbf{v}(t)$ is real. *(c)* Sketch the two signals as they would appear on a two-channel oscilloscope.

23. A signal generator with a 600-Ω output impedance is used to drive a 1-H inductor in series with a 0.01-μF capacitor. *(a)* If the inductor has a resistance of 500 Ω, what is the Q of this circuit? *(b)* What is the Q of the inductor?

24. The input to a typical oscilloscope is as shown in Figure P2.24; the voltage V_y is displayed as a vertical deflection on the oscilloscope screen. **(a)** Assuming that $V_y = 0$ displays in the middle of the screen, sketch the display when $V_{in} = 5 + 2\cos(628t)$ V with the input switch in the DC and then the AC position. **(b)** Why is there a need for the AC position? **(c)** Change the expression for V_{in} to one that would require the use of the AC switch position.

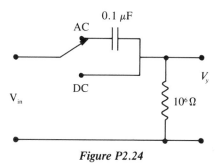

Figure P2.24

25. The two EMFs in the circuit of Figure P2.25 are independent. Show that the circuit can be replaced by a Thevenin equivalent like the one shown in Figure 2.22b, and express all three elements in terms of the ones given here.

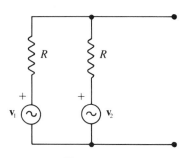

Figure P2.25

26. With reference to Figure P2.26, derive an expression for $|\mathbf{Z}_{in}|$ in terms of R, the measured amplitudes of the voltages $\mathbf{v}_A$ and $\mathbf{v}_B$, and the phase difference θ by which $\mathbf{v}_B$ leads $\mathbf{v}_A$. Reduce this expression until θ appears only in a single $\cos\theta$ term.

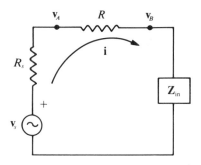

Figure P2.26

27. *(a)* Find expressions for the Thevenin equivalent circuit elements to replace the network to the left of points *AB* in Figure P2.27.   *(b)* If $R = 100\ \Omega$ and $L = 1$ H, evaluate these elements at $\omega = 100$ rad/s.   *(c)* If the circuit with these component values is used to drive a 200-μF capacitor at $\omega = 100$ rad/s, what is the phase shift in degrees between $\mathbf{i}_C$ and $\mathbf{v}_s$?

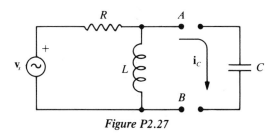

Figure P2.27

28. Redraw the circuit of Figure 2.24a with $\mathbf{Z}_L = 0$ and explain how this makes $\mathbf{Z}_{\text{in}} = \mathbf{Z}_a$.

29. Using $\mathbf{V} = V_0 e^{j\omega t}$ and $\mathbf{I} = I_0 e^{j(\omega t + \phi)}$, show that both forms of the complex power Eq. (2.130) are equal to the real Eq. (2.127).

30. A 0.1-H inductor has 100 Ω of internal resistance. Calculate the average power dissipated when a 1000-Hz signal of 1-V amplitude is applied across the inductor. [*Ans.:* 0.12 mW]

3

Signals, Filters, and Amplifiers

3.1   Introduction

When a four-terminal network is driven by a sinusoid, the output signal will be a sinusoid of the same frequency but with a generally different amplitude and phase. By resorting to a complex phasor representation for these input and output sinusoids, we are able to relate them by the complex transfer function $\mathbf{H}(j\omega)$, which multiplies the input phasor to produce the output phasor. As we saw in the last chapter, this transfer function is easily determined if schematic elements are described by complex impedances and gets its frequency dependence solely from the reactive elements. The transfer function thus specifies the amplitude and phase-changing properties of the network from which it was derived, and its frequency dependence shows the variation of these properties with the frequency of the applied sinusoid.

In this chapter, we discuss some important features of the transfer function and show how it can be used to describe the behavior of specific circuits. We begin with a qualitative discussion of some signal analysis techniques that extend the usefulness of $\mathbf{H}(j\omega)$. A more complete discussion can be found in Chapter 12.

3.2   *The Principle of Superposition and Signal Analysis*

It is always possible to decompose a time-varying signal $f(t)$ into a number of lesser signals $f_n(t)$ that will combine by addition to form $f(t)$:

$$f(t) = f_1(t) + f_2(t) + \cdots + f_n(t) \tag{3.1}$$

For this decomposition to be useful, the constituent signals $f_n(t)$ must be somehow more manageable than the total signal $f(t)$, and the system acting on the signal must obey the principle of superposition as indicated in Figure 3.1. This system acts on an input signal of either $f(t)$ or any of its constituent $f_n(t)$ to produce a corresponding output signal of $g(t)$ or $g_n(t)$. If the system is linear (describable by linear differential equations), the principle of superposition states that the $g_n(t)$ will combine by addition to form $g(t)$. Note that we have already made implicit use of this principle to separate circuit analysis into DC and AC parts.

The most useful decomposition of a signal is into a sum of sinusoidal oscillations, each with a different amplitude and frequency. For our purposes, it is best to think of a sinusoid as the projection of a phasor onto the real axis; the amplitude of the sinusoid is the magnitude of the phasor, and the angle between the phasor and the real axis at $t = 0$ is the phase angle. Because of this phase angle, the sinusoid can be a simple $\sin(\omega t)$ function, a $\cos(\omega t)$ function, or any linear combination of the two.

The general input signal to a network is not a sinusoid, but it can be described as a sum of sinusoids or phasors. Since $\mathbf{H}(j\omega)$ tells us how the network modifies each sinusoid, the output signal from a linear network is just the sum of these modified sinusoids. There are several ways to determine the sinusoids that make up a general time-varying signal, but only the Fourier series expansion will be described here; more powerful and practical methods are described in Chapter 12.

3.2.1 Fourier Series

If a real signal $f(t)$ has period T and extends in time from $-\infty$ to $+\infty$, then it can be written as the following sum of sines and cosines:

$$f(t) = a_0 + 2 \sum_{n=1}^{\infty} [a_n \cos(n\omega_0 t) - b_n \sin(n\omega_0 t)] \qquad (3.2)$$

where

$$\omega_0 = \frac{2\pi}{T}$$

Figure 3.1 *A linear system described by* $\mathbf{H}(j\omega)$ *acts on an input signal* $f(t)$ *to yield an output signal* $g(t)$.

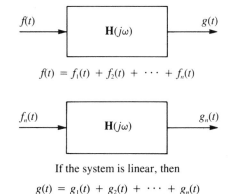

$$f(t) = f_1(t) + f_2(t) + \cdots + f_n(t)$$

If the system is linear, then

$$g(t) = g_1(t) + g_2(t) + \cdots + g_n(t)$$

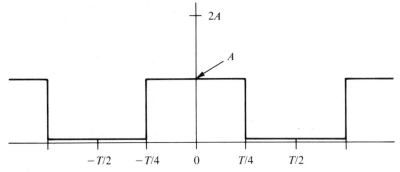

Figure 3.2   Square wave positioned as an even or symmetric function.

All frequency components represented in this Fourier series are integral multiples of the fundamental frequency, ω_0, which defines the lowest frequency and longest period component in the sum. The constant coefficients a_n and b_n can be obtained from the formulas

$$a_n = \frac{1}{T} \int_{-T/2}^{T/2} f(t)\ \cos(n\omega_0 t)\ dt \tag{3.3}$$

and

$$b_n = -\frac{1}{T} \int_{-T/2}^{T/2} f(t)\ \sin(n\omega_0 t)\ dt \tag{3.4}$$

Thus, the general periodic signal can be analyzed into a constant term and a sum of sine and cosine terms of different amplitudes and frequencies. The coefficients a_n and b_n determine the respective amplitudes of the nth cosine and sine components of angular frequency $n\omega_0$.

The cosine and sine terms in the Fourier series expansion of Eq. (3.2) respectively describe the symmetric (even) and antisymmetric (odd) parts of the general signal. Even terms $(x^0, x^2, x^4, \ldots)$ have the characteristic that $f(-x) = f(x)$, whereas odd terms $(x^1, x^3, x^5, \ldots)$ follow the expression $f(-x) = -f(x)$. Thus, since the square wave in Figure 3.2 is positioned laterally to make it an even signal, the b_n terms in its Fourier series must all be zero.

It is a useful exercise to calculate the a_n coefficients for this common periodic function. Since this $f(t)$ and $\cos(n\omega_0 t)$ are both even functions, the integral of Eq. (3.3) from $-T/2$ to $T/2$ is just twice the integral from 0 to $T/2$. So modified, the a_n integral becomes

$$a_n = \frac{2}{T} \int_0^{T/2} f(t)\ \cos(n\omega_0 t)\ dt \tag{3.5}$$

where

$$f(t) = A \qquad \text{for } 0 \leq t \leq \frac{T}{4}$$

and

$$f(t) = 0 \qquad \text{for } \frac{T}{4} < t < \frac{T}{2}$$

After substitution for $f(t)$, the integral becomes

$$a_n = \frac{2A}{T} \int_0^{T/4} \cos(n\omega_0 t) \, dt \qquad (3.6)$$

and evaluates to

$$a_n = \frac{2A}{T} \frac{\sin(n\omega_0 t)}{n\omega_0} \bigg|_0^{T/4} = \frac{2A}{n\omega_0 T} \sin\left(\frac{n\omega_0 T}{4}\right) \qquad (3.7)$$

Replacing ω_0 with $2\pi/T$ gives

$$a_n = \frac{A}{2} \frac{\sin(n\pi/2)}{n\pi/2} \qquad \text{for } n = 0, 1, 2, \ldots \qquad (3.8)$$

The functional form $\sin(\theta)/\theta$ occurs repeatedly and is described by the sinc function. [This function is usually defined as $\mathrm{sinc}(x) = \sin(2\pi x)/2\pi x$, but to avoid hidden factors of π, we will use simply $\mathrm{sinc}(\theta) = \sin(\theta)/\theta$.] Its limiting value at θ equal to zero is one.

Figure 3.3 *First four terms of the Fourier expansion of a symmetric square wave and the wave formed by the synthesis of these four terms.*

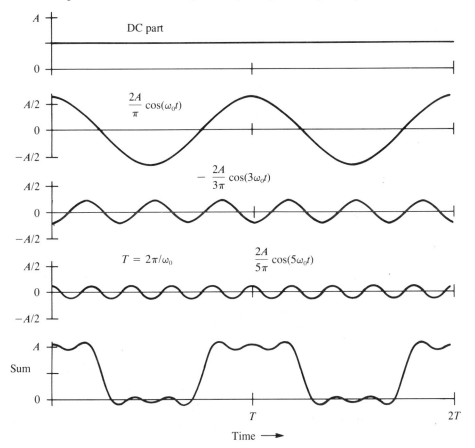

Because of the sin term, a_n will be zero for $n = 2, 4, 6, \ldots$, leaving only a_0 and the odd n terms. If we designate these nonzero terms by a new subscript m, where $n = 2m - 1$, Eq. (3.8) can be evaluated to give

$$a_m = \frac{A(-1)^{m+1}}{(2m-1)\pi} \qquad \text{for } m = 1, 2, 3, \ldots \tag{3.9}$$

The square wave can thus be represented by the infinite sum

$$f(t) = \frac{A}{2} + 2 \sum_{m=1}^{\infty} \frac{A(-1)^{m+1}}{(2m-1)\pi} \cos[(2m-1)\omega_0 t] \tag{3.10}$$

which is now over the new subscript m to eliminate the zero terms in Eq. (3.8). The square wave contains only the odd harmonics of the fundamental frequency, as can be demonstrated by evaluating a few terms of this expression:

$$f(t) = \frac{A}{2} + \frac{2A}{\pi}\cos(\omega_0 t) - \frac{2A}{3\pi}\cos(3\omega_0 t) + \frac{2A}{5\pi}\cos(5\omega_0 t) - \ldots \tag{3.11}$$

Figure 3.3 displays a graph of each of these four terms and their sum; the exact representation of the square wave requires an infinite number of terms with ever-decreasing amplitudes.

3.3   Filters and Amplifiers

The primary function of a filter circuit is to modify the frequency spectrum of a signal, producing an enhancement of some frequency components relative to others. An audio system is an everyday example: Most use filters to suppress unwanted signals at both low and high frequencies.

Schematically, a filter is a four-terminal network with transfer function $\mathbf{H}(j\omega)$, which operates on an input signal $\mathbf{v}_{in}$ to produce an output signal according to the expression

$$\mathbf{v}_{out}(j\omega) = \mathbf{H}(j\omega)\, \mathbf{v}_{in}(j\omega) \tag{3.12}$$

Because the magnitude and phase of $\mathbf{H}(j\omega)$ change with frequency, the frequency content (the Fourier coefficients) of the output signal will be different from that of the input signal.

A circuit designed to act as a filter typically produces the desired $\mathbf{H}(j\omega)$ with explicit reactive elements, but because of stray reactance, all circuits exhibit some variation with frequency and must often be treated as filters, especially at higher frequencies. Of particular importance in this regard is a class of devices known as amplifiers (see Example 2.5). When viewed as a four-terminal network described by a transfer function, an amplifier differs only slightly from a passive filter: Its distinguishing characteristic is the ability to increase the power (voltage times current) of the output signal relative to the input signal.

A frequency-independent amplifier (one characteristic of an ideal amplifier) can be described by the equation

$$\mathbf{v}_{out}(j\omega) \ = \ A \ \mathbf{v}_{in}(j\omega) \qquad (3.13)$$

where the real constant A is known as the voltage gain of the amplifier. If A is greater than one, this equation would increase the amplitude of a signal but would not modify its frequency content or shape.

Although actual amplifiers with $|A| > 1$ often approximate Eq. (3.13) over some frequency range, their actual gain always changes with frequency, at least to the extent of becoming steadily smaller above some corner or "cutoff" frequency. The response of an actual amplifier is then better described by

$$\mathbf{v}_{out}(j\omega) \ = \ \mathbf{A}(j\omega) \ \mathbf{v}_{in}(j\omega) \qquad (3.14)$$

which differs from Eq. (3.12) only in the symbol used to identify the transfer function and by the understanding that $\mathbf{v}_{in}$ may be an arbitrary voltage difference, not necessarily a voltage with respect to ground.

In the following sections, we will first explore some of the terminology and characteristics of passive filters, then employ similar mathematics to describe some of the general features of amplifiers. Amplifier networks that are designed primarily to operate as filters are known as active filters and will be discussed in Chapter 8.

3.4   Types of Filters

Filters can be classified into four main types depending on which frequency components of the input signal are passed on to the output signal: low-pass, high-pass, band-pass, and band-rejection. (A fifth type known as all-pass is used primarily for its phase-shifting properties and will not be discussed here.) For example, a low-pass filter will enhance the low frequencies relative to high frequencies: It is said to pass the low frequencies and to reject the high frequencies.

Each filter type can be further distinguished by the complexity of its transfer function and by whether it is passive or active. A more complicated circuit has additional design parameters that can be adjusted to achieve specific objectives, such as achieving a more rapid transition between the pass and rejection frequency regions.

A filter can be approximately described by the slopes of the single-term approximations to various parts of its transfer function as shown in Figure 3.4. As discussed in Section 2.10, the log-log plot of $|\mathbf{H}(j\omega)|$ can be characterized by the slopes of various straight-line segments and their intersection points—the corner frequencies of the transfer function. Each segment is proportional to ω^n, with the exponent n defining the slope of the segment on the log-log plot. The difference between the largest and smallest slope will never be greater than the number of reactances in the filter.

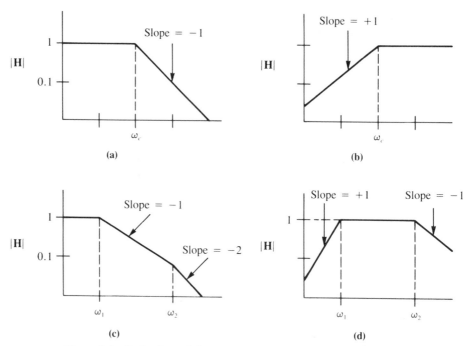

Figure 3.4   Bode plots of the single-term approximations of $|\mathbf{H}(j\omega)|$ of
various filters: (a) RC low-pass, (b) RC high-pass, (c) two-reactance
low-pass, and (d) two-reactance band-pass. All are shown as log-log
plots.

3.5   Bode Plots and Decibels

In electronics, the relative magnitude of two quantities is often given in terms of the base 10 logarithm of their ratio. The resulting unitless number is specified as being in "decibels" and so indicated by following the number with the symbol dB. Although the use of this quantity is sometimes justified by the fact that the human senses of sight and sound respond on a logarithmic scale, more often the notation is simply a technical nuisance that obscures fundamental relationships. Nevertheless, the widespread use of decibel terminology makes it necessary to make the connection between this number and the more transparent algebraic descriptions.

The decibel relationship between two powers, P_1 and P_2, is defined by the equation

$$dB = 10 \log_{10}\left(\frac{P_2}{P_1}\right) \tag{3.15}$$

Since power is proportional to the square of the voltage, this definition can also be written as

$$dB = 10 \log_{10}\left(\frac{V_2{}^2}{V_1{}^2}\right) \tag{3.16}$$

If the squares are taken out of the logarithm, we have the standard expression used to convert a voltage ratio to decibels:

$$dB = 20 \log_{10}\left(\left|\frac{V_2}{V_1}\right|\right) \tag{3.17}$$

Thus an amplifier with a gain of 1000, such that its output voltage V_2 is 10^3 times larger than its input V_1, is alternately said to have a voltage gain of 60 dB.

The decibel notation can also be used to describe the variation of a transfer function with frequency. If we apply a constant amplitude signal to the input of a four-terminal network, the amplitude of the output signal will be proportional to the magnitude of the transfer function. The output signal $\mathbf{v}_1$ and $\mathbf{v}_2$ at two frequencies is then given by

$$|\mathbf{v}_1| = |\mathbf{H}(j\omega_1)|\,|\mathbf{v}_{in}(j\omega_1)|$$

and   (3.18)

$$|\mathbf{v}_2| = |\mathbf{H}(j\omega_2)|\,|\mathbf{v}_{in}(j\omega_2)|$$

Since we are assuming that

$$\mathbf{v}_{in}(j\omega_1) = \mathbf{v}_{in}(j\omega_2) \tag{3.19}$$

the ratio of the two output voltages is just the ratio of the transfer function evaluated at two different frequencies:

$$\frac{|\mathbf{v}_2|}{|\mathbf{v}_1|} = \frac{|\mathbf{H}(j\omega_2)|}{|\mathbf{H}(j\omega_1)|} \tag{3.20}$$

Replacing the voltage ratio in Eq. (3.17) gives

$$dB = 20 \log_{10}\left(\frac{|\mathbf{H}(j\omega_2)|}{|\mathbf{H}(j\omega_1)|}\right) \tag{3.21}$$

With this equation, a high-pass filter—or amplifier—can be described as being "40 dB down at 60 Hz." Translated, this means that the ratio of the magnitude of the transfer function at 60 cycles per second to that at some higher frequency, where the transfer function is presumably flat, is 10^{-2}.

The slopes of the single-term approximations to the transfer functions shown in Figure 3.4 can also be described in decibels. Any single line on that figure is proportional to a power of the angular frequency,

$$|\mathbf{H}(j\omega)| \sim |(j\omega)^n| = \omega^n \tag{3.22}$$

and when evaluated at two frequencies, ω_1 and ω_2, this relation can be used to produce a transfer function ratio for Eq. (3.21):

$$dB = 20 \log_{10}\left(\frac{\omega_2{}^n}{\omega_1{}^n}\right) \tag{3.23}$$

If the common exponent n is brought outside the logarithm,

$$dB = n \, 20 \, \log_{10}\left(\frac{\omega_2}{\omega_1}\right) \tag{3.24}$$

the ratio is of two arbitrary frequencies, and n is a parameter of the filter.

To obtain a number from Eq. (3.24), we must introduce some standard frequency ratios. There are two common ones: the octave ($\omega_2 = 2\omega_1$) and the decade ($\omega_2 = 10\omega_1$). Using these in Eq. (3.24), we can write the slope of various regions of a transfer function as

$$dB/octave = 6.02n \text{ (normally abbreviated to } 6n)$$

or as
$$\tag{3.25}$$

$$dB/decade = 20n$$

Thus, a particular filter or amplifier can be said to "fall off by 6, 12, or 18 dB/octave," or alternatively, to "change by -20, -40, or -60 dB/decade," corresponding to single-term approximations with log-log plot slopes of -1, -2, or -3.

3.6 Passive RC Filters

A simplest filter has only one reactive circuit element, either a capacitor or an inductor. (Multiple elements of the same type connected in series and in parallel count as a single element.) Because actual inductors are bulky, expensive, and far from ideal, their use in signal applications is usually confined to frequencies above 100 kHz. To a great extent, we will therefore concentrate on filters composed of only capacitors and resistors.

3.6.1 Low-Pass Filter

The circuit shown in Figure 3.5a is easily analyzed using the method of complex impedances. This circuit is just a frequency-sensitive voltage divider, with the output signal developed across $\mathbf{Z}_C$. The general behavior of this circuit can be deduced from the observation that the capacitive reactance shorts the output signal to ground at high frequencies, and behaves as an open circuit with no effect on the output signal at low frequencies. At a more detailed level, the series current through both circuit elements is

$$\mathbf{i} = \frac{\mathbf{v}_{in}}{R + \dfrac{1}{j\omega C}} \tag{3.26}$$

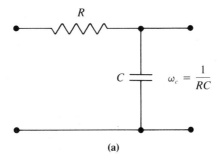

(a)

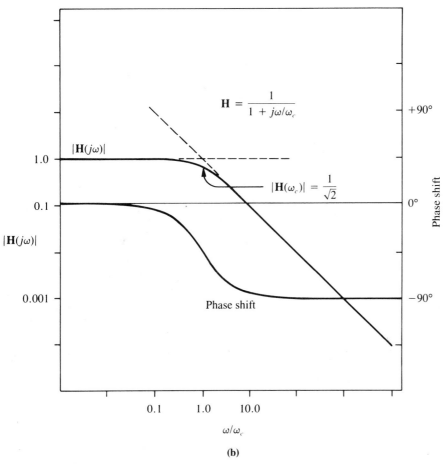

ω/ω_c

(b)

Figure 3.5 (a) The RC low-pass filter circuit, and (b) the magnitude and phase shift of its transfer function. The dashed lines show the two single-term approximations.

and develops a voltage $\mathbf{v}_{out}$ across the capacitor of

$$\mathbf{v}_{out} = \frac{\mathbf{i}}{j\omega C} \qquad (3.27)$$

Using Eq. (3.26) to substitute for the current **i** gives

$$\mathbf{v}_{out} = \frac{1/j\omega C}{R + 1/j\omega C} \, \mathbf{v}_{in} \tag{3.28}$$

which can be rearranged to yield the transfer function of this four-terminal network:

$$\mathbf{H}(j\omega) = \frac{\mathbf{v}_{out}}{\mathbf{v}_{in}} = \frac{1}{1 + j\omega RC} \tag{3.29}$$

The magnitude and phase of this function are shown as solid curves in Figure 3.5b.

The single-term approximation to this transfer function yields $\mathbf{H}_{low}$ and $\mathbf{H}_{high}$, valid at low- and high-frequency extremes. At sufficiently low frequencies, the magnitude of the second term in the denominator of Eq. (3.29) is small compared to 1 and can be dropped, leaving a single-term approximation of the form

$$\mathbf{H}_{low} = 1$$

for   (3.30)

$$\omega RC << 1$$

This agrees with the observation that at low frequencies the capacitive impedance becomes very large and can be eliminated, leaving just a resistor to connect the input to the output. At sufficiently high frequencies, the magnitude of the second term becomes large compared to 1, and the denominator can be approximated by the imaginary term alone. This yields the single-term approximation

$$\mathbf{H}_{high} = \frac{1}{j\omega RC}$$

for   (3.31)

$$\omega RC >> 1$$

In this frequency region, the capacitive impedance is small compared to the resistance.

The magnitudes of these single-term approximations are

$$|\mathbf{H}_{low}| = 1$$

and   (3.32)

$$|\mathbf{H}_{high}| = \frac{1}{\omega RC} = \left(\frac{1}{RC}\right)\omega^{-1}$$

These correspond to straight lines on a log-log plot and are shown in Figure 3.5b as dashed lines.

The curves of Eqs. (3.32) intersect at a frequency ω_c, which can be found by equating the two expressions and solving for ω_c:

$$1 = \frac{1}{\omega_c RC}$$

giving

$$\omega_c = \frac{1}{RC} \qquad (3.33)$$

This intersection defines the "corner" of the filter, and ω_c is known as the corner frequency of the filter.

The single-term approximations to the transfer function are also useful for estimating the phase shift between an input signal and an output signal. From the discussion in Section 2.10.1, a transfer function of the form $(j\omega)^n$ produces a phase shift of $n\pi/2$. Thus, Eqs. (3.30) and (3.31) imply that the phase shift across this low-pass filter approaches zero at low frequencies and $-\pi/2$ at high frequencies.

This high-frequency phase shift can be verified by using $\mathbf{H}_{high}$ to write the output signal in terms of the input signal:

$$\mathbf{v}_{out} = \frac{1}{j\omega RC}\,\mathbf{v}_{in} = \frac{-j}{\omega RC}\,\mathbf{v}_{in} \qquad (3.34)$$

If $\mathbf{v}_{in}$ is a counterclockwise spinning phasor on the complex plane, it must cross the positive real axis at periodic intervals. At the instant of one of these crossings, $\mathbf{v}_{in}$ is a real number represented by a phasor along the positive real axis as shown in Figure 3.6a. If $\mathbf{v}_{in}$ is real, Eq. (3.34) requires that $\mathbf{v}_{out}$ be proportional to $-j$. The resulting output voltage is therefore along the negative imaginary axis as shown in the figure. If the system is allowed to continue spinning, these phasors will maintain this relative phase relationship and the output voltage will lag behind the input voltage by $\pi/2$ radians, as shown in Figure 3.6b.

Although the single-term approximations are very convenient, the exact calculation of the relative amplitude and phase of the output signal must be made using $\mathbf{H}(j\omega)$. However, at a particular frequency, this function will reduce to a complex number that multiplies the input phasor to produce the output phasor.

Consider the particular frequency ω_c. In terms of ω_c, Eq. (3.29) becomes

$$\mathbf{H}(j\omega) = \frac{1}{1 + j\omega/\omega_c} \qquad (3.35)$$

and when $\omega = \omega_c$, this reduces to

$$\mathbf{H}(j\omega_c) = \frac{1}{1 + j} \qquad (3.36)$$

The magnitude of this transfer function is

$$|\mathbf{H}(j\omega_c)| = \frac{1}{\sqrt{(1 + j)(1 - j)}} = \frac{1}{\sqrt{2}} \qquad (3.37)$$

Since $|\mathbf{H}(j\omega)|$ has a maximum value of 1, this filter is said to be "down by $1/\sqrt{2}$" at its corner frequency. Note that this factor is not the same for all filters.

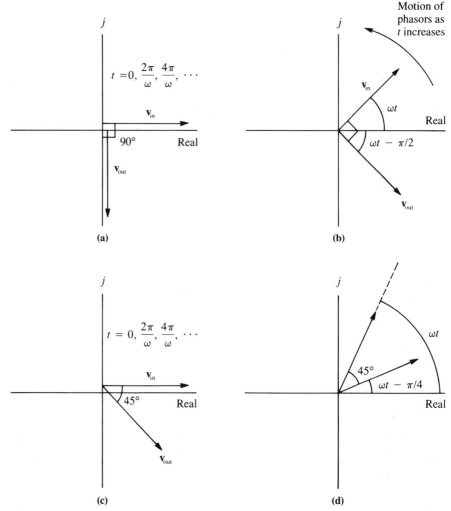

Figure 3.6 (a) At sufficiently high frequency, where v_{out} lags v_{in} by 90°, these spinning phasors represent the input and output voltages of the low-pass filter at a time when v_{in} is a real number, and (b) the same two phasors at an arbitrary time. (c) The phasor relationship when $\omega = \omega_c$ at an instant when the input is real and (d) at an arbitrary time.

The complex number of Eq. (3.36) can be put into $a + jb$ form by multiplying the numerator and denominator by $1 - j$:

$$\mathbf{H}(j\omega_c) = \frac{1}{1 + j} \frac{1 - j}{1 - j} = \frac{1}{2} - j\frac{1}{2} \tag{3.38}$$

The relationship between the input and output signals at this frequency is then given by

$$\mathbf{v}_{out} = \left(\frac{1}{2} - j\frac{1}{2}\right)\mathbf{v}_{in} = \frac{1}{\sqrt{2}} e^{-j\pi/4} \mathbf{v}_{in} \tag{3.39}$$

Viewed as two phasors spinning about the origin in the complex plane, it is clear that $\mathbf{v}_{out}$ lags behind $\mathbf{v}_{in}$ by $\pi/4$ radians as shown in Figures 3.6c–d.

Calculation of $\mathbf{H}(j\omega)$ at its corner frequencies, combined with knowledge of its single-term approximations, generally provides sufficient information to estimate the behavior of the full transfer function.

Example 3.1 Determine the phase shift introduced by the low-pass filter described by Eq. 3.29 at a frequency $\omega = 2\omega_c$.

Using the transfer function, the relationship between the input and output signals is given by

$$\mathbf{v}_{out} = \frac{1}{1 + j\omega RC}\,\mathbf{v}_{in}$$

At $\omega = 2\omega_c$, this reduces to

$$\mathbf{v}_{out} = \frac{1}{1 + j2}\,\mathbf{v}_{in}$$

Multiplying numerator and denominator by the complex conjugate of the denominator gives

$$\mathbf{v}_{out} = \left(\frac{1}{5} - j\frac{2}{5}\right)\mathbf{v}_{in}$$

making the phase shift

$$\theta = \arctan(-2) = -1.11 \text{ radians or } -63.4°$$

Plotting $\mathbf{v}_{in}$ and $\mathbf{v}_{out}$ as phasors on the complex plane at an instant when $\mathbf{v}_{in}$ is real clearly shows that $\mathbf{v}_{out}$ lags behind $\mathbf{v}_{in}$ by $63.4°$ as the minus sign suggests.

3.6.2 Approximate Integrator

The low-pass filter circuit shown in Figure 3.5a is sometimes known as an integration circuit, the implication being that the output signal is the integral of the input signal. This relationship is most readily derived by allowing the input sinusoidal signal to be represented by

$$\mathbf{v}_{in} = ve^{j\omega t} \tag{3.40}$$

For a true integration circuit, the time integral of this signal would be the output signal

$$\mathbf{v}_{out} = v\int e^{j\omega t}\,dt$$
$$= \frac{1}{j\omega}\,ve^{j\omega t} + \mathbf{v}_{out}(t=0) \tag{3.41}$$

With the understanding that $\mathbf{v}_{out}$ is only the AC part of the total signal, we can drop the last term. Since the time dependence of $\mathbf{v}_{out}$ is still the same as the input signal, Eq. (3.41) reduces to

$$\mathbf{v}_{out} = \frac{1}{j\omega}\,\mathbf{v}_{in} \tag{3.42}$$

and the transfer function of a true integrator would be

$$\mathbf{H}_I = \frac{\mathbf{v}_{out}}{\mathbf{v}_{in}} = \frac{1}{j\omega} \tag{3.43}$$

While the low-pass filter's transfer function is not that of a true integrator, its high-frequency approximation $\mathbf{H}_{high}$ differs only by a proportionality factor:

$$\mathbf{H}_{high} = \frac{1}{j\omega RC} = \frac{1}{RC} H_I \tag{3.44}$$

Thus, in the frequency region

$$\omega \gg \omega_c \tag{3.45}$$

*Figure 3.7   (a) Square wave input signal and its ideal integral.
(b) Result of passing the square wave through a low-pass filter whose
corner frequency is the same as the square wave fundamental.
(c) Result when the corner frequency is one-fifth of the square wave
fundamental. Note the change of the vertical scale between (b) and (c).*

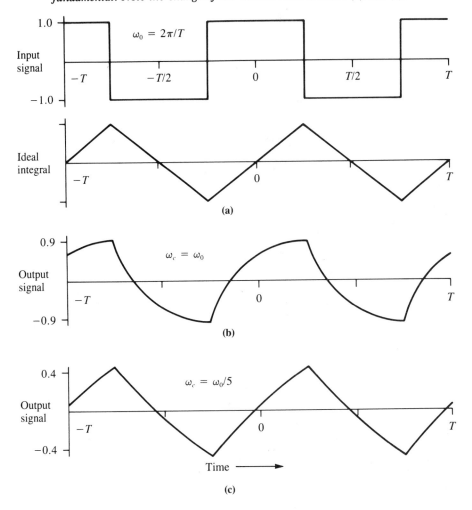

where $\mathbf{H}_{high}$ is a good approximation to the transfer function of the low-pass filter, we find that the filter's output is given by

$$\mathbf{v}_{out} = \frac{1}{RC} \mathbf{H}_I \mathbf{v}_{in} = \frac{1}{RC} \int \mathbf{v}_{in} \, dt \qquad (3.46)$$

Unfortunately, sinusoidal signals with frequencies well above ω_c are greatly attenuated by the low-pass filter; so although the filter integrates these signals, the amplitude of the integrated signal is much smaller than that of the input signal.

Consider the integration process applied to the square wave shown in Figure 3.7a. Since the average value of this signal is zero, its true integral is the triangle wave shown on this part of the figure. However, if we attempt to integrate this signal electronically by passing it through a low-pass filter whose corner frequency is the same as the fundamental (lowest) frequency component of the square wave, we get the badly distorted result shown on part (b) of the figure. By using a filter whose corner frequency is only one-fifth as large as the square wave fundamental, we are able to better satisfy the condition of Eq. (3.45) for all frequency components of the square wave; the resulting output signal is a much better approximation to the triangle wave as shown in Figure 3.7c, but note that the scale has changed between parts (b) and (c) of the figure.

3.6.3 High-Pass Filter

By interchanging the two circuit elements of the low-pass RC filter, we obtain the equivalent high-pass filter shown in Figure 3.8. Again the circuit can be viewed as a frequency-sensitive voltage divider, with the capacitive impedance being large at low frequencies and small at high frequencies. As the frequency of an input sinusoid is reduced, $\mathbf{Z}_C$ becomes larger, the series current becomes smaller, and the signal developed across the resistor shrinks. Well above the corner frequency for this filter, $\mathbf{Z}_C$ becomes much less than R, and most of the input signal will appear at the output.

Following the methods used on the low-pass filter, the transfer function for this filter is easily found to be

$$\mathbf{H}(j\omega) = \frac{j\omega RC}{1 + j\omega RC} \qquad (3.47)$$

Its single-term approximations are

$$\mathbf{H}_{low} = j\omega RC$$

and $\qquad (3.48)$

$$\mathbf{H}_{high} = 1$$

The magnitudes of these forms again intersect at a corner frequency

$$\omega_c = \frac{1}{RC} \qquad (3.49)$$

As in the case of the low-pass filter, the functions of Eq. (3.48) become better approximations to the actual transfer function for frequencies far removed from

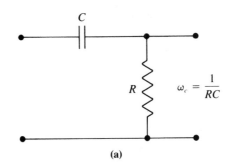

(a)

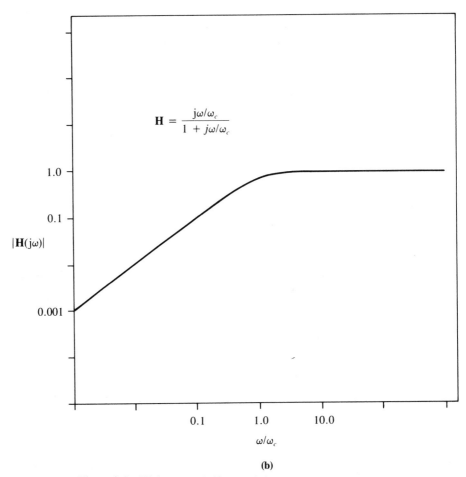

$$\mathbf{H} = \frac{j\omega/\omega_c}{1 + j\omega/\omega_c}$$

(b)

*Figure 3.8   High-pass RC filter and the magnitude of its transfer
function.*

ω_c. At sufficiently low frequencies, $\mathbf{H}_{\text{low}}$ shows that the output signal is phase
shifted ahead by $\pi/2$ radians so that it leads the input signal at these frequencies.
Equation (3.48) for $\mathbf{H}_{\text{high}}$ shows that the phase shift produced by this circuit approaches zero at high frequencies.

3.6.4 Approximate Differentiator

Just as the low-pass filter has the ability to integrate some input signals, so the high-pass filter acts like a differentiator. As before, this characteristic is limited to a certain frequency region. Consider the sinusoidal input signal

$$\mathbf{v}_{in} = v e^{j\omega t} \qquad (3.50)$$

If the output signal is to be the time derivative of this input, $\mathbf{v}_{out}$ must be given by the equation

$$
\begin{aligned}
\mathbf{v}_{out} &= \frac{d\mathbf{v}_{in}}{dt} \\
&= j\omega v e^{j\omega t} = j\omega \mathbf{v}_{in}
\end{aligned}
\qquad (3.51)
$$

making the transfer function for differentiation just

$$\mathbf{H}_D = j\omega \qquad (3.52)$$

Figure 3.9 (a) Square wave input signal and its ideal derivative, the delta function. (b) The result of passing the square wave through a high-pass filter whose corner frequency is the same as the square wave's fundamental frequency. (c) The result when the corner frequency is the same as the square wave's fifth harmonic.

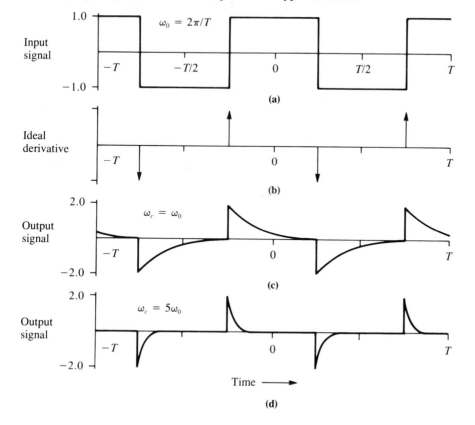

This result is proportional to the $\mathbf{H}_{low}$ for the high-pass filter,

$$\mathbf{H}_{low} = j\omega RC = RC\mathbf{H}_D \tag{3.53}$$

Of course, $\mathbf{H}_{low}$ represents the high-pass filter only when $\omega \ll \omega_c$.

Figure 3.9 shows the differentiation process applied to a square wave signal. Since the Fourier expansion of a square wave consists of an infinite series of odd harmonic terms, its faithful differentiation would require that the $\mathbf{H}_{low}$ of Eq. (3.53) correctly represent the circuit's transfer function to arbitrarily high frequencies. Since this is not possible with any circuit, the electronic differentiation of the square wave will always be approximate, as indicated by the typical results in Figure 3.9b,c.

3.7 *Complex Frequencies and the s-Plane*

The utility of the transfer function $\mathbf{H}$ is greatly enhanced by replacing the imaginary frequency variable $j\omega$ with a complex frequency variable $\mathbf{s} = \sigma + j\omega$. In this extension to the complex s-plane, the variable ω still determines the frequency of an oscillation, and the new variable σ is an inverse time constant defining the persistence of the oscillation. In terms of this new variable, a general time-dependent phasor is written

$$\begin{aligned} f(t) &= \mathbf{A}e^{st} \\ &= \mathbf{A}e^{\sigma t}e^{j\omega t} \end{aligned} \tag{3.54}$$

and can describe any of the signals shown in Figure 3.10.

3.7.1 *The Poles and Zeros of* $\mathbf{H}$

A transfer function $\mathbf{H}(j\omega)$ can be converted to the s-plane by the simple operation of replacing each occurrence of $j\omega$ with $\mathbf{s}$. This new $\mathbf{H}(\mathbf{s})$ can always be put into the general form

$$\mathbf{H}(\mathbf{s}) = \frac{\mathbf{P}(\mathbf{s})}{\mathbf{D}(\mathbf{s})} \tag{3.55}$$

where $\mathbf{P}(\mathbf{s})$ and $\mathbf{D}(\mathbf{s})$ are real coefficient polynomials in the complex variable $\mathbf{s}$. If the roots of $\mathbf{P}(\mathbf{s})$ and $\mathbf{D}(\mathbf{s})$ are $\mathbf{a}_n$ and $\mathbf{b}_m$, respectively, then the transfer function can be written in the form

$$\mathbf{H}(\mathbf{s}) = A\,\frac{(\mathbf{s} - \mathbf{a}_1)(\mathbf{s} - \mathbf{a}_2) \ldots (\mathbf{s} - \mathbf{a}_n)}{(\mathbf{s} - \mathbf{b}_1)(\mathbf{s} - \mathbf{b}_2) \ldots (\mathbf{s} - \mathbf{b}_m)} \tag{3.56}$$

where A is a real constant. Since $\mathbf{H}$ will be zero at each point $\mathbf{a}_n$ and infinite at each point $\mathbf{b}_n$, these roots are respectively known as the zeros and poles of the transfer function. It is important to realize that except for the scale factor A, these zeros and poles determine the transfer function $\mathbf{H}(\mathbf{s})$ everywhere on the s-plane, including the special case $\mathbf{H}(0 + j\omega)$, which we have already identified as the

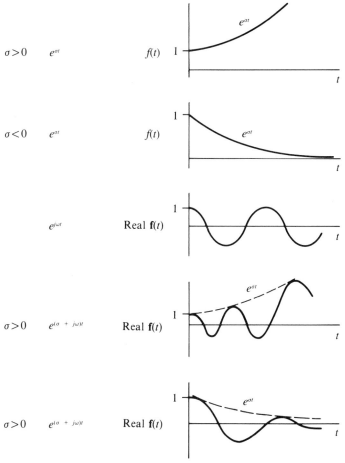

*Figure 3.10   Various forms of the function e^{st} and the corresponding
real signals.*

frequency response of a network. This relationship between $\mathbf{H(s)}$ and frequency
response is indicated by the two-pole, two-zero example in Figure 3.11.

The low-pass and high-pass transfer functions can be transformed to the s-plane
by substituting $\mathbf{s}$ for each occurrence of $j\omega$. This operation yields

$$\mathbf{H}_{\text{low}}(\mathbf{s}) = \frac{1}{1 + \mathbf{s}RC} \tag{3.57}$$

and

$$\mathbf{H}_{\text{high}}(\mathbf{s}) = \frac{\mathbf{s}RC}{1 + \mathbf{s}RC} \tag{3.58}$$

Both of these functions have a pole on the negative real axis at $\mathbf{s} = -1/RC$, but
are distinguished by the zero in $\mathbf{H}_{\text{high}}$ at $\mathbf{s} = 0$. Because their transfer functions

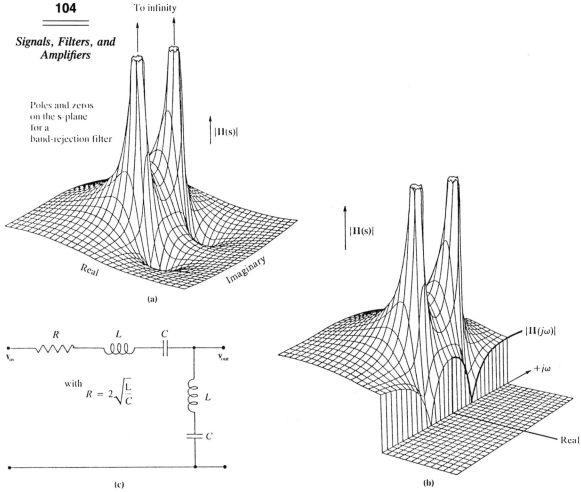

Figure 3.11 *(a) The magnitude of the transfer function of a two-pole, two-zero, band-rejection filter. (b) Same plot with a cut down the imaginary axis to show $|H(j\omega)|$. Note these scales are linear, not logarithmic. (c) A circuit described by this transfer function.*

have only one pole, these filters are often described as single-pole filters. A circuit will always have at least as many reactive elements as it has poles.

3.7.2 Frequency Response from Poles and Zeros

Because a four-terminal network is often defined in terms of its poles and zeros, it is useful to have an intuitive method for determining the frequency response $|H(j\omega)|$ from the poles and zeros of $H(s)$. The magnitude of $|H|$ follows from Eq. (3.56):

$$|\mathbf{H(s)}| = A \frac{|\mathbf{s} - \mathbf{a}_1| \, |\mathbf{s} - \mathbf{a}_2| \, . \, . \, . \, |\mathbf{s} - \mathbf{a}_n|}{|\mathbf{s} - \mathbf{b}_1| \, |\mathbf{s} - \mathbf{b}_2| \, . \, . \, . \, |\mathbf{s} - \mathbf{b}_m|} \tag{3.59}$$

Note that each factor in the numerator or denominator corresponds to the magnitude of a phasor on the complex plane, and such a magnitude is equivalent to the distance from the fixed point (pole or zero) to the variable point **s.** Since the frequency response $|\mathbf{H}(j\omega)|$ is $|\mathbf{H}(\mathbf{s})|$ evaluated along the $j\omega$ line ($\mathbf{s} = 0 + j\omega$), the determination of $|\mathbf{H}(j\omega)|$ from DC to some higher frequency corresponds to evaluating $|\mathbf{H}(\mathbf{s})|$ as **s** moves along the imaginary axis from zero to the higher frequency. The distances corresponding to typical terms from the numerator and denominator are shown in Figure 3.12.

We thus have a graphical method for determining the value of the expression given by Eq. (3.59): At a given frequency, $|\mathbf{H}(j\omega)|$ is the product of the distances from $j\omega$ to the zeros of $\mathbf{H}(\mathbf{s})$ divided by the product of the distances from $j\omega$ to the poles of $\mathbf{H}(\mathbf{s})$.

Poles or zeros located near the $+j$ axis have a dramatic effect on the frequency

Figure 3.12   Graphical method for determining $|\mathbf{H}(j\omega)|$ when the location of the poles and zeros of **H(s)** *are known. Only a typical conjugate pole pair and a typical conjugate zero pair of the general transfer function are shown.*

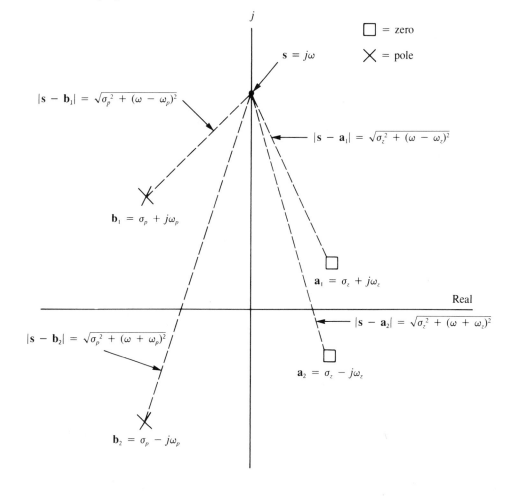

response of the filter, since Eq. (3.59) (with $\mathbf{s} = j\omega$) will then approach infinity or zero when $j\omega$ is near these points. Although complex conjugate poles and zeros in the third and fourth quadrants must be included in the products, because they are farther away from positive frequency points on the imaginary axis, they never have a dramatic effect on the frequency response.

This graphical interpretation of Eq. (3.59) greatly restricts the possible filters one can imagine constructing from linear circuit elements. For example, it would be nice to have a low-pass filter that would be constant out to some corner frequency, then drop precipitously to zero and remain there at all higher frequencies. The graphical arguments presented here show that such a filter would require an infinite number of zeros along the imaginary axis above the corner frequency. Since there must be at least one reactive circuit element for either each pole or each zero (whichever produces the larger number of elements), this ideal filter would require an infinite number of reactive circuit elements.

Example 3.2   If R is chosen to be $2\sqrt{L/C}$, the transfer function for the circuit of Figure 3.11c has zeros at ja and $-ja$, and poles at $(a/\sqrt{2})(1 + j)$ and $(a/\sqrt{2})(1 - j)$, where $a = 1/\sqrt{LC}$. Use the graphical method just discussed to determine $|\mathbf{H}(j0+)|$ and $|\mathbf{H}(j\omega)|$ as $\omega \to \infty$. (The notation $0+$ means an arbitrarily small positive frequency.)

A plot of the poles and zeros on the s-plane shows that the distance from the origin to any pole or zero is just a. Consequently, Eq. (3.59) evaluated near the origin is

$$|\mathbf{H}(0+)| = A \frac{a \cdot a}{a \cdot a} = A$$

Examination of the circuit shows that as $\omega \to 0$ from positive values, the capacitive impedance dominates both legs of the circuit, making $\mathbf{v}_{out} = 0.5\mathbf{v}_{in}$ and $A = \frac{1}{2}$ at $\omega = 0+$. The value of $|\mathbf{H}(j\omega)|$ where it crosses the vertical axis on the three-dimensional plot of Figure 3.11b is then $\frac{1}{2}$. As we move up the j axis allowing $\omega \to \infty$, the distances back to each of the poles and zeros all become large, each one arbitrarily close to ω. Evaluation of Eq. (3.59) shows that

$$|\mathbf{H}(j\omega)| \to \frac{A\omega^2}{\omega^2} = \frac{1}{2} \text{ as } \omega \to \infty$$

3.8   The Impulse Response of a Four-Terminal Network

A four-terminal network can also be characterized by its response to a short-duration pulse applied to its input. If the input signal is sufficiently narrow, the resulting output signal will be proportional to the impulse response function $h(t)$ defined in Chapter 12. In that chapter we show that $h(t)$ is related to $\mathbf{H}(\mathbf{s})$ by the Laplace transform, but for now we only need to know that there is a one-to-one relationship between $\mathbf{H}(\mathbf{s})$ and $h(t)$, and that both carry the same information about a network.

The impulse response is important because it more easily predicts a network's response to a rapidly varying input signal such as a pulse or square wave. Using $H(j\omega)$ to determine the response to a rapidly varying signal is difficult, because the Fourier series of such a signal contains many sinusoidal terms, each of which must be modified by $H(j\omega)$ and then recombined to form the output signal. Even though this procedure can be accomplished with the aid of a computer for any practical signal, its intuitive value is greatly diminished when many terms are involved.

However, the impulse response $h(t)$ is simply a sum of exponential terms e^{st}, one for each pole in the transfer function $H(s)$; and although the general form of $h(t)$ can be quite complicated, its most persistent (and most important) part is determined by the rightmost pole on the s-plane. If the rightmost pole is at $s = \sigma + j\omega$ and far removed from other poles and zeros, then the impulse response will ring with frequency ω and decay with time constant $1/\sigma$ as shown in Figure 3.13. Note that poles in the left half-plane produce damped responses, poles on the imaginary axis produce sinusoidal responses, and poles in the right half-plane produce undamped and unbounded responses. Negative frequency terms serve only to tidy up the mathematics and can be ignored in the present discussion. Passive circuits can have poles only in the left half-plane, but circuits with amplifiers can have poles anywhere.

Consider the single-pole filters described by Eqs. (3.57) and (3.58) as examples. Since both of these circuits have a single negative-real pole, the impulse response of each will exhibit a smooth exponential return to equilibrium. The square wave shown in Figures 3.7 and 3.9 is not an impulse, but since the time integral of an impulse produces a sharp step, the output response to each step can be described in terms of the time integral of the impulse response. As a result, we still expect the output response following each sudden change in the input voltage to be a smooth exponential approach toward some equilibrium with no sign of oscillation. This is indeed the case for both rising and falling edges, as can be seen in both figures.

A further example will show how the pole-zero locations can be used intuitively to relate a network's frequency response to its impulse response. Figure 3.14a shows the pole and zero locations for a hypothetical filter that will reject a narrow band of frequencies. Such a filter is known as a notch filter and is epitomized by the twin-T design given in a later section. As can be seen from the graphical technique discussed in Section 3.7.2, the two-pole, two-zero design of the figure results in a transfer function whose magnitude is 1 at DC, goes nearly to 0 at the frequencies near the transfer function zero, then again approaches a constant value of 1 at higher frequencies. Because it has only real poles, this circuit's impulse response shows no oscillation.

Now consider the similar design of Figure 3.14b, where the locations of poles and zeros have been switched. The magnitude of this transfer function is still 1 at DC and high frequencies, but becomes quite large near the frequency location of the pole. But because the poles are now complex and located near the imaginary axis, this circuit will ring for an extended time following any slight transient. With the pole locations indicated in the figure, the impulse response is formed from terms like $e^{-0.01t}e^{jt}$, each of which goes through many cycles before the decaying exponential cuts it down. The impulse response can be improved by moving the

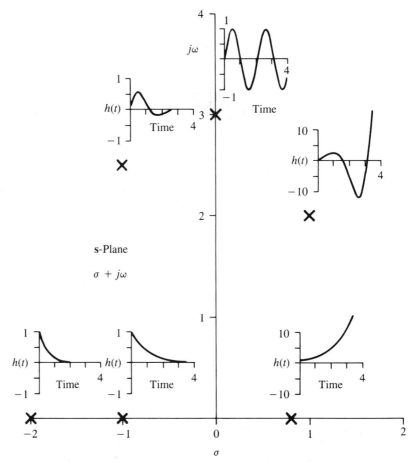

*Figure 3.13 The insets in this figure show the impulse response
functions h(t) that result from an isolated real pole or complex pole
pair.*

poles further to the left of the imaginary axis as in Figure 3.14c, but this will
reduce the sharpness of the passband of the transfer function; the two effects are
clearly inseparable.

3.9 Sequential RC Filters

Many applications require more exotic transfer functions than can be realized with
single-pole filters with their gradual transitions to 6-dB/octave slopes. Whether the
desire is for a band-pass filter, a band-rejection filter, or just a filter with steeper
slopes and a more rapid transition between pass and rejection bands, the requirement is the same: more poles and zeros and therefore more reactive circuit elements.

One way to build a circuit with more complexity is to connect two or more

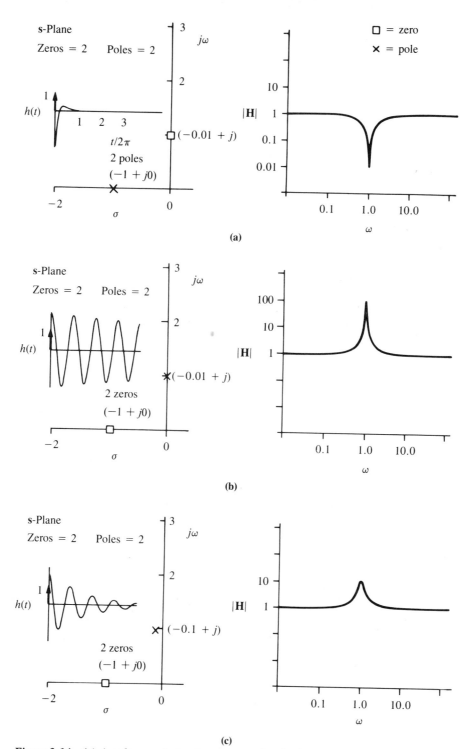

Figure 3.14 *(a) A pole-zero design for a narrow band-rejection filter resulting from a zero placed near the imaginary axis. (b) A similar design for a narrow band-pass filter resulting from a pole placed near the imaginary axis. The impulse response of this filter is most unsatisfactory. (c) A different pole placement yields a better impulse response but a broader passband. Note that the symmetrically placed $-j\omega$ member of a complex conjugate pair is not shown on these plots.*

single-pole RC filters in sequence, using the output of one to drive the next as shown in Figure 3.15. Each individual RC filter is known as a section, and if later sections draw no current from preceding ones, the overall transfer function will be the product of the individual functions. For two section filters, this convenient design rule can be approximated by choosing larger impedances for the second filter as shown in Figure 3.15. Alternatively, sections with similar impedances can be interconnected by an amplifier as shown in Figure 3.24b. Such an amplifier draws very little current from $\mathbf{H}_1$ but can provide a lot of current to $\mathbf{H}_2$.

If the multiple-section filter can be described by a transfer function that is the product of the individual transfer functions, the numerator and denominator are just the products of the individual numerators and denominators. Thus, the poles and zeros of the individual stages become the poles and zeros of the overall transfer function.

The transfer functions shown in Figure 3.15 all result from the product of two single-pole filter transfer functions. The individual single-pole sections can be described by one of the transfer functions

$$\mathbf{H}_{\text{low}} = \frac{j\omega/\omega_H}{1 + j\omega/\omega_H} \tag{3.60}$$

and

$$\mathbf{H}_{\text{high}} = \frac{1}{1 + j\omega/\omega_L} \tag{3.61}$$

where ω_L and ω_H are the corner frequencies of the low- and high-pass filters, respectively. These corner frequencies are given by $1/RC$ evaluated for the particular filter, and the single pole of these filters is located on the negative real axis at either $-\omega_L$ or $-\omega_H$.

The two-section low-pass filters shown in Figures 3.15a,c are both described by the transfer function product

$$\mathbf{H}_{\text{low}} = \frac{1}{(1 + j\omega/\omega_{L1})(1 + j\omega/\omega_{L2})} \tag{3.62}$$

where ω_{L1} and ω_{L2} are the corners of the first and second sections, respectively. Figure 3.15a is a special case, with $\omega_{L1} = \omega_{L2}$. Expanding the denominator results in three terms, each of which can dominate in a certain frequency region. Below ω_{L1} the filter is approximately flat, between the two corners it is proportional to ω^{-1} and can be represented by a -6 dB/octave line, and above ω_{L2} it is proportional to ω^{-2} producing a -12 dB/octave line. These regions are clearly displayed in Figure 3.15c, which shows a filter with $\omega_{L2} = 1000\omega_{L1}$.

The product of two high-pass transfer functions produces

$$\mathbf{H}_{\text{high}} = \frac{-\omega^2/\omega_{H1}\omega_{H2}}{(1 + j\omega/\omega_{H1})(1 + j\omega/\omega_{H2})} \tag{3.63}$$

An example of this filter type, with $\omega_{H1} = \omega_{H2}$, is shown in Figure 3.15b.

The band-pass filter of Figure 3.15d is produced by cascading one low-pass and one high-pass section. The order of the filter sections does not matter as long

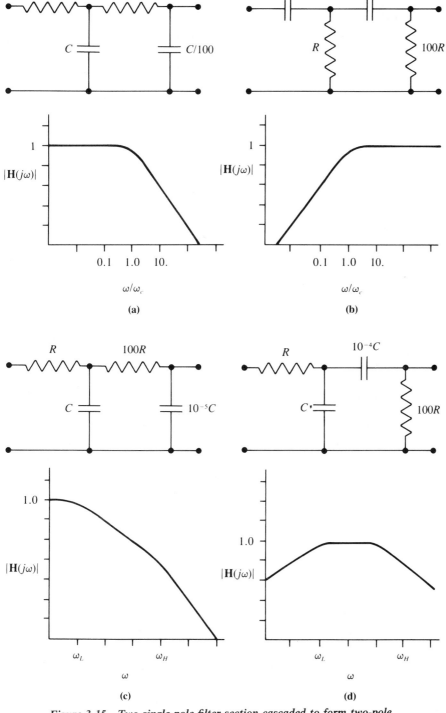

*Figure 3.15   Two single-pole filter section cascaded to form two-pole
filters of various types: (a) low-pass, (b) high-pass, (c) low-pass, and
(d) band-pass.*

as the second section is composed of larger impedances. The transfer function for this design is

$$\mathbf{H}_{pass} = \frac{j\omega/\omega_H}{(1 + j\omega/\omega_H)(1 + j\omega/\omega_L)} \qquad (3.64)$$

and the filter shown in the figure has $\omega_H = 100\omega_L$. If $\omega_H \leq \omega_L$, the transfer function would show only two straight regions.

It is not possible to produce a band-rejection filter using sections of this type.

Example 3.3 Design a band-pass filter as in Figure 3.15d so that the lower angular frequency corner is at 1000 rad/s (about 160 Hz). The largest available capacitor is 1 μF. What other components will be needed?

The corner frequencies for this filter are at $1/R_1C_1$ and $100/R_1C_1$. The largest capacitor is clearly needed in the first section of this filter, so we solve the first of these expressions for R_1 and evaluate it at the lower corner frequency with $C = 1$ μF:

$$R_1 = \frac{1}{(\omega_L C_1)} = \frac{10^6}{10^3} = 1000 \ \Omega$$

From the figure, the other components must then be $C_2 = 10^{-4} \ \mu$F $= 100$ pF and $R_2 = 100,000 \ \Omega$. The high-frequency corner is at 100,000 rad/s or nearly 16 kHz.

3.10   *Passive LCR Filters*

The two-pole *RC* filters just discussed are limited by the fact that the poles of their transfer functions are always real. Although at extremes of high and low frequency, multipole RC filters can ultimately reach a steep asymptotic slope, a filter with only real poles will always have smoothly rounded "corners," giving a slow transition between pass and rejection frequency regions. By introducing an inductor into the circuit, and thereby obtaining *LC* resonance as discussed in Chapter 2, we can develop filters with a broader range of characteristics. Such circuits with actual inductors are not widely used in current designs except at high frequencies, but their transfer functions contain some important new features and will turn up again when we discuss active filters in Chapter 8.

3.10.1   *The Series LCR Circuit*

By simple rearrangement of the elements, the series *LCR* circuit can produce any of the four filter types shown in Figure 3.16. The transfer functions of each of these filters are easily expressed as impedance ratios and can be reduced to the following expressions:

Low pass $\qquad \mathbf{H} = \dfrac{1}{(1 - \omega^2 LC) + j\omega RC}$ $\qquad (3.65a)$

High pass $\qquad \mathbf{H} = \dfrac{-\omega^2 LC}{(1 - \omega^2 LC) + j\omega RC}$ $\qquad (3.65b)$

Band pass $\qquad \mathbf{H} = \dfrac{j\omega RC}{(1 - \omega^2 LC) + j\omega RC}$ $\qquad$ (3.65c)

Band rejection $\qquad \mathbf{H} = \dfrac{1 - \omega^2 LC}{(1 - \omega^2 LC) + j\omega RC}$ $\qquad$ (3.65d)

The single-term approximations to most of these transfer functions can be obtained in the normal manner, but the band-rejection filter is complicated by the sharp null at $\omega_0 = 1/\sqrt{LC}$. This null splits the frequency spectrum in half, producing four single-term approximations: low, low-medium, high-medium, and high. At low frequencies, the numerator and denominator both approach 1, as all terms in powers of ω become much less than 1. The transfer function also approaches 1 at high frequencies, where the ω^2 terms ultimately dominate both numerator and denominator. In the middle ranges the denominator is assumed to be dominated by the $j\omega RC$ term, and the numerator can be approximated by 1 at frequencies below ω_0 and by $-\omega^2 LC$ above ω_0. These observations are summarized by the following equations:

Single-Term Approximations $\qquad$ Intersection Frequencies

$\mathbf{H}_L = 1$

$$\omega_1 = \frac{1}{RC}$$

$\mathbf{H}_{LM} = \dfrac{1}{j\omega RC}$

$$\omega_0 = \frac{1}{\sqrt{LC}} = \sqrt{\omega_1 \omega_2} \qquad (3.66)$$

$\mathbf{H}_{HM} = \dfrac{j\omega L}{R}$

$$\omega_2 = \frac{R}{L}$$

$\mathbf{H}_H = 1$

The magnitudes of these four approximations describe straight lines on a log-log plot with respective slopes of 0, -1, $+1$, and 0, and are tangent to the straight sections seen in Figure 3.16d. Note that none of the approximations approach $|\mathbf{H}|$ near the null at ω_0. As usual, the corner frequencies ω_1 and ω_2 (and in this case also ω_0) are obtained by intersecting the magnitudes of the adjacent approximations.

Although different in detail, the four *LCR* filters described by Eqs. (3.65) have several features in common. Each is completely described by two corner frequencies ω_1 and ω_2, determined by the same expressions given in Eq. (3.66) for the band-rejection filter. These corner frequencies are useful parameters for the filter, even when $|\mathbf{H}|$ is poorly represented by the single-term approximations. For example, in each case the circuit's resonant frequency ω_r and quality factor Q are given by

$$\omega_r = \sqrt{\omega_1 \omega_2} = \frac{1}{\sqrt{LC}} \qquad (3.67)$$

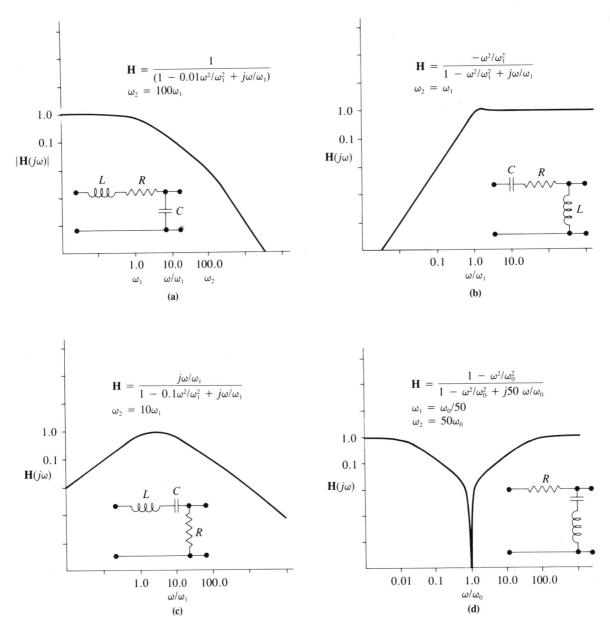

$$\mathbf{H} = \frac{1}{(1 - 0.01\omega^2/\omega_1^2 + j\omega/\omega_1)}$$
$$\omega_2 = 100\omega_1$$

$$\mathbf{H} = \frac{-\omega^2/\omega_1^2}{1 - \omega^2/\omega_1^2 + j\omega/\omega_1}$$
$$\omega_2 = \omega_1$$

$$\mathbf{H} = \frac{j\omega/\omega_1}{1 - 0.1\omega^2/\omega_1^2 + j\omega/\omega_1}$$
$$\omega_2 = 10\omega_1$$

$$\mathbf{H} = \frac{1 - \omega^2/\omega_0^2}{1 - \omega^2/\omega_0^2 + j50\ \omega/\omega_0}$$
$$\omega_1 = \omega_0/50$$
$$\omega_2 = 50\omega_0$$

Figure 3.16 Particular two-pole filters using the series LCR circuit:
(a) low-pass, (b) high-pass, (c) band-pass, and (d) band-reject.

and
$$Q = \sqrt{\frac{\omega_1}{\omega_2}} = \frac{\omega_r L}{R} \qquad (3.68)$$

Since the general shape of $|\mathbf{H}|$ is easily deduced from ω_1 and ω_2, it is useful to relate the pole locations to these corner frequencies. Rearranging the denominator of Eq. (3.65) to

$$(j\omega)^2 LC + j\omega RC + 1 = 0 \qquad (3.69)$$

then rewriting in terms of the corner frequencies and $\mathbf{s}$, we get

$$\frac{\mathbf{s}^2}{\omega_1\omega_2} + \frac{\mathbf{s}}{\omega_1} + 1 = 0 \qquad (3.70)$$

which reduces to

$$\mathbf{s}^2 + \omega_2 \mathbf{s} + \omega_1\omega_2 = 0 \qquad (3.71)$$

The roots of this quadratic correspond to the poles of the transfer function and are found directly with the quadratic formula

$$\mathbf{s} = \frac{-\omega_2 \pm \sqrt{\omega_2^2 - 4\omega_1\omega_2}}{2} \qquad (3.72)$$

Given the locations of the poles, we can now estimate the impulse response of the filter from the locations of its corner frequencies. Critical damping (two identical real roots) will occur when the radical is zero, corresponding to $\omega_2/\omega_1 = 4$; if this ratio is larger, the filter will be overdamped, with two poles on the negative real axis; if less, the filter will have a complex conjugate pair of poles in the left half-plane and be underdamped with some ringing following a transient. Thus, the impulse response can be estimated directly from the shape of $|\mathbf{H}|$.

Figure 3.17 shows the actual frequency and impulse responses of several low-pass filters as a function of the position of their corner frequencies. Note that the best response is a slightly underdamped filter somewhere between Figures 3.17b and 3.17c. The mid-frequency single-term approximation is poor for all of these figures, but it is still used to determine ω_1 and ω_2.

Example 3.4 Starting with a 100-mH inductor, choose a capacitor and resistor to produce a low-pass filter like that shown in Figure 3.16a but having $\omega_1 = \omega_2 = \omega_r = 10,000$ rad/s.

Solving Eq. (3.67) for C in terms of L and ω_r gives

$$C = \frac{1}{L\omega_r^2} = 0.1 \ \mu f$$

Setting $\omega_1 = \omega_2$ in Eq. (3.68) and solving for R gives

$$R = L\omega_r = 1000 \ \Omega$$

and completes the design of this filter.

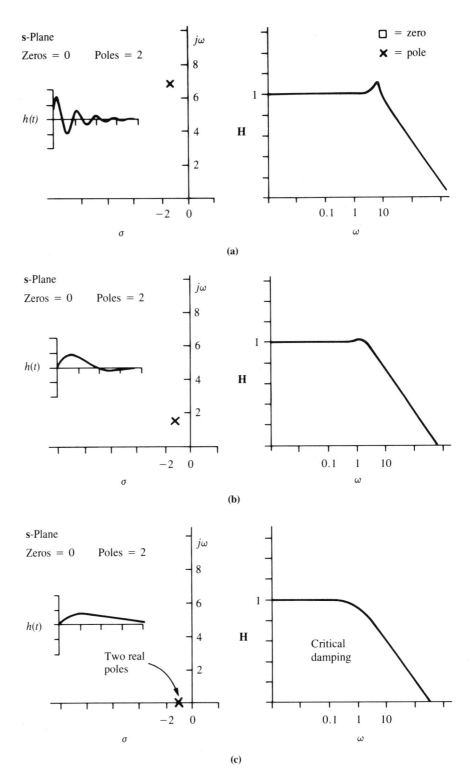

Figure 3.17 Impulse response and frequency spectrum of various low-pass series LCR filters: (a) strongly underdamped, (b) slightly underdamped, and (c) critically damped.

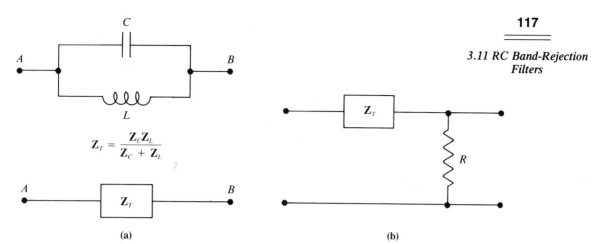

*Figure 3.18   (a) Equivalent impedance of the parallel combination of
an inductor and a capacitor: the tank circuit. (b) A band-rejection
filter using this tank circuit.*

3.10.2   The Tank Circuit

The *LC* arrangement shown in Figure 3.18a is known as a tank circuit. The equivalent impedance of this parallel combination of reactances is given by

$$\mathbf{Z}_T = \frac{j\omega L}{1 - \omega^2 LC} \tag{3.73}$$

This expression, which applies only to ideal circuit elements, goes to infinity at
$\omega = 1/\sqrt{LC}$. Because of series resistance in any actual inductor, the impedance
of a laboratory tank circuit will always remain finite. Remember that the impedance of a series *LC* combination goes to zero at this same frequency.

Figure 3.18b shows a narrow band-rejection filter made from the tank circuit.
A band-pass filter would result if the resistor and tank circuit were interchanged.
These two circuits have the same general character as the series *LCR* circuits
discussed above; the expressions for ω_1 and ω_2 are interchanged, but in terms of
these new corner frequencies, the pole locations are still described by Eq. (3.72).

3.11   RC Band-Rejection Filters

Narrow band-rejection filters can be fabricated using only resistive and capacitive
elements. Two multiple-loop networks are important but have transfer functions
that are substantially more complicated than those we have already discussed.
Although the functions can be derived by applying Kirchhoff's laws, the algebra
is lengthy and will not be given here.

3.11.1 Bridged T

The simplest band-rejection rejection filter is the bridged T shown in Figure 3.19. By assuming a voltage across the input terminals and using two circulating currents, it is possible to calculate the voltage across the output terminals. The resulting transfer function is most compactly expressed by

$$\mathbf{H}(j\omega) = 1 - \left[1 + \frac{R_1}{R_2}\left(1 + \frac{C_1}{C_2}\right) + j\sqrt{\frac{R_1 C_1}{R_2 C_2}}\left(\frac{\omega}{\omega_0} - \frac{\omega_0}{\omega}\right)\right]^{-1} \quad (3.74)$$

where

$$\omega_0 = \frac{1}{\sqrt{R_1 C_1 R_2 C_2}} \quad (3.75)$$

The magnitude of this transfer function goes through a minimum when the imaginary part vanishes at angular frequency ω_0. When rearranged, this transfer function can be expressed as a quadratic divided by a quadratic; it thus has two poles and two zeros.

The analysis of the bridged T simplifies by taking the special case of two equal capacitors. The capacitances then cancel out of Eq. (3.74), simplifying the expression. By defining $a^2 = R_1/R_2$, the new transfer function can be written

$$\mathbf{H}(j\omega) = 1 - \frac{1}{1 + 2a^2 + ja\left[\left(\frac{\omega}{\omega_0}\right) - \left(\frac{\omega_0}{\omega}\right)\right]} \quad (3.76)$$

Replacing $j\omega$ by s gives

$$\mathbf{H}(s) = 1 - \frac{\omega_0 s}{\omega_0 s(1 + 2a^2) + as^2 + a\omega_0^2} \quad (3.77)$$

and this expression reduces to

$$\mathbf{H}(s) = \frac{as^2 + 2a^2\omega_0 s + a\omega_0^2}{as^2 + (1 + 2a^2)\omega_0 s + a\omega_0^2} \quad (3.78)$$

For any resistance ratio, the roots of the denominator are both real and negative, making the impulse response well behaved. However, the transfer function still passes some signal, even at its minimum. This is most easily seen by locating the

Figure 3.19 *Bridged-T band-rejection filter.*

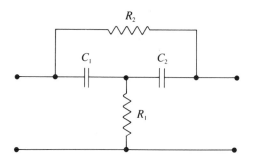

zeros of the transfer function. The two roots of the quadratic polynomial in the numerator are at

$$\mathbf{s} = -\omega_0(a \pm \sqrt{a^2 - 1}) \qquad (3.79)$$

thereby defining the zeros of $\mathbf{H(s)}$. This equation shows that the zeros always lie to the left of the imaginary axis, approaching the axis only when $a \ll 1$. Using the graphical method described in Section 3.4, it is clear that $|\mathbf{H}(\omega_0)|$ is proportional to $\omega_0 a$, the distance from $j\omega_0$ to the zero.

Since suitable filter capacitors are all less than a few microfarads, it is difficult in practice to meet the condition of a small resistance ratio a^2 while maintaining the desired null frequency as defined by Eq. (3.75). The difficulty increases as the frequency of the desired null ω_0 is decreased.

3.11.2 Twin T

The generally more desirable band-rejection circuit shown in Figure 3.20 is known as the twin T or notch filter. The transfer function for a general network of this type is very complicated, but if the resistors and capacitors are chosen as indicated in the figure, the rejection is optimized and the transfer function simplifies to

$$\mathbf{H(s)} = \frac{s^2 + \omega_0^2}{s^2 + 4\omega_0 s + \omega_0^2} \qquad (3.80)$$

where ω_0 is the notch angular frequency given by

$$\omega_0 = \frac{1}{RC} \qquad (3.81)$$

This transfer function has a complex pair of zeros located exactly on the imaginary axis at

$$\mathbf{s} = \pm j\omega_0 = \pm \frac{j}{RC} \qquad (3.82)$$

and two negative real poles. At the notch frequency the operation of the filter can be easily understood: As the voltage signal passes through each of the T sections,

Figure 3.20 The twin-T band-rejection filter.

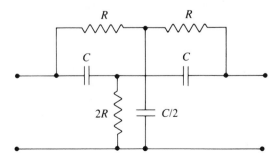

the upper signal's phase is delayed by 90° and the lower's phase is advanced by 90°; when recombined at the output, these signals cancel, producing the null result.

Do not be misled into thinking that the twin T is a particularly narrow filter; away from the null, the magnitude of the transfer function changes slowly, much like any other RC filter. Using Eq. (3.80), it can be shown (Problem 3.29) that the half-power points (frequency where $\mathbf{H}^2 = \frac{1}{2}$) of this filter occur at

$$\omega_1 \simeq \frac{\omega_0}{4}$$

and
$$\omega_2 \simeq 4.25\,\omega_0 \tag{3.83}$$

such that their difference is $\omega_2 - \omega_1 = 4\omega_0$ exactly.

Because the component values used in an actual twin-T circuit will vary somewhat from the ideal relationships indicated in Figure 3.20, the transfer function of Eq. (3.80) is only an approximation to an actual filter. Imprecise component matches will cause the zeros of the actual transfer function to move off the imaginary axis, destroying the perfect null needed in the most sophisticated applications of the twin T. This problem is reduced by using prepackaged twin-T circuits that feature precisely fabricated components arranged to minimize temperature-induced variations.

3.12   Amplifier Model

Although we are not yet ready to discuss the active circuit elements that make up an amplifier, we can extend the filter discussion to describe the overall behavior of a typical amplifier. As is often the case in actual practice, the amplifier will be treated as an active four-terminal device, totally without regard for its internal workings. Figure 3.21a shows a typical amplifier connected as a four-terminal device (the power supply connections are now shown). The symbol $\mathbf{A}$ on this figure represents the transfer function of the amplifier itself and is defined by Eq. (3.14), which we rewrite here in $\mathbf{s}$-plane notation:

$$\mathbf{v}_{out}(\mathbf{s}) = \mathbf{A}(\mathbf{s})\mathbf{v}_{in}(\mathbf{s}) \tag{3.84}$$

Note that we assume $\mathbf{v}_{out}$ is referred to ground, whereas $\mathbf{v}_{in}$ is the difference of two voltages.

One of the most useful amplifiers is the operational amplifier discussed in detail in Chapter 8. The pin connections to one of the most common such devices is shown in Figure 3.21a. Typical operational amplifiers have very large voltage gains, with $|\mathbf{A}|$ approaching 10^6 at zero frequency. However, because of various capacitances within the amplifier, this signal amplification is maintained only up to some corner frequency, above which the transfer function decreases much like a low-pass filter. Indeed, as long as the internal workings of the amplifier can be modeled by linear circuit elements, its transfer function will be algebraically similar to those of the low-pass filter circuits we have already studied, and we can use these filter functions to model the amplifier.

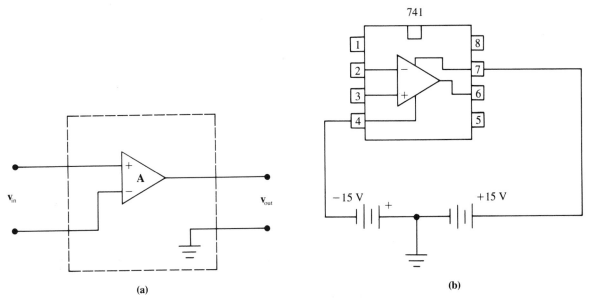

Figure 3.21 (a) The schematic representation of an amplifier showing only the signal connections. (b) The essential pin connections to the general-purpose 741 operational amplifier.

In terms of a large constant A_0, the amplifier transfer function can be written

$$\mathbf{A}(s) = A_0\mathbf{H}_{\text{low}}(\mathbf{s}) \qquad (3.85)$$

where $\mathbf{H}_{\text{low}}$ is the transfer function of a low-pass filter. The factor A_0 is the DC or zero frequency gain of the amplifier. Since the frequency variation is determined from an actual filter circuit, this transfer function correctly defines the phase shifts that must accompany any change in the transfer function's magnitude.

3.12.1 One-, Two-, and Three-Pole Amplifier Models

The simplest frequency-dependent model for the voltage amplifier is derived from the transfer function of the single-pole, low-pass RC filter of Eq. (3.57). This model describes an amplifier whose high-frequency response is determined by a single RC time constant. Multiplication of the low-pass filter transfer function by a real constant A_0 results in an amplifier transfer function expressed by

$$\mathbf{A}_1(\mathbf{s}) = \frac{A_0}{1 + \mathbf{s}/s_a} \qquad (3.86)$$

where s_a is a real positive number defining the corner frequency. Note that this amplifier has a single pole at $-s_a$ on the negative real axis and should therefore have a good impulse response, that is, decaying exponential without ringing. The frequency response for such an amplifier is shown in Figure 3.22a.

A simple model for an amplifier with two internal time constants is obtained by cascading two single-pole amplifiers together, producing the transfer function

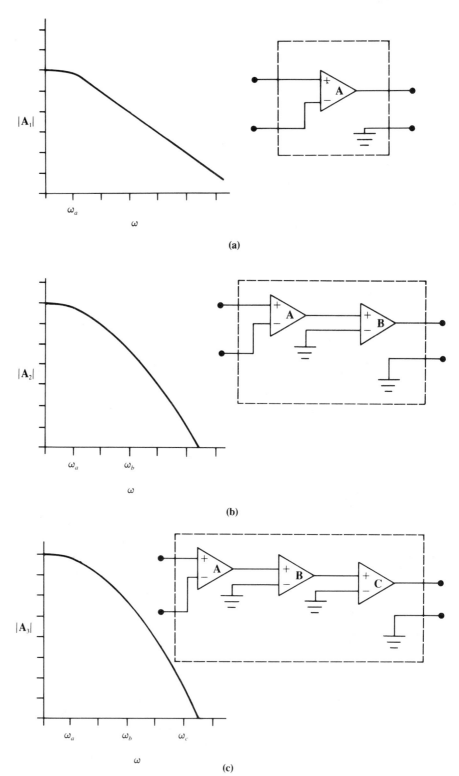

$|\mathbf{A}_1|$

ω_a

ω

(a)

$|\mathbf{A}_2|$

ω_a ω_b

ω

(b)

$|\mathbf{A}_3|$

ω_a ω_b ω_c

ω

(c)

*Figure 3.22   (a) Frequency response for a one-pole amplifier model.
(b) For a two-pole model formed by cascading two different single-pole
amplifiers. (c) For a three-pole model.*

$$A_2(s) = \frac{A_0}{(1 + s/s_a)(1 + s/s_b)} \tag{3.87}$$

where the second transfer function has corner frequency s_b. Similarly, an amplifier with three corner frequencies can be modeled by

$$A_3(s) = \frac{A_0}{(1 + s/s_a)(1 + s/s_b)(1 + s/s_c)} \tag{3.88}$$

Typical frequency response curves for these two- and three-pole amplifiers are shown in Figures 3.22b,c. Since all poles of these three amplifier models lie on the negative real axis, their impulse response will show no oscillation.

Although the nominal gain of an amplifier may be large, all are imperfect devices with gain and corner frequencies that vary, both from sample to sample and in response to changes in other variables such as temperature and power supply voltages.

3.12.2 Amplifier with Negative Feedback

In this section we will investigate negative feedback, a technique that is widely used to improve the characteristics of an imperfect amplifier. With a few extra circuit connections as shown in Figure 3.23, we can add negative voltage feedback to our amplifier. On this figure, the transfer function of the bare amplifier (its open-loop gain) is indicated by $A(s)$, and a second four-terminal network with transfer function F is used to feed the output signal of the amplifier back to its input. Note that, contrary to normal practice, the signal propagates from right to left across the F network.

Figure 3.23 Amplifier with negative voltage feedback.

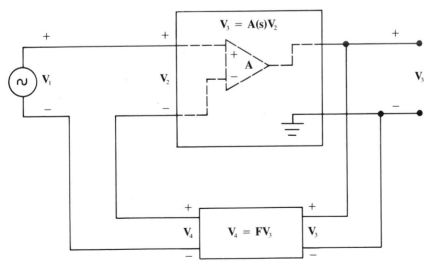

Using the voltage definitions in the figure, the overall transfer function **H** (the closed-loop gain) is

$$\mathbf{H(s)} = \frac{\mathbf{V}_3}{\mathbf{V}_1} \tag{3.89}$$

The input to network **A** is

$$\mathbf{V}_2 = \mathbf{V}_1 - \mathbf{V}_4 \tag{3.90}$$

the difference between the input signal and the feedback signal. The signal $\mathbf{V}_4$ is related to the output voltage by the feedback transfer function **F**, and the output voltage $\mathbf{V}_3$ is related to $\mathbf{V}_2$ by the amplifier transfer function **A**. Applying both of these relations, Eq. (3.90) becomes

$$\frac{\mathbf{V}_3}{\mathbf{A}} = \mathbf{V}_1 - \mathbf{F}\mathbf{V}_3 \tag{3.91}$$

Rearranging this expression to form the ratio $\mathbf{V}_3/\mathbf{V}_1$ gives the closed-loop transfer function

$$\mathbf{H(s)} = \frac{\mathbf{A}}{1 + \mathbf{AF}} \tag{3.92}$$

Figure 3.24 *(a) A 741 operational amplifier connected with F = 1 to form a unity gain amplifier. (b) A unity gain amplifier of this type used to connect two filter sections.*

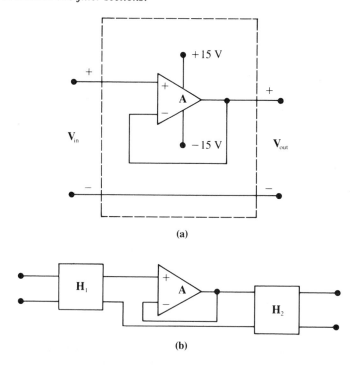

The importance of the feedback connection is easily seen by considering the situation where $|\mathbf{AF}| >> 1$. With this condition, the closed-loop transfer function reduces to

$$H(s) = \frac{1}{F}$$

(3.93)

which is independent of the amplifier's open-loop gain. Since $\mathbf{F}$ may be the transfer function of a totally passive and therefore predictable and stable network, the resulting closed-loop amplifier characteristics are also predictable and stable.

The simplest feedback network uses only resistors, resulting in an $\mathbf{F}$ that is real, positive, and less than 1. Subject to the above approximations, the closed-loop gain is then a real constant given by

$$G = \frac{1}{F}$$

(3.94)

As long as $\mathbf{A}$ is large, its exact size is now unimportant. Suppose that $\mathbf{A}$ is a real constant varying between 10^5 and 10^7, and let $\mathbf{F} = 0.1$ (see Figure 3.24). The exact closed-loop gain determined from Eq. (3.92) will vary only between 9.99900 and 9.99999, even though $\mathbf{A}$ changes by two orders of magnitude.

Problems

1. If the square wave signal of Figure 3.2 is shifted to the right by $T/4$, it can be represented by a constant plus an odd function. *(a)* Determine the first four nonzero Fourier series coefficients for the odd part alone, then *(b)* write the first five terms of the expansion that describes the complete function (constant plus odd function).

2. The unit step function $U_0(t)$ is defined as

$$U_0(t) = 0 \qquad \text{when } t \leq 0$$

and

$$U_0(t) = 1 \qquad \text{when } t > 0$$

Sketch the odd and even parts of this function with particular attention to the $t = 0$ point.

3. Use a trigonometric identity to find the odd and even parts of $y(t) = A \cos(\omega t + \theta)$.

4. In some circumstances we might wish to approximate the complex number $1 + 10j$ with $10j$. Determine the error in magnitude and phase angle introduced by this approximation. Express the magnitude error as a percent error.

5. What phase shift is introduced by the low-pass filter described by Eq. (3.29) at: *(a)* half the corner frequency, *(b)* twice the corner frequency, and *(c)* 10 times the corner frequency?

6. Derive both of the Bode plot slope Eqs. (3.25) from the decibel Eq. (3.24).

7. Show that the transfer function of Eq. (3.63) approaches an asymptote of $+12$ dB/octave at frequencies below both corners.

8. Calculate and plot the phase shift of a high-pass filter whose transfer function is given by Eq. (3.47). Take RC to be 1.

9. Write the transfer function $\mathbf{H}(j\omega)$ for the network shown in Figure P3.9 and from it find: **(a)** the corner frequency and **(b)** the value of $|\mathbf{H}|$ at the corner frequency. **(c)** How many degrees of phase shift are introduced by this network just below and just above the corner frequency?

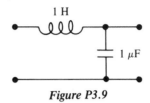

Figure P3.9

10. You have a 100-Hz square wave and a 0.1-μf capacitor. What resistance is needed to produce an approximate integrator with an output signal like that shown in Figure 3.7c?

11. If one pole in Figure 3.11 is at $(-\sigma_1, \omega_1)$ and one zero is at $(0, \omega_2)$, write an expression like Eq. (3.56) for this two-pole, two-zero transfer function.

12. If a transfer function has poles at $(-1, 2)$ and $(-1, -2)$ and a zero at $(0, 0)$, **(a)** sketch $|\mathbf{H}(j\omega)|$ on the interval $0 \le \omega < 10$. **(b)** If $|\mathbf{H}(j10)| = 1$, what is the approximate value of $|\mathbf{H}|$ at its highest point?

13. A low-pass filter has an $|\mathbf{H}(j\omega)|$ that peaks slightly at 1000 Hz and falls at 18 dB/octave at very high frequencies. On the **s**-plane, indicate the approximate location of the poles of this filter. [*Hint*: You should be able to locate the imaginary axis values to about 10%.]

14. (a) Make use of v_2 in Figure P13.14 to show that $|\mathbf{i_2}| << |\mathbf{i_1}|$. **(b)** In terms of $\mathbf{R_1}$, $\mathbf{R_2}$, $\mathbf{Z_1}$, and $\mathbf{Z_2}$, write the two loop-method equations that describe this circuit, then simplify using **(a)**. **(c)** Write an algebraic equation for $\mathbf{H(s)}$, then evaluate it at $\omega = 10^4$ rad/s and reduce to the form $1/(a + jb)$.

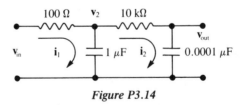

Figure P3.14

15. Show that the magnitude of the transfer function of Eq. (3.64) falls off at -6 dB/octave at both low- and high-frequency extremes.

16. Starting with a 300-mH inductor, design a low-pass series *LCR* filter as in Figure 3.16a that has an $|\mathbf{H_L}|$ to $|\mathbf{H_H}|$ corner frequency at 1000 Hz (f, not ω) and $\omega_2 = 2\omega_1$.

17. Determine the transfer function $\mathbf{H(s)}$ of the circuit shown in Figure 3.11c.

18. An *LCR* band-rejection filter like that shown in Figure 3.16d is constructed using a 1-kΩ resistor, a 1-μf capacitor, and a 1-H inductor. Unfortunately, the actual inductor has a resistance of 200 Ω. **(a)** Draw a schematic for this actual circuit and write the new transfer function. **(b)** What will be the value of $|\mathbf{H}|$ at $\omega = 1/\sqrt{LC}$?

19. If $\sigma_z = -\sigma_p$ and $\omega_z = 0.5\omega_p$ in Figure 3.12, sketch the shape of $|H(j\omega)|$ as ω runs from zero to $5\omega_p$.

20. Sketch $|H(j\omega)|$ for the *LCR* circuit shown in Figure P3.20 for the two conditions $R = 0.5\sqrt{L/C}$ and $R = 2\sqrt{L/C}$. In each case, determine the values of $|H|$ at $\omega = 0$, ∞, and ω_c, and label these points on the sketches.

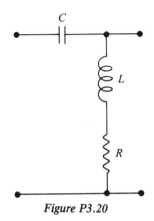

Figure P3.20

21. Determine ω_1 and ω_2 for the filter whose poles and response function are shown in Figure 3.17a. The poles are at $s = -1 + j7$ s^{-1}.

22. *(a)* Write an expression for the transfer function of the circuit shown in Figure P3.22. *(b)* What phase shift is introduced by this filter at very small and very large frequencies? *(c)* On a log-log scale, sketch $|H(j\omega)|$ and the phase shift as a function of ω.

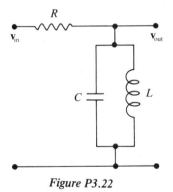

Figure P3.22

23. For the circuit of Problem 20, find the relationship between L, C, and R that will result in a critically damped response to a square wave input pulse.

24. *(a)* Using the transfer function of Eq. (2.98), show that when $Q \gg 1$, the pole positions are given by the expression $s = \omega_r(-1/2Q + j)$. *(b)* Make an s-plane plot showing these poles and identify the angle $1/2Q$. *(c)* Using this result, explain why a high-Q circuit is likely to ring when excited.

25. An actual tank circuit always must include the effects of the inherent series resistance in a real inductor. *(a)* Determine an expression for the equivalent impedance of such a circuit in terms of R_L, L, and C. **(b)** If $L = 1$ mH, $R_L = 1$ Ω, and $C = 0.1$ μf, determine the value of $\mathbf{Z}_{\text{tank}}$ at $\omega = 1/\sqrt{LC}$. **(c)** At this frequency, will the equivalent impedance produce a current-to-voltage phase shift most like an inductor, a capacitor, or a resistor?

26. A low-pass filter is constructed from an actual L and R as shown in Figure P3.26.

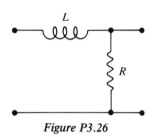

Figure P3.26

    (a) Draw a schematic for this actual circuit that includes the effects of coil resistance R_L and parasitic capacitance C_L.
    (b) Determine the transfer function of this expanded schematic.
    (c) If $R_L = 0$ while C_L is small but finite, sketch $|\mathbf{H}(j\omega)|$.
    (d) How does this sketch change if R_L is nonzero?

27. If a low-pass LCR circuit is critically damped, find a numerical value for the ratio $|\mathbf{H}(j\omega_c)/\mathbf{H}(0)|$. [The corner ω_c results from the intersection of the low-frequency and high-frequency single-term approximations.]

28. (a) Determine and plot the poles and zeros of the bridged-T transfer function of Eq. (3.74) when $a = \frac{1}{5}$. **(b)** Evaluate this equation when $\mathbf{s} = j\omega_0$.

29. After rewriting Eq. (3.80) for the twin-T filter in terms of ω, show that the full width of the rejection band at $\mathbf{H}^2 = \mathbf{H}^*\mathbf{H} = \frac{1}{2}$ is $4\omega_0$. [*Hint:* Your initial expression is a quartic, but it can be solved as a quadratic in ω^2; the answer comes up in the form $\omega_2 - \omega_1 = (\sqrt{9 + \sqrt{80}} - \sqrt{9 - \sqrt{80}})\omega_0$.]

30. The popular Butterworth condition for a filter results in a maximally flat passband. For a low-pass filter, it can be shown [Owen and Keaton, *Fundamentals of Electronics,* Vol. 1] that this condition is met by a transfer function with poles evenly spaced around a circle on the s-plane as shown in Figure P3.30. Ignoring the poles in the right half-plane, write a two-pole transfer function for a low-pass filter with poles placed as shown in quadrants II and III and show that its magnitude is given by

$$|\mathbf{H}(j\omega)| = \frac{1}{\sqrt{1 + \omega^4/\omega_c^4}}$$

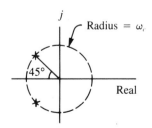

Figure P3.30

31. The general transfer function for a band-pass filter has the form

$$H(s) = \frac{As}{s^2 + Bs + \omega_0{}^2}$$

where ω_0 is at the intersection of the high- and low-frequency single-term approximations to $H(j\omega)$. For the following questions, assume that $\omega_0{}^2 \gg B^2$.

   (a) If the Q factor for this band-pass filter is defined by Eqs. (2.99) and (2.100), show that $Q = \omega_0/B$.

   (b) Also show that $Q = |H(j\omega_0)/H_L(j\omega_0)| = |H(j\omega_0)/H_H(j\omega_0)|$, where H_L and H_H are the low- and high-frequency approximations to $H(j\omega)$.

32. Use the single-term approximations of

$$A(j\omega) = \frac{A_0}{(1 - j\omega/\omega_a)(1 - j\omega/\omega_b)}$$

to determine the corner frequencies ω_1 and ω_2 in terms of ω_a and ω_b. What is the condition on ω_a and ω_b that will make $\omega_1 \simeq \omega_a$ and $\omega_2 \simeq \omega_b$?

4

Transformers

4.1 Introduction

The transformer makes use of magnetic field lines to effect a coupling between circuit loops that may be electrically isolated from each other. Although this electrical isolation is an important feature of many transformer applications, these devices are also used to modify the input or output impedance of a network so as to produce a better match to a cascaded network. The use of a transformer to transfer signals between various parts of the same circuit is especially important at high frequencies (100 kHz and above).

Faraday's law of magnetic induction states that the back EMF induced in a current loop is the result of a change in the number of magnetic field lines that the loop surrounds. These field lines are known as the magnetic flux ϕ and can be generated either by the current in the loop itself or by currents in additional loops that are electrically distinct from the first. The inductance L, which we have already discussed, is more accurately known as the self-inductance, since it describes only the part of an induced EMF that is caused by a change in the current in the same loop. The part of the induced EMF caused by the change in current in an external loop is described by a quantity M, known as the mutual inductance between the two loops. Like the capacitance and the self-inductance, the mutual inductance is a property of the physical configuration of the circuit and in our linear model is a constant, independent of currents or voltages.

The schematic symbol for a transformer is shown in Figure 4.1a and is used to indicate the presence of a circuit interaction resulting from this magnetic flux linkage. A physical transformer is constructed of two or more coils of wire closely associated in a manner that will maximize this mutual inductance effect. Ferro-

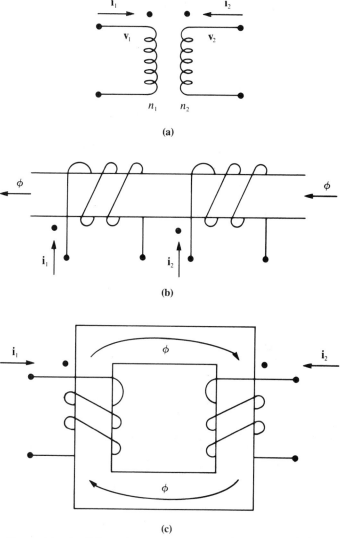

Figure 4.1 *(a) Schematic symbol for a transformer showing sign conventions. (b) A physical version of the same transformer. (c) Another version of the same transformer. To channel the flux efficiently between the coils, the square form must contain a ferromagnetic material.*

magnetic material is often used to increase the flux to current ratio in transformers, thereby increasing the magnitude of the self- and mutual inductance. This material also has the desirable property of being able to contain and channel the flux within the coils. Unfortunately, ferromagnetic materials exhibit complicated field and frequency-dependent relationships between the flux and the current, and such transformers can be expected to depart from the model presented below, especially at high frequencies. Transformers wound on air cores will have much smaller inductances, but they will be more accurately described by the relations developed below.

4.2 Transformer Equations

We can make a reasonable model for a transformer by assuming that the only time variation of the flux in the coils of Figure 4.1b is due to changes in the currents in the two coils. Thus we can describe that the flux through coil 1 by

$$\phi_1 = f(i_1, i_2) \tag{4.1}$$

and that through coil 2 by

$$\phi_2 = g(i_1, i_2) \tag{4.2}$$

Differences in the flux in the two coils may be due to different number of turns, different coil areas, or divergence of the field lines between the coils. We will come back to this later; for the moment let us concentrate on ϕ_1. The total derivative of ϕ_1 with respect to time is given by the partial differential equation

$$\frac{d\phi_1}{dt} = \frac{\partial\phi_1}{\partial i_1}\frac{di_1}{dt} + \frac{\partial\phi_1}{\partial i_2}\frac{di_2}{dt} \tag{4.3}$$

The self-inductance L of a coil is defined as the change in flux ϕ through the coil with respect to the change in current in the coil. In terms of this definition, L_1 can be written

$$L_1 = \frac{\partial\phi_1}{\partial i_1} \tag{4.4}$$

and the other partial derivative is similarly identified with the mutual inductance

$$M_{12} = \frac{\partial\phi_1}{\partial i_2} \tag{4.5}$$

With the help of these two constants, which are determined entirely by the coil configurations, Eq. (4.3) can be rewritten as a simple differential equation:

$$\frac{d\phi_1}{dt} = L_1\frac{di_1}{dt} + M_{12}\frac{di_2}{dt} \tag{4.6}$$

This equation relates the change in flux in coil 1 to the change in currents in coils 1 and 2.

Faraday's law, relating the induced EMF in a coil to the change in flux through the coil, is normally written

$$v_1 = -\frac{d\phi_1}{dt} \tag{4.7}$$

where the subscript refers to the coil number and the minus sign signifies that the induced EMF always works to generate a current that will tend to cancel the change in flux. Although this equation can be combined with Eq. (4.6) to give the induced EMF in terms of the current variations, it is a mistake to assume that the minus sign of Eq. (4.7) has a direct algebraic interpretation. Before a Kirchhoff's law loop equation can be applied to the circuit of Figure 4.2, it is necessary to establish a definite sign convention that includes the relative sense of the two windings as shown in Figure 4.1b. The sense of the windings is determined by the dots shown in this figure; if positive current is directed into the dots of both coils, then the magnetic flux generated by each current will add constructively.

If an ideal EMF drives the input loop as shown in Figure 4.2, the sign conventions shown results in the following loop equation:

$$v_1 - L_1 \frac{di_1}{dt} - M_{12} \frac{di_2}{dt} = 0 \qquad (4.8)$$

A similar argument with respect to the output loop yields the analogous result

$$v_2 - M_{12} \frac{di_1}{dt} - L_2 \frac{di_2}{dt} = 0 \qquad (4.9)$$

where L_2 and M_{12} are defined by

$$L_2 = \frac{d\phi_2}{di_2} \qquad (4.10)$$

and

$$M_{12} = \frac{d\phi_2}{di_1}$$

The self-inductance of a solenoidal coil can be determined with elementary methods. From Ampere's law, the magnetic field strength B inside the coil is given by

$$B = \frac{\mu_0 i N}{l} \qquad (4.11)$$

where i is the current, N is the number of turns in the coil, and l is its length. If the cross-sectional area of the coil is A, then the flux linking a single turn of the coil is

$$\phi' = \frac{\mu_0 i N A}{l} \qquad (4.12)$$

Figure 4.2 *Sign conventions applied to the primary or input loop of a transformer circuit.*

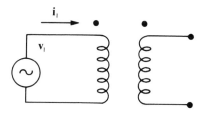

and the flux linking all N turns is

$$\phi = \frac{\mu_0 i N^2 A}{l} \tag{4.13}$$

One factor of N comes from the turns generating the flux, and the second factor comes from the turns that are linked by this flux. From Eq. (4.4), the self-inductance is given by

$$L = \frac{d\phi}{di} = \frac{\mu_0 N^2 A}{l} \tag{4.14}$$

For simple geometries, elementary methods can also be used to show that the mutual inductance between two coils is proportional to the product of the number of turns in the first coil and the number of turns in the second. If our two solenoids are wrapped on a common core (same A and l) but have different numbers of turns n_1 and n_2, then the mutual inductance is easily found to be

$$M_{12} = M_{21} = \frac{\mu_0 n_1 n_2 A}{l} \tag{4.15}$$

Although not obvious from these simple methods, more advanced formulations show that the two mutual inductances M_{12} and M_{21} are identical for any geometry and thus can always be represented by a single symbol M. In the following discussion, we will continue to assume that A/l is a common factor for both coils; the construction of actual transformers is such that this assumption is not unreasonable, and any deviation can be absorbed in the n_1 or n_2 turns parameters.

If the two coils forming a transformer's primary and secondary windings have turns n_1 and n_2, respectively, then the turns ratio for the transformer is defined as

$$n = \frac{n_1}{n_2} \tag{4.16}$$

With a common A/l for both coils and using Eq. (4.14) to define the self-inductance, this ratio can also be written as

$$n = \sqrt{\frac{L_1}{L_2}} \tag{4.17}$$

Depending on the physical configuration and the core material around which the coils are wound, some fraction of the flux generated by one coil will "leak" out of the system and not link with the other coil. The coupling coefficient k quantifies this effect and is defined by

$$k^2 = \frac{M^2}{L_1 L_2} \le 1 \tag{4.18}$$

The situation where all flux lines link both coils is known as a *perfect* transformer and is given by $k = 1$. For a perfect transformer, we can use Eqs. (4.14) and

(4.15) directly to relate the mutual inductance to the product of the two self-inductances:

$$L_1L_2 = M^2 \tag{4.19}$$

In the final equations resulting from the following arguments, we will assume a perfect transformer and use Eq. (4.19) to eliminate the need for an explicit reference to the mutual inductance. In more general treatments it is common to replace the mutual inductance with $k^2L_1L_2$. This more detailed treatment follows the same steps outlined below and leads to a more complicated transformer model.

4.2.1 Time-Independent Equations

Equations (4.8) and (4.9) can now be rewritten as

$$v_1 = L_1 \frac{di_1}{dt} + M \frac{di_2}{dt}$$

and

$$v; = M \frac{di_1}{dt} + L_2 \frac{di_2}{dt} \tag{4.20}$$

Following the normal method for AC signals, we can assume that the time dependence for the currents and voltages is of the form:

$$\mathbf{i}_1(t) = \mathbf{i}_1 e^{st}$$
$$\mathbf{i}_2(t) = \mathbf{i}_2 e^{st}$$
$$\mathbf{v}_1(t) = \mathbf{v}_1 e^{st} \tag{4.21}$$
$$\mathbf{v}_2(t) = \mathbf{v}_2 e^{st}$$

where the steady-state case would be given by $\mathbf{s} = j\omega$. Note that all of these expressions exhibit the same time dependence, but the complex amplitudes allow them to differ in phase. The last two expressions and the time derivatives of the first two can be substituted into the differential equations of (4.20). After canceling the common time dependence, we are left with a pair of simultaneous algebraic equations involving the complex amplitudes and the angular frequency variable $\mathbf{s}$; we get

$$\mathbf{v}_1 = \mathbf{s}L_1\mathbf{i}_1 + \mathbf{s}M\mathbf{i}_2$$

and

$$\mathbf{v}_2 = \mathbf{s}M\mathbf{i}_1 + \mathbf{s}L_2\mathbf{i}_2 \tag{4.22}$$

We are now ready to apply these equations to the complete transformer circuit of Figure 4.3. Note that the current directions in this figure are not arbitrary but must be chosen as shown to satisfy the sign conventions used to obtain Eqs. (4.22). The fact that the little dots shown in this figure are rarely seen on actual schematics and are not always obvious from the transformer itself should not be cause for alarm; the worst possible result is a 180° phase shift between a derived current or voltage and the physical reality. Such a condition is easily corrected by

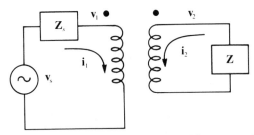

Figure 4.3 Complete transformer circuit with currents drawn to satisfy the sign conventions of Eqs. (4.22).

reversing a pair of input or a pair of output leads to the transformer. The Kirchhoff loop equations around the primary and secondary circuits are

$$\mathbf{v}_s - \mathbf{Z}_s \mathbf{i}_1 - \mathbf{v}_1 = 0$$

and

$$\mathbf{v}_2 + \mathbf{Z} \mathbf{i}_2 = 0 \tag{4.23}$$

Using Eqs. (4.22) to eliminate $\mathbf{v}_1$ and $\mathbf{v}_2$ yields

$$\mathbf{v}_s = (sL_1 + \mathbf{Z}_s)\mathbf{i}_1 + sM\mathbf{i}_2$$

and

$$0 = sM\mathbf{i}_1 + (sL_2 + \mathbf{Z})\mathbf{i}_2 \tag{4.24}$$

4.2.2 Voltage and Current Transfer Functions

The application of these equations to actual circuits is much simplified by the use of some standard relationships between the voltages and currents of the primary loop and those of the secondary loop. Using the method of determinants to solve the previous pair of equations gives

$$\mathbf{i}_1 = \frac{(\mathbf{Z} + sL_2)\mathbf{v}_s}{s^2(L_1L_2 - M^2) + (sL_1 + \mathbf{Z}_s)\mathbf{Z} + sL_2\mathbf{Z}_s}$$

and

$$\mathbf{i}_2 = \frac{-sM\mathbf{v}_s}{s^2(L_1L_2 - M^2) + (sL_1 + \mathbf{Z}_s)\mathbf{Z} + sL_2\mathbf{Z}_s} \tag{4.25}$$

Since both of these expressions have the same denominator, the current ratio $\mathbf{i}_2/\mathbf{i}_1$ is somewhat simpler:

$$\alpha = \frac{\mathbf{i}_2}{\mathbf{i}_1} = \frac{-sM}{\mathbf{Z} + sL_2} \tag{4.26}$$

For a perfect transformer we have $k = 1$, and this expression reduces to

$$\alpha = \frac{-s\sqrt{L_1L_2}}{\mathbf{Z} + sL_2} \tag{4.27}$$

or simply

$$\alpha = \frac{-n}{1 + Z/Z_2} \tag{4.28}$$

where $Z_2 = sL_2 = j\omega L_2$ for a steady-state sinusoidal signal and n is the turns ratio of the coils.

We can define an *ideal* transformer as a perfect transformer with very large self-inductances, such that Z_2 is much greater than the load impedance Z. In this approximation the current ratio will reduce to

$$\alpha = -n \quad \text{if } Z_2 \gg Z \text{ and } k = 1 \tag{4.29}$$

Since the ideal transformer condition is based on frequency-dependent imped-ances, it may be naturally satisfied at sufficiently high frequencies whenever the load impedance Z is resistive or capacitive. If inductive loads or low frequencies are to be considered, then a physical transformer can be treated as ideal only if it is constructed using relatively large inductances. Later in this chapter we will find that the perfect transformer can be easily modeled in terms of the ideal transformer and an extra inductor.

The voltage ratio, or transfer function, v_2/v_1 for a transformer can be obtained by dividing the second equation of (4.22) by the first and then removing the currents using Eqs. (4.25). The lengthy denominators of Eqs. (4.25) again cancel each other, and the resulting ratio is

$$H = \frac{v_2}{v_1} = \frac{sM(Z + sL_2) - s^2 L_2 M}{sL_1(Z + sL_2) - s^2 M^2} \tag{4.30}$$

Combining terms by powers of s gives the more compact result

$$H = \frac{sMZ}{s^2(L_1 L_2 - M^2) + sL_1 Z} \tag{4.31}$$

For a perfect transformer with $k = 1$, this voltage transfer function reduces to

$$H = \frac{\sqrt{L_1 L_2}}{L_1}$$

$$H = \sqrt{\frac{L_2}{L_1}} = \frac{n_2}{n_1} = \frac{1}{n} \tag{4.32}$$

and the ideal transformer approximation of a large L_2 provides no further simpli-fication.

4.2.3 The Input Impedance

It is often helpful to be able to represent an entire transformer circuit, such as the one in Figure 4.4a, by an equivalent impedance. This can be done by deriving an

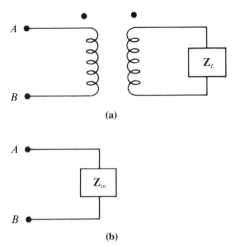

Figure 4.4 *(a) A transformer driving a load* Z_L. *(b) An equivalent input impedance.*

expression for the input impedance to that circuit and using it to replace the entire circuit. This impedance will include the effects of both the transformer and the actual load impedance.

The transformer equations most often seen in elementary texts are those corresponding to the ideal transformer, Eqs. (4.29) and (4.32). But these equations have no frequency dependence and hence fail to represent any physical transformer except in a carefully selected frequency region. The perfect transformer model, which replaces Eq. (4.29) with Eq. (4.28), is more realistic. Fortunately, as we will see from the following discussion of input impedance, the perfect transformer can be modeled in terms of an ideal transformer and an inductor.

Consider the input impedance of the circuit shown in Figure 4.4. As always, this impedance is given by $\mathbf{Z}_{in} = \mathbf{v}_{in}/\mathbf{i}_{in}$. Since $\mathbf{v}_{in}$ and $\mathbf{i}_{in}$ correspond in this case to the $\mathbf{v}_1$ and $\mathbf{i}_1$ in Figure 4.3, we can divide the first of Eqs. (4.22) by $\mathbf{i}_1$ to get

$$\mathbf{Z}_{in} = sL_1 + sM \frac{\mathbf{i}_2}{\mathbf{i}_1} \tag{4.33}$$

Using our best expression for the secondary-to-primary current ratio, Eq. (4.28), this input impedance becomes

$$\mathbf{Z}_{in} = sL_1 - \frac{s^2 M^2}{sL_2 + \mathbf{Z}L}$$

$$= \frac{s^2(L_1 L_2 - M^2) + sL_1 \mathbf{Z}_L}{sL_2 + \mathbf{Z}_L} \tag{4.34}$$

In the perfect transformer approximation with $k = 1$, the first term in the numerator vanishes, leaving

$$\mathbf{Z}_{in} = \frac{sL_1 \mathbf{Z}_L}{sL_2 + \mathbf{Z}_L} \tag{4.35}$$

But $L_1 = n^2 L_2$, and this can be used to eliminate either L_1 or L_2 from the equation. In terms of L_2, the input impedance of a perfect transformer driving a load $\mathbf{Z}_L$ is

$$\mathbf{Z}_{\text{in}} = \frac{n^2 \mathbf{Z}_L}{1 + \mathbf{Z}_L / s L_2} \tag{4.36}$$

For the ideal transformer we have $sL_2 \gg \mathbf{Z}_L$, and the expression becomes

$$\mathbf{Z}_{\text{in}} = n^2 \mathbf{Z}_L \tag{4.37}$$

Although valid only in the ideal transformation approximation, this last equation displays one of the more important uses of a transformer: It can be used to modify a load impedance, either up or down, so as to provide a better match to the driving ability of some signal source.

Examination of Eq. (4.36) will show that, aside from the factor of n^2, the right side has the form of a parallel impedance combination consisting of sL_2 and $\mathbf{Z}_L$. Thus, if we use this parallel combination as the load impedance to an ideal transformer, Eq. (4.37) will yield the same result as the original load used in Eq. (4.36). A more basic insight is obtained by considering an unloaded transformer—one with $\mathbf{Z}_L$ equal to infinity. In this case Eq. (4.36) specifies that the transformer input impedance is $n^2 s L_2 = s L_1$. The perfect transformer models of Figure 4.5 are therefore equivalent and display this result. The circuit effects of the transformer symbol in these models can be described by the simple equations of an ideal transformer, but the overall effect of the model is that of a perfect transformer.

As an example, consider the circuit shown in Figure 4.6a. Our ultimate objective will be to determine the transfer function between the signal generator voltage and the voltage developed across the capacitor C, but the first step is to write an

Figure 4.5 Perfect transformer models.

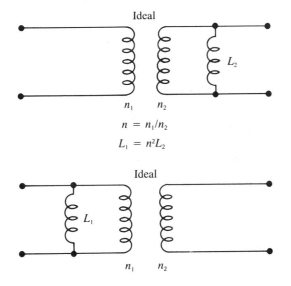

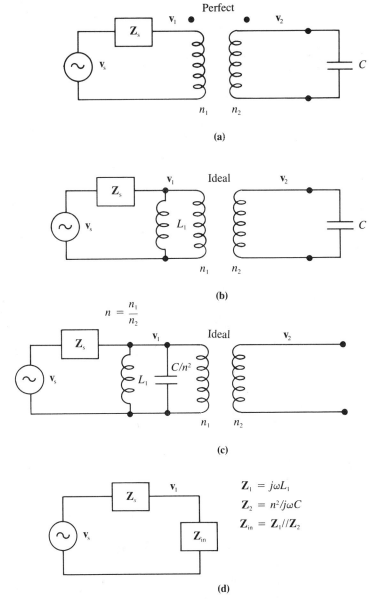

Figure 4.6 (a) An example of a transformer being used to couple a signal generator to a capacitive load. (b) The same circuit modeled using an ideal transformer. (c) The load capacitor replaced by an equivalent capacitance in the input circuit. (d) The input loop with the perfect transformer primary replaced by its equivalent input impedance.

expression for the voltage developed across the transformer primary. Given that Eq. (4.36) for the equivalent input impedance is available, the most direct route is to redraw the input loop as shown in Figure 4.6d. Using Eq. (4.36), the equivalent input impedance is just

$$\mathbf{Z}_{\text{in}} = \frac{n^2\mathbf{Z}_C}{1 + \mathbf{Z}_C/\mathbf{Z}_2} \tag{4.38}$$

where

$$\mathbf{Z}_2 = j\omega L_2$$

and

$$\mathbf{Z}_C = \frac{1}{j\omega C}$$

Since $L_1 = n^2 L_2$, this input impedance can be equally well expressed in terms of the inductance of the transformer primary:

$$\mathbf{Z}_{\text{in}} = \frac{n^2\mathbf{Z}_C}{1 + n^2\mathbf{Z}_C/\mathbf{Z}_1} \tag{4.39}$$

where

$$\mathbf{Z}_1 = j\omega L_1$$

By substituting in the expressions for the inductive and capacitive impedance, it is possible to display explicitly the frequency dependence of $\mathbf{Z}_{\text{in}}$:

$$\mathbf{Z}_{\text{in}} = \frac{j\omega L_1}{1 - \omega^2 L_1(C_2/n^2)} \tag{4.40}$$

An alternative approach to this same example is to model the perfect transformer with an ideal one as in Figure 4.6b; use the ideal transformer input impedance expression of Eq. (4.37) to determine an equivalent capacitance for the primary side; and then write the equivalent impedance of the parallel combination of L_1 and C_2/n^2. Since the input impedance of an unloaded ideal transformer is infinite, the transformer symbol in Figure 4.6c can be ignored while calculating currents in the primary loop. Equation (4.39) verifies that the equivalent impedance of this parallel combination is also the input impedance of the original circuit.

Given an equivalent version of the primary circuit loop, this part of the problem is reduced to a normal AC circuit and standard methods can be used to write expressions for either the primary loop current or the voltage developed across the transformer primary. The voltage across the primary is most useful for the present purposes and, with reference to Figure 4.6d, is given by

$$\mathbf{v}_1 = \frac{\mathbf{Z}_{\text{in}}}{\mathbf{Z}_s + \mathbf{Z}_{\text{in}}}\mathbf{v}_s \tag{4.41}$$

For purposes of simplification, we will take the signal generator output impedance to be resistive and replace $\mathbf{Z}_s$ by R. Using Eq. (4.38) to expand $\mathbf{Z}_{\text{in}}$, the input voltage becomes

$$\mathbf{v}_1 = \frac{n^2\mathbf{Z}_C\mathbf{Z}_2}{n^2\mathbf{Z}_C\mathbf{Z}_2 + R(\mathbf{Z}_2 + \mathbf{Z}_C)}\mathbf{v}_s \tag{4.42}$$

Since the voltage across the secondary winding of a perfect transformer is related to that across the primary by $\mathbf{v}_2 = \mathbf{v}_1/n$, the resulting overall transfer function is given by

$$\mathbf{H}(s) = \mathbf{v}_1/\mathbf{v}_s = \frac{nL_2s}{L_2Cs^2 + n^2L_2s + R} \tag{4.43}$$

or by

$$H(j\omega) = \frac{nj\omega L_2}{R(1 - \omega^2 L_2 C) + j\omega n^2 L_2} \quad (4.44)$$

This expression correctly predicts the bandpass behavior of this circuit and indicates a resonant behavior when $\omega_r = 1/\sqrt{L_2 C}$. A quite different transfer function, valid only at frequencies well above this resonant frequency, would be obtained from an analysis using an ideal as opposed to a perfect transformer model.

Example 4.1   Derive the transfer function v_2/v_s for the circuit of Figure 4.6a under the assumption that the transformer shown there is ideal and then compare the result with Eq. (4.44)

If the transformer is ideal, we can use Eq. (4.37) to obtain the input impedance Z_{in} needed for the equivalent circuit of Figure 4.6d:

$$\mathbf{Z}_{in} = n^2 \mathbf{Z}_L = \frac{n}{j\omega RC}$$

The voltage v_1 is then given by

$$\mathbf{v}_1 = \frac{\mathbf{Z}_{in}}{\mathbf{Z}_{in} + R} \mathbf{v}_s$$

and substituting for $\mathbf{Z}_{in}$ gives

$$\mathbf{v}_1 = \frac{n^2}{j\omega RC + n^2} \mathbf{v}_s$$

Using Eq. (4.32) to relate v_2 to v_1 gives the transfer function

$$\mathbf{H} = \frac{\mathbf{v}_2}{\mathbf{v}_s} = \frac{n}{n^2 + j\omega RC}$$

which reduces to $1/n$ at low frequency and to $n/j\omega RC$ at high frequency.

This low-frequency expression disagrees with the small ω asymptotic form of Eq. (4.44), which is $j\omega L_1/R$, but does match that equation's large ω form.

4.2.4   Output Impedance

When a problem involves several different loads or when the load is itself a complicated circuit, it is useful to have the Thevenin equivalent circuit representation of the signal generator and transformer combination shown in Figure 4.7a. Since the load impedance is infinite, the circuit can be redrawn in terms of an ideal transformer as shown in Figure 4.7b. The voltages across the primary and secondary windings are labeled $\mathbf{v}_{1o}$ and $\mathbf{v}_{2o}$ to indicate open circuit (circuit unloaded) voltages. With the output open, there is no current in either coil of the ideal transformer and the voltage $\mathbf{v}_{1o}$ is given by

$$\mathbf{v}_{1o} = \frac{\mathbf{Z}_1}{\mathbf{Z}_s + \mathbf{Z}_1} \mathbf{v}_s \quad (4.45)$$

This corresponds to an open-circuit secondary voltage of

$$\mathbf{v}_{open} = \mathbf{v}_{2o} = \frac{1}{n}\left(\frac{\mathbf{Z}_1}{\mathbf{Z}_s + \mathbf{Z}_1}\right)\mathbf{v}_s \quad (4.46)$$

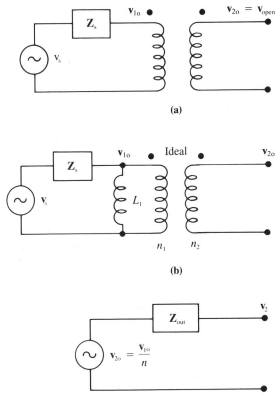

Figure 4.7 *(a) The circuit of Figure 4.6a with the load removed. (b) The equivalent circuit for this unloaded transformer. (c) The Thevenin equivalent circuit for this two-terminal network.*

Because it was determined for an unloaded circuit, this is the correct expression for the Thevenin equivalent circuit EMF shown in Figure 4.7c.

Having obtained $\mathbf{v}_{\text{open}}$, it is possible to determine an expression for $\mathbf{Z}_{\text{out}}$ from the defining equation for this quantity,

$$\mathbf{Z}_{\text{out}} = \frac{\mathbf{v}_{\text{open}}}{\mathbf{i}_{\text{short}}} \qquad (4.47)$$

The quantity $\mathbf{i}_{\text{short}}$ can be obtained directly by shorting the output terminals of each of the circuit representations of Figure 4.7 and calculating the current through the shorting wire. Since this short makes the load impedance zero, it also makes the transformer input impedance zero. The resulting current in the primary loop as a result of the shorted load is just

$$\mathbf{i}_{1s} = \frac{\mathbf{v}_s}{\mathbf{Z}_s} \qquad (4.48)$$

The corresponding secondary loop current $\mathbf{i}_{2s}$ is the desired $\mathbf{i}_{\text{short}}$ of Eq. 4.47 and is given by

$$\mathbf{i}_{\text{short}} = \mathbf{i}_{2s} = n\mathbf{v}_s/\mathbf{Z}_s \qquad (4.49)$$

Dividing the $\mathbf{v}_{\text{open}}$ expression of Eq. (4.46) by this $\mathbf{i}_{\text{short}}$ expression gives

$$\mathbf{Z}_{\text{out}} = \frac{1}{n^2}\left(\frac{\mathbf{Z}_1\mathbf{Z}_s}{\mathbf{Z}_1 + \mathbf{Z}_s}\right) \qquad (4.50)$$

For the large $\mathbf{Z}_1$ of an ideal transformer, this will reduce to

$$\mathbf{Z}_{\text{out}} = \frac{\mathbf{Z}_s}{n^2} \qquad (4.51)$$

Note that when the turns ratio n is smaller than 1, the output voltage and the output impedance of the transformer are both larger than the corresponding quantities of the driving signal generator. If the turns ratio is larger than 1, then both of these transformer output quantities will be smaller than their signal generator counterparts. This impedance-changing feature finds many applications at high frequencies, where transformer coupling between circuit sections is often more efficient than other methods.

4.3   A More Complete Transformer Model

Even if we deal only with transformers whose flux coupling is well represented by the perfect transformer model, the other properties of the coil windings must sometimes be considered. Each coil of an actual transformer is mainly an inductance but is more accurately represented by the schematic circuit of Figure 4.8a, which provides for energy loss due to wire resistance and energy storage in the electric field associated with the voltage drop across the inductor.

If we combine two of these coils to form a representation of an actual transformer, we get the schematic circuit of Figure 4.8b. The understanding of the model can be improved by redrawing the circuit using an ideal transformer as shown in Figure 4.8c. Since the resistance and capacitance on the secondary side of the transformer represent a load impedance on the ideal transformer, they can be moved to the primary side of Figure 4.8c using Eq. (4.37) for the input impedance of a loaded ideal transformer. The resulting equivalent circuit of Figure 4.8d will not reduce further unless one or more of the components can be eliminated on the basis of its size relative to other components in the circuit.

This complicated frequency-dependent model for an actual transformer, which is still assumed to have perfect flux coupling, is only one of the factors limiting the use of the transformer as a general circuit device. Other significant factors are their cost, size, and susceptibility to the pickup of unwanted signals induced by other nearby currents.

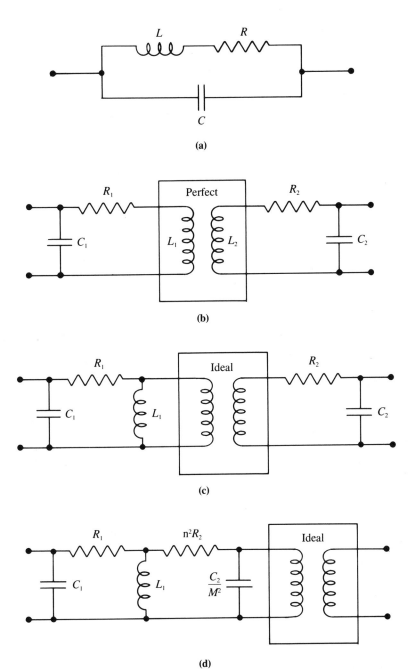

Figure 4.8 *(a) An equivalent circuit for an actual inductor. (b) The equivalent circuit of the two inductors of an actual transformer having perfect flux coupling. (c) The same circuit modeled with an ideal transformer. (d) The same circuit with the secondary circuit impedances replaced by related impedances in the primary circuit.*

1. The winding shown in Figure P4.1 is known as a center-tapped transformer. When the current **i** is increasing: *(a)* What is the sign of the voltage $\mathbf{v}_c$ with respect to ground? *(b)* of $\mathbf{v}_e$?

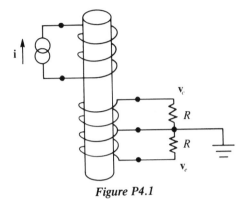

Figure P4.1

2. If $|\mathbf{v}_2|$ is to be greater than $|\mathbf{v}_1|$ on the circuit shown in Figure P4.2, should the transformer's turns ratio be greater than 1?

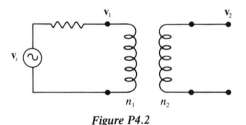

Figure P4.2

3. In order to have $R_{in} > R_L$ in the circuit shown in Figure P4.3, should the turns ratio be greater than 1?

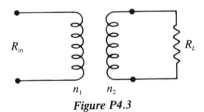

Figure P4.3

4. Using an AC signal generator $\mathbf{v}_s$, a transformer with turns ratio $n = 1$, and a battery V_B, sketch a circuit that will generate the output signal $\mathbf{v}_{out} = \mathbf{v}_s + V_B$.

5. *(a)* If the transformer shown in Figure P4.5 has turns ratio n and is ideal, determine the transfer function $\mathbf{v}_2/\mathbf{v}_s$. *(b)* What is the corner angular frequency?

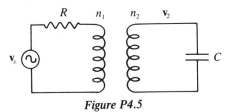

Figure P4.5

6. Show that the more general expression of Eq. (4.43) reduces to the result of the previous problem when $\omega L_2 \gg 1/\omega C$.

7. Follow the method of Figure 4.6 but let the perfect transformer drive a resistive rather than a capacitive load. **(a)** Determine the transfer function v_2/v_s. **(b)** Sketch the magnitude of this transfer function as a function of frequency when $Z_s = R_L$.

8. On a single log-log graph, sketch the magnitude of the transfer function v_2/v_s as shown in Figure P4.8 as a function of ω: **(a)** when the transformer is ideal: **(b)** when the transformer is perfect: **(c)** when the transformer's primary coil L_1 (only) is modeled as shown in Figure 4.8a.

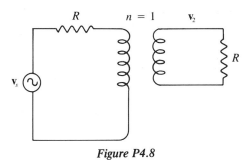

Figure P4.8

9. An ideal transformer with turns ratio $n = 10$ is used to connect two identical filter sections as shown in Figure P4.9.

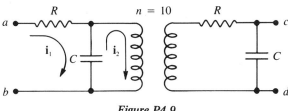

Figure P4.9

(a) Show that $|i_1/i_2|$ is greater than 100 at all frequencies (as long as the transformer is ideal).

(b) Using the result that $i_2 \ll i_1$, write an expression for the overall transfer function v_{cd}/v_{ab} of this circuit.

(c) Sketch $|v_{cd}/v_{ab}|$ versus ω and indicate how the behavior would change when a perfect transformer model is used.

10. A signal generator with output impedance R_s produces a sine wave of amplitude V_s when not loaded. If the ideal transformer shown in Figure P4.10 has $n = 2$:

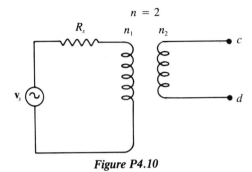

Figure P4.10

(a) Find expressions for the elements of the Thevenin equivalent circuit by using the output impedance arguments of Section 4.2.4.

(b) Show that R_{Th} can also be found by shorting $\mathbf{v}_s$ and calculating the impedance as seen from terminals cd using the method of Section 4.2.3. [*Hint:* The turns ratio is ½ when viewed from this side, since n_1 and n_2 are reversed.]

11. The perfect transformer shown in Figure P4.11 has a primary coil impedance of 100 mH and a secondary coil impedance of 25 mH. It can be modeled by a Thevenin equivalent if $\mathbf{v}_{Th}$ is allowed to be frequency-dependent.

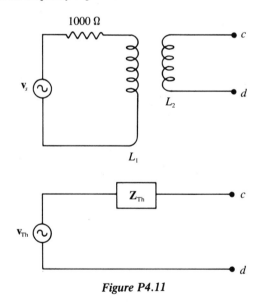

Figure P4.11

(a) Show that $\mathbf{v}_{Th} = j0.05\omega\mathbf{v}_s/(1000 + j0.1\omega)$.

(b) Show that $\mathbf{Z}_{Th} = j25\omega/(1000 + j0.1\omega)\ \Omega$.

(c) Show that the current through a shorting wire connecting terminals c and d is the same for the Thevenin equivalent as for the original circuit.

12. An ideal current source is used to drive a perfect transformer that has turns ratio n and a primary inductance of L_1. The transformer drives a resistive load R_L as shown in Figure P4.12.

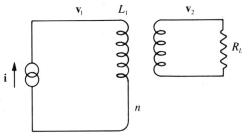

Figure P4.12

(a) Following the method of Figure 4.6, draw an equivalent circuit that shows an ideal transformer with its secondary open.

(b) From the redrawn circuit, derive an expression for the voltage $\mathbf{v}_2$ in terms of $\mathbf{i}$, ω, L, and R_L.

(c) Write an expression for the current in R_L. [*Ans.:* $\mathbf{i}_L = j\omega L_1 n\mathbf{i}/(nR_L + j\omega L_1)$]

13. When a signal generator with a large output impedance R is used to drive a small load R_L, the voltage across the load, $\mathbf{v}_L = R_L\mathbf{v}_s/(R + R_L)$, will be much less than $\mathbf{v}_s$.

(a) If an ideal transformer with turns ratio n is interposed between the signal generator and the load, show that

$$\mathbf{v}_L = \left[\frac{n(R + R_L)}{R + n^2 R_L}\right]\left(\frac{R_L}{R + R_L}\right)\mathbf{v}_s$$

(b) Find the range of n for which the term in square brackets will be greater than 1.

(c) For a special case when $R = 1000\ \Omega$ and $R_L = 50\ \Omega$, find the two integral values of n that make this term equal to 2.33. [*Ans.:* $n = 4, 5$]

14. *(a)* Write an expression for the input impedance to the perfect transformer loaded by R and C as shown in Figure P4.14.

(b) If $R = 1\ \text{k}\Omega$, $C = 0.1\ \mu\text{F}$, and $L = 100\ \text{mH}$, what is the phase relationship between $\mathbf{v}_A$ and $\mathbf{i}_A$ at resonance?

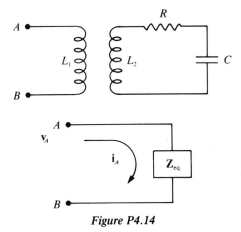

Figure P4.14

5

The **PN** *Junction and Diode Circuits*

5.1 Introduction

The development of solid state electronics has led to a modern technological explosion that can only be compared to the industrial revolution. Since the invention of the transistor in 1948, the technology has progressed from an individually prepared laboratory device to millions of components on a single silicon wafer only a few millimeters on a side. Coupled with this astounding miniaturization has been a corresponding decrease in the price per component. Allowing for inflation, a current popular microcomputer chip, containing more than a hundred thousand components, costs about the same as one of the electron tubes of earlier electronics.

As a result of this miniaturization, today's basic solid state package may equally well be a single diode or transistor, a hundred or so transistors forming an integrated (IC) circuit, thousands of transistors forming a large-scale integrated (LSI) circuit, or hundreds of thousands of transistors forming a very-large-scale integrated (VLSI) circuit. Although the catalog of such devices is large and constantly growing, there are large areas of common ground, especially with regard to the all important input and output connections. This chapter and the following two are concerned mainly with circuits formed from individual or discrete components, but the discussion is equally relevant to the behavior of integrated circuits at their input and output terminals.

The development of transistors and the large variety of other solid state devices has been based on materials known as semiconductors. The mathematical description of semiconductor behavior requires quantum mechanical methods and is well beyond the scope of this book. The following qualitative description is provided as a brief introduction to the terminology and to provide some insight into the operation of modern semiconductor devices fabricated from silicon.

151

5.2 Atoms and Energy Levels

5.2 Atoms and Energy Levels

An atom consists of a heavy nucleus and a surrounding cloud of much lighter electrons whose configuration is best described in terms of a quantum mechanical wave function. The introductory treatment of the hydrogen atom found in every modern physics textbook usually includes an energy-level diagram like Figure 5.1a. For our purposes, the most important feature of this diagram is that an electron occupies only discrete energy states described in terms of four quantum numbers: the radius-dependent principal quantum number n, the orbital angular momentum quantum number l, its projection m on some polarization axis, and a similar projection of the intrinsic spin of the electron. The two electron spin projections, which produce little or no energy shift, are not shown as separate levels in the figure, but do produce two states at each energy level shown. In general, a change in any quantum number corresponds to a change in the configuration of the atom and will usually result in a change in the atom's internal energy.

The energy-level diagram of Figure 5.1a can be divided into shells based on the radial quantum number n, and into subshells based on the angular momentum quantum number l. The first shell consists of two closely spaced states (for the two electron spin projections), the second shell consists of two subshells of two and six states, the third of three subshells of two, six, and 10 states, and so on. In the isolated hydrogen atom (one proton making $Z = 1$), the single electron will normally occupy the lowest energy state; it can be excited to a higher energy state by absorbing energy but will subsequently decay back to the lowest state by radiating one or more photons.

For higher-Z atoms with more than one electron, it is necessary to invoke the Pauli exclusion principle to locate the electrons on the energy-level diagram. This principle says that only one electron can occupy a given state; if two or more electrons are present, each must be in a different state. This simple rule can be combined with the energy-level diagram of hydrogen to yield a qualitatively correct picture of the atomic structure of the light elements. Electrons fill the states starting at the bottom—one per state (two per level shown)—and when all of the states in a shell are filled, that shell is said to be closed. Comparison of this shell structure with the periodic table of elements shows that the chemical behavior of light atoms is determined largely by those electrons that are outside closed shells. Thus hydrogen with one electron, lithium with three (one in the second shell), and sodium with 11 (one in the third shell) all have similar chemical behavior.

Electrons that are outside closed shells (sometimes closed subshells for the more complicated atoms) are known as valence electrons. Silicon ($Z = 14$), the basic material of most solid state electronic devices, has four valence electrons, as does its less commonly used sister element germanium ($Z = 32$). Two typical examples of other elements that are used to modify the characteristics of silicon or germanium crystals are arsenic ($Z = 33$) with five valence electrons and gallium ($Z = 31$) with three valence electrons. The general electronic structure of these four atoms is sketched in Figure 5.1b.

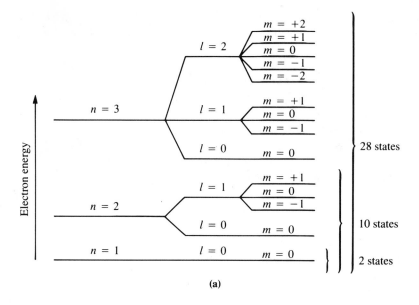

(a)

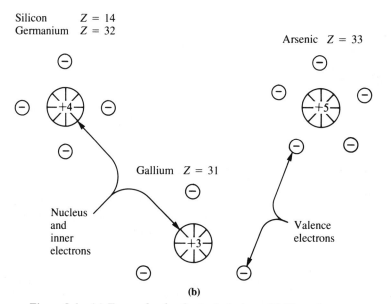

(b)

*Figure 5.1   (a) Energy levels of a typical atom. (b) The valence
electron structure of four atoms commonly used in solid state devices.*

5.3   *Energy Levels in Crystalline Solids*

To sustain an electronic current, a material must have charge carriers that are free
to move. In a crystalline solid, the core of each atom (nucleus and closed-shell
electrons) is fixed in position, but with some probability an atom may a eject a

valence electron, which is then free to move in the material. The conductivity of a material is a function of the number of free charge carriers per unit volume, and is determined by the behavior of the valence electrons. Based on this behavior, solids can be divided into three main categories: conductor, semiconductor, and insulator.

When a number of identical atoms coalesce to form a crystal, the atoms interact and bind by sharing their valence electrons. The quantum mechanical wave function of these shared electrons is no longer associated with a single atom but instead extends over the entire crystal. If we could somehow have a crystal of N atoms and at the same time neglect the interaction between the atoms, each energy level available to valence electrons in the free atom would become N identical or degenerate energy levels for the crystal as a whole. But one of the effects of the interaction between atoms is that these N degenerate levels split into N closely spaced levels. Since the number of atoms is large, it is common to refer to this set of levels as a continuous energy band describable in terms of its density of states dN/dE, the number of states per unit energy. Since a single free atom has many different energy levels, the crystal will have many different energy bands, and depending on the characteristics of the atom and its interactions within the crystal, these bands may be separate and distinct or may merge and overlap to varying degrees.

The states filled or partially filled by valence electrons in the free atom are identified with the valence band in the solid, and since each such level of the free atom splits into a number of levels equal to the number of atoms in the crystal, the valence band will always contain at least enough states to hold all of the valence electrons. If the number of valence electrons is even (both spin states filled), the valence band will contain exactly enough states to accommodate the valence electrons. The lowest unfilled energy state of the free atom also converts to an energy band, which is designated as the conduction band. Depending on the material, the valence and conduction bands may be distinct from each other or may overlap as shown by the examples of Figure 5.2.

There is no qualitative difference between the semiconductor and insulator diagrams of Figures 5.2b and 5.2c; they are distinguished only by the size of the energy gap between the valence and conduction bands. For a semiconductor this energy gap is around one electron volt, an energy increment that a valence electron can occasionally acquire from the thermal agitation of the crystal at room temperature. In an insulator the energy gap is several times larger.

For a charge to move through the crystal, it must be able to acquire a small amount of additional kinetic energy, and thus must be able to move to a new energy state. The voltages typically applied in electronics applications generate electric fields that are able to supply only relatively small amounts of energy to the atomic charges. If the valence band is completely filled with electrons, and the conduction band is distinct and completely empty, as it is for both insulators and semiconductors at absolute zero, then no electron has an attainable unoccupied energy level available and none are able to move about in the material. As the temperature of a semiconductor is increased, some electrons may be thermally excited from the valence band to the conduction band. These relatively few excited electrons now have access to many unoccupied energy levels, and consequently become free to conduct a current in the material. The common semicon-

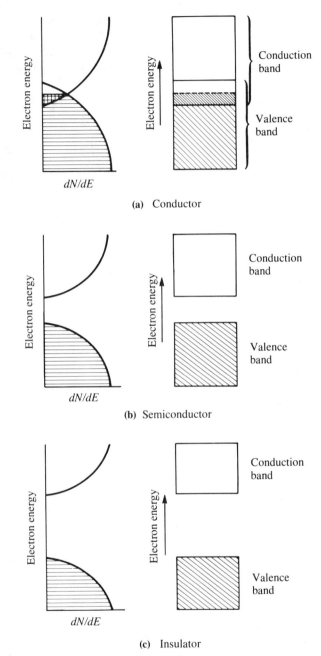

(a) Conductor

(b) Semiconductor

(c) Insulator

Figure 5.2   Possible band structures for (a) conductor, (b) semiconductor, (c) insulator. The shaded region in each figure indicates the occupation of energy levels at a temperature of absolute zero.

ductors, silicon and germanium, have an energy gap between the valence band and the conduction band that can be jumped by a significant number (in silicon about 10^{10} cm^{-3}) of thermally excited electrons at room temperature; thus they are semiconductors at room temperature.

The semiconductors, silicon and germanium, have a three-dimensional crystal-line structure in which each atom has four nearest neighbors. This structure, illus-trated schematically by the two-dimensional representation of Figure 5.3a, pro-vides a complementary model for the source of conducting electrons. The four valence electrons of each atom form covalent bonds with the four neighboring atoms as indicated. At a temperature of absolute zero, all valence electrons are involved in this covalent bonding. At room temperature, thermal excitation can break some of these bonds, thus elevating some electrons from the valence band to the conduction band. Figure 5.3b shows an electron that has been excited into the conduction band; as shown, a hole is also produced in the covalent bonding pattern of the crystal.

When an electron has been excited into the conduction band, it is free to move and conduct electric current. The hole left behind in the valence band is also free to move through the crystal, as it can migrate from one covalent bond to the next. Contrary to classical intuition, a quantum mechanical treatment of this effect puts the hole on an approximately equal footing with the electron; it behaves as a positive charge with positive mass and is free to move about in the crystal, an effect that is qualitatively different from electrons moving to fill in the holes. Each thermally excited electron thus produces two charge carriers, one positive and one negative, and these opposite charges can recombine within the crystal to form an intact covalent bond.

It is also possible to generate more charge carriers in a semiconductor by mix-ing in a few atoms of a different element. Such a process is known as doping. A typical example would be the doping of a crystal of intrinsic (pure) silicon with arsenic atoms. Each arsenic atom will replace one silicon atom at a crystal site, but the new atom will not have the correct number of valence electrons to exactly match the covalent bonding requirements. Each arsenic atom, having five rather than four valence electrons, will contribute one excess electron. This odd electron is loosely bound to the original arsenic site and is easily excited into the conduc-tion band by the thermal energy available at room temperature. Doping of this type forms an *N*-type semiconductor; the designation refers to the fact that the

Figure 5.3 (a) Simplified two-dimensional representation of the crystal structure of silicon. (b) Same crystal with a thermally excited electron, leaving behind a hole in the covalent bond structure.

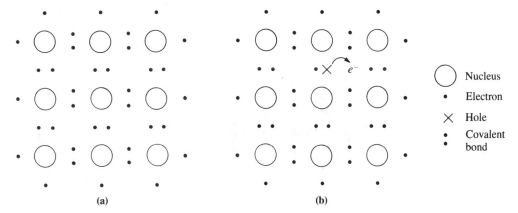

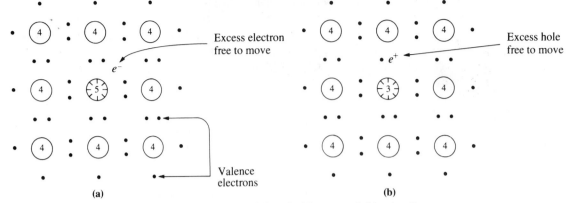

Figure 5.4 (a) Silicon crystal doped with a material having five valence electrons, an N-type semiconductor. (b) Silicon crystal doped with a material having three valence electrons, a P-type semiconductor.

majority of the free charge carriers in this material will be negative. A crystal doped in this fashion is shown schematically in Figure 5.4a.

Conversely, when intrinsic silicon is doped with a material such as indium, which has three valence electrons, each indium atom has one too few electrons to fill the covalent bond requirements. This condition is shown in Figure 5.4b. The result is that each such atom contributes one hole, which is easily excited into the electron-rich valence band where it is free to move. When doped in this manner, the majority of the charge carriers are positive, and the material is known as a *P*-type semiconductor.

Since doping increases the number of free charge carriers, it increases the conductivity of the semiconductor. However, in both *P*- and *N*-type semiconductors, a small fraction of the charge carriers will still arise from the thermal generation of electron-hole pairs. This temperature sensitive effect is indicated in Figure 5.5a by a plus-minus pair enclosed in a dashed oval. One component of this pair will add slightly to the number of majority charge carriers in the material and is of little consequence. The other component will contribute a small number of charge carriers whose sign is the opposite of the majority carriers. These are known as the minority charge carriers, and they set natural limits on the ideal performance of the devices to be described below. As indicated by the discussion above, the number of these minority charge carriers increases as the crystal temperature increases.

5.4 The PN Junction and the Diode Effect

When a *P*-type and an *N*-type semiconductor are joined or otherwise produced, an electric field is developed across the junction in a direction that repels both majority carriers away from the junction region. At first glance, this may seem to be the exact opposite of what one would expect, but it has a simple qualitative ex-

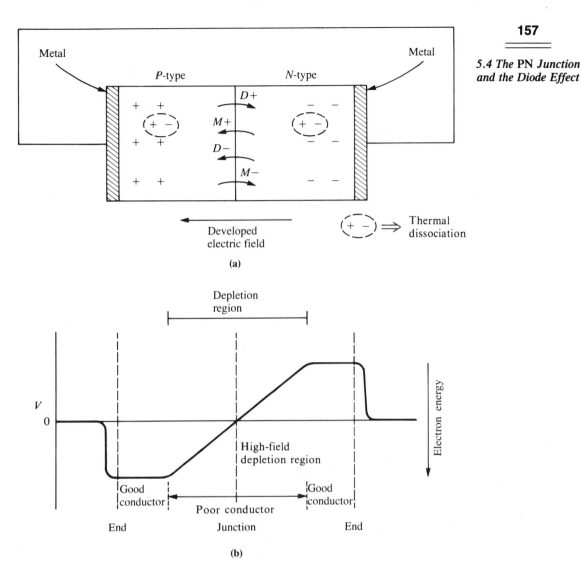

*Figure 5.5 (a) The PN junction formed from a P-type and an N-type
semiconductor showing the motion of diffusion charge $D-$ and $D+$
and the electric field-driven minority charge $M-$ and $M+$. Note that
once they cross the boundary, diffusion charge carriers quickly
recombine with majority carriers and lose their identity. (b) At
equilibrium, the electric potential developed across the junction repels
both majority carriers away from the junction, creating a "depletion"
region with few majority charge carriers.*

planation. It must be emphasized that before joining, the *P*- and *N*-type semiconductors are both electrically neutral: Each electron is exactly balanced by a positive nuclear charge fixed in the crystal lattice. A *P*-type semiconductor averages
fewer electrons per atom than an *N*-type, but it also averages fewer protons in its
nuclei.

Since the two pieces are electrically neutral when first joined, the initial effect comes only from the mobility of the charge carriers in each piece. The concentrations (number per unit volume) of positive and negative carriers are quite different on opposite sides of the junction, and the immediate effect is a simple thermal energy-powered diffusion of positive carriers into the N-type material and of negative carriers into the P-type material. (Except for the charge being carried here, this diffusion process is the same as that seen when a dyed liquid is placed next to clear liquid in a container.) As the diffusion of charge carriers progresses, the N-type material acquires an excess of positive charge in the neighborhood of the junction and the P-type acquires an excess of negative charge. This creates an electric field across the junction that points from the N-type to the P-type. The diffusion-generated field builds until an equilibrium condition is reached when the diffusion charge current is balanced by the flow of charges driven by the electric field. This condition is shown in Figure 5.5a.

The repulsion of majority carriers away from the junction, coupled with the recombination of diffusion charge when it crosses the boundary, results in the formation of a "depletion region" near the junction that has few charge carriers and behaves much like intrinsic or undoped silicon. Thus, although both of the doped semiconductors are good conductors, the depletion region itself is a poor conductor that can support an electric field and the corresponding potential gradient as shown in Figure 5.5b.

The complete semiconductor diode consists of the two semiconductor materials fabricated in the form of a *PN* junction with a semiconductor-metal junction at each end in order to connect the semiconductor device to circuit wires. Additional electric potential changes occur at these semiconductor-to-metal junctions, with the result that both ends of the device are at the same potential and can be connected as shown in Figure 5.5a without affecting the potential drop across the *PN* junction.

5.4.1 Current in the Diode

If the diode is connected to a constant EMF as shown in Figure 5.6a, the sense of the applied voltage is such that the potential of the N-type semiconductor becomes more positive with respect to the P-type. More mobile charges are pulled away from the *PN* junction, creating a thicker depletion region and increasing the voltage change across the region. The only charge carriers able to support a net current across the *PN* junction under these conditions are the minority carriers of each semiconductor. As a result of the high impedance thus created in the junction region, the current in this direction is small. It is known as the reverse current and is designated in Figure 5.6a by I_r.

When the diode is slightly forward-biased by a low-voltage EMF as shown in Figure 5.6b, the potential of the N-type semiconductor becomes less positive with respect to the P-type. Under these conditions, the depletion region becomes somewhat thinner and the potential difference between the P- and N-type materials becomes smaller. The combined effect is to allow more of the majority carriers to diffuse across the depletion region of the junction, producing a forward current I_f.

As the applied voltage is increased, the reverse potential across the *PN* junction and the thickness of the depletion region both continue to shrink, and the *PN*

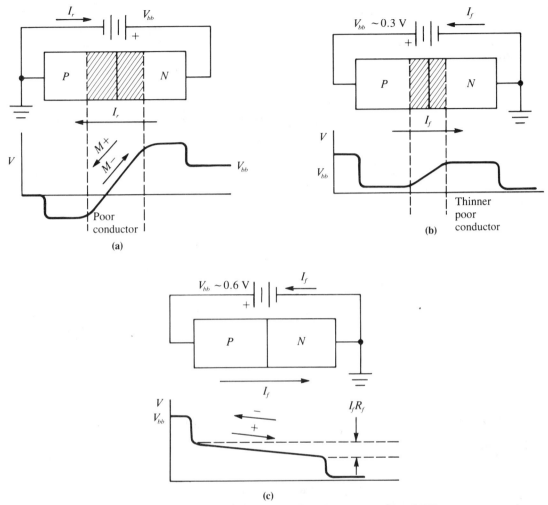

Figure 5.6 (a) Circuit connection resulting in a reverse-biased PN junction showing electric potential at various points. (b) Slightly forward-biased PN junction. (c) Strongly forward-biased PN junction. Note that any applied voltage in excess of that needed to collapse the junction is dropped across the small resistance, R_f, of the doped semiconductor materials.

junction becomes less and less of a factor impeding current through the diode. When the applied voltage exceeds that needed to effectively eliminate the depletion region, the junction itself becomes a negligible effect: Further increases in the voltage result in an increased current according to the equation $\Delta V = R_f \, \Delta I$, where R_f is the dynamic or effective forward resistance of the diode and depends on the geometry and conductivity of the semiconducting materials themselves.

We can make a more quantitative argument by considering the energy distribution of free electrons and holes in the region of the junction. To simplify the discussion, we will use a classical argument and discuss only the motion of the positive charges; the motion of the negative charges follows the same rules, and the currents generated by the two components can be added to form the total current. The thermal agitation of a group of particles results in a distribution of particle energies whose high-energy tail can be approximated by the exponential

$$N = N_0 e^{-(E-E_0)/kT} \tag{5.1}$$

where N is the number of particles with energy greater than E, k is the Boltzmann constant, T is the absolute temperature, and N_0 is the number of particles with energy greater than some reference energy E_0. In considering the equilibrium diffusion of positive charges across the junction from P to N, we need to identify the energy E with that needed to overcome the potential difference between the two materials. If V is a positive applied voltage in a direction to cause a forward current from P to N across the junction, then the energy needed to overcome the potential decreases as V increases, making $E - E_0 = -qV$. Substituting into Eq. (5.1) gives

$$N = N_0 e^{qV/kT} \tag{5.2}$$

If we define a new quantity V_T as

$$V_T = \frac{kT}{q} \tag{5.3}$$

Eq. (5.2) can be written as

$$N = N_0 e^{V/V_T} \tag{5.4}$$

where V_T is 25.3 mV at room temperature (20°C). Since the motion of these charges across the junction constitutes a current, we can rewrite this equation as

$$I_{D+} = I_0 e^{V/V_T} \tag{5.5}$$

where the notation I_{D+} identifies this as a thermally generated diffusion current. This function is sketched in Figure 5.7a.

The diagrams in Figure 5.7b–5.7d show the application of this equation to the three diode bias conditions: shorted with $V = 0$ and I_{D+} balanced by the minority current $-I_{M+}$; reverse-biased so that $I_{D+} \approx 0$, leaving only the minority current $-I_{M+}$; and forward-biased with the total current equal to $I_{D+} - I_{M+}$. If we assume that I_{M+} remains constant for all three voltage bias conditions whereas I_{D+} changes with the applied voltage according to Eq. (5.5), we can derive a single equation that describes all three biasing conditions. The assumption that the minority current I_{M+} is independent of the applied voltage is only an appxoximation, but it is made plausible by the fact that this current depends mainly on the concentration of minority charges in the vicinity of the junction; this concentration will certainly vary with the applied voltage, but not as rapidly as the diffusion current.

In the shorted condition defined by Figure 5.7a, the total positive charge current must be zero. Since this current is composed of two parts, we have

$$I = I_{D+} - I_{M+} = 0 \tag{5.6}$$

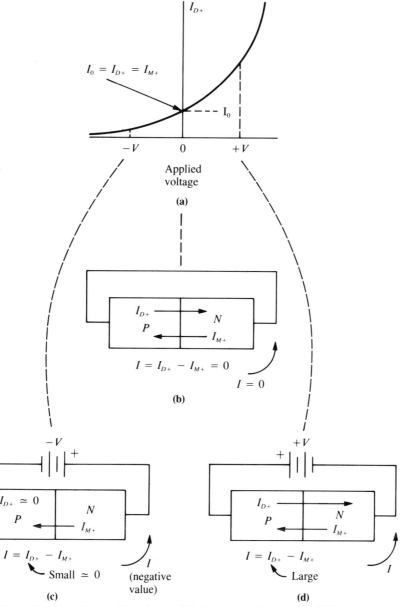

*Figure 5.7   (a) The positive charge diffusion current across the PN
junction as a function of the applied voltage V. The two components of
the positive charge current in the diode under the condition of (b) no
applied voltage, (c) reverse applied voltage, and (d) forward applied
voltage.*

and using Eq. (5.5) with $V = 0$, this becomes

$$I_{M+} = I_0 \tag{5.7}$$

We can now combine this equation with (5.5) and (5.6) to give a general expression for the current,

$$I = I_0(e^{V/V_T} - 1) \tag{5.8}$$

which remains valid even after the negative charge contribution has been added. This equation is plotted in Figure 5.8c.

Both I_0 and V_T are temperature-dependent: V_T from its definition in Eq. (5.3), and I_0 because is must represent a reverse current of minority charges when the diode is reverse-biased (applied V is negative). Near room temperature, the reverse current I_0 increases by about 7%/°C as a result of the increasing number of broken covalent bonds.

This equation gives a reasonably accurate prediction of the current-voltage relationship of the *PN* junction itself—especially the temperature variation—and can be improved somewhat by choosing I_0 and V_T empirically to fit a particular diode; nominal values for a silicon signal diode would be $I_0 = 1$ nA and $V_T = 50$ mV. However, for a real diode, other factors are also important: In particular, edge effects around the border of the junction cause the actual reverse current to increase slightly with reverse voltage, and the finite conductivity of the doped semiconductors ultimately restricts the forward current to a linear increase with increasing applied voltage V. A better curve for a real diode is shown in Figure 5.8d and will be discussed in the following section.

5.4.2 The PN Diode as a Circuit Element

The schematic symbol of a diode is shown in Figure 5.8a. The arrow points from anode to cathode in the direction of the forward-biased current: from the *P*-type material to the *N*-type. The characteristic current-voltage *(IV)* curve of an idealized diode is shown in Figure 5.8b. This idealized device will conduct current without impedance in the forward direction and allows no current in the reverse direction. For many circuit applications, this ideal diode model is an adequate representation of an actual diode and simply requires that the circuit analysis be separated into two parts: forward current and reverse current. For other applications, a more accurate description may be needed.

The operation of an actual diode can be summarized by the *IV* curve shown in Figure 5.8d. The curve is characterized by two linear regions of constant slope connected by an exponential curve in the region of small forward bias. The exponential region of this curve corresponds to an applied voltage across the diode, which has reduced but not eliminated the depletion region of the junction. The *PN* diode is thus our first example of a nonlinear circuit element: The current through the device is not always proportional to the voltage across the device. Some circuits make use of the nonlinear region of this curve, but for most applications this region of the curve can be avoided and the device can be modeled as a "piecewise" linear circuit element.

Noting that the slope at any point on this *IV* curve has the units of inverse

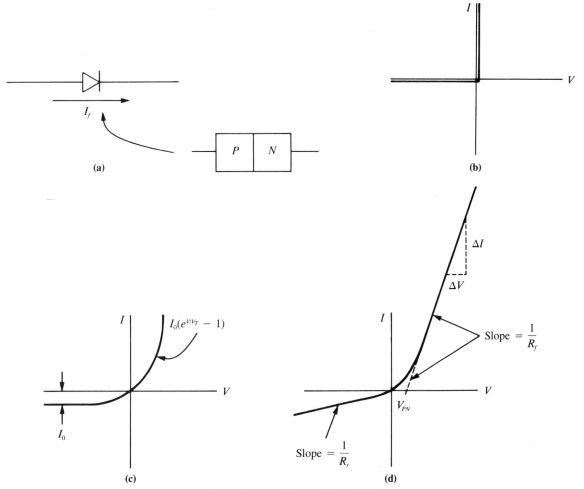

Figure 5.8 (a) Schematic symbol for a diode. (b) Current-versus-voltage curve for an ideal diode. (c) Current-versus-voltage curve for the PN junction region. (d) Current-versus-voltage curve for an actual PN diode.

resistance, we can assign a "dynamic" resistance to the diode in each of its linear regions: R_f in the forward-biased region and R_r in the reverse-biased region. These resistances are defined by

$$R_f = \frac{\Delta V}{\Delta I}$$

and   (5.9)

$$R_r = \frac{\Delta V}{\Delta I}$$

in their respective regions as shown in Figure 5.8d. Typical signal diodes have a forward resistance of a few ohms and a reverse resistance around a megohm.

The equivalent circuit model of an actual diode forward-biased into the linear region is shown in Figure 5.9a. This model reproduces the linear portion of the *IV* curve of Figure 5.8d but extends to intersect the *V* axis at V_{PN}. The voltage, V_{PN}, represents the effective voltage drop across a forward-biased *PN* junction (the turn-on voltage) and is represented in the model by a constant voltage source. For a germanium diode, V_{PN} is approximately 0.3 V; for a silicon diode it is close to 0.6 V. The reverse-bias equivalent circuit of Figure 5.9b is somewhat simpler and does not require a voltage source, since the linear portion of the actual diode curve passes very close to the origin.

Using two ideal diodes to restrict the current to the appropriate direction through each branch, these two linear element models can be combined as shown in Figure 5.9c. The most important diode feature not included in this model is the switching time between forward current I_f and reverse current I_r. Actual diodes require a time ranging from a picosecond to a microsecond to switch in this direction, even when the applied voltage across the diode is assumed to switch instantly. This represents a significant high-frequency limitation for many diodes.

In the above discussion we noted that the constants I_0 and V_T used in Eq. (5.8) to parameterize the diode junction current-voltage curve are both functions of temperature. The 7%/°C change in the reverse saturation current I_0 (generated by thermally produced minority charge carriers) is the major effect at room temperature. In addition to the obvious effects on a reverse-biased diode, Eq. (5.8) can be used to predict the change in the "turn-on" voltage V_{PN} for a forward-biased diode. If we pick a point on the forward part of the *IV* curve and attempt to hold *I* constant as the temperature is increased, Eq. (5.8) indicates that we must decrease the applied voltage *V* to compensate for the increase in I_0. Some of this

Figure 5.9   (a) Model for the linear portion of a physical forward-biased diode. (b) Model for a reverse-biased diode. (c) Combined model for an actual diode in terms of the PN voltage drop, the forward and reverse resistances, and ideal diode schematic elements.

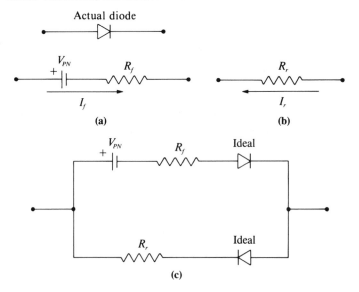

effect is canceled by the slow increase of V_T with temperature, but the combined effect is that the forward *IV* curve shifts (to the left) by about -2 mV/°C for both germanium and silicon. Thus the turn-on voltage of a *PN* junction decreases as the temperature increases; this effect is also seen in transistors, and for silicon devices is the most significant temperature variation.

Example 5.1   Use Eq. (5.8) to find the change in V_{PN} with temperature for a silicon signal diode. Take $V_{PN} = 0.6$ V at $T = 20°C$.

Dropping the 1 from Eq. (5.8) and then taking the natural logarithm gives

$$V = V_T \ln\left(\frac{I}{I_0}\right)$$

and in order to have $V = 0.6$ V at 20°C with $V_T = 50$ mV, the logarithm must be

$$\ln\left(\frac{I}{I_0}\right) = \frac{600 \text{ mV}}{50 \text{ mv}} = 12$$

Since V_T is proportional to the absolute temperature, at 21°C it increases to

$$V_T = 50 \text{ mV}\left(\frac{294 \text{ K}}{293 \text{ K}}\right) = 50.17 \text{ mV}$$

and the 7% increase in I_0 per degree changes the logarithm to

$$\ln\left(\frac{I}{1.07I_0}\right) = \ln\left(\frac{I}{I_0}\right) - \ln(1.07)$$

or

$$\ln\left(\frac{I}{1.07I_0}\right) = 12 - 0.068 = 11.93$$

At this one degree higher temperature, the original equation for *V* now becomes

$$V = 50.17 \text{ mV}(11.93) = 598.5 \text{ mV}$$

The change in V_{PN} per degree change in temperature is then -1.5 mV.

5.4.3   The Zener Diode

When a diode is reverse-biased, the current I_r across the junction region is related to the number of minority charge carriers available to cross the depletion region. As the reverse voltage is increased, two new effects can dramatically increase the number of these carriers: Zener breakdown, where the electric field near the junction becomes large enough to excite valence electrons directly into the conduction band; and avalanche breakdown, where the minority carriers are accelerated in the electric field near the junction to sufficient energies that they can excite valence electrons by collision. Both processes can result in the sudden onset of increased current in the *reverse* direction beyond a certain breakdown voltage. For most diodes, effects of this type will result in permanent damage to the diode if the *peak inverse voltage* limit is exceeded.

The Zener diode is constructed so that this breakdown can occur in a nonde-structive way. The current-voltage curve of a Zener diode is similar to an ordinary diode except in the reverse-biased region. As shown in Figure 5.10a, the Zener diode curve exhibits an additional and abrupt slope change at a particular reverse-bias voltage. An equivalent circuit model for the Zener diode in this portion of its operating region is shown in Figure 5.10b. Since the slope in this region is very steep, the dynamic resistance of the diode in this region, R_Z, is small—in the range 5–50 Ω. Zener diodes are available in a wide range of breakdown voltages, ranging from a few volts to several hundred volts. The best characteristics—small R_Z and little change of V_Z with temperature—are found on diodes with V_Z in the range 6–7 V.

A common application of the Zener diode is in a circuit designed to provide a constant voltage reference or voltage source. A typical circuit is shown in Figure 5.11a, where a variable load R_L has been added to the basic circuit. Note that the current I_Z across the Zener diode is in the reverse direction, opposite to the usage of a normal diode. This circuit can be analyzed using the graphical technique of Figure 5.11b. First consider the circuit with the Zener diode removed: As different load currents are required, the voltage drop across resistor R will vary according to the relationship

$$V_L = V_s - RI_L \tag{5.10}$$

This "load line" resulting from this equation is plotted as a solid line of slope $-R$ in the figure. When the Zener diode is put back into the circuit, the behavior at high values of I_L is unchanged, because the reverse voltage across the Zener,

Figure 5.10 *(a) The circuit symbol and current-voltage curve of a Zener diode. (b) A model for the Zener diode in the linear portion of its breakdown region.*

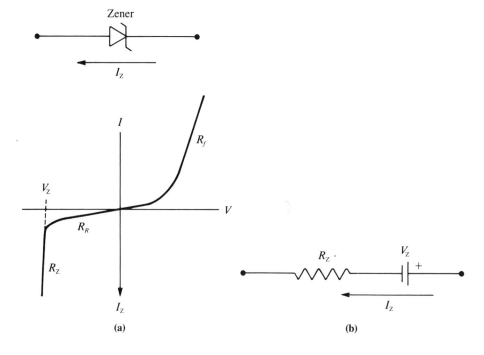

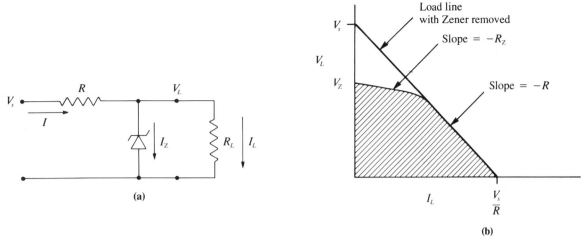

Figure 5.11 *(a) A Zener diode used as a voltage reference. (b) Load line analysis of this circuit.*

V_L here, is less than the Zener's breakdown voltage. As the load current is decreased, V_L increases as prescribed by the load line curve until it reaches the Zener breakdown voltage V_Z. At this point the Zener begins to conduct whatever current is needed to keep the voltage from rising appreciably above this point. The line bounding the crosshatched region describes the behavior with the Zener diode in the circuit.

For a more analytical treatment, we need an equivalent circuit model of the reverse-biased Zener diode. In the low-voltage region, before breakdown, the Zener diode can be treated like any other reverse-biased diode. Above the breakdown voltage, the Zener can be represented by the resistor–voltage source combination shown in Figure 5.12a. This is the normal operating region of this type of diode. With this model replacing the Zener, the voltage reference circuit can be described in the breakdown region by the equivalent circuit shown in Figure 5.12b.

Figure 5.12 *(a) An equivalent circuit model of the voltage reference of Figure 5.11a in the Zener breakdown region. (b) The Thevenin equivalent circuit for this voltage reference.*

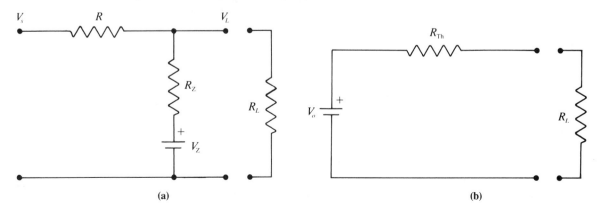

The value of the Zener diode to this voltage reference circuit can now be determined analytically by calculating the elements of a Thevenin equivalent circuit representation. The current in this unloaded circuit is just

$$I = \frac{V_s - V_Z}{R - R_Z} \tag{5.11}$$

The open circuit output voltage is

$$V_o = V_s - IR \tag{5.12}$$

Combining these two equations gives the Thevenin equivalent voltage in terms of circuit parameters:

$$V_o = V_Z \left(\frac{R}{R + R_Z} \right) + V_s \left(\frac{R_Z}{R + R_Z} \right) \tag{5.13}$$

Note that for small R_Z this voltage is close to V_Z and is very insensitive to changes in V_s. The Thevenin equivalent impedance is easily found by shorting both voltage sources; the two resistors are now in parallel, and the expression is

$$R_{\text{Th}} = \frac{RR_Z}{R + R_Z} \tag{5.14}$$

If $R_Z \ll R$, this resistance reduces to R_Z, giving the voltage source a reasonably small output impedance. The combined result shows a voltage reference that is insensitive to voltage changes in the original EMF and to changes in the load current. This result is of course valid only when the diode is operating in the linear portion of its reverse-biased breakdown region.

Example 5.2 On the Zener regulator circuit of Figure 5.11a, assume that $V_s = 15$ V, that the desired V_L is 10 V, and that the load resistance can vary from 500 Ω to infinity. Choose a reasonable value for R, and determine the worst-case power dissipation in R and in the Zener.

With V_L at 10 V, the load current I_L will vary from zero to 20 mA as the load resistor varies. For the Zener to be beyond the knee of its *IV* curve, it must carry at least a few milliamperes of reverse current; let's choose 5 mA. Since $I = I_Z + I_L$, this means that under the worst-case condition we will need 25 mA of current I through the resistor R, 5 mA for the Zener and 20 mA for the load. With the voltage at both ends of R fixed, this current fixes the values of R needed:

$$R = \frac{(15 - 10) \text{ V}}{25 \text{ mA}} = 200 \ \Omega$$

Under normal conditions, with $V_L = 10$ V, the power dissipated by R is given by $P = (5 \text{ V})(25 \text{ mA}) = 125$ mW. But if the output was accidentally shorted, the power dissipated in R would then be $P = (15 \text{ V})^2/(200 \ \Omega) = 1.125$ W. This is outside the design limits posed by the question, but a resistor at least as large as 1 W would be a wise choice.

The worst-case power dissipation in the Zener occurs when the load is open, making $R_L = \infty$ and $I_L = 0$. For this condition the full 25 mA flows in the Zener, making its power dissipation $P = (10 \text{ V})(25 \text{ mA}) = 250$ mW.

The design thus requires a 1-W, 200-Ω resistor and a ¼-W, 10-V Zener.

5.4.4 Light-Emitting Diodes

The light-emitting diode or LED emits light in proportion to the forward current I_f through the diode. These devices are used in place of small incandescent lamps and have the advantage of lower voltage operation and longer life. In addition, the light output from an LED responds quickly to variations in the current through the diode and can be used to generate modulated or pulsed light signals at frequencies as high as 10 MHz. A variation on the LED, known as the diode laser, is able to produce a narrow spectrum of coherent red or infrared light that is very well collimated and is particularly useful for optical fiber communications applications. Commercial light-emitting diodes are available in a variety of colors, ranging from infrared to green, and operate at continuous currents up to about 30 mA.

Although LEDs are made using semiconducting materials formed from various compounds of gallium (arsenide, phosphide, or combinations), they are still doped to form a *PN* junction and behave electrically much like the silicon *PN* junction we have already discussed. If the *PN* junction is forward-biased as shown in Figure 5.6b, holes in the *P* material and electrons in the *N* material are both impelled into the *PN* junction region, where they recombine. As an electron in the conduction band recombines with a hole in the valence band, the electron is making a transition to a lower-lying state and releases energy in the amount of the band gap. Normally this energy goes into heating the material near the junction, but in an LED a significant fraction is emitted as infrared or visible light. The circuit connection needed to produce this result is shown in Figure 5.13a; the only requirements are a voltage source and a series resistor to limit the current I_f.

5.4.5 Light-Sensitive Diodes

When a normal silicon diode is reverse-biased as shown in Figure 5.13b, a depletion region is formed in which the current is carried entirely by thermally generated pairs of charge carriers. As we have already seen, this typically results in a large reverse resistance and a small current I_r. However, if light of the proper wavelength is incident on the depletion region, the absorption of light photons by the crystal can produce additional electron-hole pairs. Like all pairs created in this

Figure 5.13   (a) A typical circuit application of a light-emitting diode. (b) This circuit reverse-biases the photodiode, thereby making it sensitive to increased minority carriers produced by the incident light.

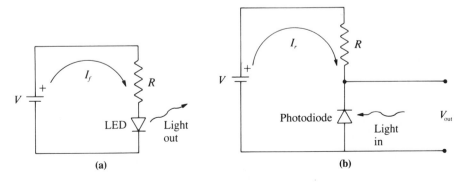

high-field region, they are swept apart and out of the junction region by the applied potential. The incident light thus increases the reverse current I_r, and with the diode connected as shown in Figure 5.13b, this increased current will cause the output voltage V_{out} to decrease.

Diodes designed to optimize the photosensitive effect are known as photodiodes or photocells. Their characteristics vary widely, from the silicon photocells used to convert sunlight directly into electrical energy to high-speed devices designed to receive frequency-modulated light signals. One variation on the signal-detecting photodiode is the light-sensitive *NPN* device. Although such devices have the form of a three-terminal transistor, they are still most often used as two-terminal *PN* diodes.

Figure 5.14 *(a) Semiconductor arrangement for the PIN photodiode showing the spatially extended region of potential gradient that results from the P-type, high-resistance silicon, N-type sandwich. (b) A circuit model for the photodiode. The diode shown in this figure represents a nonideal, but light-insensitive diode; the light-sensitive current is generated by the current source I_p. (c) An LED and a photodiode packaged together to form an optocoupler.*

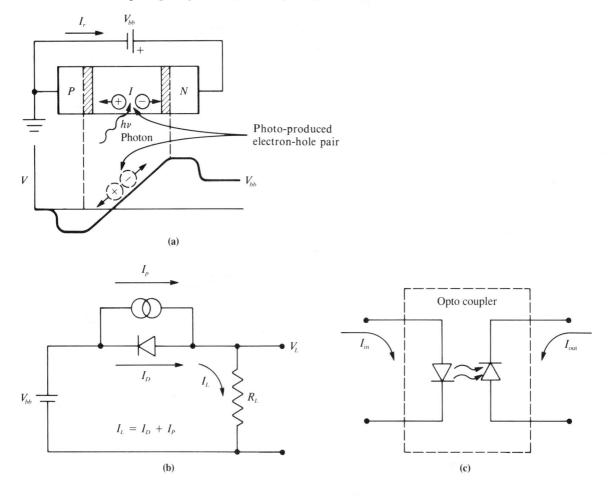

(a)

$I_L = I_D + I_P$

(b)

(c)

An important variation on the *PN* device is the *PIN* photodiode, where the *I* stands for intrinsic (undoped and therefore high-resistance) silicon. This structure is shown in Figure 5.14a; the sandwiched layer of pure silicon increases the thickness of the high-resistance depletion region and broadens the spectrum of light that can be efficiently absorbed. Also, the onset of the photocurrent (rise time) following a light impulse is much faster in a *PIN* diode than in the *PN* diode: less than 1 ns versus several microseconds. This allows the *PIN* diode to be used in much higher frequency applications.

The circuit model for a photodiode is shown in Figure 5.14b. In this figure the diode symbol stands for a normal signal diode having a forward *PN* voltage drop and a reverse resistance R_r. When an external reverse bias is applied, this resistance gives rise to the dark current I_d. The current source I_p represents the photo-induced current and is a true external EMF that is present even if the reverse-biasing voltage V_b is zero. This fact allows the photodiode to be used in two modes: photovoltaic with zero external bias, or photocurrent with the diode back-biased by an external source. The photovoltaic mode is convenient for some applications, since it can directly produce a voltage signal across a large impedance, but the best characteristics (linearity, speed, and lack of temperature sensitivity) are obtained in the photocurrent mode.

An LED and a photodiode are often used together to couple one part of a circuit to another optically. They can be connected by a long optical fiber for communication, or they may be encapsulated in the same package to replace transformers in low-power signal transfer applications. Like a transformer, these devices permit one part of a circuit to be electrically isolated from another while still allowing signals to be transferred between the circuits. Optical coupling is widely used in medical applications to protect the patient from dangerous effects that may result from electronic failure. Even the low voltages present in modern electronic equipment can be fatal in medical applications, because parts of the circuit are often connected to the patient in a way that circumvents the protection afforded by the body's normally high-resistance skin.

5.5 *Circuit Applications of Ordinary Diodes*

Ordinary diodes can be divided into two general classes: power diodes and signal diodes. Power diode applications are generally limited to the various rectifier circuits such as those described below; their function is to convert alternating current power into direct current power. The more important characteristics of diodes of this type are their ability to withstand large currents, dissipate power, and withstand large peak inverse voltages (PIV). Power diode packages are generally made of plastic or metal, and package size varies dramatically with the current carrying and power dissipation capability of the diode. Signal diodes are generally used in low-voltage and low-current applications where speed and a large backward-to-forward resistance ratio are of major importance. Their packages are generally the size of a quarter-watt resistor and are made of glass, ceramic, or plastic.

Some signal diode applications make explicit use of the *PN* voltage drop or other of the more subtle characteristics of the diode; one common example is to use the *PN* voltage drop to set voltage levels for other circuit elements. For such applications a suitable model of the diode, such as we have discussed above, can be used to describe the circuit behavior. For many circuits, however, only the basic diode effect is of any significance, and these circuits can be analyzed under the assumption that the diode is an ideal device. Except for the initial discussion of the half-wave rectifier, the circuits below will be treated in this simplified manner.

5.5.1 Rectifiers

The basic half-wave rectifier circuit is shown in Figure 5.15a. The input signal to the rectifier is assumed to be a purely AC signal with a time-average value of zero. Since current passes through an ideal diode only when the input signal is positive, the signal V_O across the load resistor will be as shown. This signal is exactly the top half of the voltage signal, and for an ideal diode does not depend

Figure 5.15   (a) The half-wave rectifier circuit and its idealized effect on a sine wave input signal. (b) The equivalent circuit during the negative part of the wave when the diode is reverse-biased. (c) The equivalent circuit during the positive part of the wave. (d) The solid curve displays the combined output signal resulting from this model; the positive and negative parts of the curve are discontinuous because V_{PN} appears in only one equivalent circuit. The smoothing shown by the dashed curve arises from the nonlinear portion of the IV curve of a real diode.

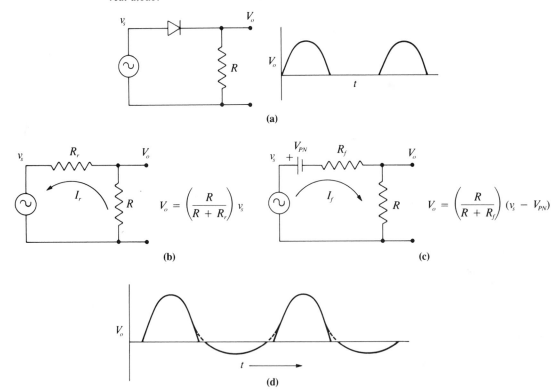

at all on the size of the load resistor. The rectified signal is now a combination of an AC signal and a DC component. Generally, it is the DC part of a rectified signal that is of interest, and the unwelcome AC component is described as ripple. We will see below how this signal can be converted to one with a much smaller ripple.

When diodes are used in small-signal applications of a few volts, their behavior is not closely approximated by the ideal model because of the *PN* turn-on voltage. The equivalent circuit model developed above can be used to evaluate the detailed action of the rectifier under these conditions. The equivalent rectifier circuit during the negative part of the wave is shown in Figure 5.15b. In this reverse-biased state the diode behaves like a resistor of magnitude R_r, and the output signal is just a smaller-amplitude version of the input signal. During the positive half of the wave the circuit can be represented by the equivalent circuit shown in Figure 5.15c. During the part of the wave when the input is positive but less than the *PN* turn-on voltage, the model predicts no loop current and the output signal voltage is therefore zero. When the input exceeds this voltage, the output signal becomes proportional to $v_s - V_{PN}$. The predicted output resulting from a low-voltage input sine wave is shown in Figure 5.15d. This sketch exaggerates the effect actually observed; usually the curved portion of the *IV* curve of an actual diode will smooth the corners as shown dashed in the figure, thereby making them less obvious.

In power supply applications it is common to use a transformer to isolate the power supply from the 110-V AC line. A half-wave rectifier can be connected to the transformer secondary as shown in Figure 5.16a to generate the typical half-wave output signal. As can be seen from the figures, the half-wave rectifier produces an output signal whose fundamental frequency is the same as the input AC signal.

By using a transformer with a center tap in conjunction with a second diode, it is possible to produce a full-wave rectifier as shown in Figure 5.16b. This circuit produces the signal shown; both the positive and negative halves of the AC waveform now appear across the load resistor. This rectified wave has a fundamental frequency that is twice that of the corresponding half-wave rectifier, moving the ripple to a higher frequency where it is easier to remove by a low-pass filtering process.

When a center-tapped transformer is not available, the bridge circuit of Figure 5.16c will also act as a full-wave rectifier. The diodes act to route the current from both halves of the AC wave through the load resistor in the same direction, and the voltage developed across the load resistor becomes the rectified output signal. The diode bridge is a commonly used circuit and is available as a four-terminal component in a number of different power and voltage ratings.

5.5.2 Clamping

When a signal drives an open-ended capacitor as shown in Figure 5.17a, the average voltage level on the output terminal of the capacitor is determined by the initial charge on that terminal and may therefore be quite unpredictable. Thus, even though the high-pass filter function may not be desired, it is necessary to connect the output to ground (or some other voltage reference) via a large resistor.

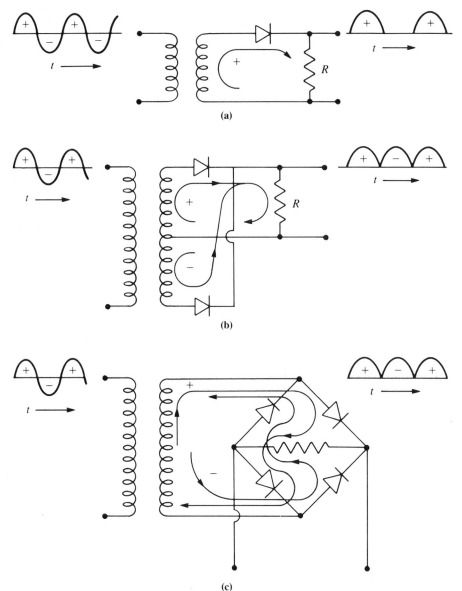

*Figure 5.16 (a) A half-wave rectifier circuit using a transformer. (b)
A full-wave rectifier using a center-tapped transformer. (c) A full-wave
rectifier using the bridge circuit.*

This action drains any excess charge, and for the circuit in Figure 5.17b results in
an average or DC output voltage of zero.

A simple alternative method of establishing a DC reference for the ouput voltage is by use of the diode clamp. This circuit is shown in Figure 5.17c. By
conducting whenever the voltage at the output terminal of the capacitor goes neg-

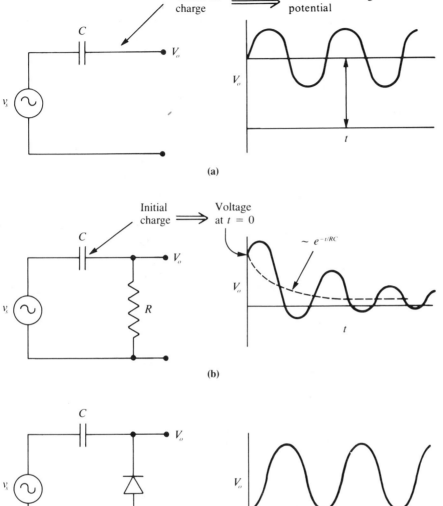

*Figure 5.17 (a) The DC signal level is undefined after a capacitor. (b)
A resistor used to establish the DC voltage after a capacitor. (c) A
diode clamp circuit that allows no part of the wave to go negative.*

ative, this circuit builds up an average charge on the terminal that is sufficient to
prevent the output signal from ever going negative. Positive charge on this terminal is effectively trapped and only leaks to ground slowly through the reverse
resistance of the diode. The output voltage in an equilibrium condition is shown
in Figure 5.17c.

5.5.3 Clipping

When it is necessary to limit the voltage swing of a signal, a diode clipping circuit can be used. An example that restricts both positive and negative voltage swings is shown in Figure 5.18. When the signal voltage is between the limits set by the reference voltages, neither diode conducts and the signal is unaffected. If the signal voltage swings beyond either limit, the appropriate diode will conduct, causing an additional voltage drop across the series resistor R. The clipped signal is shown in Figure 5.18b.

5.5.4 The Charge Pump

A two-diode circuit known as the charge pump is shown in Figure 5.19a. In the ideal case the output voltage of this circuit is proportional to the total number of pulses generated by the input signal. It can thus be used to count the number of input pulses occurring in some time interval. This is a transient circuit, and a switch is needed to discharge capacitor C_2 between counting applications and establish a zero-count voltage reference.

A qualitative description of the operation of this nonlinear circuit is as follows. Diode D_1 acts like the diode in a clamp circuit and supplies charge as needed to keep the voltage V_1 from going negative. During the positive slope of the input

Figure 5.18 *(a) A typical clipping circuit and (b) the resulting signal.*

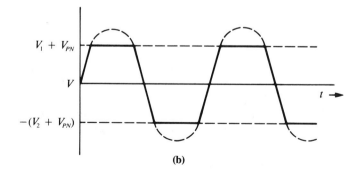

(a)

(b)

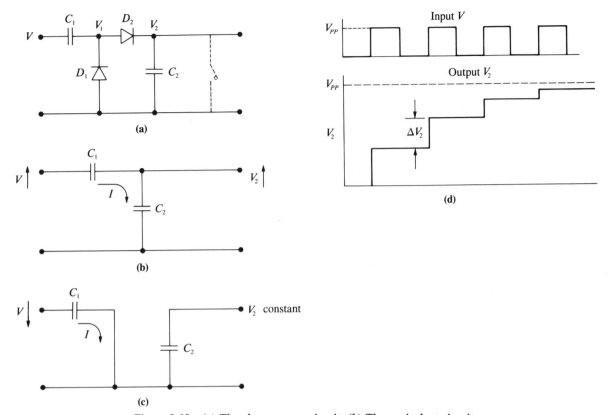

Figure 5.19 *(a) The charge pump circuit. (b) The equivalent circuit when dV/dt > 0. (c) The equivalent circuit when dV/dt < 0. (d) The output signal resulting from a string of pulses.*

signal V, positive charge on the right side of C_1 is pushed across diode D_2 onto capacitor C_2. While V is constant, there is no current in the circuit; but when V falls, the charge at C_1 is replaced by current through D_1. The charge on C_2 is trapped by diode D_2, and during this part of the wave V_2 may exceed the other voltages in the circuit. When D_2 is conducting, the equivalent circuit is as shown in Figure 5.19b, and with D_1 conducting it is as shown in Figure 5.19c.

A more quantitative description can be obtained by observing that in Figure 5.19b, charge conservation requires that the positive charge deposited on the top of C_2 must be equal to the magnitude of the negative charge induced on the right side of C_1 by the applied signal. The resulting voltage loop equation is

$$V - \frac{\Delta q}{C_1} - \frac{q_2 + \Delta q}{C_2} = 0 \qquad (5.15)$$

Solving for Δq gives

$$\Delta q = \left(\frac{C_1 C_2}{C_1 + C_2}\right)\left(V - \frac{q_2}{C_2}\right) \qquad (5.16)$$

And since $V_2 = q_2/C_2$,

$$\Delta q = \left(\frac{C_1 C_2}{C_1 + C_2}\right)(V - V_2) \tag{5.17}$$

The voltage increase on C_2 is therefore

$$\Delta V_2 = \frac{\Delta q}{C_2} = \left(\frac{C_1}{C_1 + C_2}\right)(V - V_2) \tag{5.18}$$

If the step size ΔV_2 is to be small compared with V so that an output signal like that shown in Figure 5.19d can develop, then C_1 must be much less than C_2. This is the basic design requirement for a diode pump.

5.5.5 *The Frequency-to-Voltage Converter*

When a load resistor is added to the charge pump, the circuit becomes a frequency-to-voltage converter as shown in Figure 5.20. After each charging impulse from the input square wave, C_2 will discharge across R for a time equal to the

Figure 5.20   A frequency-to-voltage converter made from a charge pump and a load resistor.

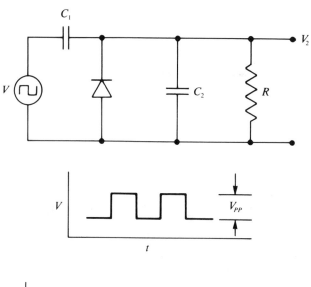

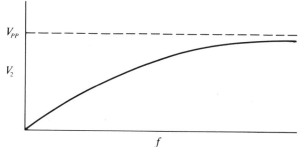

period of the input signal. During this part of the cycle the dwindling charge on C_2 is given by

179

5.5 Circuit Applications of Ordinary Diodes

$$q = q_2 e^{-t/RC_2} \qquad (5.19)$$

where q_2 is the initial charge on C_2, and t is measured from the charging edge of the square wave. The stored charge lost into R in a time $1/f$ is given by the initial charge minus the stored charge after a time $1/f$ and is expressed by

$$\Delta q_2 = q_2(1 - e^{-1/RC_2 f}) \qquad (5.20)$$

If $fRC_2 \gg 1$, we can approximate the exponential in Eq. (5.20) by the first two terms of its Taylor series expansion ($e^x = 1 + x + x^2/2! + x^3/3! + \ldots$). The one in Eq. (5.20) cancels the first term of the expansion, leaving

$$\Delta q_2 = \frac{q_2}{fRC_2} \qquad (5.21)$$

When this is set equal to the charge gained per cycle as given by Eq. (5.17), the resulting expression can be solved for V_2 to give

$$V_2 = \frac{C_1 C_2 Rf V}{C_1 + C_2 + C_1 C_2 Rf} \qquad (5.22)$$

With the further requirement that $C_1 C_2 Rf \ll C_1 + C_2$, a generally desirable linear relationship between voltage and frequency is obtained:

$$V_2 = \left(\frac{C_1 C_2}{C_1 + C_2}\right) RVf \qquad (5.23)$$

Since the charge pump itself required that $C_1 \ll C_2$, the design requirements just imposed can be simplified to

$$\frac{1}{C_2} \ll Rf \ll \frac{1}{C_1} \qquad (5.24)$$

Within this restricted range of frequencies the output voltage of the circuit will be proportional to the frequency of the input signal. This result was derived under the assumption of a square-wave input signal, but a similar expression would be obtained for a sinusoidal signal.

5.5.6 The Voltage Doubler

A voltage doubler is a kind of rectifier that produces a DC signal of approximately twice the voltage of the corresponding half- or full-wave rectifier. In one form it is constructed from a charge pump as shown in Figure 5.21a. In this application of the pump, the capacitors C_1 and C_2 are typically the same size; and with the circuit unloaded, the DC voltage developed on C_2 will reach the peak-to-peak voltage of the input AC signal after one or two cycles. As can be seen, the first capacitor and diode act as a clamp, and the second diode charges C_2 whenever the voltage V_1 exceeds V_2. When a load is connected, capacitor C_2 loses charge to the load, causing the voltage V_2 to fall during the part of the cycle where V_1 is

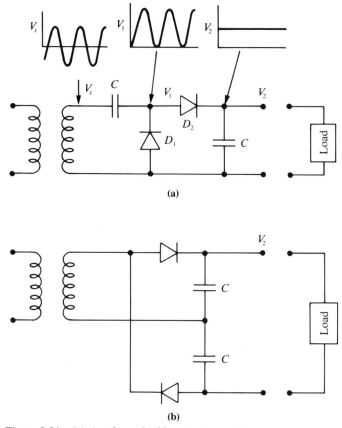

Figure 5.21 *(a) A voltage doubler circuit made from a charge pump.*
(b) Another common form of the voltage doubler circuit.

less than V_2. Thus, there will be ripple on the output voltage at the same frequency as the input AC signal, like a half-wave rectifier.

Another common form of the voltage doubler circuit is shown in Figure 5.21b. This circuit develops equal and opposite voltages across the capacitors (with respect to their common point), and the output voltage is taken across the sum of the two capacitors. Normally one of the output terminals will be grounded, making the other voltage equal to the peak-to-peak AC signal voltage. Note that in the circuit shown neither side of the transformer secondary is at ground potential.

5.5.7 *Inductive Surge Suppression*

An important use of diodes is to suppress the voltage surge present when an inductive load is switched out of a circuit. In the absence of the diode shown in Figure 5.22, the current in the inductor would generate a large transient voltage across the switch whenever it is opened. In the case of a mechanical switch, this can lead to a spark, radiofrequency noise, and rapid oxidation of the switch contacts; the voltage surge can be large enough to destroy a solid state transistor switch. When the switch is closed, the diode shunting the inductor does not con-

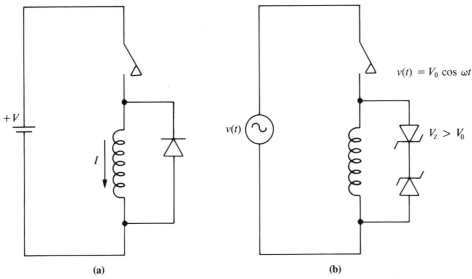

Figure 5.22 *(a) Diode inductive surge suppressor useful for a DC current in the inductor. (b) Alternative circuit for an AC current.*

duct, since it would be reverse biased. However, when the switch is opened, the diode provides an alternative path for the inductor current, and the inductor will not develop a voltage higher than V_{PN}.

When the inductor is driven by an AC source, a diode cannot normally be used. In this case a better solution is to use two back-to-back Zener diodes as shown in Figure 5.22b. Each Zener must have a breakdown voltage slightly in excess of the amplitude of the AC voltage swing expected across the inductor. Normally one diode will be reverse-biased, but not sufficiently for it to enter its breakdown region. If the switch is opened while the inductor is carrying a significant current, the induced EMF will be great enough to push one diode into its breakdown region, thereby providing an alternative path for the inductive current.

Depending on the Q of the inductor and other circuit considerations, it may be necessary to include a series resistor in either of these diode bypass circuits in order to dissipate the inductor's energy more quickly.

5.6 Power Supplies

Although schematics often show a DC voltage source as a battery, in practice the use of batteries is limited to special-purpose applications such as stable reference voltages, power for portable equipment, or backup power in case of AC line failures. Most often, the actual DC voltage source will be a power supply. The diode rectifier circuits discussed above serve to convert AC power to DC power, but a complete power supply has several other important parts. A block diagram of a typical power supply is shown in Figure 5.23a. This diagram shows the four main parts of a typical power supply as separate units: the transformer, the recti-

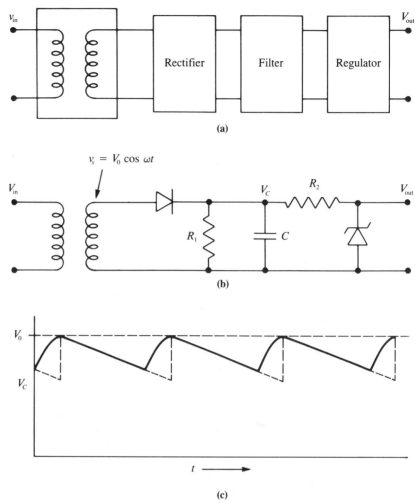

Figure 5.23 *(a) Block diagram of a typical power supply. (b) Simple power supply circuit. (c) The signal present across the capacitor. The dashed curve is the approximate signal assumed in Example 5.3.*

fier, the filter, and the regulator. While power supplies generally have all four of these operational units, they are not always as physically distinct as this diagram indicates.

The simple power supply circuit shown in Figure 5.23b makes use of circuits already discussed: a transformer and rectifier, a low-pass filter circuit whose only physical component may be a large electrolytic capacitor, and a Zener diode regulator. The low-pass filter circuit serves to reduce the ripple on the rectified DC signal and could certainly be improved by replacing the resistor with an inductor. Since inductors are large and expensive, a more practical solution is to add a voltage regulator. The regulator serves both to reduce the ripple on the output voltage of the power supply and to reduce the output impedance of the supply, thereby making the output voltage more independent of the load.

The Zener regulator shown is very basic and makes use of a variable voltage drop across the series resistor R_2 to allow V_{out} to show much less ripple than that present on the filtered voltage V_f. When designing such a regulator, careful consideration must be given to the power dissipated in each of the circuit elements. Since a power supply will generally be operated under widely varying load conditions, the power dissipation in each component should be calculated under "worst-case" assumptions for that particular component. For example, the series resistor R_2 will dissipate the most power when the output terminals of the supply are shorted, and it should therefore have a power rating in excess of that calculated under this condition. At the other extreme, the Zener diode will dissipate the most power when the power supply's output terminals are open, since this will produce the maximum current through the Zener. If the power supply is ever to be operated in the unloaded condition—and most will be sooner or later—the Zener's power rating must exceed that calculated under this worst-case condition.

Example 5.3 On the regulated power supply circuit of Figure 5.23b, assume that $v_s = 15 \cos(2\pi 60 \text{ Hz } t)$ V, $R_1 = 100 \ \Omega$, $R_2 = 10 \ \Omega$, $C = 10,000 \ \mu\text{F}$, and $V_Z = 10$ V. Further assume that the reverse resistance of the Zener is $1 \ \Omega$. Calculate the peak-to-peak ripple voltage that is present on the capacitor voltage V_C and on the output voltage V_{out}.

The voltage signal on the capacitor is shown in Figure 5.23c. If the rectifying diode is ideal, the maximum voltage on the capacitor will reach the full amplitude of the driving AC signal, V_0 as shown in the figure and 15 V in this example. By approximating the actual signal with the dashed triangle wave shown, the mathematics of this problem is much reduced. In this approximation, the rectifying diode charges the capacitor to V_0 over a very short time, after which the capacitor undergoes a constant-current discharge into R_1 and R_2 for a time T equal to the period of the AC wave. This current is actually given by

$$I_C = \frac{V_C}{R_1} + \frac{(V_C - V_Z)}{R_2}$$

where V_C is a function of time. If the peak-to-peak ripple is not a large fraction of V_T, we can make a reasonable approximation to this current by replacing V_C with V_0; for our example this gives

$$I_C = \frac{15 \text{ V}}{100 \ \Omega} + \frac{5 \text{ V}}{10 \ \Omega} = 650 \text{ mA}$$

When this current discharges the capacitor for a time that is equal to the AC period T (1/60 s), it changes the voltage on the capacitor by

$$\Delta V_C = \frac{-\Delta Q}{C} = \frac{-I \ \Delta t}{C} = \frac{-(0.65 \text{ A}) \ (1/60 \text{ s})}{(10^{-2} \text{ F})} = -1.1 \text{ V}$$

The voltage signal at the capacitor thus has about 1 V of peak-to-peak ripple. Assuming a Zener diode model like that shown in Figure 5.12a, we can use Eq. (5.13) to determine the open circuit output voltage V_{out}. That equation becomes

$$V_{out} = V_Z\left(\frac{R}{R + R_Z}\right) + V_C\left(\frac{R_Z}{R + R_Z}\right)$$

which with our numbers is

$$V_{out} = (10 \text{ V})\left(\frac{10}{11}\right) + \left(15 \text{ V} - \frac{1.1t}{T}\text{ V}\right)\left(\frac{1}{11}\right)$$

and reduces to

$$V_{out} = \left(9.09 + 1.36 - \frac{0.1t}{T}\right)\text{V} = \left(10.45 - \frac{0.1t}{T}\right)\text{V}$$

The peak-to-peak ripple voltage at the output is thus only 0.1 V, with the improvement factor being equal to the resistance ratio R_Z/R_2.

Note that the use of a full-wave rectifier would decrease this ripple by a factor of 2, since the capacitor would then discharge for only 1/120 s before recharging.

5.6.1 *Integrated Circuit Voltage Regulators*

A much better DC power supply can be constructed using a common and inexpensive three-terminal regulator. These regulators are integrated circuits consisting of several solid state devices and are designed to provide the desirable attributes of temperature stability, output current limiting, and thermal overload protection. The most popular units can provide up to 1 A of output current and are designated 79xx or LM320-xx for negative voltages and 78xx or LM340-xx for positive voltages. The xx designates the voltage magnitude; available devices are limited to the values 05, 06, 10, 12, 15, 18, and 24. The circuit symbol is just a box with three terminals as shown in Figure 5.24a.

The values of C_1 and C_2 vary with application. In the least demanding case when V_{in} is already a nominally constant voltage, C_1 should be about 0.1 μF. Since the high-frequency response of the regulator is limited, the capacitor C_1 is used to remove induced high-frequency signals from V_{in}. In a similar manner C_2 helps to hold V_{out} constant under rapidly changing load conditions; a typical value for this capacitor is 1 μF.

Although the internal circuit of the regulator is fairly complicated, its use in normal applications is quite simple, and only a few operating principles must be considered. A typical power supply circuit is shown in Figure 5.24b. The input voltage to the regulator will be a DC voltage but may have a substantial ripple; the essential requirement is that at any instant the regulator input voltage must exceed the output voltage by at least 2.5 V. Typically only about 2% of an input voltage variation will appear at the output. Depending on the average voltage across the regulator and the current through it, the regulator may be required to dissipate substantial amounts of power in the form of heat. This is minimized by a design where the input voltage is as close as possible to the output voltage. A 5-V drop across the regulator is a reasonable design goal, although input voltages as high as 35 V are acceptable.

The 78xx and 79xx series of regulators will dissipate up to 2 W of power into room-temperature air; with proper heat sinks attached, this can be increased to 15 W. Additional devices, circuits, and design details can be found in voltage regulator handbooks published by several manufacturers, including Texas Instruments and National Semiconductor.

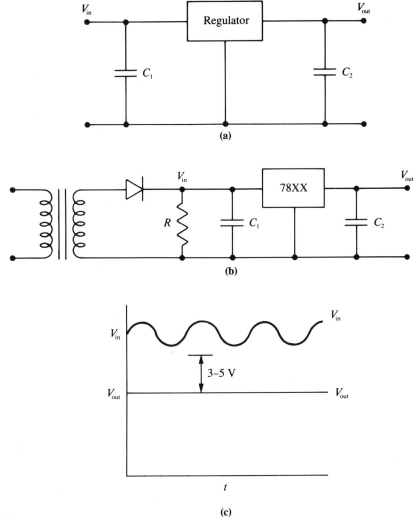

Figure 5.24 *(a) Typical connection of a three-terminal IC voltage regulators. (b) Improved power supply using a three-terminal regulator.*

Problems

1. If a silicon crystal has 10^{22} atoms/cm^3 and 10^{10} atoms/cm^3 have broken covalent bonds at room temperature, what fraction of doping atoms must be added to double the conductivity of the material? [*Ans.*: $2:10^{12}$]

2. A certain diode has a cross-sectional area of 10^{-8} m^2 and in the shorted condition develops a depletion region that is 0.6 μm across. *(a)* Using the parallel-plate capacitance formula $C = \epsilon A/d$ with $\epsilon = 16 \times 10^{12}$ F/m, calculate the capacitance of this junction. *(b)* If a forward voltage is applied across the diode, will this capacitance increase or decrease?

3. If a physical diode is modeled as shown in Figure 5.9a with $V_{PN} = 0.6$ V and $R_f = 10$ Ω: *(a)* What voltage is developed across the diode when $I_f = 100$ mA? *(b)* How much power is dissipated in the diode at this current? [*Ans.*: not 100 mW]

4. Assume that the diode in the circuit shown in Figure P5.4 can be modeled by $V_{PN} = 0.6$ V, $R_f = 5$ Ω, and $R_r = 5 \times 10^4$ Ω. If the input signal is $v_s = 5\cos(\omega t)$, sketch the output signal v_d and label the positive and negative amplitudes as accurately as possible.

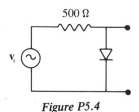

Figure P5.4

5. Given that you have a signal generator producing a sine wave with no DC component, a capacitor, a diode, and a 15-V battery with the negative side unalterably connected to ground, design a circuit that will give a sine wave output clamped so that it is always above 15 V.

6. A simple ohmmeter can be constructed from an ammeter as shown in Figure P5.6. In this design the measured resistance is given by the equation $R = 10/I - 1$ kΩ. If this ohmmeter is used to measure the forward resistance of a signal diode, the measurement will typically be about 100 Ω when R_f is really closer to 1 Ω. Explain with a current versus voltage diagram.

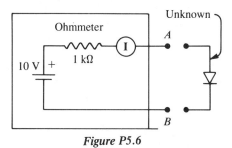

Figure P5.6

7. Sketch the current versus voltage curve for a typical Zener diode and identify the following quantities in the sketch: V_{PN}, V_Z, R_Z, R_f, R_r.

8. In the circuit shown in Figure P5.8, the 20-V Zener is rated at 1 W and the load resistor R_L is sometimes switched out of the circuit as shown.

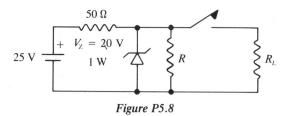

Figure P5.8

(a) If $R = \infty$ and R_L is out of the circuit, how much power is dissipated in the Zener?
[*Ans.*: 2 W]

(b) What value of R makes $P_{Zener} = 1$ W? [*Ans.*: 400 Ω]

(c) With this value of R in place, what is the minimum value of R_L that may be used before V_{out} falls below 20 V? [*Ans.*: 400 Ω]

9. *(a)* In the circuit of Figure P5.9, how much power is dissipated in the Zener? *(b)* How much power is dissipated in the 100-Ω resistor? *(c)* How much power would be dissipated in each element if the output were shorted?

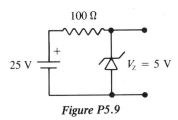

100 Ω

25 V

$V_Z = 5$ V

Figure P5.9

10. On the circuit of Figure 5.11, take $V_s = 25$ V, $V_Z = 15$ V, and allow R_L to be variable.

(a) What value of R is required if the breakpoint on the curve is to be at $I_L = 500$ mA? [*Ans.*: 20 Ω]

(b) How much power will be dissipated in the Zener when $R_L = 50$ Ω? [*Ans.*: 3 W]

11. For each circuit shown in Figure P5.11, sketch the output voltage as a function of time if $v_s(t) = 10 \cos(2000\pi t)$ V. Assume that the circuit elements are ideal.

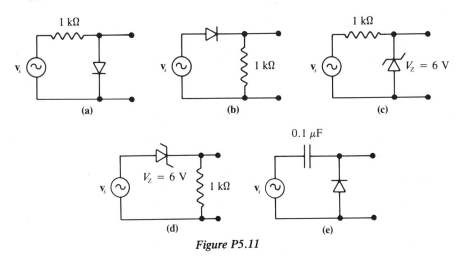

1 kΩ

v_s

(a)

1 kΩ

v_s

(b)

1 kΩ

v_s

$V_Z = 6$ V

(c)

v_s

$V_Z = 6$ V

1 kΩ

(d)

0.1 μF

v_s

(e)

Figure P5.11

12. Assuming that the AC voltage source v_s as shown in Figure P5.12 (top of page 188) offers zero resistance to DC current, write an expression for the voltage v_{ab} when $v_s = 10 \cos(100\pi t)$.

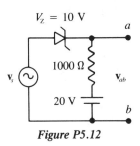

Figure P5.12

13. Using a voltmeter for readout, devise a circuit and method for measuring R_Z of a Zener diode.

14. Assuming that the diodes in the circuit shown in Figure P5.14 are ideal, write expressions for the voltages at points *A* and *B*. [*Ans.:* $V_0 + V_0 \cos(\omega t), 2V_0$]

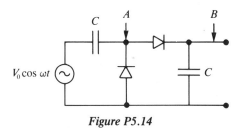

Figure P5.14

15. Using the results of the previous problem, write expressions for the voltages at points *D* and *E* as shown in Figure P5.15.

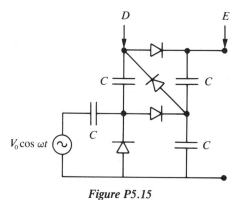

Figure P5.15

16. (a) Use an integrated circuit voltage regulator to design a 5-V power supply for use in an automobile. *(b)* If the DC voltage available from the car varies between 12 and 16 V (depending on engine RPM), what voltage variation is to be expected from your supply? *(c)* Under worst-case conditions and assuming a maximum power dissipation of 2 W, how much current can this device supply without overheating? [*Ans.:* 180 mA]

6

Single Transistor Circuits

6.1 Introduction

From the Thevenin and Norton models for a two-terminal network, it is clear that a signal source can be characterized by its open circuit voltage V_{open} and its short circuit current I_{short}. Taken together, these parameters specify the strength of the source and dictate that the power dissipated in a resistive load will be always less than $0.25 V_{\text{open}} I_{\text{short}}$. A passive network introduced between source and load can only reduce this limit, due to resistive losses within the network. To recoup these losses or to increase the maximum power ($0.25 V_{\text{open}} I_{\text{short}}$) available from a source, the network must contain an active element: a device other than the signal source that can add energy to the electrical domain. When such a device is included as part of a four-terminal network, the network is known as an *active* network.

know for the final (this derivation)

The active element is not a true energy source like a battery; rather, it controls or modulates the flow of energy from such a source. A good analogy is the spill gate on the dam of a large water storage reservoir: The small energy expenditure of moving a barrier can cause tons of water to give up its potential energy. In this chapter we will introduce several types of transistors, all of which have the ability to act as electronic spill gates controlling the flow of charge from a power supply into a circuit.

Transistors are of two general types: bipolar and field effect. The former control the current by varying the number of charge carriers in a critical region, and the latter by varying the shape of a conducting volume. Bipolar transistors have the widest range of application to analog circuits, but field effect transistors (FETs) are being steadily improved and excel in many important areas. The majority of modern transistors are fabricated from silicon, although germanium bipolar types continue to be available.

In this chapter we will develop design and analysis techniques for single transistor circuits: first for the bipolar types, then for the somewhat easier field effect types.

Because of increasing circuit complexity, we will adhere to the following convention with regard to subscripts on voltages, currents, and elements: (1) A single subscript and two identical subscripts both identify a voltage measured with respect to ground, but the latter form identifies a power supply voltage. (2) Two different subscripts define a voltage difference $V_{AB} = V_A - V_B$. The two-subscript notation may occasionally be used to identify an element C_{AB} or current I_{AB} between points A and B. (3) There is also a three-subscript convention for three-terminal devices such as transistors. The first two subscripts identify two terminals of the device and have the two-subscript meaning; the third is either O or S for open or shorted and refers to the unnamed third terminal. When shorted, the third terminal is connected to the terminal identified by the middle subscript.

6.2   *Bipolar Transistors*

The bipolar transistor consists of three regions of doped semiconductor material as shown in Figure 6.1 and comes in two types, known as *NPN* or *PNP* depending on the geometry. Both types can be fabricated from either germanium or silicon, but the physics of germanium favors the *PNP* type whereas for silicon it favors the *NPN* structure. The thin central region, known as the base, has majority charge carriers of opposite polarity to those in the surrounding material. A transistor contains two *PN* junctions and can be visualized as two connected diodes as shown in the figure. However, the "transistor action" requires the thin base region, and two connected but distinct diodes do not exhibit all the properties of a transistor.

The two outer regions are known as the emitter and the collector. Under the proper operating conditions the emitter will emit or inject majority charge carriers into the base region, and because the base is very thin, most will ultimately reach the collector. The base region derives its name from an early manufacturing technique whereby the collector and emitter materials were deposited onto a piece (base) of opposite polarity material. In modern terminology this initial piece of material is known as the substrate and is not necessarily the electrical base of the transistor.

The schematic symbols for both types of bipolar transistors are shown at the bottom of Figure 6.1. On each symbol, the arrow indicates the forward-biased (small resistance) direction of the *PN* junction between base and emitter; in normal operation this junction is conducting. The similar *PN* junction between base and collector is not explicitly indicated on the symbol because in normal operation this junction is reverse-biased (large resistance) and nonconducting.

Note that DC current is oppositely directed in the two transistor types, and that the arrow on the schematic symbol also indicates the direction of both I_B and I_C. Some multiple-transistor applications make use of this current reversal and are

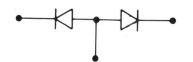

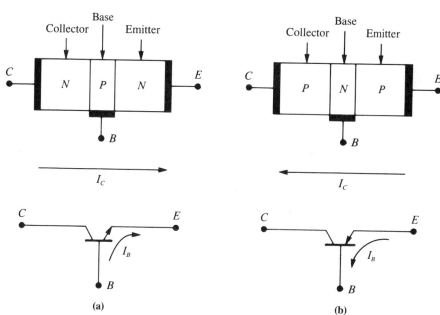

Figure 6.1 *(a) NPN bipolar transistor and circuit symbol. (b) PNP bipolar transistor and circuit symbol.*

possible only as a consequence of the availability of both *NPN* and *PNP* transistors. In the discussion of bipolar transistors in this chapter, we will concentrate on the *NPN* type because it is the most common. The analysis of *PNP* types is no different, except that a few additional minus signs appear in the algebra.

6.2.1 The NPN Transistor: Principles of Operation

The operation of a bipolar transistor can be described qualitatively in terms of the flow of charges across the base and its surrounding depletion regions. If the collector, emitter, and base of an *NPN* transistor are shorted together as shown in Figure 6.2a, the diffusion process described earlier for diodes results in the formation of two depletion regions that surround the base as shown. The diffusion of negative carriers into the base and positive carriers out of the base results in a relative electric potential as shown in Figure 6.2b.

When the transistor is biased for normal operation as in Figure 6.2c, the base terminal is slightly positive with respect to the emitter (about 0.6 V for silicon), and the collector is positive by several volts. As indicated, the external EMFs generate two loop currents, I_C and I_B; when properly biased, the transistor acts to make $I_C \gg I_B$. The depletion region at the forward-biased base-emitter junction becomes thinner as shown, just like a forward-biased diode. The depletion region

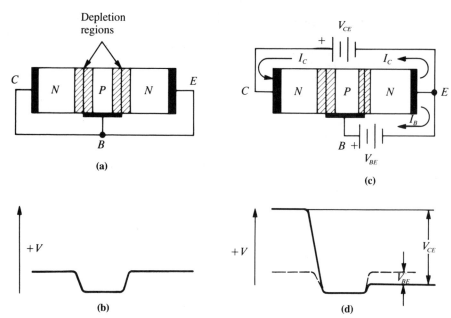

Figure 6.2 *(a) NPN transistor with collector, base, and emitter shorted together. Depletion regions are formed at each PN junction. (b) The voltage levels developed within the shorted semiconductor. (c) The NPN transistor biased for operation with $V_{CE} > V_{BE}$. (d) The voltage levels developed within the biased semiconductor. Note that the electric force on a negative charge is toward $+V$.*

at the reverse-biased base-collector junction grows and is able to support the increased electric potential change indicated in Figure 6.2d.

As a result of the narrowed depletion region and small potential step at the base-emitter junction, negative carriers from the emitter are able to diffuse into the base region. A small fraction of these negative charges work their way out the base terminal through recombination with holes as shown in Figure 6.3a, but most drift across the narrow base and are swept into the collector by the potential gradient of the reverse-biased base-collector junction. For a typical transistor, 95% to 99% of the charge carriers from the emitter make it to the collector and constitute almost all of the collector current I_C. The holes lost to recombination in the base are replenished by the external base current I_B.

The transistor operation is also influenced by the thermal dissociation of covalent bonds in the N-type material of the collector. This dissociation produces a flow of positive minority carriers into the base as shown in Figure 6.3a. The number of such carriers increases with temperature, doubling for each 10°C change. The effects of this current are magnified since it has exactly the same effect as external base current: It increases the number of holes in the base, raises the potential of the base, and consequently increases the flow of negative charges across the base-emitter junction. This part of the collector-to-base current is known as I_{CBO} and is defined by the circuit of Figure 6.3b. At room temperature, this current is generally in the range 10^{-11} to 10^{-10} A for silicon, whereas for germanium it is a much more troublesome 10^{-6} A. This large temperature-depen-

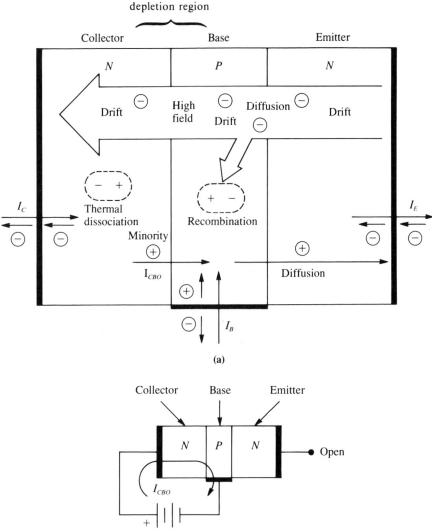

(a)

(b)

Figure 6.3   (a) The flow of charge in a biased NPN semiconductor. The most significant effects are caused by the flow of negative majority carriers from the N-type semiconductors of the collector and emitter. (b) The definition of I_{CBO} (leakage current from Collector to Base with the emitter Open) caused by the flow of minority charge carriers from collector to base.

dent leakage current is sufficient reason to avoid germanium devices whenever possible.

The behavior of a transistor can be summarized by the characteristic curves shown in Figure 6.4a. These curves plot the collector current I_C as a function of the collector-emitter voltage V_{CE} at various constant base currents I_B. Each curve

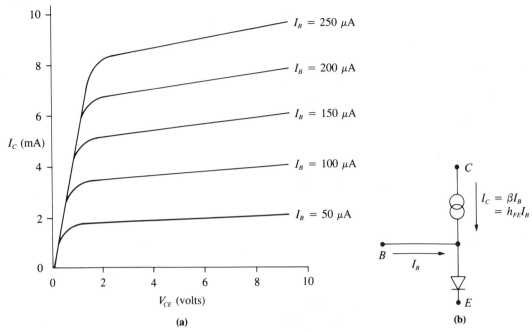

Figure 6.4   The characteristic curves of an NPN transistor. (a) In normal use the transistor is biased to operate in the region to the right of the knee of these curves. In this region the collector current is determined mainly by the base current and is only slightly influenced by the collector to emitter voltage. (b) The transistor operation is reasonably well modeled by an ideal current source and a physical diode.

starts from zero in a nonlinear fashion (not seen on the scale of this figure), rises smoothly, then rounds a knee to enter a region of essentially constant I_C. This flat region corresponds to the condition where the depletion region at the base-emitter junction in Figure 6.2c has essential disappeared; the slight rise in collector current with increasing V_{CE} can be explained by a decrease in the thickness of the base drift region due to an increase in the thickness of the depletion layer around the base-collector junction. To be useful as a linear amplifier, the transistor must be operated exclusively in the flat region, where the collector current is determined by the base current.

When biased to operate in the flat region, the transistor can be approximated by the simple model of Figure 6.4b. If the diode symbol in the base leg is taken to represent a real diode with an *IV* curve like Figure 5.8d, this model can describe the temperature-dependent *PN* voltage drop between base and emitter, the small forward resistance between the base and emitter, the large resistance between the base and collector, and even the nonlinear portions of the curves shown in Figure 6.4a. The current source in the collector leg has an infinite impedance (effectively disconnecting base and collector) and is taken to be a linear function of the base current related by the *static forward current transfer ratio,* which is variously expressed as β or h_{FE}.

$$I_C = h_{FE}I_B \tag{6.1}$$
$$= \beta I_B$$

(Note that we will later introduce a similar AC parameter h_{fe}, distinguished notationally only by the lowercase subscripts.) Even with the assumption of an idealized diode, this model describes the essential features of transistor operation: the base-emitter connection is like that of a forward-biased diode, low-impedance and showing the standard V_{PN} voltage drop; the base-collector connection is nonconducting; and the collector current is proportional to the base current.

For many applications V_{PN} can be taken to be a constant (0.3 V for germanium and 0.6 V for silicon) or assumed small and ignored completely. In more sophisticated applications, especially where temperature stability is required, it may be necessary to consider the voltage drop between the base and emitter more carefully. Figure 6.5a shows the relationship between the base-emitter voltage V_{BE} and the collector current I_C. This curve has the general shape of the current-voltage curve of a forward-biased diode and can be described by a similar equation:

$$I_C = I_D(e^{V_{BE}/V_T} - 1) \tag{6.2}$$

The constant V_T is the same threshold voltage used in the diode expression (Eq. 5.1), and I_D depends on the construction details of the transistor. As in the case of diodes, both I_D and V_T are functions of temperature, and if the collector current I_C is held constant, these temperature effects combine to produce a change in V_{BE} of about -2 mV/°C. If the circuit design is such that V_{BE} is held constant, as it would be with the simple voltage bias of Figure 6.2c, then I_C must increase with temperature. This is the major temperature-dependent effect on the operation of silicon bipolar transistors. For germanium devices, the increase of minority carrier current I_{CBO} with temperature is of greater importance. One additional temperature dependent parameter is h_{FE} itself. For silicon, this quantity increases by approximately 1%/°C.

All of the temperature effects discussed tend to increase the collector current I_C as the temperature increases. Since the internal heating of the transistor is proportional to this current (which in turn increases the internal heating), the overall effect can produce a "thermal runaway" heating condition. Proper circuit design and choice of components can minimize this effect and in critical designs it is common to use additional temperature-sensitive circuit components to compensate for changes in the transistor. At power dissipation levels under 1 W, circuits constructed with silicon transistors are usually temperature-stable if they are designed to be electrically stable—that is, insensitive to changes in transistor parameters. Since transistor parameters vary widely, even for identically labeled devices, we will stress circuits that are relatively insensitive to these parameters. Consequently, temperature stability will not be discussed in detail.

6.2.2 Basic Circuit Configurations

The transistor is a three-terminal device that we will use to form a four-terminal circuit. Unlike the passive four-terminal networks of the previous chapters, the transistor circuit is an active device with power input from the constant EMF that

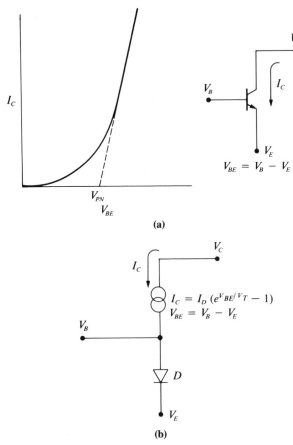

$V_{BE} = V_B - V_E$

(a)

$$I_C = I_D \left(e^{V_{BE}/V_T} - 1\right)$$
$$V_{BE} = V_B - V_E$$

(b)

Figure 6.5   (a) The nonlinear relationship between the collector current and the base-emitter voltage. (b) This model uses a physical diode to approximate the nonlinear behavior. This model is helpful in understanding some of the more sophisticated, temperature-insensitive DC biasing techniques.

supplies the collector-emitter voltage bias. The resulting circuit has a preferred direction of signal propagation: Small voltage changes at the base-emitter junction produce large current changes in the collector and emitter, whereas small changes in the collector-emitter voltage have very little effect at the base. The result is that the base must always be a part of the input of a four-terminal arrangement. The three possible circuit configurations are shown in Figure 6.6. These sketches have been drawn to emphasize the common element in the naming convention. In an actual circuit the distinction between the common emitter *(CE)* and common collector *(CC)* configurations is not always obvious from this common feature. A more reliable indication is that the output voltage signal (with respect to ground) is taken from the collector of the common emitter circuit and from the emitter of a common collector circuit.

Transistor circuits are complicated by the need to provide the DC bias currents and voltages necessary for linear operation. Every linear design must have a pro-

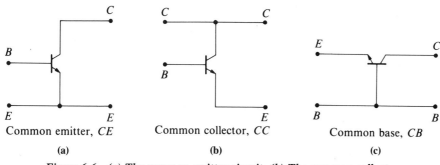

Figure 6.6 *(a) The common emitter circuit. (b) The common collector circuit. (c) The common base circuit.*

vision for establishing the DC operating point (also called the Q or quiescent point) defined by the constant values of I_B, V_{BE}, I_C, and V_{CE} in the absence of an input signal. The design of a transistor circuit thus has two parts: DC biasing and AC analysis. Before beginning the investigation of specific circuits with the additional problems introduced by their biasing components, we will consider some AC circuit models of the bare transistor.

6.2.3   Small-Signal Models

In the simple transistor model expressed by Eq. (6.1), the collector current was taken to be proportional to the base current. A more general statement capable of describing the family of curves displayed in Figure 6.4a would be of the form

$$I_C = I_C(I_B, V_{CE}) \tag{6.3}$$

where I_C is some smooth but transistor-dependent function. For AC analysis we are interested only in the time derivative of this equation:

$$\frac{dI_C}{dt} = \frac{\partial I_C}{\partial I_B}\frac{dI_B}{dt} + \frac{\partial I_C}{\partial V_{CE}}\frac{dV_{CE}}{dt} \tag{6.4}$$

When the first partial derivative is evaluated at a particular I_B, V_{CE} point on the curves of Figure 6.4a, it becomes a constant h_{fe} known as the forward current transfer ratio. This parameter describes the vertical spacing $\Delta I_C/\Delta I_B$ between the curves of Figure 6.4a. Evaluated at this same point, the second partial becomes a constant h_{oe} known as the output admittance (inverse of resistance). This parameter is just the slope $\Delta I_C/\Delta V_{CE}$ of one of the curves as it passes through the specified point.

The second subscript e on the h parameters indicates that voltages are being referenced to the emitter terminal (terminal E on the common emitter configuration of Figure 6.6a). In terms of these h parameters, Eq. (6.4) becomes

$$\frac{dI_C}{dt} = h_{fe}\frac{dI_B}{dt} + h_{oe}\frac{dV_{CE}}{dt} \tag{6.5}$$

These h's are functions of the variables I_B and V_{CE}, both of which normally have an AC component caused by the driving input signal. This differential equation is linear only in the limit of small AC signals, where the h's are effectively constant.

Equation (6.3) is not a complete description of the transistor, since it does not include the input signal V_{BE}. This voltage must also be related to I_B and V_{CE} and could be expressed by

$$V_{BE} = V_{BE}(I_B, V_{CE}) \tag{6.6}$$

where V_{BE} is another smooth transistor-dependent function. Following the same argument used above, the time derivative of this equation could be written as

$$\frac{dV_{BE}}{dt} = h_{ie}\frac{dI_B}{dt} + h_{re}\frac{dV_{CE}}{dt} \tag{6.7}$$

where h_{ie} and h_{re} are the partial derivatives of V_{BE} with respect to I_B and V_{CE}, respectively.

In this formulation we chose I_B and V_{CE} as the independent variables in Eqs. (6.3) and (6.6); we could equally well have chosen any other pair of the four variables V_{BE}, V_{CE}, I_B, and I_C. Because current and voltage variables are mixed, the h's are known as hybrid parameters. Other choices produce similar models with different parameterization; see Section 6.8.4.

The typical current or voltage signal will have both DC and AC components, as for example,

$$I_B = I_B(\text{DC}) + i_B(t) \tag{6.8}$$

The time derivative involves only the AC part, just di_B/dt. If we restrict ourselves to sinusoidal AC signals and introduce the complex notation, then $i_B(t)$ is replaced by

$$\mathbf{i}_B(t) = \mathbf{i}_B e^{j\omega t} \tag{6.9}$$

which has the simple time derivative $j\omega\mathbf{i}_B$. The other variables have the same form with a common frequency ω. If these complex forms are then used to replace the time derivatives in Eqs. (6.5) and (6.7), the $j\omega$ factor appears in each term and will cancel to produce

$$\mathbf{i}_C = h_{fe}\mathbf{i}_B + h_{oe}\mathbf{v}_{CE} \tag{6.10}$$

and

$$\mathbf{v}_{BE} = h_{ie}\mathbf{i}_B + h_{re}\mathbf{v}_{CE} \tag{6.11}$$

For the present we have no need for the complex voltage and current notation and can simplify the discussion by taking only the real (resistive) part of each equation. The complex notation will be reintroduced as needed to describe circuits with reactive elements.

The relationship between the voltages and currents of Eqs. (6.10) and (6.11) and the transistor is shown in Figure 6.7a. A similar set of equations, with different h's and different voltage differences, can be developed for a transistor in either of the other two configurations shown in Figure 6.6. However, we will use only the common emitter parameters, even when describing a transistor in the common base or common collector configuration.

The transistor in Figure 6.7a can be modeled by the Thevenin and Norton equivalent circuits shown in Figure 6.7b; the model is only a variation on Figure

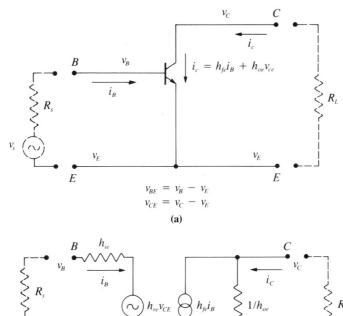

$$i_c = h_{fe}i_B + h_{oe}v_{ce}$$

$$V_{BE} = V_B - V_E$$
$$V_{CE} = V_C - V_E$$

(a)

(b)

Figure 6.7 (a) An NPN transistor in the common emitter configuration. (b) The hybrid-parameter, small-signal model for the transistor.

2.23c. The identification of h parameters with elements in this transistor model follows directly from Eqs. (6.10) and (6.11). With the labeling used in the figure, Eq. (6.10) is seen to be a Thevenin point equation at the collector terminal C, and Eq. (6.11) is a partial loop equation from the base terminal B to the emitter terminal E. These two equations thus define a four-terminal linear element model for the transistor. The four parameters listed in order of importance to the typical circuit analysis are given in Table 6.1; typical values for a small-signal transistor are also given.

Although manufacturers list the nominal or average values of the hybrid parameters for specific transistor types, the variation between samples of the same type is large (greater than a factor of 2). As a result, detailed calculations based on this

Table 6.1 Hybrid Parameters

Symbol	Name	Typical Values for a Small-Signal Transistor
h_{fe}	Forward current ratio	10^2
h_{ie}	Input impedance	$2 \times 10^3 \ \Omega$
h_{oe}	Output admittance	$2 \times 10^{-5} \ \Omega^{-1}$
h_{re}	Reverse voltage ratio	10^{-4}

model are justified only in those special cases where the parameter values of a particular transistor have been determined.

Rather than use this full hybrid parameter model in our discussion of transistor circuits, we will introduce several approximations that greatly simplify the calculations. A couple of examples based on the hybrid parameter model will introduce the level of approximation to be used in the remainder of this chapter. Using the load resistor R_L indicated in Figure 6.7, we can relate the output voltage v_{CE} to the collector current i_C:

$$v_{CE} = -R_L i_C \tag{6.12}$$

Substituting this into Eq. (6.10) gives an expression that can be put in the form

$$i_C = \frac{h_{fe} i_B}{1 - h_{oe} R_L} \tag{6.13}$$

Now we can see that if R_L is such that $h_{oe} R_L \ll 1$, the collector and base currents can be related by the expression

$$\boxed{i_C = h_{fe} i_B} \tag{6.14}$$

which is the AC equivalent of Eq. (6.1). Since h_{oe} is of the order of $10^{-5}\ \Omega^{-1}$ (mho), the approximation is good to about 10% if R_L is less than $10^4\ \Omega$. This argument is exactly equivalent to making the requirement that $i_2 \ll i_C$ in the output loop of Figure 6.7b.

By a similar argument, we can show that the term h_{re} can also be dropped. Using either Eq. (6.11) or the equivalent circuit model of Figure 6.7b, the current in the input loop of the equivalent circuit is found to be

$$i_B = \frac{v_{BE} - h_{re} v_{CE}}{h_{ie}} \tag{6.15}$$

With Eq. (6.14), we can relate the output voltage v_{CE} to the base current i_B:

$$v_{CE} = -R_L i_C = -R_L h_{fe} i_B \tag{6.16}$$

Substitution of i_B from Eq. (6.15) and a little algebra yields the voltage ratio

$$\frac{v_{CE}}{v_{BE}} = -\frac{R_L h_{fe}}{h_{ie} - R_L h_{fe} h_{re}} \tag{6.17}$$

For a load resistor of $10^4\ \Omega$, substitution of the nominal values of h_{fe} and h_{re} from Table 6.1 shows that the second term in the denominator is about $10^2\ \Omega$. Since h_{ie} is about 10^3, the second term can again be dropped with about 10% error to yield an approximate expression for the AC voltage gain of

$$\frac{v_{CE}}{v_{BE}} = -\frac{h_{fe}}{h_{ie}} R_L \tag{6.18}$$

These approximations have effectively eliminated the $1/h_{oe}$ resistance in the output loop of Figure 6.7b and the $h_{re} v_{CE}$ voltage source from the input loop.

Example 6.1 Equation 6.17 can be made exact by replacing h_{fe} with $h_{fe}/(1 + h_{oe}R_L)$. For $R_L = 100\ \Omega$, $1\ k\Omega$, $10\ k\Omega$, and $100\ k\Omega$, make a table showing v_{CE}/v_{BE} from this exact equation and the approximation of Eq. (6.18). Use the nominal h parameter values from Table 6.1.

R_L	$h_{fe}/(1 + h_{oe}R_L)$	v_{CE}/v_{BE}(exact)	v_{CE}/v_{BE}(approx.)
$100\ \Omega$	99.8	-5.00	-5.00
$1\ k\Omega$	98.0	-49.2	-50.0
$10\ k\Omega$	83.3	$-435.$	$-500.$
$100\ k\Omega$	33.3	$-1998.$	$-5000.$

Thus for voltage gains as high as several hundred volts, the simplest approximation works quite well.

6.2.4 Ideal and Perfect Bipolar Transistor Models

By making use of the preceding approximations we can define a simplified AC model for the transistor that is independent of circuit configuration. By ignoring the effects of h_{oe} and h_{re}, we can rewrite Eqs. (6.10) and (6.11) as

$$i_C = h_{fe}i_B \tag{6.19}$$

and

$$v_{BE} = h_{ie}i_B \tag{6.20}$$

Note that there is no *PN* voltage drop in these equations; that is a DC effect and these equations describe only AC signals. The first of these equations defines the relationship between base and collector currents as shown in Figure 6.8a. Since i_C is typically 100 times larger than i_B, we can make the further approximation that $i_E = i_C$ (also, $I_E = I_C$).

Simplifying further, we define an "ideal" transistor as one that has $h_{ie} = 0$; Eq. (6.20) then says that $v_B = v_E$ and leads to the equivalent circuit model of Figure 6.8b. When the effects of h_{ie} cannot be ignored, we can use the "perfect" transistor model of Figure 6.8c; this model satisfies both of the above equations and is needed whenever there is no other impedance to limit the base current significantly. Figure 6.8d shows an alternative form of the perfect model where h_{ie} in the base leg has been replaced by an equivalent, but much smaller, resistance in the emitter leg.

With some experience, it is not necessary to replace the transistor symbol with the current source model of Figure 6.8b; the transistor symbol itself can take on the meaning of this "idealized" equivalent circuit. When the perfect transistor model is needed, the resistor h_{ie} or h_{ie}/h_{fe} can be added directly to the ideal transistor symbol as shown in Figure 6.13b. In the following sections we will adopt the rule that the transistor symbol appearing in an AC equivalent circuit follows the ideal model; if the perfect model is required, h_{ie} will be shown explicitly.

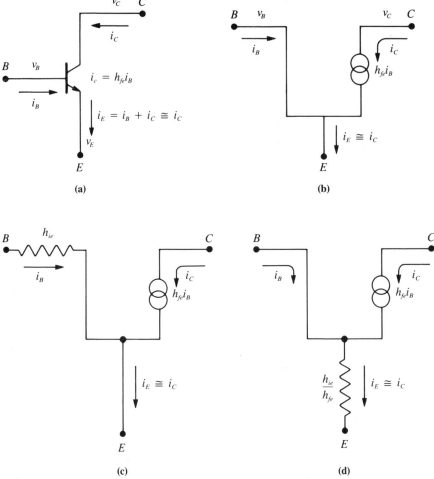

Figure 6.8 *(a) The currents in an NPN transistor. (b) The model for an ideal transistor. (c) The perfect transistor model includes the effects of h_{ie}. (d) An equivalent version of the perfect model uses a smaller emitter resistor instead of h_{ie} in the base.*

6.2.5 Ideal Transistor Working Rules

The DC model of Figure 6.5b, sporting a real diode D, is consistent with the ideal AC model of Figure 6.8b: We need only assume that the diode is sufficiently forward-biased to show a constant *PN* voltage drop with an insignificant forward resistance. This combined model yields the two basic rules governing the operation of the ideal transistor in a circuit:

1. The base and emitter are at the same AC voltage and differ only by a constant DC potential V_{PN}.

2. The collector current is equal to the emitter current and proportional to the base current.

When applying these rules to an AC equivalent circuit, remember that the AC potential v_{PN} is zero as long as the transistor is biased for linear operation.

It may appear that we have discarded too much in these approximations, but remember that the goal is to facilitate the estimation of voltage gains, current gains, input impedances, output impedances, and corner frequencies. These minimal models (ideal and perfect) still include the all-important parameter h_{fe} and generally make reasonably accurate predictions. In practice, the small-signal requirement, $i_C \ll I_C(DC)$, is often stretched to the limit, and it makes little sense to worry excessively over hybrid parameter details.

6.2.6 Transconductance Model

An alternative description of transistor operation uses the forward transconductance $g_m = h_{fe}/h_{ie}$, which is the hybrid parameter ratio appearing in the transfer function of Eq. (6.18). Using the small-signal model of Figure 6.8b, we can derive the basic relationship between g_m and the voltage and current signals. The base current in Figure 6.8b is

$$i_B = \frac{v_B - v_E}{h_{ie}} \tag{6.21}$$

and the collector current is

$$i_C = h_{fe}i_B \tag{6.22}$$

Substitution of i_B from the previous expression gives

$$i_C = \frac{h_{fe}}{h_{ie}} (v_B - v_E)$$

or just

$$i_C = g_m(v_B - v_E) = g_m v_{BE} \tag{6.23}$$

The transconductance parameter thus relates the current between collector and emitter to the applied voltage between base and emitter: The input voltage is controlling the conductance (inverse resistance) of the high-current path through the transistor. Using the nominal h-parameter values from Table 6.1, we get a g_m of $50 \times 10^{-3}\ \Omega^{-1} = 50$ mmhos. We will not make any immediate use of this relation, but since transconductance is used to describe field effect transistors, it is sometimes convenient to apply the same parameter to bipolar transistors.

6.3 The Common Emitter Amplifier

Of the three transistor circuit configurations, the common emitter is the most versatile. It features a low input impedance, a moderate output impedance, and can produce both a voltage gain and a current gain between the input and output terminals. The following sections will first outline the additional circuit connections necessary to establish the DC bias for the common emitter amplifier, then apply the simple transistor model to the AC analysis of various amplifiers. Two types of negative feedback will be discussed: current feedback from collector to base, and voltage feedback from emitter to base.

6.3.1   DC Biasing

The simplest circuit for a common emitter amplifier is shown in Figure 6.9a. In this circuit, the only path for DC bias current into the base is through R_B. Since the base voltage is only a diode drop V_{PN} above the grounded emitter, the current through R_B and into the base is

$$I_B = \frac{V_{CC} - V_{PN}}{R_B} \qquad (6.24)$$

Figure 6.9   *(a) Simplest form of CE amplifier uses R_B to provide DC bias current and set the DC operating point. (b) This alternative circuit gives an operating point that is less dependent on h_{FE}. (c) Another variation that works with smaller values of R_F.*

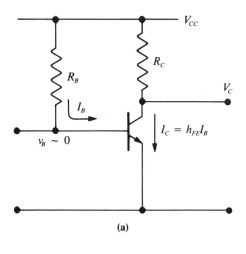

(a)

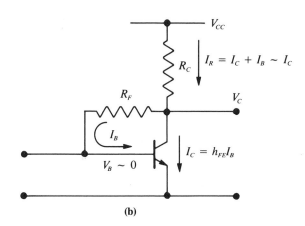

(b)

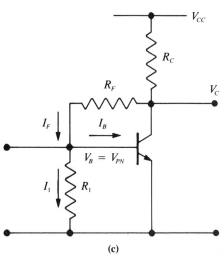

(c)

Since V_{CC} is a power supply voltage generally greater than 10 V, V_{PN} can often be dropped to give

$$I_B = \frac{V_{CC}}{R_B} \qquad (6.25)$$

If h_{FE} is known for the transistor, then this base current can be used to predict the collector current following

$$I_C = h_{FE}I_B \qquad (6.26)$$

The DC voltage at the collector is determined by the power supply voltage and the Ohm's law voltage drop across resistor R_C, and can be written as

$$V_C = V_{CC} - R_CI_C \qquad (6.27)$$

or
$$V_C = V_{CC} - R_Ch_{FE}I_B \qquad (6.28)$$

or
$$V_C = V_{CC}\left(\frac{1 - h_{FE}R_C}{R_B}\right) \qquad (6.29)$$

This average collector voltage should be large enough to provide at least a 2-V drop between collector and emitter and clearly must be less than V_{CC}. In the absence of other circuit requirements, a convenient algebraic choice for V_C is $V_{CC}/2$. This results in a particular relationship

$$R_B = 2h_{FE}R_C \qquad (6.30)$$

which then establishes the relative sizes of the two resistors.

Example 6.2   On the circuit of Figure 6.9a, let $V_{CC} = 10$ V and $h_{FE} = 100$. Determine R_C and R_B such that $V_C = 6$ V and $I_C = 1$ mA.
   When the collector current is 1 mA, the voltage drop across R_C must be 4 V. Independent of any other condition, this requires a resistor given by

$$R_C = \frac{10 \text{ V} - 6 \text{ V}}{1 \times 10^{-3} \text{ A}} = 4 \times 10^3 \, \Omega = 4 \text{ k}\Omega$$

With $h_{FE} = 100$, a collector current of 1 mA requires a base current of 10 μA. Solving Eq. (6.25) for R_B gives

$$R_B = \frac{10 \text{ V} - 0.6 \text{ V}}{10 \times 10^{-6} \text{ A}} = 9.4 \times 10^5 \, \Omega = 940 \text{ k}\Omega$$

   In practice, this will be only a first guess for R_B, because h_{FE} will not be known exactly. For example, if, after the circuit is built, V_C turns out to be too high, it means that I_C and I_B are too small and a smaller value for R_B is needed.

   Although this circuit works reasonably well for a silicon transistor in a temporary application where the resistor R_B can be chosen to match the h_{FE} of a particular transistor, it has several problems as a general-purpose design. Foremost is the fact that h_{FE} is quite variable among samples of the same transistor type. Since the transistor may need replacing at some later date, a well-designed circuit should have an operating point that is less dependent on this parameter. There are several ways this can be achieved.

One method is to connect the base-biasing resistor to the collector instead of to V_{CC}. This connection is shown in Figure 6.9b, where the bias resistor is labeled R_F. Since an increase in the base current to the transistor results in a decrease in the collector voltage and a corresponding decrease in the base current through R_F, this connection amounts to negative feedback. With the base-biasing resistor connected in this manner, and again neglecting V_{PN}, we get a base current of

$$I_B = \frac{V_C}{R_F} \tag{6.31}$$

The DC current in the collector resistor I_R is the sum of the base current and the collector current,

$$I_R = I_C + I_B = (h_{FE} + 1)I_B \tag{6.32}$$

but since $h_{FE} \gg 1$, this expression can be approximated by the collector current alone, $I_C = h_{FE}I_B$. The voltage at the collector is thus given by

$$V_C = V_{CC} - \frac{R_C h_{FE} V_C}{R_F} \tag{6.33}$$

Solving for V_C gives

$$V_C = \frac{R_F}{R_F + h_{FE}R_C} V_{CC} \tag{6.34}$$

To attain the nominal DC operating point of $V_C = V_{CC}/2$, it is clear that we should have $R_F = h_{FE}R_C$. Under this condition, the change in V_C for a given change in h_{FE} is only half that of the previous design.

Example 6.3 Assuming that $V_{CC} = 15$ V, $R_C = 5$ kΩ, and $h_{FE} = 100$ in Figure 6.9b, use Eq. (6.34) to find an R_F that will make $V_C = 7.5$ V; then use an iterative procedure to find the actual operating point when $V_{PN} = 0.6$ V.
From the discussion after Eq. (6.34), we estimate that

$$R_F = h_{FE}R_C = 500 \text{ k}\Omega$$

If the base is really at a potential of 0.6 V while $V_C = 7.5$ V, the current into the base through R_F is

$$I_B = \frac{V_{CC} - V_{PN}}{R_F} = \frac{7.5 - 0.6}{5 \times 10^5} = 13.8 \text{ }\mu\text{A}$$

However, this base current predicts a higher V_C:

$$V_C = V_{CC} - R_C h_{FE} I_B = 15 - 5000(100)(13.8 \times 10^{-6}) = 8.1 \text{ V}$$

Splitting the difference between the first estimate and this one gives us a collector voltage of $V_C = 7.8$ V, and this yields a base current of

$$I_B = \frac{7.8 - 0.6}{5 \times 10^5} = 14.4 \text{ }\mu\text{A}$$

which in turn predicts that $V_C = 7.8$ V. The iteration has converged more quickly than we had a right to expect, and shows that neglecting V_{PN} has a small effect on the calculated operating point. Normal uncertainties in h_{FE} have a much larger effect.

A more common bias stabilization technique employs a series resistor between the emitter and ground as shown in Figure 6.10a. The base-to-ground connection through the transistor can be replaced by an equivalent resistor as shown in Figure 6.10b. The value of this resistor is just the DC input impedance of the transistor between the base and ground. Remember that the base-collector junction is reverse-biased and presents a high impedance as viewed from the base and therefore need not be considered in this analysis. The input impedance to the base is just V_B/I_B, and taking the base and emitter to be at the same voltage (neglecting V_{PN}), we have

$$R_{in} = \frac{V_B}{I_B} = \frac{V_E}{I_B}$$

$$= \frac{R_E I_C}{I_B}$$

(6.35)

But the collector current is related to the base current by h_{FE}, giving for the input impedance

$$R_{in} = h_{FE}R_E$$

(6.36)

This is an extremely important observation, because it allows us to replace the transistor in the base circuit with an equivalent resistance. As a result, the base voltage can be calculated from a resistance network involving only the currents on the base side of the transistor as shown on the schematics of Figure 6.11. Since the base and emitter are at the same DC voltage (neglecting the V_{PN} diode drop), the base voltage calculated from these networks is also the emitter voltage developed by I_C across R_E on the corresponding circuit.

Using the equivalent resistance drawing of Figure 6.11b, the base voltage can be written

$$V_B = \frac{h_{FE}R_E}{R_B + h_{FE}R_E} V_{CC}$$

(6.37)

Figure 6.10 (a) A transistor circuit that includes an emitter resistor. (b) The equivalent input resistance viewed from the base is the emitter resistor times the DC current gain.

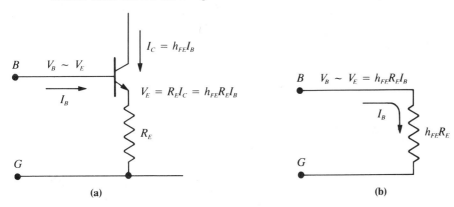

(a)   (b)

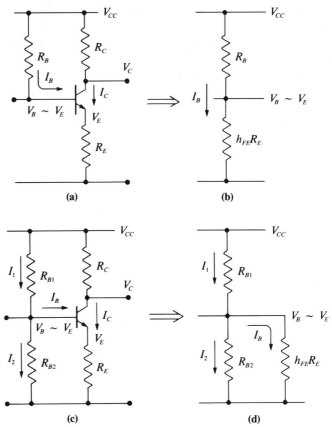

Figure 6.11 *(a) The CE amplifier with emitter resistor and (b) the equivalent circuit seen from the base. (c) An improved biasing scheme using two base resistors, and (d) its equivalent circuit seen from the base.*

Since this is also the emitter voltage developed by the current through R_E, the emitter and collector current can be written as

$$I_C = \frac{h_{FE}}{R_B + h_{FE}R_E}V_{CC} \tag{6.38}$$

The resulting voltage drop across R_C is given by

$$V_{CC} - V_C = R_C I_C \tag{6.39}$$

and if Eq. (6.38) is used to replace I_C, this expression can be solved for V_C to give

$$V_C = V_{CC}\left(1 - \frac{h_{FE}R_C}{R_B + h_{FE}R_E}\right) \tag{6.40}$$

If the resistors are chosen to again produce $V_C = V_{CC}/2$, this circuit has about the same sensitivity to changes in h_{FE} as the circuit of Figure 6.9b.

A further improvement can be made by introducing a second base bias resistor R_{B2} as shown in Figure 6.11c. It is clear from the equivalent resistor circuit of Figure 6.11d that if $R_{B2} \ll h_{FE}R_E$, then the base voltage is determined almost entirely by the two base resistors R_{B1} and R_{B2}. Using the equivalent resistance drawing of Figure 6.11d, which is a voltage divider consisting of R_{B1} and the parallel combination of R_{B2} and $h_{FE}R_E$, the base voltage of this circuit is found to be

$$V_B = \frac{R_{B2}h_{FE}R_E}{R_{B1}(R_{B2} + h_{FE}R_E) + h_{FE}R_ER_{B2}}V_{CC} \qquad (6.41)$$

and if $R_{B2} \ll h_{FE}R_E$, this reduces to

$$V_B = \left(\frac{R_{B2}}{R_{B1} + R_{B2}}\right)V_{CC} \qquad (6.42)$$

In this limit the base voltage is no longer dependent on h_{FE}, making the emitter voltage, the collector current, and the collector voltage also independent of h_{FE}.

The DC biasing methods just outlined for the common emitter amplifier work equally well for transistors in the common collector or common base configuration.

Example 6.4 The circuit of Figure 6.11c has $R_{B1} = R_{B2} = 50\ \text{k}\Omega$ and $R_C = R_E = 1\ \text{k}\Omega$. Determine the operating voltage V_E if $V_{CC} = 6$ V and $h_{FE} = 100$.

If R_{B2} and $h_{FE}R_E$ are combined to make a parallel resistor $R_p = 33.33\ \text{k}\Omega$, then the equivalent circuit of Figure 6.11d can be evaluated as a voltage divider to give

$$\begin{aligned}
V_E &= \frac{R_pV_{CC}}{R_p + R_{B1}} \\
&= \frac{(33.3 \times 10^3\ \Omega)(6\ \text{V})}{33.3 \times 10^3\ \Omega + 50 \times 10^3\ \Omega} \\
&= 2.4\ \text{V}
\end{aligned}$$

But this result ignores the base-emitter voltage drop, an effect that is made more significant by the small size of V_{CC}.

The base-emitter voltage drop V_{PN} can be included by drawing the somewhat more complicated equivalent circuit of Figure 6.12. Using the currents shown on that figure, we can write the equations

$$V_B = V_{PN} + I_Bh_{FE}R_E$$

$$V_B = (I_1 - I_B)R$$

and
$$V_B = V_{CC} - I_1R$$

Combining these to eliminate I_1 and I_B, we obtain

$$V_B = \frac{RV_{PN} + R_{FE}R_EV_{CC}}{R + 2h_{FE}R_E}$$

and with V_{PN} taken as 0.6 V, this expression evaluates to

$$V_B = \frac{30 \times 10^3 + 600 \times 10^3}{50 \times 10^3 + 200 \times 10^3} = 2.52\ \text{V}$$

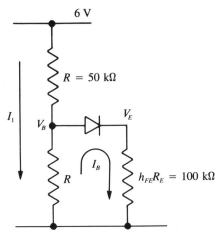

Figure 6.12   The DC equivalent circuit for Example 6.4.

The emitter voltage is then $V_E = 1.92$ V, implying an emitter (and collector) current of 1.9 mA and a collector voltage of 4.1 V. Thus, the careful calculation predicts that $V_{CE} = 2.2$ V, whereas the same argument applied to the first estimate of V_E gives only 1.2 V.

6.3.2   Equivalent Circuit for AC Signals

The mathematical analysis of a transistor amplifier's response to an AC input signal is greatly facilitated by the use of the AC equivalent circuit. The most important point in deriving this AC equivalent circuit is that all constant voltage sources are replaced with wires and all constant current sources with open circuits. This is an obvious consequence of the fact that such constant sources do not develop any AC signal and would seem to be a reasonably simple step. In practice, this replacement so dramatically alters and simplifies the circuit configuration that it may be mistaken for some more profound operation.

6.3.3   Approximate AC Model

The circuit shown in Figure 6.13a is the basic common emitter amplifier using the simplest biasing method. Lowercase letters indicate the AC voltages and currents; the total signal at any point is of course the sum of the DC and AC components. Because it is constant, the power supply voltage V_{CC} is an AC ground indistinguishable from the normal ground of the circuit. We can therefore relocate the upper end of R_B and R_C to the common ground line as shown in Figure 6.13b.

Figure 6.13c shows the common emitter amplifier redrawn as an AC equivalent circuit; note that V_{CC} and the ground now share a common wire. In this example the physical transistor is being modeled as a perfect transistor: The transistor symbol is ideal and h_{ie} is shown explicitly. It must be emphasized that in general the signal current i_s is different from the base current i_B, because some of i_s flows directly to ground through R_B. However, for this particular circuit the biasing resistor R_B will always be much larger than h_{ie}, making i_B and i_s essentially the

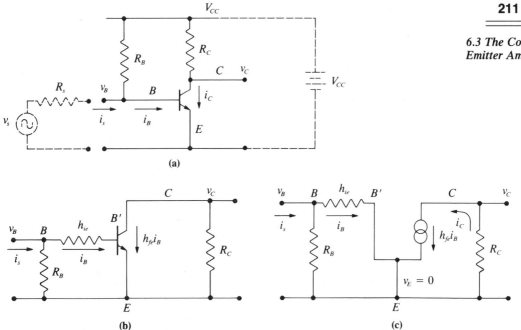

Figure 6.13 *(a) The basic CE amplifier. (b) The AC equivalent circuit
drawn using an ideal transistor symbol with h_{ie} shown explicitly. (c)
The same AC equivalent using the perfect transistor equivalent circuit
of Figure 6.8c.*

same. Following the arguments used in Figure 6.8, this AC equivalent circuit can
also be drawn using only linear circuit elements as shown in Figure 6.13c. On
both of the AC equivalent figures, $B'CE$ is an ideal transistor and BCE is a perfect
transistor.

The four-terminal networks of Figures 6.13b and 6.13c are now cast in the
general form. With our ideal and perfect transistor approximations, the backward
transfer function **B** is always zero, but we will have to calculate the other three
elements: input impedance, output impedance, and forward transfer function
$\mathbf{v_C}/\mathbf{v_B}$. So far in this chapter we have been dealing only with resistive impedances
and real voltages, thereby making the transfer function real as well. However, no
generality has been lost; resistors can be replaced by complex impedances in any
of the AC equations.

6.3.4 The Basic CE Amplifier

We can now use either of the AC equivalent circuits shown in Figures 6.13b and
6.13c to calculate the AC voltage gain between the base and collector of the
common emitter circuit in Figure 6.13a. The base voltage is developed across the
input resistor h_{ie} and is seen to be

$$v_B = h_{ie}i_B \qquad (6.43)$$

The collector voltage can be similarly expressed as the voltage drop across the resistor R_C:

$$0 - v_C = R_C h_{fe} i_B \tag{6.44}$$

Note that by carefully writing this equation as a voltage difference, we naturally obtain the minus sign in the following equation. Using Eq. (6.43) to eliminate i_B, we can write the amplifier's voltage transfer function between the base and collector as

$$\frac{v_C}{v_B} = -\frac{h_{fe} R_C}{h_{ie}} \tag{6.45}$$

The minus sign in this expression indicates that the voltage signal at the collector is inverted (180° out of phase) with the signal on the base. As expected, this result agrees with Eq. (6.18), which was obtained directly from the traditional hybrid model with the effects of h_{re} and h_{oe} ignored.

The input and output impedances of the common emitter amplifier can be deduced directly from Figure 6.13b and 6.13c. The input impedance R_{in} to this amplifier circuit is just the parallel combination of R_B and h_{ie}, and since h_{ie} is usually much smaller than R_B, R_{in} generally reduces to just the input impedance of the transistor itself, namely, h_{ie}. These figures also show that the circuit's output impedance is the collector resistor R_C. Note that the voltage gain given by Eq. (6.45) is the open circuit gain of the amplifier, and that v_C is the open circuit output voltage of the amplifier. This output voltage and the output impedance of the amplifier thus form the elements of a Thevenin equivalent circuit.

When this circuit is driven by a signal source v_s that has an output impedance R_s as shown in Figure 6.13a, some of the signal voltage v_s will be lost due to the Ohm's law drop in R_s. This means that the overall voltage gain $|v_C/v_s|$ will generally be less than the bare transistor's gain $|v_C/v_B|$. If the effect of R_B on the input impedance can be ignored, then the current i_s from the signal source is approximately equal to the current i_B into the base, and the base-to-signal source voltage ratio is

$$\frac{v_B}{v_s} = \frac{h_{ie}}{R_s + h_{ie}} \tag{6.46}$$

Combining this ratio with Eq. (6.45) gives the overall transfer function between the signal source and collector,

$$H = \frac{v_C}{v_s} = -\frac{h_{fe} R_C}{R_s + h_{ie}} \tag{6.47}$$

the magnitude of which is indeed less than the magnitude of Eq. (6.45). Note that this expression has the same form as Eq. (6.45) if the denominator is interpreted as the total series impedance between the input signal (v_B in the first case, v_s in the last) and the base. This substitution of $h_{ie} + R_s$ for h_{ie} to obtain the overall voltage gain will work for any circuit where $i_s \simeq i_B$. When the base bias resistors cannot be ignored, they can be treated as part of the signal source. The circuit can then be reduced to the simple form used here by developing a Thevenin equivalent circuit for the signal source plus bias resistors.

6.3.5   *High-Frequency Operation*

The high-frequency operation of the common emitter amplifier is limited by the parasitic capacitance between the collector and base as shown in Figure 6.14. This capacitor provides a path by which the large and inverted signal at the collector drives a feedback current into the base. Other examples of negative feedback

Figure 6.14 *(a) Because of the physical proximity of the collector and base connections within the transistor, a parasitic capacitance C_F must be included in the high-frequency analysis. (b) The base-to-collector voltage gain of this amplifier looks like a low-pass filter, and (c) its input impedance also varies with frequency.*

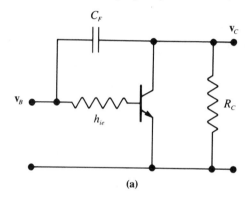

(a)

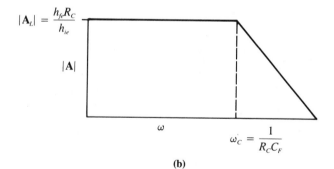

(b)

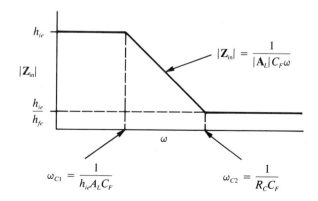

(c)

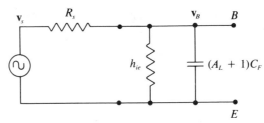

*Figure 6.15   The equivalent input circuit for a signal source driving a
common emitter transistor and its base-to-collector parasitic
capacitance C_F.*

between collector and base are shown in Figures 6.9b,c. The AC analysis of the
general case is given in Appendix C.

However, the corner frequency of the typical *CE* amplifier can be estimated
from the Miller effect derived for a general inverting amplifier in Appendix B.
The low-frequency pass band is most restricted when a large voltage gain,

$$A = \left| \frac{v_C}{v_B} \right| \tag{6.48}$$

exists between the base and collector. Because of this gain, the normal h_{ie} input
impedance of the amplifier is reduced by a parallel Miller impedance $\mathbf{Z}_M$,

$$\mathbf{Z}_M = \frac{\mathbf{Z}_F}{1 + \mathbf{A}} \tag{6.49}$$

as described in Appendix B and shown in Figure 6.15. The parasitic capacitance
between collector and base provides the impedance

$$\mathbf{Z}_F = \frac{1}{j\omega C_F} \tag{6.50}$$

If R_F in Eq. (C.3) is replaced by this $\mathbf{Z}_F$, we find that the voltage gain $|\mathbf{A}|$ has the
frequency dependence shown in Figure 6.14b. At low frequencies the magnitude
of the voltage gain is a constant,

$$A_L = \frac{h_{fe} R_C}{h_{ie}} \tag{6.51}$$

but above the corner frequency ω_c, the gain declines at 6 dB/octave.

Assuming that ω_c is at a relatively high frequency, we can use A_L in place of
$\mathbf{A}$ in Eq. (6.49) to get

$$\mathbf{Z}_M = \frac{1}{j\omega A_L C_F} \tag{6.52}$$

where we have also taken $A_L \gg 1$. The total input impedance is the parallel
combination of $\mathbf{Z}_M$ and h_{ie}, and displays the frequency dependence shown in Figure 6.14c. The corner frequency ω_{c1} is given by

$$\omega_{c1} = \frac{1}{h_{ie} A_L C_F} \tag{6.53}$$

or
$$\omega_{c1} = \frac{1}{h_{fe}R_C C_F} \qquad (6.54)$$

and is seen to be at a lower frequency than the corner in $|\mathbf{A}|$.

The capacitive component of the *CE* transistor's input impedance forms a low-pass filter when it is driven as shown in Figure 6.15. If $A_L \gg 1$, the transfer function between the source and base is

$$\mathbf{v}_B/\mathbf{v}_s = \frac{h_{ie}}{h_{ie} + R_s(1 + j\omega\, h_{ie} A_L C_F)} \qquad (6.55)$$

and is seen to have a corner frequency at

$$\omega_1 = \frac{h_{ie} + R_s}{h_{ie} R_s A_L C_F} \qquad (6.56)$$

If $R_s \gg h_{ie}$, this expression reduces to

$$\omega_1 = \frac{1}{h_{ie} A_L C_F} \qquad (6.57)$$

a frequency that is lower than might be expected because of the Miller-enhanced capacitance $A_L C_F$.

Example 6.5   A transistor with $h_{ie} = 2000\ \Omega$ and a base-to-collector stray capacitance of 5 pF is used to build a *CE* circuit like Figure 6.13a. If the amplifier's gain $|v_C/v_B|$ at low frequency is measured to be 1000, determine the high-frequency corner when the amplifier is driven by a 100-kΩ output impedance source.

The equivalent input circuit to this transistor is as shown in Figure 6.15. It forms a low-pass filter with a corner frequency given by Eq. (6.56). Evaluating that expression with $R_s = 10^5\ \Omega$, $C_F = 5$ pF, and $A_L = 1000$ gives

$$\omega_1 = \frac{2000 + 10^5}{(2 \times 10^3)(10^5)(10^3)(5 \times 10^{-12})} = 10^5 \text{ rad/s}$$

or

$$f_1 = 10.6 \text{ kHz}$$

Use of a signal generator with a smaller output impedance will yield a higher corner frequency. If R_s is 2000 Ω, the corner moves up to over 30 kHz, and would go to infinity if R_s were zero. As we shall see in the next chapter, most of the Miller effect can be eliminated by using the two-transistor cascode circuit.

6.3.6   *CE Amplifier with Emitter Resistor*

The *CE* amplifier is often constructed with an emitter resistor R_E as in Figure 6.11. This resistor provides a form of negative feedback that can be used to stabilize both the DC operating point and the AC gain. If the base-biasing resistors are omitted (considered as part of the signal source), the circuits of Figure 6.11 have the common AC equivalent shown in Figure 6.16a. This AC equivalent can be analyzed with the same methods used in the DC analysis. The voltage at the

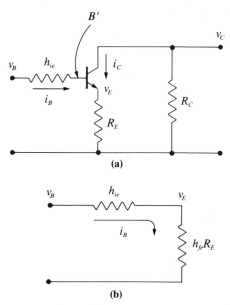

Figure 6.16 *(a) The AC equivalent of the CE amplifier with an
emitter resistor added for stability. The base-biasing resistors have
been omitted. (b) The AC equivalent circuit using the transistor symbol
and h_{ie}.*

emitter is the same as the voltage at B' on the base of the idealized transistor
symbol shown in this figure. As a result, the input impedance looking into the
transistor at B' is

$$R_{\text{in}} = \frac{v_E}{i_B} \tag{6.58}$$

Since the AC current through R_E is much larger than the base current i_B, the
voltage v_E will also be much larger than would result if there was base current
alone in R_E. The input impedance is thus much larger than R_E.

Neglecting, as usual, the direct contribution of the base current, the voltage on
the emitter can be written in terms of the collector current,

$$v_E = R_E i_C \tag{6.59}$$

or the base current,

$$v_E = R_E h_{fe} i_B \tag{6.60}$$

Putting this last result into Eq. (6.58) reduces the input impedance to circuit parameters only:

$$R_{\text{in}} = h_{fe} R_E \tag{6.61}$$

Including the effects of h_{ie} makes the input impedance look like the equivalent
circuit of Figure 6.16b. Since h_{ie} is small, it can usually be neglected whenever
the emitter impedance is more than a few hundred ohms.

Using this expression for R_{in} and recalling that $v_E = v_B'$, it is a simple exercise to find that the voltage transfer function across this transistor is

$$\frac{v_C}{v_B} = -A = -\frac{h_{fe}R_C}{h_{ie} + h_{fe}R_E} \tag{6.62}$$

If $h_{fe}R_E >> h_{ie}$, the gain becomes independent of the hybrid parameters:

$$\frac{v_C}{v_B} = -A = -\frac{R_C}{R_E} \tag{6.63}$$

Because it is unaffected by variations in the hybrid parameters, this result is valid even for a large-amplitude signal.

For small input signals it is often desirable to retain the large voltage gain of the basic *CE* amplifier even though an emitter resistor is used for DC stability. This combination results if a large capacitor C_E is used to bypass the AC signal around the emitter resistor as shown in Figure 6.17. Equation (6.62) will yield the

Figure 6.17 (a) A bypass capacitor on the emitter resistor is used to recover the full AC gain of the circuit at high frequencies. (b) The AC equivalent circuit. (c) The transfer function of this amplifier has two corner frequencies.

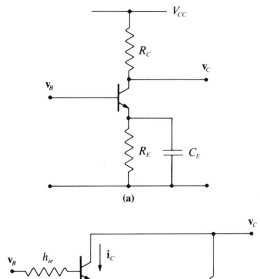

(a)

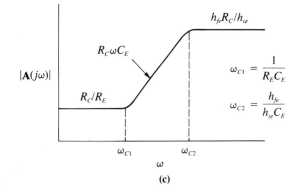

(b)

(c)

correct frequency-dependent transfer function if R_E is replaced by a complex impedance:

$$Z_E = \frac{R_E}{1 + j\omega R_E C_E} \tag{6.64}$$

The magnitude of the resulting transfer function is shown in Figure 6.17c; it is similar to a high-pass filter, with R_E setting the gain at low frequencies and the capacitor maximizing the gain at high frequencies.

6.4 The Common Collector Amplifier

The common collector amplifier is sometimes called an emitter follower because the output voltage signal at the emitter is nearly identical to the voltage signal input on the base. Although this amplifier's voltage gain is thus always less than unity, it has a large current gain and is normally used to match a high-impedance source to a low-impedance load. The common collector amplifier can be DC-biased by any of the methods described above for a common emitter circuit with an emitter resistor.

6.4.1 Voltage Gain

A typical common collector amplifier is shown in Figure 6.18a. Note that although the output is taken from the emitter terminal, the "commonness" of the collector is more obvious on the AC equivalent circuit of Figure 6.18b. The hybrid parameters used to model the transistor in this configuration normally have a c subscript replacing the e used with the common emitter circuit. However, the differences in the parameters are quite small, and we will continue to use the common emitter parameters.

The input impedance circuit of Figure 6.18c is unchanged from those used earlier for a common emitter amplifier with emitter resistor R_E. Using this equivalent circuit, the voltage gain from base to emitter can be written as

$$\frac{v_E}{v_B} = \frac{h_{fe}R_E}{h_{ie} + h_{fe}R_E} \tag{6.65}$$

The gain is thus seen to be in phase and slightly less than 1. For a typical circuit, R_E will be somewhere between 100 Ω and 10 kΩ, giving a gain between 0.8 and 1.0.

6.4.2 Input and Output Impedances

The input impedance of the circuit in Figure 6.18a is easily obtained from the equivalent circuit of Figure 6.18c. As can be seen from the figure, the input impedance to the circuit is nothing more than the parallel combination of the

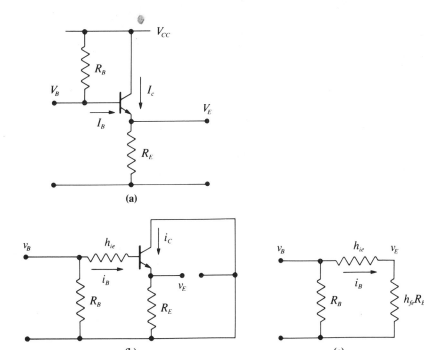

Figure 6.18 *(a) The basic common collector amplifier. (b) An AC
equivalent circuit for this amplifier. (c) The equivalent input impedance
of this circuit.*

input impedance to the transistor $(h_{ie} + h_{fe}R_E)$ in parallel with the base bias resistor. The complete expression is

$$R_{in} = \frac{R_B(h_{ie} + h_{fe}R_E)}{R_B + h_{ie} + h_{fe}R_E} \qquad (6.66)$$

If R_E is greater than 100 Ω, it is clear that h_{ie} can be neglected with little effect. One of the main advantages of this circuit is its large input impedance, but unfortunately the base bias resistor R_B reduces the overall input impedance.

A more complete circuit with the signal generator included is shown in Figure 6.19. From Figure 6.19b it is clear that the circuit's input impedance R_{in} looking into the *AB* terminals is given by Eq. (6.66) with R_B replaced by the parallel combination of R_{B1} and R_{B2}. If R_{in} is used to draw the equivalent circuit shown in the right part of Figure 6.19b, we can write

$$\frac{\mathbf{v}_B}{\mathbf{v}_s} = \frac{R_{in}}{R_{in} + R_s + \mathbf{Z}_C} \qquad (6.67)$$

Thus, even though $\mathbf{v}_E/\mathbf{v}_B$ is approximately 1, the overall voltage gain $\mathbf{v}_E/\mathbf{v}_s$ may be much less than 1 when R_s is large.

The output impedance of the common collector amplifier can be substantially less than the output impedance of the driving signal source and is an important feature of this amplifier. Depending on the relative sizes of the various resistors,

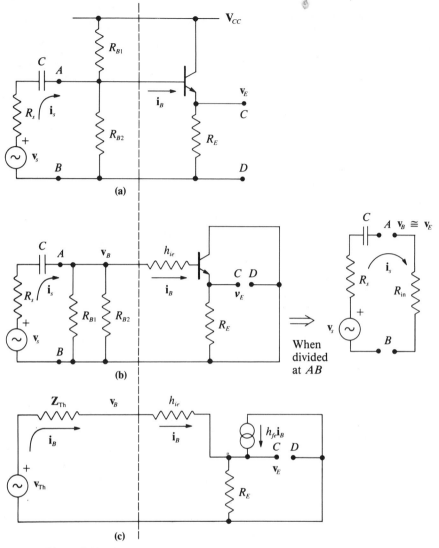

Figure 6.19 *(a) A complete CC circuit driven by a signal generator of output impedance R_s. (b) The AC equivalent circuit. (c) A Thevenin equivalent circuit for the signal generator and biasing resistors simplifies the appearance of the AC equivalent circuit.*

the small-signal output impedance can be either R_E or essentially independent of R_E.

To obtain the most useful output impedance expression, we must divide the circuit of Figure 6.19 along the dashed line and associate the base bias resistors with the signal generator. The left portion of the circuit can be replaced by a Thevenin equivalent as shown in Figure 6.19c, but since we are here only concerned with the circuit's performance above the corner frequency, we will replace $\mathbf{Z}_{Th}$ with R_{Th}. Note that v_{Th} will be less than v_s due to the effects of the base bias resistors. The input impedance of the circuit to the right is now simply $h_{ie} + h_{fe}R_E$.

As always, the output impedance of the circuit is given by the ratio $v_E(\text{open})/i_E(\text{short})$. Since the circuit of Figure 6.19c does not include a load resistor across terminals C and D, the output voltage v_E shown is the open circuit voltage. Since the circuit to the right of the dashed line can be replaced with the equivalent resistance $h_{ie} + h_{fe}R_E$ as shown in Figure 6.16b, the output voltage is easily found to be

$$v_E(\text{open}) = \frac{h_{fe}R_E v_{\text{Th}}}{R_{\text{Th}} + h_{ie} + h_{fe}R_E} \qquad (6.68)$$

The short circuit current is given by $h_{fe}i_B(\text{short})$, and since the short circuit in this case means that $v_E = 0$, the shorted base current is

$$i_B(\text{short}) = \frac{v_{\text{Th}}}{R_{\text{Th}} + h_{ie}} \qquad (6.69)$$

The short circuit emitter current is therefore

$$i_E(\text{short}) = \frac{h_{fe}v_{\text{Th}}}{R_{\text{Th}} + h_{ie}} \qquad (6.70)$$

Dividing (6.68) by (6.70) gives

$$R_{\text{out}} = \frac{R_E(R_{\text{Th}} + h_{ie})}{R_{\text{Th}} + h_{ie} + h_{fe}R_E} \qquad (6.71)$$

From this expression we see that if $R_{\text{Th}} + h_{ie} \ll h_{fe}R_E$, the output impedance reduces to

$$R_{\text{out}} = \frac{R_{\text{Th}} + h_{ie}}{h_{fe}} \qquad (6.72)$$

If the above inequality is reversed by constructing a circuit with a sufficiently small R_E, the output impedance is just R_E itself. In either case the output impedance can be much smaller than either R_s or R_{Th}.

Example 6.6 A common collector circuit like Figure 6.18a is capacitively coupled on the input to a high-frequency signal source with an adjustable output resistance. At its output, the circuit is capacitively coupled to a long coaxial cable that is terminated in 50 Ω; the effective load is 50 Ω. In order to minimize reflected signals in the coaxial cable, we want our driving circuit to have an output impedance of 50 Ω also. If the signal amplitude never exceeds 1 mV and assuming transistor parameters like those in Table 6.1, determine R_E, R_B, and the source output impedance R_s.

We first need to choose a reasonable value for I_C: 1 mV into 50 Ω is 20 μA, so $I_C = 2$ mA is large enough that we are well within the small-signal domain. A V_E of 10 V is convenient and requires that $R_E = 5000$ Ω. Since $h_{FE} = 100$, we need $R_B = 500$ kΩ to establish this DC operating point. Both of these resistors are large enough to neglect in the AC analysis. Solving Eq. (6.72) for R_s (R_{Th} in the equation) gives

$$R_s = h_{fe}R_{out} - h_{ie}$$

and evaluation using our component values gives the required source resistance,

$$R_s = 100(50 \ \Omega) - (2000 \ \Omega) = 3000 \ \Omega$$

Note that, unlike many of our examples, the actual value of h_{ie} has a significant effect on this result.

6.5   The Common Base Amplifier

The common base amplifier can produce a voltage gain but generates no current gain between the input and output signals. It is normally characterized by a very small input impedance and an output impedance like the common emitter amplifier. Because the input and output currents are of similar size, the stray capacitance of the transistor is of less significance than for the common emitter amplifier. As a result, the common base amplifier is often used at high frequencies, where it provides more voltage amplification than the other one-transistor circuits.

6.5.1   DC Biasing

Again, the DC biasing can be accomplished by the same techniques used for the common emitter circuit. The common base amplifier in Figure 6.20a has the same DC configuration as the common emitter amplifier of Figure 6.11b; only the orientation and input connections have been changed to put the transistor into the common base configuration.

6.5.2   Voltage Gain

Above a corner frequency, the capacitor between base and ground on the circuit of Figure 6.20a provides an effective AC ground at the transistor's base, and allows us to draw the simplified AC equivalent circuit shown in Figure 6.20b. The base current in this circuit is given by

$$i_B = -\frac{v_E}{h_{ie}} \tag{6.73}$$

making the voltage gain

$$\frac{v_C}{v_E} = \frac{h_{fe}R_C}{h_{ie}} \tag{6.74}$$

which is the same as the common emitter amplifier except for the lack of the voltage inversion.

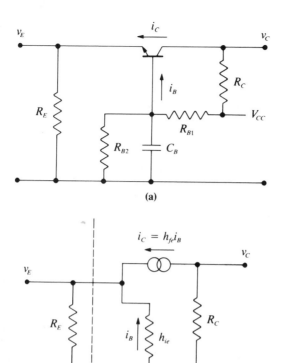

Figure 6.20 *(a) A common base circuit and (b) its AC equivalent above the corner frequency.*

6.5.3 Input and Output Impedances

The input impedance looking into the emitter of the transistor in Figure 6.20b (the part to the right of the dashed line) is defined by

$$R_{in} = \frac{v_E}{-i_C} \tag{6.75}$$

and can be rewritten in terms of the base current,

$$R_{in} = -\frac{v_E}{h_{fe}i_B} \tag{6.76}$$

Substituting for i_B using Eq. (6.73) gives

$$R_{in} = \frac{h_{ie}}{h_{fe}} \tag{6.77}$$

The input impedance looking into the emitter is quite small, in the range of a few tens of ohms, and the overall input impedance to the circuit is just the parallel combination of R_E and this R_{in}.

Because of its high-frequency response and small input impedance, this circuit is often used to receive high-frequency signals transmitted via a coaxial cable (as in Example 6.6). For this purpose the input impedance of the amplifier is adjusted to match the distributed impedance of the coaxial cable—usually 50 to 75 Ω. If h_{ie}/h_{fe} is too small, it can be easily increased by adding a resistor in series with h_{ie} in the transistor base. The output impedance of this circuit is never greater than R_C.

6.6   The PNP Transistor

The methods developed above for the *NPN* transistor work equally well with the *PNP* type. The basic *PNP* common emitter circuit of Figure 6.21a uses a negative power supply and consequently has its DC current reversed as shown in the figure. In all other respects it is identical to the *NPN* circuit used earlier. The AC equivalent circuit shown in this figure is also unchanged from the *NPN* example, except that the direction of assumed positive current has been reversed to match the actual DC currents.

The reversed sense of DC current in the *PNP* transistor adds an important dimension to the design of multitransistor circuits. When used with multiple voltage power supplies, this feature can be used to eliminate the need for coupling capacitors between transistor stages.

Figure 6.21   (a) Simple circuit using a PNP transistor. (b) The AC equivalent is similar to that used for the NPN transistor. The positive sense of AC current is chosen here to match the positive DC current, but this agreement is rarely needed.

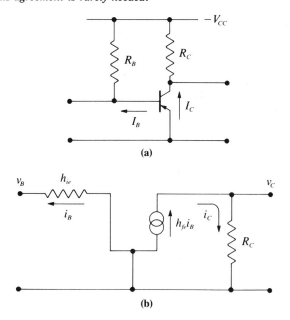

6.7 Current Sources

The common collector configuration can also be used as a current source (or sink) in multitransistor circuits. A constant current source uses a stable base circuit to develop a constant voltage at the base and across the emitter resistor. Figure 6.22a shows an *NPN* transistor used as a current sink with the base voltage determined by a voltage divider resistor network. The voltage at the emitter is then fixed at $V_B - V_{PN}$, which in turn establishes a fixed collector current through R_E. This design produces a source current that is independent of the power supply voltage.

Figure 6.22b shows a *PNP* transistor used to form a current source. The voltage across the emitter is developed in this example by the reverse current across a Zener diode, and the voltage drop across the emitter resistor is $V_Z - V_{PN}$.

The third example, shown in Figure 6.22c, is a current sink similar to the first, but two diodes have been used in the base circuit to obtain a base voltage that is itself determined by *PN* voltage drops. This circuit arrangement will be stable against temperature-induced changes in V_{PN} as long as the transistor and diodes have similar characteristics and are maintained at the same temperature. Note that the diode placement in the two examples produces an I_C that is independent of power supply voltage.

An ideal current source (a Norton equivalent circuit with an infinite output impedance) provides an output current that is entirely independent of the load impedance, and our single transistor model unrealistically predicts this type of behavior. A more exact treatment would need to consider the more subtle effects that arise as a result of voltage changes at the collector and would need to include the bare transistor's output impedance $1/h_{oe}$. Typically, the actual output impedance of a single-transistor current source is less than a megohm.

Figure 6.22 *(a) A current source (sink) with the reference voltage determined by a resistive voltage divider. (b) A current source using a Zener diode as a reference. (c) A improved current source (sink) providing temperature stabilization against changes in V_{PN}.*

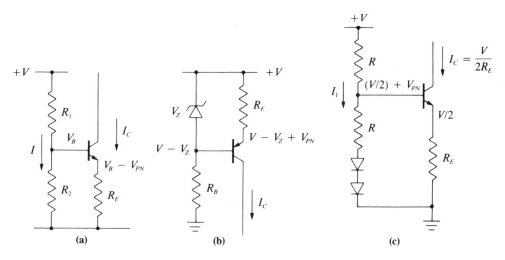

6.8   The Junction Field Effect Transistor (JFET)

Bipolar transistors have several significant faults that are not present in the field effect transistor: (1) low impedance between base and emitter; (2) a temperature characteristic that tends to increase the collector current with increasing temperature; (3) small high-frequency gains for types with large power-handling ability; and (4) highly nonlinear operation when $|V_{CE}|$ is less than 2 V. To a large degree these faults are all caused by the passage of the main charge current through a series base region whose majority charge carriers have opposite polarity. The field effect transistor avoids these problems by using an entirely different mechanism to control the current through the device.

The JFET comes in two types, *N*-channel and *P*-channel, as shown in Figure 6.23. The designation refers to the polarity of the majority charge carriers in the bar of semiconductor that connects the drain terminal *D* to the source terminal *S*. Since the channel is formed from a single-polarity material, its resistance is a function only of the geometry of the conducting volume and the conductivity of the material. The gate terminal *G* of the JFET is connected to a small implant of opposite-polarity semiconductor shown on the drawing as two separate but electrically connected strips. Unlike the bipolar transistor, the JFET operates with all *PN* junctions reverse-biased, producing a very high input impedance into the gate. In contrast to the bipolar transistor, virtually all of the current in FET is thus in a semiconducting material of a single polarity.

6.8.1   Principles of Operation

Again we will concentrate on a JFET of a particular type, the *N*-channel; the operation of a *P*-channel device is similar except for the polarity of the DC bias voltages. The sketch in Figure 6.24 shows the depletion region formed when the *PN* junction is reverse-biased by the applied voltages. Just as for a simple diode,

Figure 6.23   (a) The basic geometry of an N-channel JFET and its circuit symbol. The two P regions are electrically connected. (b) A P-channel JFET and its circuit symbol.

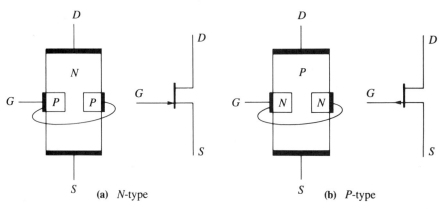

(a) *N*-type   (b) *P*-type

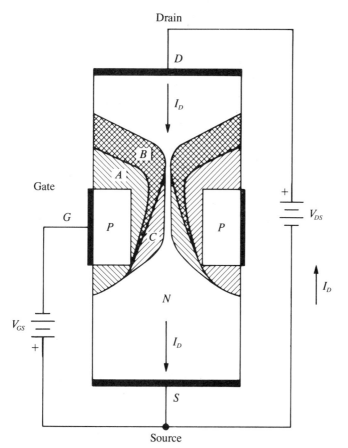

*Figure 6.24   An N-channel JFET with DC bias voltages applied. The
depletion region (shown cross-hatched) grows as the reverse voltage
across the PN junction is increased and reduces the volume of the
conducting N region. The depletion region is thicker near the drain
because the ohmic voltage drop across the N material connecting drain
to source makes the reverse PN junction voltage larger in this area.*

this depletion region grows as the reverse bias across the *PN* junction is increased,
thereby constricting the cross section of conducting *N*-channel material and increasing the resistance of the channel. The major current I_D in the channel is
caused by the applied voltage between drain and source, V_{DS}, and is controlled by
the applied voltage between gate and source, V_{GS}. The resistive voltage drop between drain and source causes the reverse bias across the *PN* junction to be larger
near the drain, thus causing the depletion region to be larger in this area.

The JFET has two distinct modes of operation: the variable-resistance mode,
and the pinch-off mode. In the variable-resistance mode, indicated in the figure
by depletion region *A,* charge carriers are available in a large resistive channel of
the *N*-channel material. The cross-sectional area of this channel changes as a result
of the depletion region generated by applied voltage V_{GS}, but for small changes in
V_{DS} the overall shape is effectively unchanged. In this mode the JFET thus be-

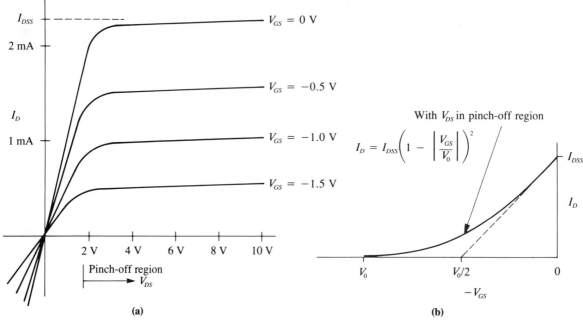

Figure 6.25 *(a) The characteristic curves of a typical N-channel JFET. The pinch-off mode of operation corresponds to the region of this graph where the curves attain a constant I_D as a function of V_{DS}. (b) The relationship between I_D and V_{GS} in the pinch-off region.*

haves like a resistor whose value is controlled by V_{GS}. In the pinch-off mode, indicated by depletion regions B and C, the channel has been heavily constricted with most of the drain-source voltage drop occurring along the narrow and therefore high-resistance part of the channel near the depletion regions. Under these conditions, an increase in V_{DS} produces a lengthening of the pinched-off region and a corresponding increase in the effective resistance between drain and source. The combined effect of increased voltage and increased resistance leaves I_D essentially constant in this operating mode.

The characteristic curves of a typical JFET are shown in Figure 6.25a. At small values of V_{DS} (in the range of a few tenths of a volt), the curves of constant V_{GS} show a linear relationship between V_{DS} and I_D. This is the variable-resistance region of the graph. As V_{DS} increases, each of the curves of constant V_{GS} enters a region of nearly constant I_D. This is the pinch-off region, where the JFET can be used as a linear voltage and current amplifier. Note that the pinch-off region is attained more quickly for large values of V_{GS} but still occurs even when the applied gate source voltage is zero. At $V_{GS} = 0$, the current through the JFET reaches a maximum known as I_{DSS}, the current from *Drain* to *Source* with the gate *Shorted* to the source. If V_{GS} goes positive for this N-channel JFET, the *PN* junction becomes conducting, and the JFET becomes just a forward-biased diode.

In the pinch-off region the relationship between drain-source current I_D and gate-source voltage V_{GS} can be approximated by the relation

$$I_D = I_{DSS}\left(1 - \left|\frac{V_{GS}}{V_0}\right|\right)^2 \qquad (6.78)$$

where V_0 is the gate source voltage for which the drain current is zero. This curve is shown in Figure 6.25b. These parameters are normally given on the manufacturer's specification sheets: V_0 is listed as $V_{GS}(\text{off})$ or V_p for JFET devices and as $V_{GS}(\text{threshold})$ or V_t for the similar MOSFET devices discussed in Section 6.11. Equation (6.78) can be used as the basic equation for determining the DC operating point of a FET, but the algebra can be cumbersome with this quadratic equation, so it is sometimes more convenient to use the linear relation

$$I_D = I_{DSS}\left(1 - B\left|\frac{V_{GS}}{V_0}\right|\right) \qquad (6.79)$$

with B chosen somewhere between 1.25 and 2 to match the characteristics of a particular JFET. With $B = 2$, this line is tangent to the curve of Eq. (6.11) at V_{GS}, as shown by the dashed line in Figure 6.25b.

6.8.2 Variable-Resistance Operation

The operation of a JFET as a variable resistor is shown by the circuit of Figure 6.26a. Since there are no *PN* junctions between drain and source, this device requires no "turn-on" drain-source voltage and has a simple resistance character for small voltages of either polarity across the drain and source terminals. The variable-resistance dependence on V_{GS} can be approximated by the expression

$$R_{DS} \simeq \frac{r_0}{(1 - |V_{GS}/V_0|)^2} \qquad (6.80)$$

where r_0 is the on-resistance from drain to source in the *N*-channel material when V_{GS} is zero and V_{DS} approaches zero. This parameter is given on the specification sheets of some JFET types and typically is in the range of 100 to 1000 Ω. When not given, it can be approximated by $2/I_{DSS}$. By contrast, the on-resistances of the power MOSFET devices described in Section 6.11.3 can be less than an ohm.

On the characteristic curves of Figure 6.25a, the variable resistance R_{DS} appears as an inverse slope near the origin. Since these curves extend into the third quadrant, a bipolar signal can be applied across the drain-source connection. However, for real devices these curves become increasingly nonlinear as $|V_{DS}|$ increases, thereby restricting the variable-resistance application to signal amplitudes that are no more than a few tenths of a volt. For linear operation, the range of R_{DS} variation is not large, typically from r_0 to about $25r_0$, corresponding to applied gate-source signals of 0 to $0.8V_0$. The linearity can be improved somewhat by employing a feedback circuit like the one shown in Figure 6.26b.

6.8.3 DC Biasing

The self-bias circuit for a JFET is shown in Figure 6.27. Since the gate current is only a few picoamperes, the gate resistor R_G can assume a wide range of values

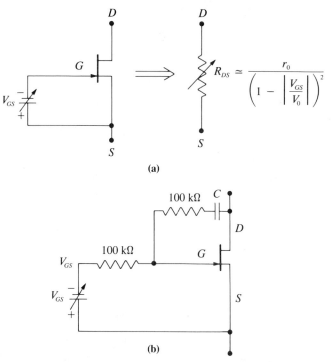

(a)

(b)

Figure 6.26 (a) Basic circuit for operating a JFET as a variable resistor, and (b) a more linear variable-resistance design. A small DC or AC voltage can be applied across the drain-source channel.

without having any effect on the operating point. Its only purpose is to establish the gate at a fixed voltage, zero in this case. With the gate voltage set to zero, the drain current through the source resistor R_S raises the emitter to a voltage $R_S I_D$, producing the gate-to-source voltage

$$V_{GS} = -R_S I_D \tag{6.81}$$

Figure 6.27 (a) The essential features of the self-biasing circuit for a JFET. For a small signal application, typical values are $R_G = 1\ M\Omega$ and $R_S = 1\ k\Omega$.

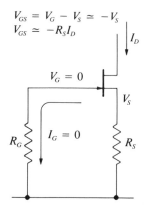

If this expression is substituted for V_{GS} in Eq. (6.78), the result can be solved to give R_S in terms of a target I_D and the JFET parameters V_0 and I_{DSS}:

$$R_S = \frac{V_0}{I_0}\left(1 - \sqrt{\frac{I_D}{I_{DSS}}}\right) \tag{6.82}$$

Since the two transistor parameters may each vary by much as a factor of five from sample to sample, the I_D operating point actually obtained with this R_S may be far from the intended value.

6.8.4   Small-Signal AC Model

The JFET characteristic curves of Figure 6.25a can be described by an equation of the general form

$$I_D = I_D(V_{GS}, V_{DS}) \tag{6.83}$$

where the function on the right varies with the particular transistor. Following the method outlined in Section 6.2.3 for bipolar transistors, this expression yields the AC relationship

$$\mathbf{i}_D = \frac{\partial I_D}{\partial V_{GS}}\mathbf{v}_{GS} + \frac{\partial I_D}{\partial V_{DS}}\mathbf{v}_{DS} \tag{6.84}$$

where the AC currents and voltages are complex, but the partial derivatives evaluate to real numbers. In the pinch-off region where this analysis is effective, the curves of constant V_{GS} are essentially flat; this makes the second partial derivative equal to zero and allows the equation to be rewritten as

$$\mathbf{i}_D = g_m\mathbf{v}_{GS} \tag{6.85}$$

where $g_m = \partial I_D/\partial V_{GS}$. Thus the AC drain-to-source current would seem to be simply proportional to $\mathbf{v}_{GS}$. This g_m parameter is known as the transconductance and has units of ohm^{-1} or mhos. For a typical JFET, g_m will be a few times 10^{-3} mho.

Although this model will work at low frequencies and can be represented schematically by a simple current source as shown in Figure 6.28b, the gate-source and gate-drain parasitic capacitances present in all FETs require a slightly different treatment. Since these capacitive elements can be described by complex impedances, we could simply expand on this method by allowing the function in Eq. (6.83) (and a similar one for $\mathbf{I}_G$) to be complex; this in turn would make the partial derivatives complex and double the number of parameters.

This approach leads to the **y**-parameter model shown in Figure 6.28c. (The **y** notation identifies an admittance defined as $1/\mathbf{Z}$.) Although we will not make use of the **y**-parameter model, some specification sheets define the AC characteristics of an FET in terms of these eight (real and imaginary) parameters. For our purposes we need to know only that g_m is the real part of $\mathbf{y}_{fs}$ (the forward transadmittance). This should be obvious from a comparison of Figures 6.28b and 6.28c.

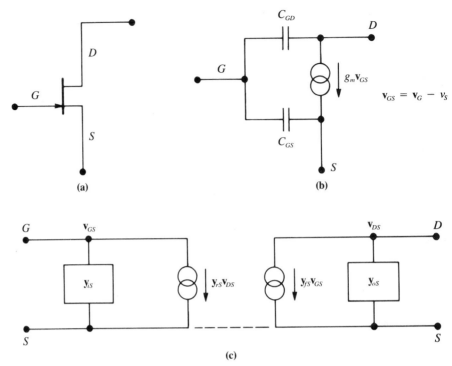

Figure 6.28 *(a) The circuit symbol for a JFET shown in relation to (b) a simple AC model for its operation in the pinch-off region, and (c) a more complete y-parameter model. Note that the y parameters are complex.*

Equation (6.85) can be expanded into a reasonable model of the field effect transistor by the inclusion of two coupling capacitors as shown in Figure 6.28b. This model displays the major capacitive elements explicitly, and as a result the three parameters needed here (g_m, C_{GS}, and C_{GD}) are all real numbers. This alternative description of the AC operation of an FET is also found on some specification sheets. The capacitors C_{GS} and C_{GD} (related to the imaginary parts of the admittances y_{is} and y_{os} shown in Figure 6.28c) are typically around 10 pF. Although these stray capacitance values are actually slightly smaller than their bipolar transistor counterparts, their significance is increased by the lack of an internal resistive connection between the gate and either drain or source. Moreover, small-signal FET circuits tend toward much larger resistors and smaller currents than their bipolar equivalents, with the result that the same-size capacitors charge slower and produce longer time constants.

6.9 JFET Common Source Amplifier

The common source configuration for an FET is similar to the common emitter bipolar transistor configuration. The basic circuit is shown in Figure 6.29a. As with the bipolar configuration, the common source amplifier can provide both a voltage and current gain. Since the input resistance looking into the gate is ex-

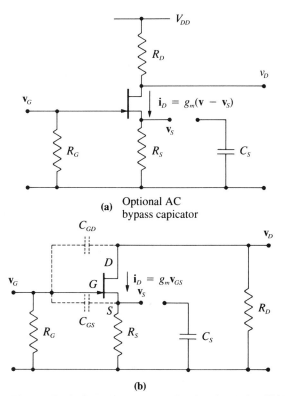

Figure 6.29 *(a) Typical common source circuit using DC self-bias and
(b) its AC equivalent circuit. The addition of the bypass capacitor C_S
will increase the AC gain but make it dependent on g_m.*

tremely large (10^7 to 10^9 Ω), the current gain available from the FET amplifier
can be quite large, but the voltage gain is generally inferior to that available from
a bipolar device.

Because of the large input resistance to the gate, the DC input impedance to
the transistor is much larger than that of any bipolar configuration. The overall
input impedance to the amplifier is usually determined by other components: either
the resistors necessary to obtain the DC operating point or the parasitic capacitances. Thus, FET amplifiers are most useful with high output-impedance signal
sources where a large current gain is the primary requirement. The lack of gate
current at low frequencies also results in a major simplification in the analysis of
circuits involving FET devices.

6.9.1 Low-Frequency Analysis

The common source amplifier can be analyzed using the AC equivalent circuit
shown in Figure 6.29b. The voltage gain between the gate and drain can be found
from the transconductance expression

$$i_D = g_m(v_G - v_S) \qquad (6.86)$$

which is just an expansion of Eq. (6.85). Since at low frequencies we can ignore the capacitors shown in the figure, the source voltage is given by

$$v_S = R_S i_D$$
$$= g_m R_S (v_G - v_S) \tag{6.87}$$

Solving for v_S gives

$$v_S = \frac{g_m R_s v_G}{1 + g_m R_S} \tag{6.88}$$

In the absence of the capacitors, the current through R_S and R_D is the same, and the drain voltage is given by

$$v_D = \frac{-v_S R_D}{R_S} \tag{6.89}$$

Substituting v_S from Eq. (6.88) and rearranging gives

$$\frac{v_D}{v_G} = -A = -\frac{g_m R_D}{1 + g_m R_S} \tag{6.90}$$

where $-A$ is again being used to represent the transfer function of an inverting amplifier. If $g_m R_S \gg 1$, this reduces to

$$\frac{v_D}{v_G} = -A = -\frac{R_D}{R_S} \tag{6.91}$$

Since g_m is typically a few times 10^{-3} mho, the resistor R_S must be several thousand ohms for this approximation to be valid. A large gain will thus require a drain resistor R_D on the order of 10^5 Ω, and this can limit the DC drain current to a fraction of a milliampere unless a relatively large power supply voltage V_{DD} is used.

At low frequencies, the only drain-to-source connection in the FET model of Figure 6.28b is through the infinite-impedance current source. With this model replacing the FET in the AC equivalent of Figure 6.29b, it is easy to show that the output impedance of this circuit is just the drain resistor R_D.

This circuit produces its largest AC voltage gain when a capacitor C_S is used to bypass the AC current around the source resistor R_S as shown in Figure 6.29. This converts R_S in Eq. 6.90 to a $\mathbf{Z}_S$ that becomes smaller as the frequency increases. The complete transfer function is easily found by replacing R_S in Eq. (6.90) with the impedance of the parallel combination of R_S and $\mathbf{Z}_C$. The resulting transfer function $-\mathbf{A}$ is now complex,

$$\frac{\mathbf{v}_D}{\mathbf{v}_G} = -\mathbf{A}(j\omega) = -\frac{g_m R_D (1 + j\omega C_S R_S)}{1 + j\omega C_S R_S + g_m R_S} \tag{6.92}$$

but above the low-frequency corners it reduces to a real constant,

$$\frac{\mathbf{v}_D}{\mathbf{v}_G} = -A = -g_m R_D \tag{6.93}$$

A plot of the magnitude of the transfer function of Equation 6.92 would result in a frequency response much like that of Figure 6.17c. The higher corner frequency in this case is given by

$$\omega_{c2} = \frac{g_m}{C_S} \tag{6.94}$$

Example 6.7   Given a JFET with parameters $V_0 = -3$ V, $I_{DSS} = 2.5$ mA, and $g_m = 3$ mmhos, determine the resistor values needed in the common source circuit of Figure 6.29a if the power supply is 20 V and a gain of -10 is required. Assume that a bypass capacitor C_S is not used.

Since we need to have a gain of -10, Eq. (6.91) indicates that $R_D/R_S \simeq 10$. With a common current I_D in both resistors, the voltage drop across R_D will also be about 10 times greater than the voltage drop across R_S. Choosing $V_S = 0.8$ V should produce $V_D \simeq 12$ V, an 8-V drop from the power supply. From Eq. (6.79), we can then estimate the drain current:

$$I_D = (2.5 \text{ mA})\left(1 - \frac{0.8}{3}\right)^2 = 1.34 \text{ mA}$$

To get $V_S = 0.8$ V, we must have

$$R_S = \frac{0.8 \text{ V}}{1.34 \text{ mA}} = 595 \text{ } \Omega$$

Equation (6.90) and the requirement that $v_D/v_G = -10$ allows us to solve for R_D:

$$R_D = \frac{10[1 + (3 \text{ mmho } 595 \text{ } \Omega)]}{3 \text{ mmhos}} = 9283 \text{ } \Omega$$

Since $g_m R_S$ is only about 1.5 in this design, Eq. (6.90) does not reduce to (6.91) and R_D is more than 10 times R_S. This shows up now as an operating point for V_D, which is not 12 V as expected, but

$$V_D = (20 \text{ V}) - (9283) \text{ } \Omega)(1.34 \text{ mA}) = 7.6 \text{ V}$$

If important, V_D could be increased by assuming that $V_S = 0.9$ V initially.

Because we did not attain $g_m R_S \gg 1$ in this design, the AC gain of this design will vary with the FET sample. This situation is much improved if we start with a larger power supply, perhaps 30 V, and continue with $V_D = 12$ V. We would then need V_S to be about 1.8 V, yielding $I_D = 0.4$ mA, $R_S = 4.5$ kΩ, and $R_D = 45$ kΩ. With this start, $g_m R_S = 13.5$, and Eq. (6.91) is a good approximation to the AC gain.

In either design the value of R_G is unimportant to the problem as stated; a choice in the range of 1 to 10 MΩ is typical and makes the overall gain insensitive to a signal generator's output impedance.

6.9.2   *High-Frequency Operation*

At high frequencies the effects of the stray capacitances C_{GD} and C_{GS} must be considered. These capacitances are shown on the AC equivalent circuit in Figure 6.29b, but if R_G and C_S are both large, this circuit can be redrawn as shown in Figure 6.30a. These are reasonable assumptions; the effects of R_G can always be

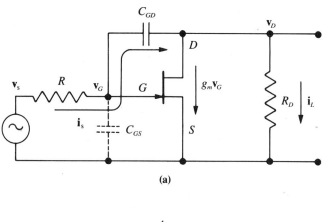

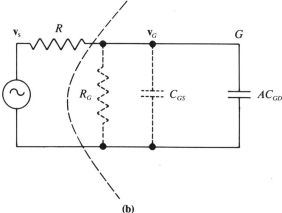

Figure 6.30 *(a) The high-frequency AC equivalent of the common source circuit of Figure 6.29a, and (b) the equivalent circuit seen by the signal generator.*

included later, and the source bypass capacitor C_S is normally chosen large enough to effectively ground the source at the lowest intended operating frequency.

With the source end of the parasitic capacitance C_{GS} now at ground, its only effect is to reduce the input impedance to the FET at high frequencies, and even though C_{GS} and C_{GD} are about the same size, C_{GD} has a much larger effect on the circuit performance at high frequencies because it provides a negative feedback path between gate and drain. This is an example of the Miller effect discussed in Appendix B. When viewed from the gate, this capacitor looks much larger because a small increase in $\mathbf{v}_G$ produces a much larger decrease in $\mathbf{v}_D$.

We start the high-frequency analysis of this circuit by deriving an expression for the gate-to-drain transfer function:

$$-\mathbf{A} = \frac{\mathbf{v}_D}{\mathbf{v}_G} \tag{6.95}$$

Because of the capacitances, this function will now be complex. This transfer function can be determined by first writing the three-current sum at the drain terminal,

$$\frac{\mathbf{v}_D - \mathbf{v}_G}{\mathbf{Z}_{GD}} - \frac{\mathbf{v}_D}{R_D} - g_m\mathbf{v}_G = 0 \tag{6.96}$$

then rearranging to yield the desired voltage ratio,

$$\frac{\mathbf{v}_D}{\mathbf{v}_G} = -\mathbf{A} = -\frac{(g_m\mathbf{Z}_{GD} + 1)R_D}{\mathbf{Z}_{GD} + R_D} \tag{6.97}$$

At low frequencies, where $|\mathbf{Z}_{GD}| \gg R_D$, this transfer function reduces to a real number

$$-A_L = -g_mR_D \tag{6.98}$$

and at medium high frequencies, where $|\mathbf{Z}_{GD}| \ll R_D$ and $g_m|\mathbf{Z}_{GD}| \gg 1$, it reduces to

$$-\mathbf{A}_H = \frac{-g_m}{j\omega C_{GD}} \tag{6.99}$$

At even higher frequencies, where $g_m|\mathbf{Z}_{GD}| \ll 1$, Eq. (6.97) reduces to unity as a result of capacitance C_{GD} effectively shorting the gate to the drain. Intersecting $|\mathbf{A}_L|$ with $|\mathbf{A}_H|$ shows that this low-pass transfer function will have a high-frequency corner at

$$\omega_{c2} = \frac{1}{R_DC_{GD}} \tag{6.100}$$

For a typical FET design with $R_D = 5 \times 10^4$ Ω and $C_{GD} = 5$ pF, we find a high-frequency cutoff of about 300 kHz, but as we will see below, it is usually superseded by a lower-frequency corner caused by the reactive input impedance.

Since the resistive impedance into the FET gate is quite large (>100 MΩ), the input impedance to the common source circuit is determined by the current through the parasitic capacitances. By far the most important of these is C_{GD} because of its enhancement by the Miller effect. Below the corner frequency of Eq. (6.100), where $\mathbf{A}$ is a real constant A_L, Eq. (B.4) for the Miller equivalent impedance can be rewritten as

$$\mathbf{Z}_M = \frac{\mathbf{Z}_{GD}}{1 + A_L} \tag{6.101}$$

and using Eq. (6.98) for $A_L \gg 1$, it becomes

$$\mathbf{Z}_M = \frac{\mathbf{Z}_{GD}}{A_L} = \frac{1}{j\omega g_m R_D C_{GD}} \tag{6.102}$$

This impedance corresponds to an enhanced capacitance $A_L C_{GD}$ connected between the gate and ground as shown in Figure 6.30b. The input impedance to the common source amplifier is actually the parallel combination of all three elements to the right of the dashed line, but for most circuits R_G and C_{GS} are of little consequence.

The overall transfer function of the amplifier is

$$\mathbf{H}(j\omega) = \frac{\mathbf{v}_D}{\mathbf{v}_s} \tag{6.103}$$

or

$$\mathbf{H}(j\omega) = \frac{-A_L\mathbf{v}_G}{\mathbf{v}_s} \tag{6.104}$$

Note that the s subscript refers to the signal generator, not the source. Ignoring the effects of R_G and C_{GS}, Figure 6.30b becomes a single-loop voltage divider, giving

$$\frac{\mathbf{v}_G}{\mathbf{v}_s} = \frac{\mathbf{Z}_{GD}}{A_L R + \mathbf{Z}_{GD}} \tag{6.105}$$

The transfer function has a high-frequency corner at

$$\omega_{c1} = \frac{1}{R A_L C_{GD}} \tag{6.106}$$

which for most applications is well below the ω_{c2} of Eq. (6.100). When the signal generator impedance is large, the bandpass can be quite small: For example, $R = 10^5 \, \Omega$, $A_L = 50$, and $C_{GD} = 5$ pF results in an amplifier with a high-frequency corner only slightly over 6 kHz.

We will see in the next chapter that it is possible to eliminate the gain factor A_L in Eq. (6.106) by adding a second transistor in the *cascode* connection.

6.9.3   *Reactive Loads and Resonant Circuits*

In the typical high-frequency (above 10 MHz) amplifier, the drain resistor is replaced by an inductor or a transformer. As we shall see, this inductive load on the drain can compensate for the signal loss due to stray capacitance in the circuit and result in a high-gain amplifier, even at high frequencies. In most circuits a drain capacitance is also present, either as a separate, often variable, capacitor or as stray capacitance between wire leads. A typical circuit that will function as a bandpass amplifier is shown in Figure 6.31a. Such amplifiers are known as tuned circuit amplifiers or radiofrequency (RF) amplifiers.

The DC bias of this circuit is different from any circuit we have previously investigated. If we assume that the inductor is ideal, with no resistive part, then it cannot support a DC voltage drop. As a consequence, $V_D = V_{DD}$ for any choice of I_D. This has the surprising effect that a combined AC and DC signal will exceed V_{DD} over half its cycle.

The high-frequency AC equivalent circuit for this RF amplifier is shown in Figure 6.31b, but the generalized form shown in Figure 6.31c is a better choice for analysis. The three-current sum at the drain of the JFET can be written

$$\frac{\mathbf{v}_G - \mathbf{v}_D}{\mathbf{Z}_1} - g_m \mathbf{v}_G - \frac{\mathbf{v}_D}{\mathbf{Z}_2} = 0 \tag{6.107}$$

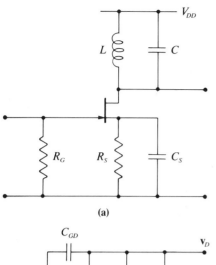

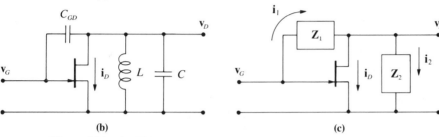

Figure 6.31 *(a) This common source circuit has a tank circuit
replacing the drain resistor, a common high-frequency application.
(b) Above the corner frequency caused by C_S, the source is at an AC
ground, but the equivalent circuit must include the stray feedback
capacitance between the drain and gate. (c) The same circuit with
generalized impedances.*

and rearranging the terms gives the general transfer function

$$\frac{\mathbf{v}_D}{\mathbf{v}_G} = \frac{\mathbf{Z}_2}{\mathbf{Z}_1 + \mathbf{Z}_2} - g_m \frac{\mathbf{Z}_1 \mathbf{Z}_2}{\mathbf{Z}_1 + \mathbf{Z}_2} \tag{6.108}$$

In the example of Figure 6.31b,

$$\mathbf{Z}_1 = \frac{1}{j\omega C_{GD}} \tag{6.109}$$

and

$$\mathbf{Z}_2 = \frac{j\omega L}{1 - \omega^2 LC}$$

Substituting these into (6.108) gives

$$\frac{\mathbf{v}_D}{\mathbf{v}_G} = -\left(\frac{j\omega L}{1 - \omega^2 L(C + C_{GD})}\right)(g_m - j\omega C_{GD}) \tag{6.110}$$

For a typical high-frequency JFET, the relative sizes of g_m and C_{GD} are such
that the last parenthetical factor will reduce to just g_m at frequencies below 100

MHz. Series resistance in an actual inductor will of course keep the gain from being infinite at the resonant frequency:

$$\omega_r = \frac{1}{\sqrt{L(C + C_{GD})}} \tag{6.111}$$

Thus we find that with a resonant amplifier, the stray capacitance does not seriously degrade the gain below 100 MHz, but it does change the location of the pass band.

6.10   JFET Common Drain Amplifier

The common drain FET amplifier is similar to the common collector configuration of the bipolar transistor. A general common drain JFET amplifier, self-biased using the same methods discussed above for the common source amplifier, is shown in Figure 6.32a. This configuration, which is sometimes known as a source follower, is characterized by a voltage gain of less than 1, and features a large current gain as a result of having a very large input impedance and a small output impedance. In this configuration, the stray capacitances shown in Figure 6.32b have much less effect on the circuit operation, with C_{GD} having the major effect on high frequency operation.

6.10.1   Voltage Gain

From the AC equivalent circuit of Figure 6.32b, the output voltage at the source is

$$\mathbf{v}_S = g_m R_S(\mathbf{v}_G - \mathbf{v}_S) \tag{6.112}$$

Solving for $\mathbf{v}_S$ gives the voltage gain between the gate and source:

$$\frac{\mathbf{v}_S}{\mathbf{v}_G} = \frac{g_m R_S}{1 + g_m R_S} \tag{6.113}$$

Since g_m is small, R_S must be in the neighborhood of 10 kΩ before this equation will predict unity gain.

6.10.2   Input and Output Impedances

The input impedance of the common drain circuit of Figure 6.32a is determined largely by the parallel combination of R_G and C_{GD} as can be seen from the AC equivalent in Figure 6.32b. The gate source capacitance C_{GS} has less effect on the input impedance, since the voltage developed across C_{GS} is typically less than $0.1\mathbf{v}_G$ whereas the voltage developed across G_{GD} is the full $\mathbf{v}_G$.

The output impedance can be obtained by finding the short circuit current in terms of $\mathbf{v}_G$. With $\mathbf{v}_S$ shorted to ground, the drain-to-source current is

$$\mathbf{i}_S(\text{short}) = g_m \mathbf{v}_G \tag{6.114}$$

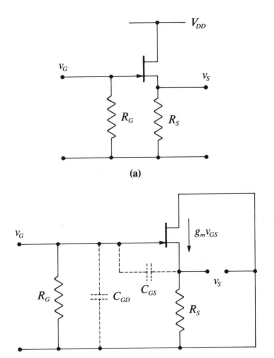

Figure 6.32 *(a) The common drain circuit. (b) Its AC equivalent circuit.*

Dividing the output voltage, given by Eq. (6.113), by this result gives

$$R_{out} = \frac{R_S}{1 + g_m R_S} \tag{6.115}$$

which reduces to $1/g_m$ for sufficiently large values of R_S.

6.11 The Insulated-Gate Field Effect Transistor (IGFET)

The insulated-gate FET, also known as a metal oxide semiconductor field effect transistor (MOFSET), is similar to the JFET but exhibits an even larger resistive input impedance due to the thin layer of silicon dioxide (SiO_2) that is used to insulate the gate from the semiconductor channel. This insulating layer forms a capacitive coupling between the gate and the body of the transistor as shown in Figure 6.33. The consequent lack of an internal DC connection to the gate makes the device more versatile than the JFET, but it also means that the insulating material of the capacitor can be easily damaged by the internal discharge of static charge developed during normal handling. This is true even for the otherwise rugged power MOSFET devices discussed below.

The MOSFET is widely used in large-scale digital integrated circuits where its high input impedances can result in very low power consumption per component. Many of these circuits feature bipolar transistor connections to the external terminals, thereby making the devices less susceptible to damage. Another common use of the MOSFET is to switch the current path for analog signals, like a relay with no mechanical parts; the enhancement types discussed below are particularly useful in this application. It is also possible to obtain integrated circuit devices that contain several MOSFETs and are specifically designed to perform this switching function in many variations and with minimal analog signal loss.

6.11.1 Enhancement and/or Depletion: Principles of Operation

The MOSFET comes in four basic types, *N*-channel or *P*-channel, and depletion or enhancement. The configuration of an *N*-channel, depletion MOSFET is shown in Figure 6.33a. Its operation is similar to the *N*-channel JFET discussed previously: A negative voltage placed on the gate generates a charge depleted region in the *N*-type material next to the gate, thereby reducing the area of the conducting channel between the drain and source.

However, the mechanism by which the depletion region is formed is different from the JFET. The shaded area in the figure shows the normal depletion region associated with any *PN* junction. As the gate is made negative with respect to the source, more positive carriers from the *P*-type material are drawn into the *N*-channel, where they combine with (produce covalent bonds) and eliminate the free

Figure 6.33   (a) A depletion or depletion-enhancement type of MOSFET. (b) An enhancement-type MOSFET. The cross-hatched area shows the depletion region of an unbiased device. The terminal U is connected to the central section of semiconducting material known as the substrate.

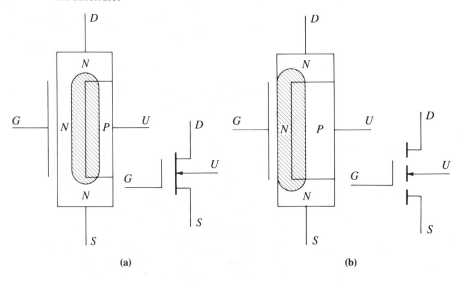

(a)   (b)

negative charges. This action enlarges the depletion region toward the gate, reducing the area of the *N*-channel and thereby lowering the conductivity between the drain and source. For negative applied gate-source voltages the observed effect is much like a JFET, and g_m is also about the same size, typically a few times 10^{-3} mho.

However, since the MOSFET gate is insulated from the channel, positive gate-source voltages may also be applied without losing the FET effect. Depending on construction details, the application of a positive gate-source voltage to a depletion-type MOSFET can repel the minority positive carriers in the depleted portion of the *N* channel back into the *P*-type material as discussed below, thereby enlarging the channel and reducing the resistance. If the device exhibits this behavior, it is known as an enhancement-depletion MOSFET.

A strictly enhancement MOSFET results from the configuration shown in Figure 6.33b. Below some threshold of positive gate-source voltage, the connecting channel of *N*-type material between the drain and source is completely blocked by the depletion region generated by the *PN* junction. In this depleted *N*-type material, most of the charges are paired in covalent bonds. However, as a result of thermal action, these bonds are constantly breaking and re-forming, so that at any instant there are a few charges of both signs that are free to move. As the gate-source voltage is made more positive, the minority positive carriers are repelled back into the *P*-type material, leaving free negative charges behind. The effect is to shrink the depletion region and increase the conductivity between the drain and source.

Figure 6.34 shows the drain current versus gate-source voltage curves for MOSFET devices of the depletion, depletion-enhancement, and enhancement types. The curves are all very similar and can be described by a variation of Eq. (6.78) used for the JFET. For depletion-type devices (all JFETs are depletion

Figure 6.34   The drain current as a function of the applied gate-source voltage for depletion, depletion-enhancement, and enhancement MOSFETs.

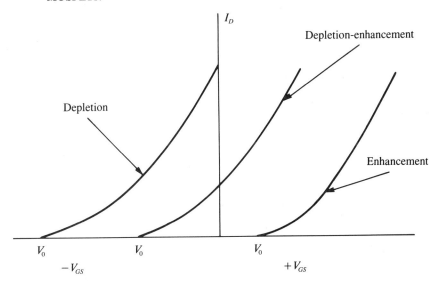

types), the drain current at $V_{GS} = 0$ is a convenient parameter for describing a particular device. For enhancement types this point is of no use, and is replaced by $I_D(\text{on})$ at some particular V_{GS}. The zero drain current gate-source voltage labeled V_0 here is variously listed as $V_{GS}(\text{off})$ or $V_{GS}(\text{pinch-off})$ for depletion types and as $V_{GS}(\text{threshold})$ for enhancement types. If V_0 is known and $I_D(\text{on})$ is available at some particular $V_{GS}(\text{on})$, then the drain current is given by

$$I_D = \left(\frac{I_D(\text{on})}{\{[V_{GS}(\text{on})/V_0] - 1\}^2} \right) \left(\frac{V_{GS}}{V_0} - 1 \right)^2 \qquad (6.116)$$

The first expression in large parentheses is just a constant (similar to I_{DSS}) for a particular device, but is expressed here in terms of the quantities normally available on manufacturers' specification sheets. At $V_{GS} = V_0$, the last term in parentheses vanishes to produce $I_D = 0$; and when $V_{GS} = V_{GS}(\text{on})$, the squared terms cancel, making $I_D = I_D(\text{on})$.

The base or substrate material (the *P*-type material in an *N*-channel device) of the MOSFET may be brought out to an external pin *U* as in Figure 6.33, internally connected to the source, or simply left floating. These options, *P*-channel devices, and multiple gate configurations produce a wealth of device options, some of which are indicated by the schematic symbols shown in Figure 6.35.

6.11.2 DC Biasing and AC Analysis

The DC bias circuit for an *N*-channel depletion-type MOSFET is identical to the self-bias technique used on the JFET. An *N*-channel enhancement type is some-

Figure 6.35   Circuit symbols for the various field effect transistors.

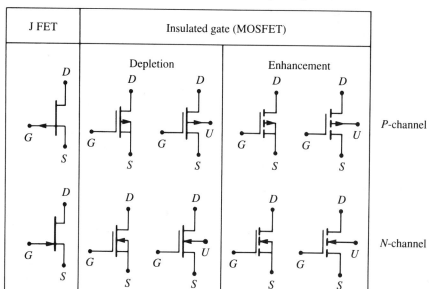

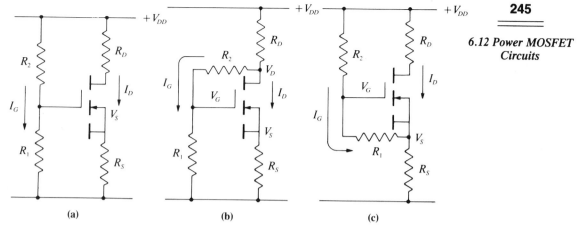

Figure 6.36 *Three additional DC biasing techniques that can be used with any type of FET. The devices indicated here are N-channel, enhancement type MOSFETs with the body of P-type material connected to the source.*

what different, since the quiescent gate voltage must then be positive with respect to the source. Any of the circuits shown in Figure 6.36 can be biased either way.

The circuit of Figure 6.36a can develop the needed bias while retaining a relatively large input impedance given by the parallel combination of R_1 and R_2, but the actual operating point for this design depends on the sample dependent FET parameters of Eq. (6.116). The circuit of Figure 6.36b offers some stability against sample variation but has a lower input impedance given by the parallel combination of R_1 and $R_2/(1 + A)$, where A is the magnitude of the AC gain of the amplifier. The arrangement of Figure 6.36c does not provide any added stability, but since V_S follows V_G, the input impedance is only slightly less than R_2.

The AC analysis of any MOSFET type follows that developed for the JFET above. The devices have a broader range of application due to the addition of enhancement and depletion-enhancement types. In low-frequency applications, the extremely large input resistance of the capacitively coupled MOSFET gate (10^{10} to 10^{15} Ω) may be of significance.

6.12 Power MOSFET Circuits

Traditionally, MOSFET devices have had the drain-to-source current confined to a thin planar volume of silicon lying parallel to the gate as shown in Figure 6.33. The limited cross-sectional area of material thus available for conduction effectively limited the power-handling capability of MOSFET devices to less than 1 W. More recently, new designs and manufacturing techniques have been developed to produce a more complicated, three-dimensional gate structure. These transistors are identified by various manufacturers as HEXFET, VMOS, or DMOS, depending on the geometry of the gate structure: respectively hexagonal, V-shaped, or

D-shaped. They feature "on" state resistances of less than an ohm, drain-to-source voltages as high as 500 V, power dissipations exceeding 100 W, and excellent high-frequency operation. In contrast to the normal MOSFET, these devices have a much larger forward transconductance g_m, typically a few mhos. These devices thus feature very high current gain at both high frequency and high power, a combination that is hard to obtain with traditional bipolar power transistors.

Power MOSFETs are currently available only as enhancement devices, and because of the relatively poor conductivity of *P*-type silicon, they are mostly *N*-channel devices. In a circuit like those shown in Figure 6.36, the power MOSFET makes an excellent high-power amplifier. There is one significant difference between DC bias of these devices and the lower-power MOSFETs: The gate-source turn-on voltage (V_0 on the enhancement curve of Figure 6.34 and in Eq. [6.116]) is quite large, ranging from 4 V to 10 V and generally increasing with the maximum drain-source voltage allowed by the device. Power supplies of relatively high voltage are thus required for effective operation.

Although the DC input impedance to the gate of the power MOSFET is still quite large, dynamic effects produce a large effective capacitance between gate and source; for some devices this can be as high as 1000 pF, and at high frequencies greatly reduces the input impedance of the device. Power MOSFETs are well suited to making high-current on-off current switches as discussed in the next section.

6.13 Switching an Inductive Load

Transistors of various kinds can be used to switch the current through an inductive load such as a motor or solenoid. Examples using the enhancement-type power MOSFETs are shown in Figure 6.37. However, if a conducting transistor is sud-

Figure 6.37 (a) A diode across the inductor load protects against inductive surge. (b) A Zener diode across the transistor can perform the same function.

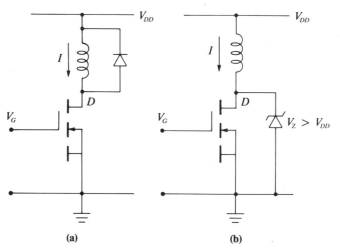

denly turned off, the collapsing magnetic field will continue to drive current through the coil, building up the charge at the drain connection. If no alternate path is provided, this charge and its associated positive voltage can cause a destructive breakdown (spark) within the transistor.

Two methods are available to protect the transistor. The first, shown in Figure 6.37a, uses a diode to limit the drain voltage to no more than the power supply. The diodes used in this application will normally be power diodes, not signal diodes, because they must be able to carry the full operating current of the coil, at least for a short time. The second method shown in this figure uses a Zener diode across the transistor. The breakdown voltage V_Z of this diode can be any value less than the breakdown voltage of the transistor and greater than the power supply voltage V_{DD}.

Problems

1. When the transistor of Figure 6.4b is operated with 6 V between the collector and emitter, the curve indicates that a base current of 100 μA will produce a collector current of 3.5 mA. *(a)* What is the h_{FE} of this transistor at this operating point? *(b)* What is the exact value of the emitter current?

2. Based on the solid state models of diode and transistor operation, do you expect the I_D of Eq. (6.2) to increase or decrease with temperature?

3. In most of the transistor analysis discussed in this chapter, we have assumed that h_{oe} is zero. How would this assumption change the curves shown in Figure 6.4a?

4. Sketch a four-terminal network like Figure 6.7b that has $h_{re} = h_{oe} = 0$. What is the input impedance looking into the *BE* terminals?

5. Using the nominal parameters given in Table 6.1, determine a typical value of g_m for a small-signal bipolar transistor.

6. Determine values for R_B and R_C that will yield the operating point shown in Figure P6.6 under the assumption that: *(a)* $V_{BE} = 0$ and *(b)* $V_{BE} = 0.6$ V. *(c)* Repeat both calculations for $h_{FE} = 80$.

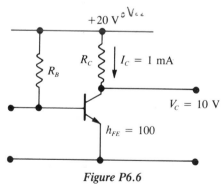

Figure P6.6

7. Derive Eq. (6.41) using the equivalent circuit of Figure 6.11d.

8. (a) If $h_{FE} = 100$, determine an expression for the base-biasing resistor R_B that will result in a DC operating point $V_C = V_{CC}/2$ for the circuit in Figure P6.8. *(b)* If the circuit is built with the R_B just found, what will be the operating point if $h_{FE} = 50$?

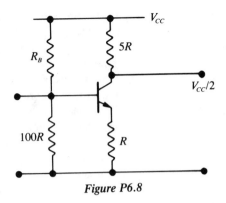

Figure P6.8

9. Equation (6.34) was determined under the assumption that $V_{BE} = 0$. If V_{CC} in Figure 6.9b is 15V, $V_{CE} = 7.5$ V, $R_C = 10$ kΩ, and $h_{FE} = 100$, this equation says that R_F must be 10^6 Ω. *(a)* If V_{BE} actually equals 0.6 V, what will be the true operating point? *(b)* If V_{BE} equals 0.6 V but h_{FE} actually equals 120, what will be the true operating point? Don't change the value of R_F.

10. If the h_{fe} of each of the transistors is 100 and all other hybrid parameters are negligible, what is the approximate AC voltage gain v_{out}/v_s of each of the circuits in Figure P6.10? To simplify things, the DC bias resistors have been omitted; assume that the circuits work without them.

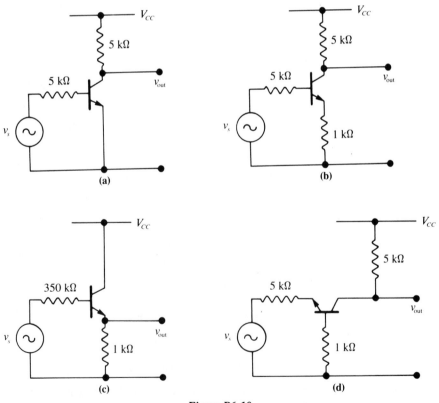

Figure P6.10

11. Retain the transistor symbol and draw an AC equivalent of the circuit shown in Figure P6.11.

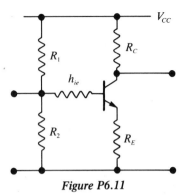

Figure P6.11

12. If the h_{fe} of each of the transistors is 100 and all other hybrid parameters are neglected, what is the input impedance of each of the circuits shown in Figure P6.12? In some cases the DC bias resistors have been omitted; assume that the transistors operate without them.

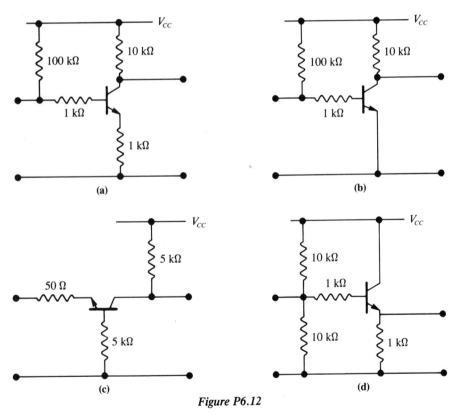

Figure P6.12

13. The transistor used in the circuit shown in Figure P6.13 has $h_{fe} = h_{FE} = 100$; all other hybrid parameters and V_{BE} are to be neglected.

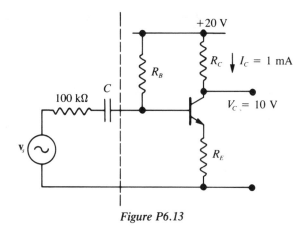

Figure P6.13

(a) Determine the resistors R_C, R_E, and R_B such that the AC voltage gain $v_C/v_B = -10$ when the DC collector voltage and current are as shown.

(b) What is the input impedance of the circuit to the right of the dashed line?

(c) If an AC voltage source is attached as shown, what value of C will result in a corner angular frequency of 5 rad/s?

14. If $\mathbf{Z}_F$ is the impedance of a capacitor C and $A = 50$, show how the Miller effect described by Eq. (B.4) magnifies the effective capacitance to $50C$.

15. A circuit similar to Figure 6.9c also has a resistor R_E in the emitter leg. If $|v_C/v_B| = A$, use the Miller effect to determine the input impedance to this circuit. (*Hint:* The input impedance is made up of three parallel resistive elements.)

16. (*a*) Using the v_{open}/i_{short} method, derive an expression for the output impedance of the simplified *CC* circuit shown in Figure P6.16. Assume that the transistor works without DC bias resistors. Neglect all hybrid parameters except h_{fe}, but do not assume that $h_{fe}R_E >> |\mathbf{Z}_s|$.

(*b*) With $\mathbf{Z}_s = R_s$, sketch the resulting R_{out} as a function of R_s.

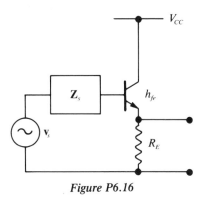

Figure P6.16

17. (*a*) Derive an expression for the transfer function v_C/v_B of the circuit in Figure 6.17.

(*b*) Find the three single-term linear approximations to this transfer function and verify the corner frequency expressions given in the figure.

18. **(a)** What is the collector current I_C of the circuit shown in Figure P6.18? **(b)** Sketch the two-terminal equivalent circuit that best describes the circuit inside the dashed lines. Label the elements with calculated or estimated values.

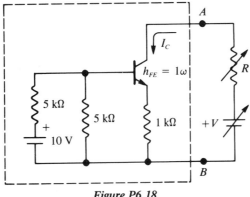

Figure P6.18

19. In the absence of an applied input signal, calculate the power dissipated in the resistor R_C and in the transistor of Figure 6.12.

20. Assume that the JFET shown in Figure P6.20 follows the characteristic curves given in Figure 6.25a.

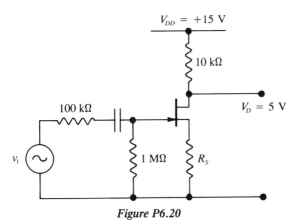

Figure P6.20

(a) Find the value for R_S that will result in a DC operating point $V_D = 5$ V.
(b) What is the AC voltage gain v_D/v_1 of this circuit above the low-frequency corner? [*Ans.*: −4.5]

21. Estimate the on-resistance r_0 for the JFET described by Figure 6.25a. [*Ans.*: 1 kΩ]

22. Sketch the behavior of R_{DS} versus V_{GS} for a JFET operated as a variable resistor. Label r_0 and V_0 on your graph.

23. **(a)** If $I_{DSS} = 2$ mA and $V_0 = 1.5$ V for the circuit shown in Figure P6.23, choose R_D and R_S such that $I_D = 1$ mA and $V_D = 10$ V. **(b)** If $g_m = 10^{-3}$ mho, what is the AC gain of your resulting amplifier? [*Ans.*: −3.47]

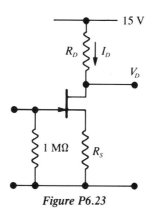

Figure P6.23

24. What is the input impedance to the amplifier of the previous question?

25. *(a)* If $R_S = 10$ kΩ, $R_D = 50$ kΩ, $C = 1$ μF, and $g_m = 10^{-2}$ mho, sketch $|v_D/v_G|$ from Eq. (6.92). *(b)* Find the corner frequencies of this transfer function.

26. Find an expression for the input impedance to the circuit shown in Figure P6.26. [*Ans.:* $R_S + R(1 + g_m R_S)$]

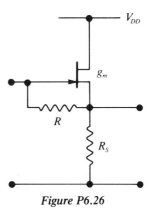

Figure P6.26

27. If $R_s = 100$ kΩ and $C_{GD} = 2$ pF, find f_c for the designs of Example 6.7.

28. Describe the operation of the circuit shown in Figure P6.28 in terms of a Norton equivalent circuit. Label the Norton components in terms of whichever JFET parameters you need: g_m, I_{DSS}, V_0, C_{DG}, C_{GS}.

29. Calculate the ratio v_2/i_1 for the two circuits shown in Figure P6.29. These transistors are MOSFETs, but for AC analysis they behave just like JFETs.

30. Suppose that we build a circuit like that of Figure 6.31a and for our specific choice of components it turns out to have AC gain $|v_D/v_G| = 100$ at a resonant frequency of 16 MHz. The manufacturer's specification sheet says that the JFET has $C_{DG} = 5$ pF.
(a) What is the magnitude of the input impedance to this amplifier at resonance?
(b) If the circuit is to be driven by a signal source with a 1000-Ω output impedance, what will be the expected overall gain $|v_D/v_s|$ at the resonant frequency? [*Ans.:* 2]

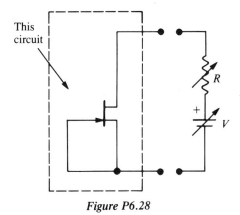

Figure P6.28

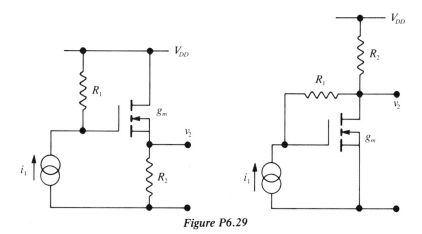

Figure P6.29

31. For the FET circuit shown in Figure P6.31, choose the smallest possible values for C_1 and C_2 consistent with a flat response above a corner frequency of 100 Hz. [*Ans.*: C_1 = 1450 pF; C_2 = 1.6 μf]

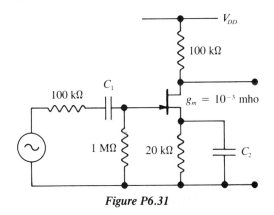

Figure P6.31

32. If on the MOSFET circuits of Figures 6.36a and 6.36b, we have $V_{DD} = 20$ V, $R_D = 8$ kΩ, $R_S = 1$ kΩ, and $R_1 = R_2 = 100$ kΩ:

(a) Plot V_{GS} versus the drain current I_D, and label the intersections on both axes.

(b) If the nominal DC operating point for a particular MOSFET gives $I_D = 1$ mA, use these plots to explain why the circuit of Figure 6.36b will show less sample-to-sample variation as a result of changing transistor parameters in Eq. (6.116).

7

Multiple Transistor Circuits

7.1 Introduction

Integrated circuit technology has made it possible to manufacture a complex circuit of interconnected transistors on a single silicon chip, and the quality and diversity of these prefabricated devices are rapidly eliminating the need to fabricate such circuits from discrete transistors. Only occasionally, for reasons of availability, power requirements, low noise, or high-frequency response, is it truly necessary to fabricate a circuit involving more than a couple of discrete transistors. Rather than undertake the design of a complicated circuit using discrete components, it is far better to spend a few hours reviewing a well-stocked library of handbooks and data manuals (catalogs) issued by IC manufacturers. Not only can this save circuit development time, but the resulting circuit will benefit from the many hours of engineering work that went into the IC's design.

For this reason, the circuits discussed in this chapter are limited for the most part to simple two- or three-transistor examples that perform a specialized circuit function. In many cases the circuits shown, or variations on them, are used as building blocks of much larger integrated circuits. They are presented here not only as circuits to be used independently, but also as aids to the detailed understanding of the operation of integrated circuits, especially as that operation relates to the IC's input and output characteristics.

7.2 Coupling Between Single Transistor Stages

Quite often the single-transistor amplifier discussed in the previous chapter does not provide enough gain for an application, or more often, it does not combine gain with the desired input and output impedance characteristics. Perhaps the most obvious solution (but generally not the best) is to connect several single-transistor amplifiers, or stages, in tandem one after the other. Because of DC biasing considerations, it is usually not practical to connect the output of one stage directly to the input of another; some kind of coupling device must be used that permits a change in the DC level between two stages. Possible coupling devices include capacitors, transformers, additional transistors, and occasionally batteries or Zener diodes.

7.2.1 Capacitive Coupling

A favorite circuit of yesteryear featured capacitor coupling between single-transistor stages as shown in Figure 7.1 Although for some applications this is still a reasonable approach, the coupling capacitors are likely to be many times more expensive than a complete IC amplifier. Cost aside, these amplifiers are easily analyzed using the methods introduced in previous chapters and serve as a useful introduction to the general problems of interstage coupling.

The interstage coupling capacitor C_2 serves the same purpose as C_1: It allows the following transistor to be DC biased without concern for the DC component of the driving signal. Thus, the DC analysis of the two transistor stages is separated into two distinct problems, which can each be handled by the techniques of the last chapter.

The AC analysis of circuits of this type is best approached using the Thevenin equivalent circuit elements of input impedance, open circuit voltage gain, and output impedance of each stage. Figure 7.1b shows a three-section AC equivalent circuit for the two-stage *CE-CE* amplifier shown. The transistor action is approximately modeled by the two hybrid parameters h_{ie} and h_{fe}, and it is assumed that h_{fe} is large enough that the open circuit voltage gain of each transistor can be derived from the ratio of its collector and emitter resistors as discussed in Section 6.3.6.

Since each section has the basic form shown in Figure 7.1c, these models are useful for determining the high-pass filter characteristics resulting from the coupling capacitors C_1 and C_2. Analysis of Figure 7.1c shows that the corner frequency resulting from each capacitor is given by

$$\omega_c = \frac{1}{(R_A + R_B)C} \tag{7.1}$$

where R_A and R_B represent the total resistance in series with a particular capacitor. To retain amplifier gain at low frequencies (10–100 Hz), these coupling capacitors typically must be larger than 0.1 μF.

Occasionally the configuration of the individual stages may permit DC coupling between some stages as shown in Figure 7.2a. The circuit is in a *CC-CE-CC* configuration. The only voltage gain occurs between the base and collector of Q_2; the remainder of the circuit is designed to minimize the voltage loss due to circuit loading. The first *CC* stage has Q_1 biased to develop a 2-V DC signal at its emitter; this also serves to bias the emitter of the *CE* stage to (2 V − V_{PN}) or about 1.5 V. The 1.5-kΩ emitter resistor at Q_2 thus must carry 1 mA of current—

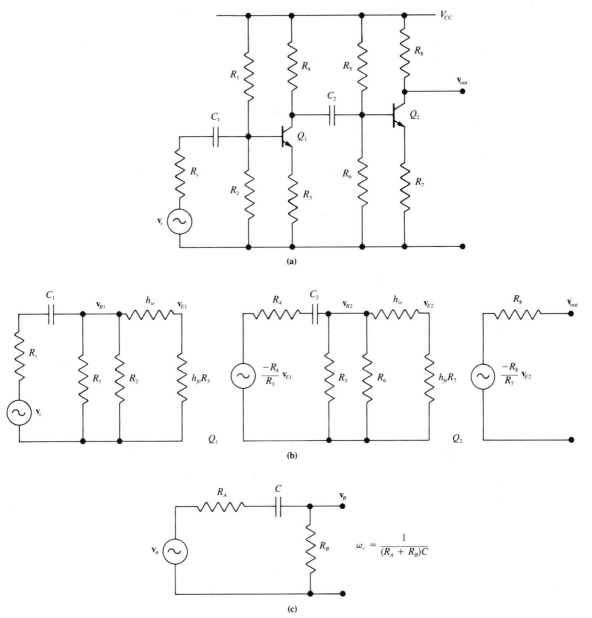

Figure 7.1 *(a) The basic two-stage capacitively coupled CE-CE amplifier and (b) an AC equivalent expressed in terms of the open circuit voltage gain of each stage and the input and output impedances of each stage. (c) A simplified circuit showing the high-pass filter effects of each of the coupling capacitors C_1 and C_2.*

current that also must pass through the 5-kΩ collector resistor. This collector current drops the collector to 5 V below the power supply V_{CC}, setting the base of Q_3 to 10 V and its emitter to (10 V $-$ V_{PN}) or about 9.5 V. Any change of the bias resistors at the base of Q_1 will change all of these quiescent currents and voltages.

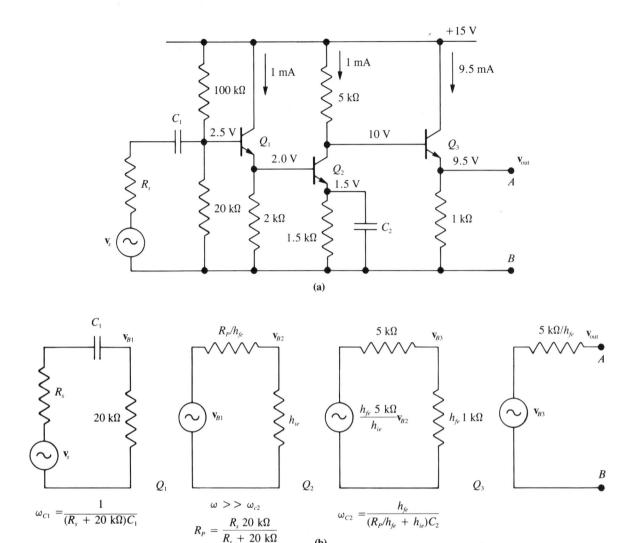

Figure 7.2 *A three-stage capacitively coupled CC-CE-CC amplifier featuring high input impedance, low output impedance, and one stage of voltage gain. (b) The AC equivalent circuit above the corner frequency ω_{c2}.*

Again we can make the AC analysis in terms of Thevenin equivalent circuits—four sections this time, as shown in Figure 7.2b. The formulas for the equivalent circuit components are shown in the figure and can be slightly simplified by assuming that $R_s \ll 20$ kΩ; this makes R_P on the second loop of the equivalent circuit equal to the signal generator output impedance R_s.

The *CC-CE-CC* arrangement results in a large input impedance and a small output impedance for the overall circuit; it also features good impedance matching between stages, allowing the full voltage gain potential of the *CE* stage to be realized. In the pass region, the voltage gain of the *CE* stage is $5000h_{fe}/h_{ie}$, and the surrounding *CC* stages isolate this high-gain stage. In this example, the overall input impedance is determined by the fairly small base biasing resistors of Q_1. The first common collector stage features a small output impedance, which can drive the second stage with little loss of signal between $\mathbf{v}_{B1}$ and $\mathbf{v}_{B2}$. Above the corner frequency, the AC input impedance to Q_2 is small (just h_{ie}), since this common emitter stage uses a bypassed emitter resistor for maximum AC gain. The moderately large output impedance at the collector of Q_2 is only required to drive the even larger input impedance of the last common collector stage, again maintaining the signal.

Most of the resistors shown on this circuit were chosen to obtain the DC coupling between the stages. As noted above, only the 5-kΩ collector resistor of Q_2 is directly related to the overall voltage gain of this amplifier. It is possible to increase this resistor to obtain more gain, but only if the emitter resistor of Q_2 is also increased to retain the DC bias condition.

As in the previous example, the input coupling capacitor C_1 results in a high-pass filter with a corner frequency given here by

$$\omega_{c1} = \frac{1}{(R_s + 20\ \text{k}\Omega)C_1} \tag{7.2}$$

The bypass capacitor C_2 also has a high-pass filter effect, this time with a corner frequency given by

$$\omega_{c2} = \frac{h_{fe}}{(R_P/h_{fe} + h_{ie})C_2} \tag{7.3}$$

These capacitors must be chosen to assure that these poles lie well below the minimum operating frequency of the amplifier. If the amplifier is to operate in the low audiofrequency range, C_2 may need to be relatively large, sometimes requiring a polarized electrolytic. The high-frequency corner will be determined by the Miller-enhanced stray capacitance C_{BC} of Q_2, but since Q_2 is being driven by a low-impedance signal (R_s/h_{fe}), this corner can easily be over 1 MHz.

7.2.2 Transformer Coupling

In high-frequency applications, transformers or a combination of transformers and capacitors are commonly used to provide frequency-selective coupling between transistor stages. The necessary inductances are small at these frequencies, making for relatively compact circuits, and they can even be used to improve the high-frequency gain of the transistor stages. The circuit of Figure 7.3a features two N-channel JFETs in the common source configuration. Transformer coupling is particularly convenient for JFETs, since the DC gate voltages of the FETs are automatically held to zero by the secondary coils of transformers T_1 and T_2. The AC equivalent shown in Figure 7.3b includes capacitors C_{S1} and C_{S2}, which represent the stray capacitances between the gate and drain of each of the transistors. The resonances available from the combination of inductances and capacitances can be used to enhance the gain in a narrow frequency range.

As shown in Section 4.2.3, a capacitance C in parallel with the secondary of a transformer develops a transformer input impedance of

$$\mathbf{Z}_{\text{in}} = \frac{j\omega L}{1 - \omega^2 LC/n^2} \tag{7.4}$$

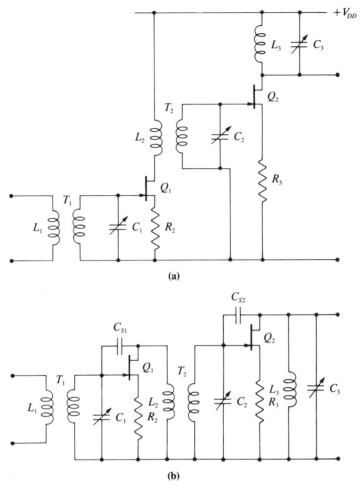

Figure 7.3 *(a) Transformer T_2 provides inductive coupling between two JFETs. (b) AC equivalent circuit showing the stray capacitances C_{S1} and C_{S2}.*

where n is the turns ratio of the transformer and L is the impedance of the primary coil. Note that this is just the parallel combination of L and C/n^2. For idealized (resistanceless) coils this impedance goes to infinity at a resonant frequency given by

$$\omega_r = \frac{n}{\sqrt{LC}} \tag{7.5}$$

As discussed in Section 6.9.3, this behavior results in a resonance shape for the transfer function of each of the transistors in this circuit.

The abundance of reactive elements makes the analysis of even the simplest of these circuits fairly involved; fortunately, a qualitative analysis of the equivalent circuit of Figure 7.3b will be sufficient to associate inductors with capacitors for the purpose of identifying the resonant frequencies. For notational simplicity, we will consider a design where all of the resonances occur at the same frequency. Starting with the second transistor Q_2, the elements L_3 and C_3 form a tank circuit modified by the effects of the feedback capacitor C_{S2}. Because of the voltage gain of this stage, the AC voltage on the gate side of C_{S2} is much smaller than on the drain side. The gate side is thus effectively at an AC ground, placing C_{S2} in parallel with C_3. At a resonant frequency given by

$$\omega_r = \frac{1}{\sqrt{L_3(C_3 + C_{S2})}} \tag{7.6}$$

this stage develops a maximum voltage gain of magnitude A_2, which is limited by the transconductance g_m of the FET and the Q (or series resistance) of the inductance L_3.

As seen from the secondary of transformer T_2, the gate-drain capacitance of Q_2 is enhanced by the Miller effect to become $(A_2 + 1)C_{S2}$ in parallel with C_2 across the secondary of transformer T_2; the resulting impedance loading T_2 is

$$\mathbf{Z}_2 = \frac{1}{j\omega[C_2 + (A_2 + 1)C_{S2}]} \tag{7.7}$$

which is just the impedance of an effective capacitor $C_{L2} = [C_2 + (A_2 + 1)C_{S2}]$. The load on transistor Q_1 is then L_2 in parallel with $C_{L2}/n_2{}^2$, and by the same argument used for Q_2 this will combine with the feedback capacitor C_{S1} to produce a maximum gain of magnitude A_1 at

$$\omega_r = \frac{n_2}{\sqrt{L_2(C_{L2} + C_{S1})}} \tag{7.8}$$

The Miller effect-enhanced C_{S1} combines with C_1 to produce a load impedance for transformer T_1,

$$\mathbf{Z}_1 = \frac{1}{j\omega[C_1 + (A_1 + 1)C_{S1}]} \tag{7.9}$$

which defines another effective capacitance $C_{L1} = [C_1 + (A_1 + 1)C_{S1}]$. If the circuit is driven from a high-output impedance source, this capacitance will result in a third frequency-selective stage with its resonant frequency at

$$\omega_r = \frac{n_1}{\sqrt{L_1 C_{L1}}} \tag{7.10}$$

In practice, these three resonant frequencies will often be adjusted to slightly different values, thereby producing a bandpass filter with a wider and relatively constant gain pass band.

At high frequencies (10 MHz and up), where circuits of this type are normally used, the inductors and transformers are usually small air core coils, often of only a few turns. Since the capacitors are also small values (pF), the circuit can be easily modified by the presence of stray capacitance and inductance associated with circuit wires, probes, or even fingers. The circuit shown is already a six-pole system but is still only an idealization of the actual circuit. Although the circuit shown is stable against oscillation, since all of the poles are on the left side of the s-plane, consideration of other possible stray capacitance connections (e.g., the drain of Q_2 to the gate or source of Q_1) can easily result in poles in the right half plane and oscillation rather than amplification.

7.3   *Darlington and Sziklai Connections*

Two bipolar transistors in either the Darlington or Sziklai connection can be used as a single, high gain transistor. These circuits have AC current gains on the order of h_{fe}^2, providing high amplification in a single stage. Of course the DC current gain will also be large, something like h_{FE}^2, and must be taken into account when developing the DC bias circuit for those devices.

The arrangement of two *NPN* (or two *PNP*) transistors shown in Figure 7.4a is known as the Darlington connection. The three unconnected terminals, the base of Q_1 and the collector and emitter of Q_2, behave like a single transistor with current gain $h_{fe1}h_{fe2}$ and an overall base to emitter DC voltage drop of $2V_{PN}$. The DC collector current in transistor Q_1 is less than optimal in the basic Darlington connection, and the h_{fe} of this transistor can generally be increased by the addition of a small resistor (the order of h_{ie}) between the base and emitter of Q_2 as shown in Figure 7.4b. Two transistors in the Darlington connection can be purchased in a single three-terminal package, usually with the resistor R included in the package.

A similar circuit result can be obtained with the Sziklai connection of Figure 7.5. Because it uses one transistor of each polarity (*NPN* and *PNP*), this connection is also known as the complementary Darlington. The combination again results in a three-terminal device that behaves like a single high-current-gain transistor. Since the base of Q_1 is separated from the collector of Q_2 only by a

Figure 7.4   (a) The Darlington connection of two transistors to obtain higher current gain. (b) An improved version of this connection.

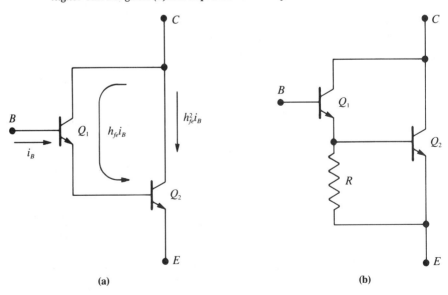

(a)   (b)

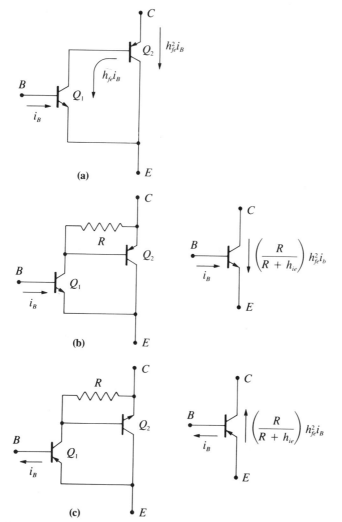

*Figure 7.5 (a) The Sziklai connection of an NPN and a PNP
transistor to obtain the equivalent of a high-current-gain NPN
transistor. (b) An improved version of the same connection. (c) This
Sziklai connection behaves like a high-current-gain PNP transistor.*

forward-biased diode, the collector of Q_2 becomes the effective emitter of the
composite device. Likewise, the emitter of Q_2 functions as the effective collector
of the composite, since it is separated from the base of Q_1 by a reverse-biased
diode. Thus, the overall circuit behaves like a transistor of the same polarity as
Q_1, an *NPN* in the example of Figures 7.5a and 7.5b or a *PNP* in Figure 7.5c.

As in the case of the Darlington connection, the operation of the Sziklai combination is improved by the addition of a resistor from the base to emitter of Q_2.
Unlike the Darlington, the Sziklai arrangement exhibits a single V_{PN} voltage drop
between the base and the effective emitter.

Because of the higher conductivity of *N*-type silicon compared to *P*-type, *NPN*
transistors (and *N*-channel FETs) are the preferred high-power devices. As a con-

sequence, a common use of the Sziklai connection is to form a high-power effective *PNP* transistor using a low-power *PNP* at Q_1 and a high-power *NPN* at Q_2. Because fabrication techniques further limit the quality of *PNP* transistors in integrated circuits, this arrangement is often found on the push-pull output stage of IC operational amplifiers.

7.4   Current Source as an Emitter Load

In some circuits, additional transistors serve mainly to enhance the performance of an otherwise normal one-transistor amplifier. The circuit of Figure 7.6a shows an ideal current source replacing the source resistor of a common drain (source follower) amplifier. If the constant current I_{SS} is chosen to be approximately $I_{DSS}/2$, the source voltage will automatically assume the correct DC bias. Since the

Figure 7.6 *(a) A constant current source makes a high-impedance active load for the CC amplifier. (b) The AC equivalent of this circuit shows the simplifying effect of a constant current source load. (c) A second transistor in the CC configuration provides the constant current.*

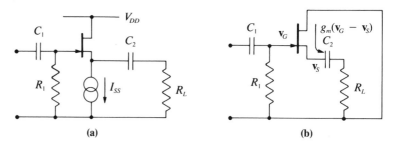

(a)   (b)

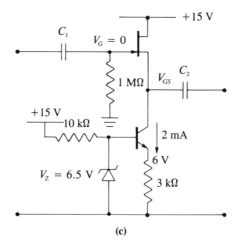

(c)

ideal current source is presumed to have an infinite impedance, none of the AC drain current i_D can pass through the current source, and the current source does not appear in the AC equivalent circuit of Figure 7.6b. Above the corner frequencies associated with capacitor C_2, the voltage gain is just

$$v_S/v_G = \frac{g_m R_L}{1 + g_m R_L} \tag{7.11}$$

The gain of this amplifier is thus affected only by R_L, without modification by the normal source resistor R_S.

Figure 7.6c shows an implementation of this circuit using a bipolar transistor current source. The presumption in this circuit is that the FET chosen will have an I_{DSS} of 4 mA or greater. The circuit is easily modified for other I_{DSS} by changing the 3-kΩ emitter resistor to some other value. A Zener diode of approximately 6 V is preferred because of its inherent temperature stability.

7.5   Cascode Connection

Although the cascode connection shown in Figure 7.7a bears a superficial resemblance to the circuit just discussed (see Figure 7.6b), its operation is quite different. Transistor Q_2 in this circuit is in the common base configuration: Its base is at a fixed voltage and the signal path is from the emitter to the collector. Looking into the emitter of Q_2 from the collector of Q_1, we see a very small AC input impedance given by h_{ie}/h_{fe}. Transistor Q_2 thus acts to keep the collector of Q_1 at a fixed voltage: a *PN* voltage drop below the constant base voltage of Q_2. An AC signal applied to the base of Q_1 generates an AC collector current through both transistors and will cause a change in the voltage across R_4, but the collector of Q_2 is effectively at an AC ground.

The main function of this circuit arrangement is to reduce the Miller effect derived in Appendix B. The effective Miller capacitance is given by

$$C_M = C_{CB}(|A_L| + 1) \tag{7.12}$$

where A_L is the low-frequency voltage gain of the transistor. This capacitance appears as a parallel addition to the input impedance, and limits the high-frequency response of the amplifier by introducing a low-pass filter stage into the base circuit. The cascode connection eliminates the $(|A_L| + 1)$ multiplication by holding the collector of Q_1 to an effective AC ground (making $A_L = 0$). In the AC equivalent circuit of Figure 7.7b, the stray capacitance C_{CB} runs directly from the base to the AC ground.

The AC equivalent for the cascode circuit consists of two sections, with Q_1 acting as an AC current source for the second section. The cascode connection actually only moves the Miller effect from transistor Q_1 to transistor Q_2, for even though the base of this transistor is at a constant voltage, the collector voltage is changing. However, the output impedance of the constant voltage source driving the base of Q_2 is tiny, minimizing the effect of stray capacitance at this transistor.

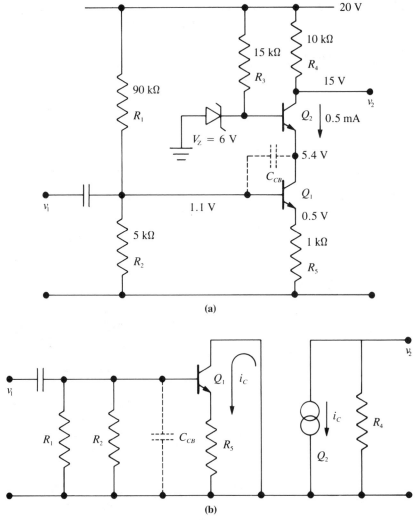

(a)

(b)

*Figure 7.7   (a) Two NPN transistors in the cascode connection. (b)
The AC equivalent circuit if Q_2 is treated as an ideal current source.*

The corner frequency of the low-pass filter formed at the base of Q_1 in Figure
7.7b moves to lower frequencies as the signal generator output impedance and/or
the input impedance to Q_1 gets larger. For this reason the Miller effect is particularly limiting for high-impedance FET common source circuits, even though the
stray capacitance of these devices is generally smaller than that of bipolar transistors. A cascode connection can be formed using two FETs, but the most convenient arrangement is the FET and bipolar combination shown in Figure 7.8. This
circuit retains both the high input impedance of the FET and the convenient voltage biasing of the bipolar transistor.

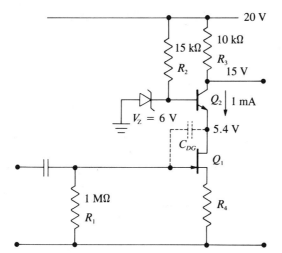

Figure 7.8 *The cascode connection of a bipolar transistor and an FET
is very useful at high frequencies.*

7.6 MOSFET Biasing with a Current Source

The N-channel, enhancement-type MOSFET must be biased with the gate several
volts above the source. Although this can be done with a gate resistor network as
shown in the last chapter, variations on the current source form of biasing shown
in Figure 7.9a are also seen. The constant current I_{GG} in resistor R_1 develops the
gate-to-source voltage needed to bias the MOSFET into conduction. If the current
source has a large enough impedance, it can be neglected as in the AC equivalent

Figure 7.9 *(a) An idealized current source used to bias an N-channel,
enhancement-type MOSFET. (b) The AC equivalent circuit has an
input impedance much greater than R_1.*

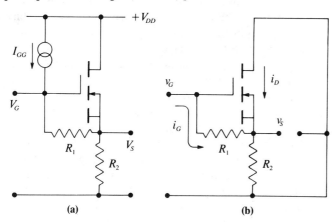

circuit of Figure 7.9b. Because the source voltage closely follows the voltage applied to the gate, the input impedance to this circuit is $(1 + g_m R_2)R_1$, generally much greater than R_1.

7.7 Matched Transistor Pairs

Many useful circuits can be formed using a matched pair of nearly identical transistors. Such pairs can, in principle, be found by a careful selection process starting with a large transistor sample. A far better approach is to begin with a package containing two matched transistors fabricated on a single silicon chip. Since such transistors are in close proximity, they operate at nearly the same temperature and retain their matched characteristics much better than two identical but physically separated transistors. Several of the following circuits make use of one or more matched pairs.

7.8 Current Mirror

The current mirror is an elegantly simple circuit that is usually used to bias or interconnect parts of a larger circuit. The basic configurations using a pair of *PNP*s or a pair of *NPN*s are shown in Figure 7.10a and 7.10b. The signal is an applied current source, shown as I_P in the figure, and known as the programmed current. At transistor Q_1 this current divides, with small and equal amounts passing through the base emitter junctions of the two transistors. The magnitude of this base current is just sufficient to bias Q_1 into the conduction needed to pass the remainder of I_P, and since Q_2 is presumed to be identical to Q_1, it responds with a collector current that is also equal to I_P, to first-order independent of the voltage at the collector of Q_2.

To understand the effect Q_1 has on the operation of Q_2, we must depart from our usual practice of taking the magnitude of the base emitter voltage drop V_{PN} to be a constant. As can be seen from Figure 6.5a, this voltage is in fact a function of I_C (or I_B), and resembles the current-voltage curve of a forward-biased diode. As I_P increases, $|V_{PN}|$ also increases, following a curve like that shown in Figure 6.5a. If two transistors are connected as a current mirror, this same voltage will appear across the base emitter junction of Q_2. To the extent that Q_2 follows the same curve, its collector current must then also be I_P.

One problem with the basic mirror circuit is that for a constant collector current, V_{BE} varies somewhat with the collector-to-emitter voltage V_{CE}. Thus, when the load on Q_2 presents a changing voltage, the collector current in Q_2 will not exactly mirror that in Q_1. A linear analysis of this current source would show a fairly small output impedance, or in more general language, this current source is said to have a poor compliance. An improved version of the current mirror, shown in Figure 7.10c, uses matched emitter resistors to increase the fraction of the base voltage, which is strictly a function of the collector current. If the emitter resistors are large enough that the voltage dropped across them is several times the nominal

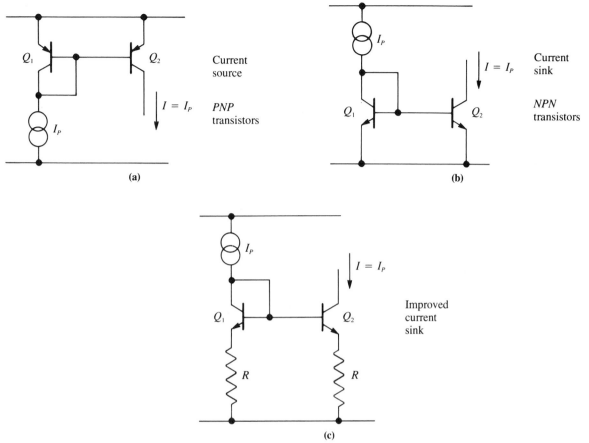

Figure 7.10 *(a) A current mirror using a pair of matched PNP transistors provides a programmable current source. (b) A current mirror using a pair of matched NPN transistors. (c) An improved design.*

PN voltage drop (0.5 V), then small changes in V_{BE} will have a diminished effect on the collector current and the compliance of the current mirror will be greatly improved.

7.9   Differential Amplifier

Perhaps the most common feature of modern analog designs is the differential amplifier constructed from a matched pair of bipolar or field effect transistors. Variations on the circuit discussed here are universally used to form the input stage of the extremely versatile operational amplifier devices to be treated in the next chapter.

A basic circuit using bipolar transistors is shown in Figure 7.11a. This circuit is presumed to operate from two separate (but not necessarily equal-magnitude)

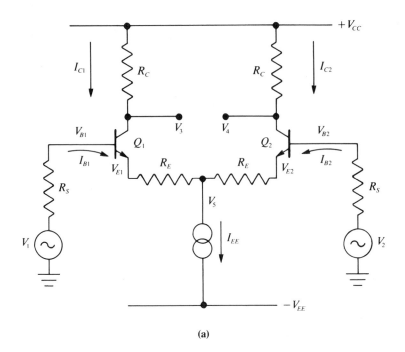

(a)

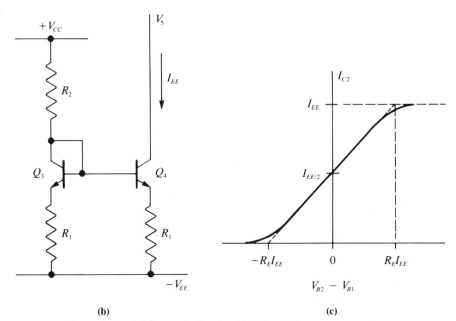

(b)

(c)

*Figure 7.11 (a) A matched pair of NPN transistors connected as a
differential amplifier. (b) A current mirror provides the constant
current sink. (c) The collector current in Q_2 as a function of the
difference in base voltages.*

power supplies, one above ground and one below. This is not actually necessary, but it greatly simplifies the DC biasing problem and is the most common situation. No explicit DC biasing circuit is shown on this schematic; we may assume, as is usually the case in practice, that the voltage sources V_1 and V_2 provide a DC path to ground for the base bias current I_{B1} and I_{B2}. This is an important requirement, but so common that is easily overlooked when dealing with the operational amplifiers.

A high-impedance current source, shown as an ideal device I_{EE} in Figure 7.11a, is essential to the operation of the differential amplifier. One possible circuit for this current source is shown in Figure 7.11b. This circuit sets the source current at

$$I_{EE} = \frac{V_{CC} - V_{EE} - V_{PN}}{R_1 + R_2} \tag{7.13}$$

A single-transistor current source such as the one shown in Figure 7.6c can also be used in place of the current mirror.

The DC operating point can be found by setting both voltage sources V_1 and V_2 to zero. If the transistors, Q_1 and Q_2, and all resistors are carefully matched, half of the source current I_{EE} will pass through each transistor:

$$I_{C1} = I_{C2} = \frac{I_{EE}}{2} \tag{7.14}$$

The collector voltages must then be given by

$$V_3 = V_4 = V_{CC} - \frac{R_C I_{EE}}{2} \tag{7.15}$$

It is evident that this condition persists as long as I_{EE} divides evenly between Q_1 and Q_2 and therefore does not really require that the source voltages be zero, only that $V_1 = V_2$.

Since the differential amplifier shown has no coupling capacitors, the amplifier will respond to both AC and DC signals and there is no need for an explicitly AC analysis. However, it is still convenient to distinguish between the bias voltages and an applied signal by using lowercase letters for the applied signal and discussing it as an AC signal. We will therefore stick with the AC notation and point out that components of this input signal can now extend down to zero frequency. Assume, then, that v_1 and v_2 are any two time-varying signals. Taking a clue from the above discussion of the operating point, it is useful to write these two signals in terms of two other signals v_C and v_D, defined by

$$v_1 = \frac{v_C + v_D}{2}$$

$$\tag{7.16}$$

and
$$v_2 = \frac{v_C - v_D}{2}$$

where v_C and v_D are known as the common mode signal and the difference signal, respectively. Inverting these equations points up the distinction between these new composite signals:

$$v_C = v_1 + v_2$$

and

$$v_D = v_1 - v_2 \tag{7.17}$$

Since the application of identical signals (common mode only with the difference equal to zero) to the bases of the two transistors continues to result in an even split of I_{EE} through the two collector resistors, this type of input ideally produces no change in the current or voltage at either collector. Thus,

$$i_{C1} = i_{C2} = 0$$

and

$$v_3 = v_4 = 0 \tag{7.18}$$

must hold for any common-mode input signal. Since $i_C = 0$ implies that $i_B = 0$, the base and emitter voltages of both transistors must follow the common input voltage. With base and collector currents constant, there is no AC current in either R_E, with the consequence that the voltage v_5 at the top of the current source must also equal the common-mode input signal. If the current source is not infinite impedance (poor compliance), this voltage change will change I_{EE}, destroying the argument and producing a common-mode output signal at the collectors. The suppression of the common-mode part of the input signal before it reaches the collector output terminals (common-mode rejection) is therefore only as good as the current source.

The application of a pure difference signal to the two bases results in a change in the sharing of I_{EE} between Q_1 and Q_2 as shown in Figure 7.11c and can produce a voltage response v_3 and v_4 at the collectors. If we assume identical transistors and resistive components, then for equal and opposite input signals of the form

$$v_1 = \frac{v_D}{2}$$

and

$$v_2 = \frac{-v_D}{2} \tag{7.19}$$

the base currents are given by

$$i_{B1} = \frac{v_1 - v_5}{R_S + h_{fe}R_E}$$

and

$$i_{B2} = \frac{v_2 - v_5}{R_S + h_{fe}R_E} \tag{7.20}$$

Remembering that the current source I_{EE} behaves like a very large impedance to some negative voltage supply, it is clear that V_5 will remain constant and v_5 will be zero as long as no attempt is made to change the current through this impedance. Since equal and opposite changes in the collector currents do not affect I_{EE}, V_5 will not change as a result of the application of the pure difference signal of Eqs. (7.19). Substituting the difference signals for v_1 and v_2 and setting v_5 to zero yields

$$i_{B1} = \frac{v_D}{2(R_S + h_{fe}R_E)}$$

and

$$i_{B2} = \frac{-v_D}{2(R_S + h_{fe}R_E)} \tag{7.21}$$

Multiplication by h_{fe} gives the respective collector currents, and the resulting collector voltages (including the DC part) are

$$v_3 = V - \frac{v_D h_{fe} R_C}{2(R_S + h_{fe} R_E)}$$

(7.22)

and

$$v_4 = V + \frac{v_D h_{fe} R_C}{2(R_S + h_{fe} R_E)}$$

The difference of these two output signals is then

$$v_3 - v_4 = \frac{-v_D h_{fe} R_C}{R_S + h_{fe} R_E}$$

(7.23)

The output of an operational amplifier can be taken from either collector, or both collectors can be used to feed a second differential amplifier stage. Although any of these outputs from an ideal differential amplifier will be sensitive only to the difference signal v_D, the input signal can be a combination of difference and common-mode signals. A common application is to fix the second input at some constant voltage v_2 (often ground) and apply a time-varying signal v_1 at the other input. The resulting signal v_3 will be inverted (it will go down when v_1 goes up), whereas the signal v_4 will be noninverted. Each of these output signals will be proportional to $|v_1 - v_2|$ but will be superimposed on the DC bias voltages given by Eq. (7.15).

A simplified version of the differential amplifier with the emitter resistors omitted is often seen on integrated circuit schematics. An example is shown in Figure 7.12a. Although this circuit will function, it does not have the large input impedance usually associated with differential amplifiers. The input impedance is easily obtained from the AC equivalent of this circuit, shown in Figure 7.12b. The base current i_1 into Q_1 produces a collector current i_C as shown; these combine at the emitter connection (which can assume any voltage), then divide again in the matched transistor Q_2. If we ground the input v_2 as indicated by the dashed connection, it is clear that the input impedance at v_1 is just $2h_{ie}$.

One way to avoid explicit resistors and still get a large input impedance is to replace each transistor with a Darlington pair as shown in Figure 7.12c. By the same argument used above, the input impedance to this circuit must be $2h_{fe}h_{ie}$.

7.9.1 Differential Amplifiers with Active Loads

Integrated circuit schematics often use additional transistors to augment the performance of a differential amplifier, and the current mirror appears frequently. In addition to its use for the constant current I_{EE}, this circuit finds various applications as an active collector load. The circuit of Figure 7.13 uses a current mirror as a high-impedance load, effectively developing an output current that is the difference of the two collector currents. The analysis is straightforward: The base current into Q_1 determines I_1, which becomes the program current of the mirror. Since this current is mirrored in Q_4, and I_2 is determined by the condition $I_1 + I_2 = I_{EE}$, the difference ($I_1 - I_2 = 2I_1 - I_{EE}$) must flow in the output circuit. The reverse polarity version of this circuit, shown in Figure 7.13b, is often used

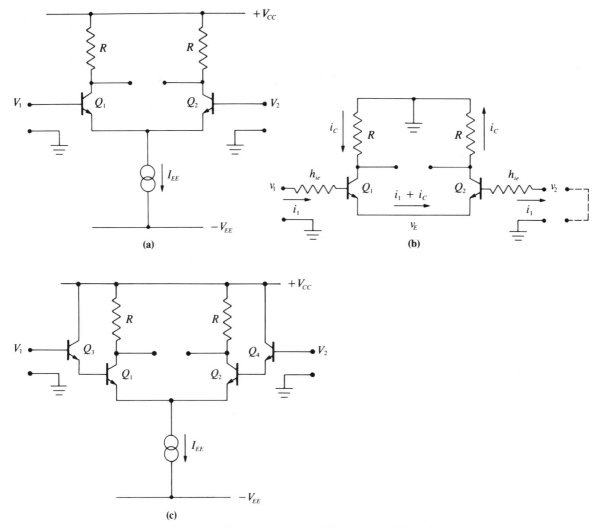

Figure 7.12 *(a) Simplest form of a differential amplifier. (b) Equivalent AC circuit assuming a perfect current source for I_{EE}. (c) Higher input impedance and higher gain is obtained with a double Darlington connection.*

as the input stage to operational amplifiers designed for use with a single-voltage power supply.

Still another use of the current mirror in conjunction with a differential amplifier is shown in Figure 7.14a. This time the current mirror produces an output current equal to I_2 or $I_{EE} - I_1$. Note that while the AC part of I_2 will be proportional to the difference in the two base voltages, the value of I_2 with both inputs grounded is $I_{EE}/2$. The advantage of this circuit lies in its ability to develop an output voltage that is offset by any voltage from $-V_{EE}$ to nearly V_{CC}, as shown by the examples in Figure 7.14b. This circuit is therefore used as a DC coupling between integrated circuit amplifier stages.

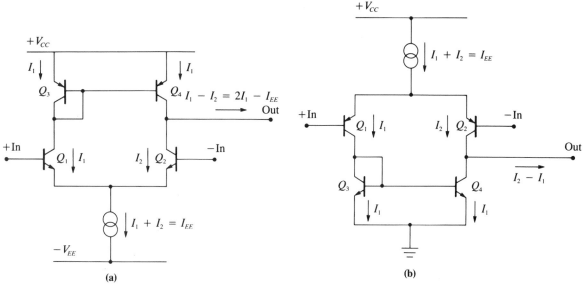

Figure 7.13 (a) A PNP current mirror can replace the collector resistors of an NPN differential amplifier. (b) The same circuit with transistor polarities reversed is a common input stage to single-voltage operational amplifiers.

7.10 Push-Pull Output Stages

The last stage of an amplifier is often required to supply relatively large currents to the load and must therefore have a small output impedance. In some applications a single transistor in the common collector (or common drain) configuration can be used as the output stage; this arrangement is shown in Figure 7.2a. However, although the small-signal output impedance of this common collector is small as shown in Figure 7.2b, for large voltage swings this stage exhibits an asymmetric current-driving ability.

A simplified common collector amplifier and its response to a large-amplitude square wave are shown in Figure 7.15. Following the rising edge of the square wave, the transistor becomes conducting and charges the capacitive load C through a relative small transistor-dependent resistance, thereby producing an output signal with a relatively short rise time. By contrast, at the falling edge of the input pulse, the transistor stops conducting, and the capacitor C must slowly discharge through R_E. The output signal's fall time can be reduced only by using a smaller R_E; for a fixed quiescent input voltage V_s, this translates directly into a higher DC collector current and increased heat dissipation in the transistor and the emitter resistor.

This effect becomes even more apparent when the output stage is included in an overall negative feedback loop. As a result of feedback, the output stage can be overdriven as needed to make the output signal properly track the input signal (see Figures 3.23 and 8.19a). Such a positive overdrive can modify the rise of the output signal of Figure 7.15b by decreasing the transistor's conductivity to the

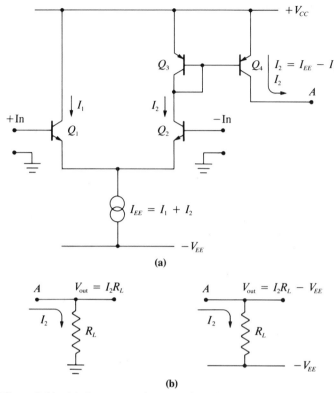

Figure 7.14 *(a) A current mirror used in place of one collector resistor is often used as a level-shifting device. (b) The average voltage of the output signal depends on the load connection.*

positive supply and allowing the capacitor to charge more rapidly. On the other hand, a negative overdrive can only reduce the collector current to zero, leaving the capacitor to discharge through R_E.

The push-pull amplifier eliminates this problem by using two transistors to provide both a source and a sink for output current. Any amplifier stage, but most often the push-pull type, can be classified according to its DC biasing characteristics as shown in Table 7.1. Up to this point we have considered only class A operation of transistors.

For each of these classes of transistor operation, the collector current resulting from an applied sine wave is shown in Figure 7.16. Only class A operation results

Table 7.1 Amplifier Class

Class	Transistor action
A	Transistor conducts full-cycle
AB	Transistor conducts more than half-cycle
B	Transistor conducts half-cycle only
C	Transistor conducts less than half-cycle

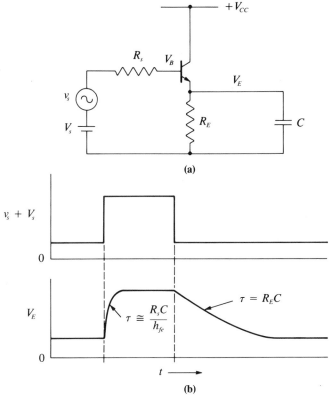

Figure 7.15 *(a) An NPN common collector amplifier driving a capacitive load. (b) The response to a large-amplitude square pulse.*

in current over the entire waveform, and only this class can reproduce an applied voltage swing with a single transistor. When transistors are used in the push-pull configuration, classes AB and B can also give a good approximation to the input signal and have the advantage of lower power loss in the transistors themselves. Extreme examples of power-efficient class C operation are normally limited to high-power, tuned circuit amplifiers (such as radio transmitters), which are used mostly to excite oscillations in *LCR* circuits.

Most present designs for push-pull output stages use bipolar transistors for all but the lowest-power applications. Even though field effect transistors can easily provide the large current gains necessary for the output stage and are free of the thermal runaway problems common in bipolar designs, until recently these devices have been limited to output currents of less than 100 mA. Recent advances in high-power MOSFET technology have made available high-current and high-voltage versions of field effect transistors that are extremely useful in nonlinear high-power devices such as current switches, but because of biasing problems and device variability, their application to linear circuits requires techniques that are beyond the scope of this discussion.

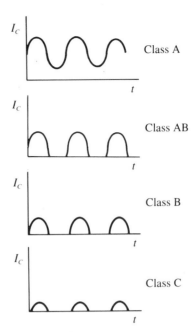

*Figure 7.16   Amplifier classification by collector current response to
an input sine wave signal. The amplifier class is normally determined
by the DC biasing.*

7.10.1   Push-Pull with Bipolar Transistors

The design of push-pull stages using bipolar transistors is greatly simplified by the availability of complementary pairs of matched *NPN* and *PNP* bipolar transistors. A simple example is shown in Figure 7.17a, where each of the transistors is in the common collector, or emitter follower, configuration. If the *PN* voltage drops are ignored, this circuit will function as a class B amplifier with Q_1 acting as a current source for positive-voltage input signals and Q_2 acting as a current sink for negative signals. Since the transistors conduct alternately, there is never a current directly from V_{CC} through Q_1 and Q_2 to V_{EE}. When driving a purely AC signal, this push-pull circuit will consume only half as much power in transistor heating as an equivalent class A push-pull amplifier.

In reality, the *PN* voltage drop between the base and emitter of these transistors means that both transistors will be off when $-0.6 \text{ V} < V_{in} < 0.6 \text{ V}$. This operation is therefore slightly into class C: When the input voltage approaches zero, neither output transistor conducts. The resulting distortion of the output signal is known as "crossover" distortion. For some applications negative feedback can reduce this distortion back to acceptable limits, and this simple circuit will produce an acceptable output signal.

The base-to-emitter resistor *R* improves the operation of the circuit in the crossover region, especially when a feedback connection allows the input to be overdriven as needed to produce the correct V_{out}. By providing a path for current directly from the input signal to the load, this resistor keeps the input from being

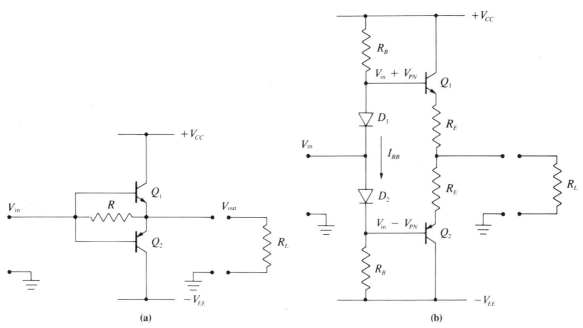

Figure 7.17 *(a) A matched complementary pair consisting of an NPN and a PNP transistor can be connected to form a class C push-pull amplifier. (b) The same amplifier biased for class B operation.*

decoupled from the output even when both transistors are off. For example, if $R = R_L$, then V_{out} will be $V_{\text{in}}/2$ rather than zero in the crossover region. Even though R may be a relatively small resistor, its effective impedance as seen by the driving signal V_{in} is reasonably large and given by

$$R_{\text{eff}} = R\left(1 + \frac{h_{fe}R_L}{h_{ie}}\right) \tag{7.24}$$

under the assumption that $R \ll h_{ie}$. For typical values of h_{fe} and h_{ie} and $R_L = 50\ \Omega$, this effective impedance will evaluate to approximately about $3.5R$.

A better version of this circuit is shown in Figure 7.17b. The identical resistors R_B are chosen to produce a DC current I_{BB} that is large compared to the base currents needed to drive Q_1 and Q_2. The PN voltage drops developed across the diodes D_1 and D_2 then serve to bias Q_1 and Q_2 just to the point of conduction. The resistors R_E are small (a few ohms), and serve both to stabilize the DC bias against variations in Q_1 and Q_2 and to limit the output current under short circuit conditions.

A higher-current gain output stage can be obtained by replacing each transistor with a Darlington pair as shown in Figure 7.18a. Four diodes are now required to bias the circuit for class B operation. The version of this circuit shown in Figure 7.18b uses the Sziklai connection for transistors Q_2 and Q_4. With this arrangement, both of the high-power transistors Q_3 and Q_4 can be of the *NPN* type. Because of the better conductivity of *N*-type silicon, a transistor with a given

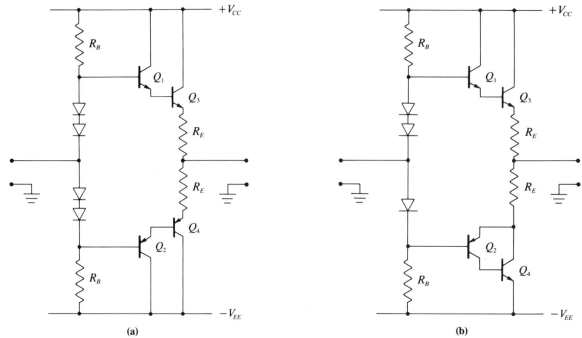

Figure 7.18 *(a) A high-current gain class B push-pull amplifier using transistors in the Darlington connection. (b) The PNP Darlington is often replaced by the Sziklai connection so that Q_4 can be an NPN power transistor.*

power-handling ability can be made smaller in the *NPN* type. For this reason the Sziklai connection often forms part of the output stage of even relatively low-power integrated circuit amplifiers.

The design of high-power stages using bipolar transistors is complicated by the temperature variation of the transistor parameters. All of the major effects tend to increase the collector current as the temperature increases, and this increase in collector current results in more power dissipation, more heating, and a further increase in temperature. If not compensated by other circuit components that depend on temperature or collector current, this cycle can lead to "thermal runaway" and the destruction of circuit components. In applications where the transistor power dissipation exceeds a few watts, the temperature on the silicon chip can be in the range of 100°C to 200°C. Under these conditions, the problems with bipolar design become more complex because of local hot spots developing within the chip itself.

Although the design techniques used in very-high-power stages (more than a few watts) are beyond the scope of this discussion, for modest power handling the transistor temperature can be kept within reasonable limits and the problems are not too severe. For silicon, the most important effects result from changes in h_{FE} and V_{PN} with temperature. The emitter follower-type circuits shown here are desensitized to these parameter changes by the addition of emitter resistors large enough to drop approximately a volt: As temperature effects cause I_C to increase, the increased voltage drop across R_E reduces V_{BE}, thereby decreasing I_C.

7.10.2   Totem Pole Output Stage

Complementary pairs of transistors are not easily fabricated on silicon chips, and a variation on the push-pull circuit is commonly used at the output of IC devices where the highly linear response of the complementary push-pull is not required. The push-pull variation shown in Figure 7.19 uses two like-polarity bipolar transistors and is known as the totem pole connection; a similar circuit can be fabricated from MOSFET devices. The totem pole connection is a common output stage on most of digital IC devices discussed in Chapter 9.

The use of a single-polarity transistor complicates the drive requirements, since one must be on while the other is off; their bases must therefore be driven by signals that are 180° out of phase. This "phase splitting" operation is accomplished by transistor Q_1. If the emitter and collector resistors on this stage are identical ($R_1 = R_2$), then the emitter and collector signals will both have the amplitude of the input signal V_{in}, but the collector signal will be inverted. The circuit is easily analyzed as a two-state on/off device. With V_{in} low, Q_1 will be nonconducting, allowing R_1 to pull the base of Q_2 high while R_2 pulls the base of Q_3 low; Q_2 will thus be conducting while Q_3 is not, and V_{out} will be near V_{CC}. When V_{in} is high, the collector current through Q_1 will turn Q_3 on and reduce the voltage on the base of Q_2; V_{out} is then pulled to ground.

7.11   Reading an Integrated Circuit Schematic

The typical linear IC schematic shown in Figure 7.20 is that of the 741 operational amplifier. The schematic shows 20 transistors labeled Q_1 through Q_{22} (Q_{19} and Q_{21} are not shown explicitly). Without attempting a detailed analysis, it is instructive to identify the functions performed by various groups of transistors in this

Figure 7.19   *The totem pole arrangement of two like-polarity output transistors is commonly used as the output stage on digital circuits. The circuit shown is highly simplified.*

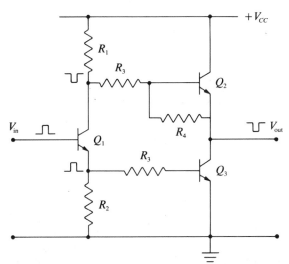

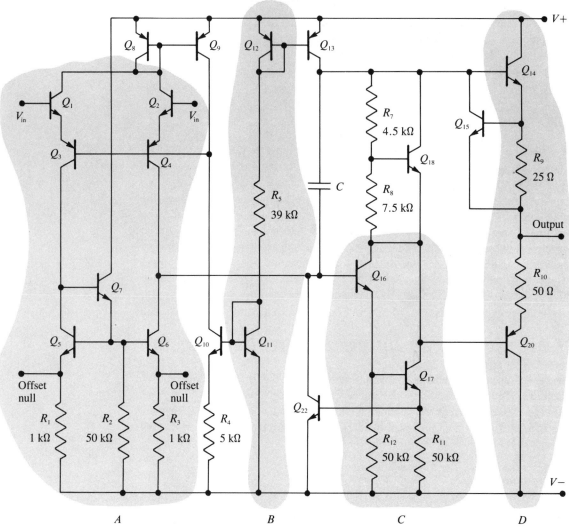

Figure 7.20 *The schematic of a 741 integrated circuit operational amplifier. (Courtesy Signetics Corp.)*

circuit. For this purpose the basic circuit found in several manufacturers' data books has been shaded to show the major functional blocks of this circuit. Although they are not shown linked together in separate blocks, there are also several current mirrors in this schematic: Q_8 and Q_9, Q_{10} and Q_{11}, Q_{12} and Q_{13}, and less obviously, Q_5, Q_7, and Q_6. Except for the last triplet, these current sources (transistors Q_9, Q_{10}, and Q_{13}) are used mostly as large resistors; this application is common on IC chips, since a large resistor takes more physical space on the chip than a transistor current source.

The various transistors that make up group *A* function as a differential amplifier, but the configuration is different from our earlier examples. Since the voltage at the collectors of Q_1 and Q_2 is held fixed by the base-to-collector connection around Q_8, these *NPN* input transistors are in a common collector configuration with their emitters driving *PNP* transistors Q_3 and Q_4, which are in the common base configuration. (The wire passing through the base of Q_4 does not indicate a new device; it is just a convenient way to draw the schematic.) The high impedances generated by current sources Q_9 and Q_{10} serve to bias these *PNP* transistors. This *CC* and *CB* combination input stage provides both current and voltage gain much like a common emitter transistor, and the current mirror action of Q_8 and Q_9 changes the voltage at the bases of Q_3 and Q_4 in a manner that acts to keep the current through Q_8 constant: The form of this differential input is different, but the effect is much the same as our earlier examples.

The current mirror formed by Q_5, Q_6, and Q_7 presents a low impedance to the collector current from Q_3 but causes Q_6 to appear as a large impedance to the current from Q_4. The action of this transistor triplet is much like the current mirror shown in Figure 7.13a.

The two transistors of group *B* develop the primary bias current through resistor R_5. Since the voltage across this resistor is just $V_+ - V_- - 2V_{PN}$, this current will be different depending on the choice of power supply voltages. The two current mirrors driven by these transistors distribute this reference current to the rest of the circuit, with the result that the entire amplifier circuit will be properly DC biased for almost any choice of power supply.

The Darlington pair of transistors in group *C* are the high-gain part of this circuit. The base of Q_{16} is driven by the single-sided output from the differential amplifier (group *A*), and the output voltage signal is developed at the collector of Q_{17}. This output signal is applied directly to the base of Q_{20} and through the DC level-shifting transistor Q_{18} to the base of Q_{14}. (In an AC analysis, Q_{18} can be replaced with a wire, but for a DC analysis its collector will be about 1 V above its emitter.) The capacitor C_1 provides a negative-feedback path around the high-gain Darlington stage and defines a low-pass filter transfer function for the overall amplifier (the dominate pole). The associated transistor Q_{22} is normally off and becomes conducting only when the current through R_{11} approaches 0.1 mA. When conductive, it reduces the gain of the Darlington pair, effectively setting a maximum on the current through R_{11}. This in turn limits the base (and collector) current of Q_{20} and thus serves to protect this output transistor.

Finally, the two transistors of group *D* form a complementary push-pull output amplifier. The output terminal will normally be connected to ground through an external load resistor R_L, allowing the currents developed by Q_{14} and Q_{20} to develop an output voltage across R_L. The associated transistor Q_{15} performs the same function as Q_{22} discussed earlier; it is normally off, and by sensing the current through R_9 is able to limit the maximum current through Q_{14}, thereby protecting this output transistor.

All stages of the 741 amplifier are DC-coupled, so the device will function for DC as well as AC signals. The overall DC voltage gain $V_{out}/(V_2 - V_3)$ is about 10^5, and the device is so widely used that its cost is about the same as a single transistor.

Problems

1. Derive Eq. (7.1) for the corner frequency of the circuit of Figure 7.1c.

2. The resistors in Figure 7.1a are: $R_1 = R_5 = 40$ kΩ, $R_2 = R_6 = 10$ kΩ, $R_3 = R_7 = 1$ kΩ, $R_4 = R_8 = 2$ kΩ, and $R_s = 8$ kΩ. Assume that $h_{ie} = 2$ kΩ and that $h_{fe} = 100$.

(a) Determine the value of each coupling capacitor such that all corner frequencies are 100 rad/s.

(b) What is the AC gain v_{out}/v_s at frequencies well above the corner frequency?

3. If the signal generator of Figure 7.2 has a 500-Ω output impedance and both transistors have $h_{fe} = 100$ and $h_{ie} = 2$ kΩ, what size capacitors are necessary if both corner frequencies are to be at 1000 rad/s?

4. Work out the DC quiescent voltages in Figure 7.2a if all PN voltage drops are 0.7 V instead of the 0.5 V shown. Assume that $h_{FE} = \infty$.

5. The DC quiescent voltages shown in Figure 7.2a assume that h_{FE} is very large. Work out these voltages exactly if $h_{FE} = 50$ and all $V_{PN} = 0.5$ V.

6. The circuit shown in Figure P7.6 uses two PNP transistors in CC current amplifier configurations and an NPN in the CB voltage amplifier configuration. Assume that $h_{fe} = h_{FE} = 100$ and $V_{BE} = 0.5$ V. Assume arbitrarily large capacitors and neglect small currents whenever possible.

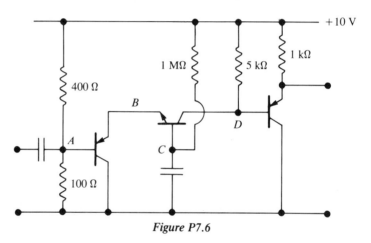

Figure P7.6

(a) Taking the base-emitter voltage drops into account, determine the DC operating voltages at points A, B, C, D, and E.

(b) After building this circuit, suppose that the actual voltage at point E turns out to be about 10 V. Should R_B be made larger or smaller?

(c) Draw an AC equivalent circuit retaining the transistor symbols and determine the overall AC voltage gain if h_{ie} of each transistor is 2000 Ω.

(d) Estimate the output impedance of this circuit.

7. (a) Evaluate each of the resonant frequencies given by Eqs. (7.6), (7.8), and (7.10), under the assumption that the stray capacitances C_{S1} and C_{S2} are each 10 pF, the tuning capacitors are each zero, the transformers all have turns ratio $n = 1$, all inductances are 1 mH, and each FET develops a voltage gain A of 500. (b) Determine the value of a tuning capacitance that will maximize the overall gain at one frequency.

8. If both transistors in Figure 7.4b have $h_{ie} = R$ and $h_{fe} \gg 1$, show that the overall current gain of the circuit is $0.5h_{fe}^2$.

9. If $R = h_{ie}$ and $h_{fe} = 100$, use the result of the previous problem to determine the input impedance of the circuit to the right of the dashed line in Figure P7.9.

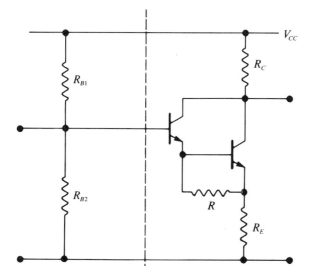

Figure P7.9

10. Show that the current gain of the Sziklai connection shown in Figure 7.5b is $i_C/i_B = Rh_{fe}^2/(R + h_{ie})$.

11. (*a*) If all transistors in the circuit shown in Figure P7.11 have $V_{PN} = 0.5$ V, what are the quiescent operating point values of I_C and V_C? (*b*) Repeat this using a Sziklai pair instead of the Darlington pair.

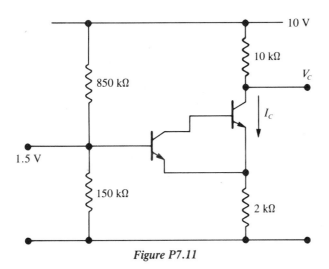

Figure P7.11

12. Design a circuit like that of Figure 7.6a using a second *N*-channel JFET as a current source. Assume that JFET Q_1 develops 2 mA of drain current when $V_{GS} = -2$ V, whereas JFET Q_2 develops 2 mA when $V_{GS} = -4$ V. Draw a circuit diagram identifying the transistors as Q_1 and Q_2 and indicate the quiescent voltage at the point where the drain of one connects to the source of the other.

13. Write an expression for the AC input impedance to ground looking into the emitter of Q_2 in Figure 7.7a. Express your result in terms of h_{ie}, h_{fe}, and the reverse Zener resistance R_Z. Assume that the transistors Q_1 and Q_2 are identical.

14. At sufficiently high frequencies the input capacitance of the circuit in Figure 7.7a can be ignored. *(a)* Under this condition, what is the AC voltage gain of this circuit? *(b)* Locate an additional capacitor on this circuit so as to increase the AC gain at high frequencies.

15. If the MOSFET transistor of Figure 7.9b has transconductance g_m, determine an expression for the input impedance of the circuit.

16. *(a)* If the *PN* voltage drops of both transistors in Figure P7.16 are 0.6 V, what is the current I into a small but variable load? *(b)* Will this result hold if the load impedance is greater than 100 kΩ?

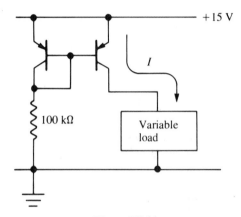

Figure P7.16

17. If the *PN* voltage drops in Figure P7.17 are all 0.5 V:

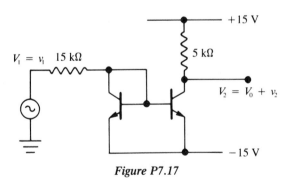

Figure P7.17

(a) What is the value of the DC voltage V_0?
(b) Determine a value for the ratio v_2/v_1.

18. If the *PN* voltage drops in Figure P7.18 are all 0.5 V:
(a) What is the value of the DC voltage V_0?
(b) Determine a value for the ratio v_2/v_1.

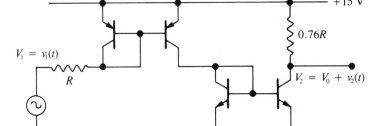

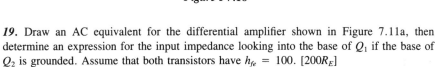

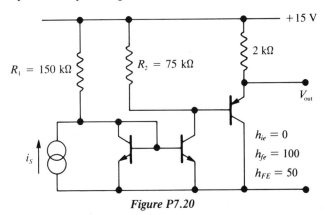

Figure P7.18

19. Draw an AC equivalent for the differential amplifier shown in Figure 7.11a, then determine an expression for the input impedance looking into the base of Q_1 if the base of Q_2 is grounded. Assume that both transistors have $h_{fe} = 100$. [$200R_E$]

20. Assuming that the *PN* voltage drops in Figure P7.20 are zero:
(a) Determine the quiescent output voltage.
(b) What is the ratio V_{out}/i_s if $h_{ie} = 0$?
(c) If $V_{PN} = 0.6$ V and R_2 is unchanged, what value of R_1 would be needed to produce the same quiescent output voltage?

Figure P7.20

21. When the input signal V_{in} is positive, show that the input impedance to the circuit of Figure 7.17a is $(h_{ie}R + h_{ie}R_L + h_{fe}RR_L)/(h_{ie} + R)$.

22. If the input and output terminals on the circuit of Figure 7.18b are grounded and all *PN* voltage drops are 0.5 V, what is the collector current through Q_3 and Q_4?

23. Write an expression for V_{GS} in Figure 7.9a.

24. In the push-pull circuit of Figure 7.17b, let $R_E = 25 \ \Omega$ and set the power supplies to $+15$ V and -15 V. Ground the output terminal and apply a signal $V_{in} = 2 \cos(1000t)$ V to the input terminal. Assume that the *PN* voltage drops are all 0.5 V.
  (a) What is the peak instantaneous power dissipated in R_E connected to Q_1?
  (b) At this same instant, what power is being dissipated in each transistor?

25. Consider the level-shifting transistor Q_{18} in the 741 schematic of Figure 7.20. If a V_{BE} of 0.6 V is required to get the transistor to its DC operating point, determine the level-shifting voltage V_{CE}.

8

Operational Amplifiers

8.1   Introduction

The DC-coupled operational amplifier (op amp) has become a common feature of modern analog electronics. The most popular versions can be purchased for the price of a couple of resistors, and are well suited to applications at or below audio frequencies (DC to 20 kHz). Higher-performance models are also available, some with frequency response in excess of 50 MHz. The operational amplifier is constructed of several transistor stages, usually featuring a differential amplifier input stage (either bipolar or FET), an intermediate-gain stage, and some form of push-pull output stage. Commercially available op amps are fabricated from discrete components or integrated on a single silicon chip. The former are used in the most demanding applications; they usually feature slightly better specifications and significantly higher prices.

Although operational amplifier performance varies considerably depending on design and intended application, most have DC open-loop voltage gains in the range of 10^3 to 10^6, with the devices intended for high-frequency operation having the lower gain. The voltage gain invariably falls off at higher frequencies, with the Bode plot showing one or more turning points like a low-pass filter. The most popular devices (such as the 741) feature a dominate-pole design resulting in an open-loop transfer function that falls off at 6 dB/octave above a corner frequency of about 5 Hz.

Because of their high open-loop gain, operational amplifiers are intended to be used in circuits that develop substantial negative feedback as discussed in Chapter 3. This feedback allows the designer to fabricate a circuit to obtain a specific transfer function, with little concern about the variable nature of the parameters in the individual active circuit elements. On the negative side, feedback also generates the major and sometimes exasperating problem of oscillations at high frequency. This problem is relatively benign when using dominate-pole amplifiers such as the 741 (the main reason for their popularity), but becomes a primary concern when using higher-performance devices that feature a unity gain point above 1 MHz.

At various times in this chapter we will use three different symbols to label the transfer function of an amplifier: $A(j\omega)$, $G(j\omega)$, and $H(j\omega)$. The symbol A will always be used to represent the gain of the amplifier itself without any feedback connections (open-loop gain). The symbols G and H will both be used to represent the overall gain of the amplifier with feedback (closed-loop gain). When G appears, it will always refer to an approximate form of H obtained under the assumption of infinite and frequency-independent A (Section 8.3) . Other transfer function symbols will be used to describe specific networks in a circuit; the transfer function symbol $F(j\omega)$ will be reserved to describe a feedback network, usually but not necessarily passive.

8.2   Open-Loop Amplifier

An unusually complete schematic representation of an operational amplifier is shown in Figure 8.1a. In practice this gets reduced to the bare triangle, sometimes even without the $+/-$ input terminal indicators. The power supply connections needed for proper operation are shown explicitly in this diagram, and it is important to remember that although they are generally omitted for reasons of simplification, the connections do provide the major source or sink for current at the output terminal of the amplifier. Although devices are available that operate from a single supply as low as 5 V, most require two supplies (typically ± 15 V) on opposite sides of the ground as shown.

Depending on the precision needed for the intended application, the equivalent circuit for a real operational amplifier can be drawn with various degrees of complication. An approximate description will often suffice, and we will use the amplifier model shown in Figure 8.1b. As we shall see, most applications do not require even this level of detail.

The symbol $A(j\omega)$ represents the open-loop voltage gain of this ideal amplifier and is defined by the equation

$$\mathbf{V}_{out} = \mathbf{A}(j\omega)\,(\mathbf{V}_+ - \mathbf{V}_-) \tag{8.1}$$

Although this complex, frequency-dependent parameter A can be usefully approximated by a real constant A_o at sufficiently low frequencies, a more complete description can be obtained by replacing $A(j\omega)$ with

$$\mathbf{A}(j\omega) = \mathbf{A}_o\,\mathbf{H}_{low}(j\omega) \tag{8.2}$$

where $\mathbf{H}_{low}$ is the complex transfer function of some passive low-pass filter and A_o is the DC open-loop gain of the amplifier. For a dominant-pole op amp such

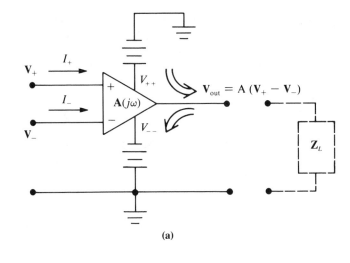

(a)

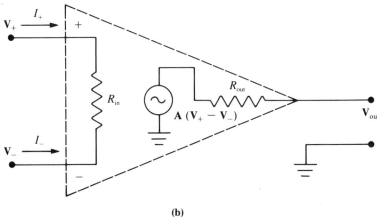

(b)

Figure 8.1 (a) Schematic diagram of an ideal operational amplifier.
(b) First-order equivalent circuit of the operational amplifier.

as the 741, the basic one-pole filter model is sufficient, and we get the result first discussed in Section 3.12.1:

$$\mathbf{A}(j\omega) = \frac{A_o}{1 + j\omega/\omega_c} \tag{8.3}$$

With a suitable choice of parameters A_o and ω_c, this expression provides a complete algebraic description of the open-loop gain and phase shift of a typical dominant-pole operational amplifier over the frequency range DC to 1 MHz as shown in Figure 8.2.

In some of the more subtle problems associated with operational amplifier applications we will find it necessary to use this type of complete description of the open-loop transfer function. We will always need to consider it at high frequencies, and sometimes when we are concerned with the details of input or output impedances. However, for most applications it is possible and desirable to use simplifying approximations.

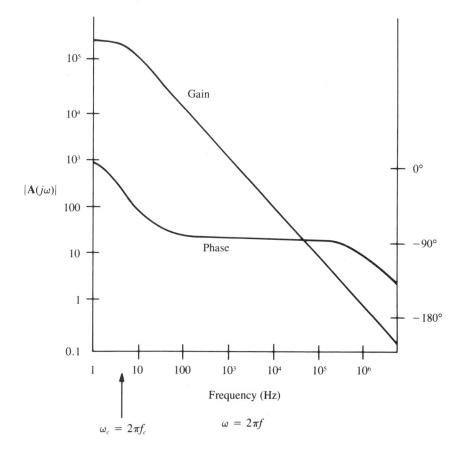

Figure 8.2 Bode plot of the open-loop gain and phase shift of a 741 operational amplifier. For the 741, the DC open-loop gain $A_o = 2 \times 10^5$, and the corner frequency $\omega_c = 25$ rad/s or about 4 Hz.

8.3 *Ideal Amplifier Approximation*

Operational amplifiers are nearly ideal devices with large forward transfer functions, virtually nonexistent reverse transfer functions, large input impedances, and small output impedances. As a result, if we use the amplifier in a circuit with a negative feedback connection, we can identify two simple rules that greatly simplify the circuit's design:

Rule 1: The input currents I_+ and I_- are zero.

Rule 2: The voltages V_+ and V_- are equal.

Note carefully that these rules require linear circuit operation with negative feedback and only apply when the output is neither voltage nor current limited.

The first rule comes directly from the fact that an operational amplifier's inputs are actually transistor bases or gates, usually in a high-input-impedance differential-amplifier configuration. As such, these inputs require very little current for normal operation, rarely more than a microampere and as little as a few picoamperes if the amplifier has FET inputs. Exceptions to this rule are small and are discussed further in Section 8.4.5.

The second rule requires a specific assumption about the amplifier's open-loop gain **A**, which is defined by Eq. (8.1). If we take this gain to be a real constant that is infinite at all frequencies, then Eq. (8.1) can produce a finite output signal V_{out} only when the voltage difference $\mathbf{V}_+ - \mathbf{V}_-$ is zero. (In Section 8.4 we will relax this second rule, and allow **A** to take on its actual, finite and frequency dependent character.)

Because these two rules allow neither current nor voltage signals at the operational amplifier's input terminals, they suspend the cause and effect relationship between the operational amplifier's input and output signals. It is best to sidestep this problem by simply assuming that the amplifier drives its output terminal with whatever voltage and current is necessary to validate the rules, sending the needed signal back to the input via the negative feedback connection.

Operational amplifiers rarely distinguish between AC and DC signals; therefore, it is convenient to represent most of the signals on these circuits with uppercase letters. When the frequency variation of a circuit is an issue, we will resort to boldface notation, which should be interpreted as a single frequency sinusoid when appropriate.

8.3.1   Noninverting Amplifiers

The simplest operational amplifier circuit is the unity-gain, noninverting amplifier in Figure 8.3a. Only Rule 2 is required for the analysis of this circuit, since from the connections V_{in} appears on one input terminal and V_{out} appears on the other. Rule 2 requires that they be equal! Note that the open-loop gain of the op amp is not used explicitly, and that cause and effect are therefore not considered. Using Rule 1, we see that no input current is required to drive this circuit, and we can therefore conclude that this idealized amplifier has an infinite input impedance.

Figure 8.3   (a) Noninverting, unity-gain amplifier. (b) Noninverting amplifier with gain. According to Rule 1, both amplifiers have an infinite input impedance.

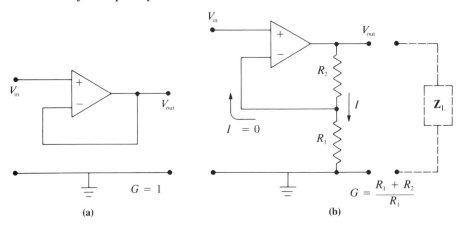

By including two resistors we can produce a noninverting amplifier with gain as in Figure 8.3b. Rule 1 specifies that no charge flows into either op amp input terminal and makes it possible to describe this circuit's operation with only one current. The voltage at the negative terminal is set by the voltage divider and given by

$$V_- = \frac{R_1 V_{out}}{R_1 + R_2} \qquad (8.4)$$

Applying Rule 2 to this particular circuit gives

$$V_{in} = \frac{R_1 V_{out}}{R_1 + R_2} \qquad (8.5)$$

Dividing by V_{out} and inverting gives the standard transfer function form:

$$\mathbf{G}(j\omega) = \frac{R_1 + R_2}{R_1} \qquad (8.6)$$

where we use **G** to represent the closed-loop gain (transfer function of the amplifier with feedback).

Note that the addition of a load impedance does not change this analysis, since we have not placed any limits on the current at the output of the operational amplifier. In this approximation, the operational amplifier will supply current (in or out) to the load as necessary to satisfy the condition imposed by Rule 2. We will consider the more subtle consequences of the operational amplifier's small but nonzero output impedance in Section 8.4.1.

8.3.2   Inverting Amplifiers

The inverting amplifier configuration in Figure 8.4a has many variations and a wide range of applications. The signal input to this amplifier is through a resistor to the negative op amp terminal, and for circuit-drawing convenience, the amplifier symbol is usually flipped so that the minus input is on top; the feedback connection continues to be from the output to the negative input.

It is a simple matter to again use Rule 1 to determine that only one current is needed to describe this circuit. Since the positive input terminal is tied to ground in this example, application of Rule 2 means that the negative input terminal must also be at ground potential. This condition of the negative terminal is sometimes referred to as a "virtual ground" and leads immediately to the following equation:

$$\frac{V_{in} - 0}{R_1} = \frac{0 - V_{out}}{R_2} \qquad (8.7)$$

The resulting closed-loop transfer function for this circuit is

$$\mathbf{G}(j\omega) = -\frac{R_2}{R_1} \qquad (8.8)$$

where the minus sign means that the output is inverted with respect to the input. Again, note that the addition of a load resistor does not change the analysis.

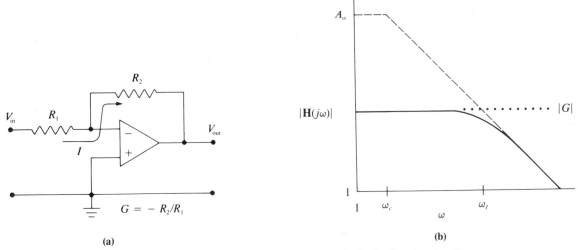

Figure 8.4 *(a) Inverting amplifier with gain. (b) Bode plot showing the gain of the open-loop amplifier $|\mathbf{A}|$ (dashed), the approximate closed-loop gain $|\mathbf{G}|$ (dotted), and the actual closed-loop gain $|\mathbf{H}|$ (solid).*

Most practical applications of this circuit will also feature a third resistor replacing the wire between the positive input terminal and ground. The addition of this resistor serves a useful purpose as discussed in Section 8.4.5, but it does not affect our first-order analysis since under the present assumptions it carries no current and therefore develops no potential drop. When included in the design, this third resistor should have a value equal to the parallel combination of R_1 and R_2.

The Bode plot of Figure 8.4b shows the relation of the open-loop gain of the operational amplifier to the closed-loop gain for both the inverting and noninverting amplifiers just discussed. Note that the approximations that produced the frequency-independent gain must fail when our constant $|\mathbf{G}|$ intersects the open-loop gain $|\mathbf{A}(j\omega)|$ of the operational amplifier. The actual closed-loop gain curve $|\mathbf{H}(j\omega)|$ must lie under both curves and is shown as a solid line in this figure.

8.3.3 Current Summing Junction

A variation of the inverting amplifier configuration yields the current-to-voltage converter shown in Figure 8.5a. This circuit produces an output voltage that is proportional to the input current. The voltage drop caused by the current $\mathbf{I}$ in the feedback resistor requires that

$$V_{\text{out}} = -RI \tag{8.9}$$

The figure shows an ideal current source driving this circuit, but since both input terminals are at the same potential, the operation of the circuit is unchanged if the ideal source is replaced with a real one having a shunt impedance.

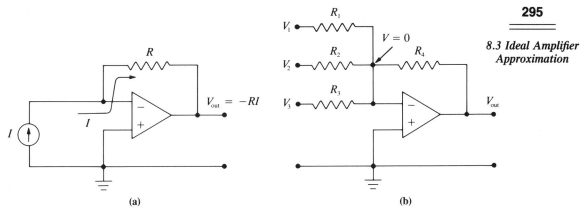

Figure 8.5 *(a) Current-to-voltage converter. (b) Multiple inputs to the current summing junction.*

If several different current sources simultaneously drive the negative input of an inverting amplifier, the output voltage will be proportional to the algebraic sum of the input currents. On the voltage summing amplifier of Figure 8.5b, the current inputs are generated by connecting input voltage signals to the virtual ground point through resistors. The expression for the output voltage is easily found from the current equality:

$$\frac{V_{out}}{R_4} = -\frac{V_1}{R_1} - \frac{V_2}{R_2} - \frac{V_3}{R_3} \tag{8.10}$$

If the three input resistors have the same value, the output will be proportional to the sum of the input voltage signals.

A two-input version of this circuit, with a constant reference voltage applied to one of the inputs, provides a convenient method for obtaining an output signal with any required voltage offset.

Example 8.1   Use an op amp to improve the frequency-to-voltage converter of Figure 5.20.

A simple improvement on that passive design would be to connect a unity-gain, noninverting amplifier to the output terminal. This amplifier would serve as a buffer and isolate the converter from the effects of varying loads.

However, we can do better by incorporating the op amp directly into the basic charge pump design. In the discussion of the charge pump in Figure 5.19, we found that the build-up of charge on C_2 gradually reduced the voltage step ΔV_2 that was added to C_2 on successive cycles, resulting in a fundamental nonlinearity in that circuit's operation. This problem is eliminated by the op amp circuit of Figure 8.6. With $R = \infty$, this is just a charge pump and the charge Δq_2 passing through D_2 ends up on C_2 just as on the passive design. However, because the negative input terminal of the op amp is fixed at a virtual ground, the output of the op amp must change by $-\Delta q_2/C_2$. Since nothing changes at the input, all cycles deposit the same amount of charge on C_2, and the nonlinearity has been removed.

From Eq. (5.15) with V_2 always zero, we find that the charge added to C_2 each cycle is

$$\Delta q_2 = \frac{C_1 C_2 V}{C_1 + C_2}$$

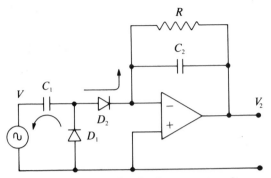

Figure 8.6 An active design for the frequency-to-voltage converter.

with no restriction on the relative sizes of the capacitors. With the resistor in place, the capacitor loses charge between cycles. Following the argument of Section 5.5.5, this loss is given by

$$\Delta q_2 \simeq \frac{-V_2}{fR}$$

under the condition that $RC_2 f \gg 1$; the minus sign is due to the inverting op amp configuration. Equating charge gained to charge lost gives the steady-state equation

$$V_2 = -\frac{C_1 C_2}{C_1 + C_2} RVf$$

which is essentially the same as Eq. (5.23) but without the additional design limitations imposed by (5.24).

This design is also buffered by the op amp and its operation is unaffected by load variations.

8.3.4 Differentiation Circuit

If one or more of the gain-setting resistors in the inverting amplifier just discussed are replaced by generalized impedances, the resulting closed-loop transfer function will show additional frequency dependence. Because of size, cost, and stray signal pickup, inductors are rarely used in operational amplifier circuits. The simplest modification is therefore to replace one resistor with a capacitor. The only modification needed in the analysis is to use the standard complex expressions for impedances, currents, and voltages.

When R_1 is replaced by a capacitor as in Figure 8.7a, the amplifier's transfer function is

$$\mathbf{G}(j\omega) = -\frac{R}{\mathbf{Z}_C} = -j\omega RC \qquad (8.11)$$

As noted in Chapter 3, this is the transfer function of the ideal differentiator multiplied by a constant factor $-RC$. If the input signal is a pure sine wave, this transfer function would retard its phase by $\pi/2$ rad, thereby converting it into a

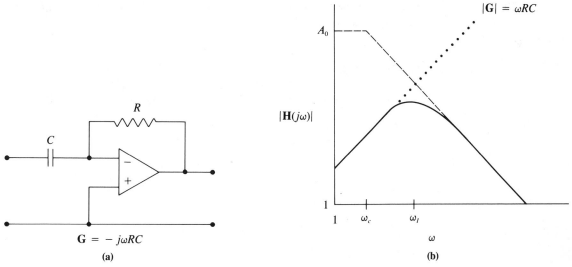

Figure 8.7 (a) Operational amplifier differentiation circuit. (b) Bode plot for the differentiator.

cosine wave as required. In the limit of infinite open-loop gain **A,** there is no restriction on the value of RC; it is simply a gain-adjusting factor.

The Bode plot of the magnitude of this transfer function shows a kind of high-pass filter with linearly increasing closed-loop gain as the signal frequency increases. As shown in Figure 8.7b, this increasing function $|\mathbf{G}|$ must sooner or later intersect the open-loop gain curve of the operational amplifier. The infinite open-loop gain approximation clearly fails as we approach this crossover frequency, with the result that the actual amplifier will have a bandpass characteristic as shown by the solid line in the figure. Obviously, the differentiation feature applies only to signals with frequencies below the intersection region. Note that Figure 8.7b is a log-log plot; changing RC changes the vertical position of the closed-loop curve $|\mathbf{G}|$ but does not alter its slope. Thus smaller values of the RC product will permit differentiation of higher-frequency signals. This observation emphasizes the need to remain aware of the limitations of the infinite open-loop gain approximation.

The failure of the differentiation at high frequencies, when $|\mathbf{G}|$ intersects $|\mathbf{A}|$, means that this circuit must yield incorrect results for signals with significant high-frequency components; the square wave is an obvious example. If a square wave is applied to the input of this circuit, the output signal will show the requisite spikes at the rising and falling edges, but in most designs each spike will exhibit substantial damped oscillation—ringing at a frequency near the intersection of $|\mathbf{G}|$ with $|\mathbf{A}|$.

The increase in gain with frequency also means that this circuit will naturally accentuate any high-frequency noise components present in the input signal, a characteristic that often limits the usefulness of this circuit and may necessitate further high-frequency compromises. The addition of a resistor in series with C or a capacitor in parallel with R will further limit the high-frequency gain and result in a more predictable and therefore more practical circuit design. We will investigate this modification in Section 8.6.2.

Example 8.2 If a 741 op amp, a 0.1-μF capacitor, and a 10-kΩ resistor are used to build a differentiation circuit, at what frequency will the differentiation fail?

The high-frequency limit occurs near the point where $|\mathbf{G}|$ intersects the open-loop gain curve $|\mathbf{A}|$. In Figures 8.2 and 8.7b, the unity-gain point of the open-loop amplifier occurs at about 10^6 Hz. The corresponding angular frequency is $\omega_1 = 6.28 \times 10^6$ rad/s, and the -6 dB/octave part of $|\mathbf{A}|$ can be written as

$$|\mathbf{A}| = \frac{\omega_1}{\omega} = \frac{6.28 \times 10^6 \text{ rad/s}}{\omega}$$

From Eq. (8.11), the magnitude of $\mathbf{G}$ is

$$|\mathbf{G}| = RC\omega = (10^{-7} \text{ F})(10^4 \ \Omega)\omega = 10^{-3}\omega$$

and its intersection with $|\mathbf{A}|$ is given by

$$10^{-3}\omega = \frac{6.28 \times 10^6 \text{ rad/s}}{\omega}$$

or
$$\omega = 7.9 \times 10^4 \text{ rad/s}$$

which is labeled in Figure 8.7b as ω_l and in this example corresponds to a limiting frequency of 12,600 Hz.

8.3.5   *Integration Circuit*

By reversing the resistor and capacitor in the differentiation circuit, we get the integrator configuration in Figure 8.8a. The transfer function of this circuit is the algebraic and operational inverse of the differentiator, and is given by

$$\mathbf{G}(j\omega) = -\frac{\mathbf{Z}_C}{R} = -\frac{1}{j\omega RC} \tag{8.12}$$

Once again, the ideal amplifier approximation has led to a transfer function that approaches an infinite gain, this time as the frequency decreases. Although we certainly expect a real amplifier to depart from this equation at sufficiently low frequencies because of its finite open-loop gain, the voltage and current offsets discussed in Section 8.4.5 also complicate the problem in this large-time-constant regime.

A more predictable circuit, known as a leaky integrator, is obtained by adding a resistor in parallel with the charge storage capacitor as shown in Figure 8.9b. The resulting transfer function is simply that of a single-pole, low-pass filter and integrates properly only above its corner frequency. This type of design, where the closed-loop transfer function never crosses the open-loop response, can be accurately analyzed by the simple techniques outlined here and often results in a more practical circuit. This useful practice is nothing more than an electronic equivalent of the old "know your enemies" adage.

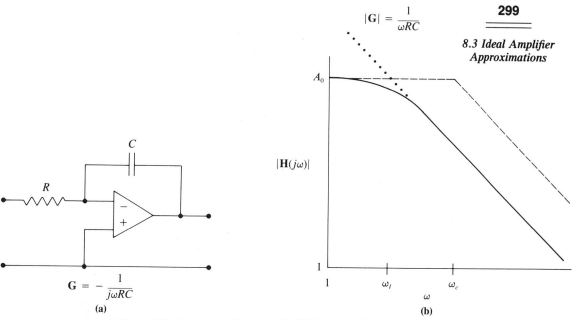

$$|\mathbf{G}| = \frac{1}{\omega RC}$$

$$\mathbf{G} = -\frac{1}{j\omega RC}$$

(a)

(b)

Figure 8.8   (a) Integration circuit. (b) Bode plot for the integrator.

Operational amplifier differentiators and integrators can be combined to form a device known as an analog computer whose function is to solve a set of simultaneous differential equations. With this capability, the analog computer can be used to simulate a mechanical system. For a time, analog computers competed with digital computers in this special application, but because of their superior speed, accuracy, and flexibility, digital computers are now used for all but the smallest simulations. Differentiators and integrators are now used mainly for direct signal modification and are just one of the many ways a signal can be "conditioned" for use by another electronic system.

Figure 8.9   (a) A single-pole, buffered high-pass filter. (b) An inverting amplifier design for a single-pole, low-pass filter.

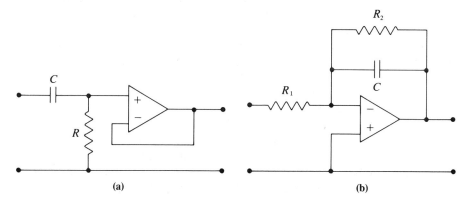

(a)

(b)

8.3.6 *Active Filters*

One way of improving the performance of a filter is shown in Figure 8.9a. The passive high-pass filter in this example is coupled to a high-impedance, noninverting amplifier, which effectively isolates the filter from any possible load. An amplifier used in this way is generally described as a buffer. Several such filters can thus be cascaded together (the output from one driving the next) without modifying the operation of any of the individual filters. Under these conditions the overall transfer function is just the product of the individual transfer functions. Three identical single-pole, high-pass filters would give

$$\mathbf{G}(j\omega) = \frac{j\omega}{1 + j\omega/\omega_c} \cdot \frac{j\omega}{1 + j\omega/\omega_c} \cdot \frac{j\omega}{1 + j\omega/\omega_c} \tag{8.13}$$

or just

$$\mathbf{G}(j\omega) = \frac{-j\omega^3}{(1 + j\omega/\omega_c)^3} \tag{8.14}$$

Such a filter will show an asymptotic slope of 18 dB/octave at sufficiently low frequencies, but the transfer function changes only slowly during the transition between this region and the flat, high-frequency region; it does not have a sharp knee. A sharp knee on the transfer function requires complex poles, not just more real ones.

The differentiator and integrator circuits also function as high- and low-pass filters, respectively. When modified by the third circuit element as described in the two previous sections, these circuits have transfer functions that are again similar to those of single-pole, passive filters, modified only by the voltage gain derived from the amplifier. A low-pass example is shown in Figure 8.9b; two high-pass examples are shown in Figures 8.28a and 8.28b. The transfer function for the circuit of Figure 8.9b is

$$\mathbf{G}(j\omega) = -\frac{R_2}{R_1(1 + j\omega R_2 C)} \tag{8.15}$$

Because of the imbedded amplifier, these designs are known as active filters and are superior to their passive counterparts because of the voltage and current gain available from the amplifier. In addition, with a single amplifier and only capacitive reactive circuit elements, it is possible to build a more versatile two-pole circuit that has the filter characteristics of a series *LCR* circuit. Two general designs for low-pass, two-pole, active filters are shown in Figures 8.10a and 8.10b. Equivalent high-pass designs can be obtained by simply interchanging resistors and capacitors. Unlike passive *RC* filters, the poles of these active filters can be complex, with the result that a wider variation in filter performance is possible. Active filters with more than two poles can be formed by cascading several two-pole designs.

With the addition of extra circuit elements, the designer has available more parameters, and the choice of closed-loop gain and corner frequency does not completely determine the passive circuit element values. Several standard design choices are available that optimize specific filter characteristics: the Chebyshev,

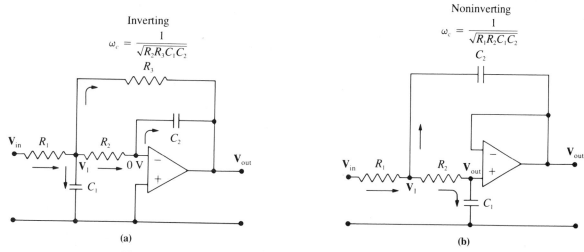

Figure 8.10 *(a) A two-pole, low-pass, inverting active filter. (b) This two-pole, low-pass, noninverting design has unity gain at DC.*

which produces a reasonably sharp cutoff condition; the Butterworth, which yields a maximally flat bandpass characteristic at the expense of a slower roll-off with frequency; and the Bessel, which has an even slower roll-off but strives for a linear phase-shift characteristic in the pass band, resulting in minimal signal distortion. The general design of multipole active filters is beyond the scope of this text, but the literature provides tabular techniques for designing low-pass, high-pass, and bandpass filters for each of these types.

The analysis of the two-pole filters in Figure 8.10 can proceed at several different levels. Using the inverting filter as an example, it is easy to get the asymptotic forms of the transfer function directly from the schematic. At low frequency, both $|\mathbf{Z}_C|$ can become much greater than any of the resistors, with the result that in this regime,

$$\mathbf{G}(\text{low}) = -\frac{R_3}{R_1} \qquad (8.16)$$

At high frequency, both $|\mathbf{Z}_C|$ become much less than any resistor, resulting in

$$\mathbf{V}_1 = \frac{\mathbf{V}_\text{in}}{1 + j\omega R_1 C_1} \qquad (8.17)$$

and

$$\mathbf{V}_\text{out} = -\frac{\mathbf{V}_1}{j\omega R_2 C_2} \qquad (8.18)$$

The combination of (8.17) and (8.18) gives

$$\mathbf{G}(\text{high}) = -\frac{1}{j\omega R_2 C_2 (1 + j\omega R_1 C_1)} \qquad (8.19)$$

which can be written approximately as

$$G(\text{high}) = \frac{1}{\omega^2 R_1 R_2 C_1 C_2} \qquad (8.20)$$

Intersecting the low-frequency asymptotic form given in Eq. (8.16) with this high-frequency asymptote gives an expression for a corner frequency of this filter:

$$\omega_c = \frac{1}{\sqrt{R_2 R_3 C_1 C_2}} \qquad (8.21)$$

A more complete analysis of this circuit produces the transfer function

$$G(s) = -\frac{R_3}{R_1 R_2 R_3 C_1 C_2 s^2 + C_2(R_1 R_2 + R_1 R_3 + R_2 R_3)s + R_1} \qquad (8.22)$$

The derivation of this expression is fairly lengthy, but is simplified by first writing an expression for V_1 in terms of V_{in} and V_{out}, then eliminating V_1 by using the integrator transfer function of Eq. (8.12). This transfer function has the same general form and behavior as that of the low-pass series LCR circuit discussed in Chapter 3.

It is clear from Eq. (8.22) that by choosing the proper component values, a filter could be designed such that each of the three terms in the denominator would dominate in a different frequency region: low frequency for the constant term, middle for the term in s, and high for the term in s^2. If such an approximation is valid, the Bode plot of $|G(j\omega)|$ can be represented by three straight lines, and there will be two additional corner frequencies that can be used to describe the filter. In this manner, any multipole filter can therefore be represented by several straight-line approximations on the Bode plot. As we have noted before, the slope of these straight lines is related directly to the phase shift over the corresponding frequency interval, although in practice the integral multiple of 6 dB/octave thus obtained is valid only at frequency extremes and near the center of an intermediate interval.

Multipole filters are generally designed using tables of component value ratios generated specifically for Chebyshev, Butterworth, or Bessel types. However, two-pole filters can be easily designed using the critical damping condition as discussed in Chapter 3. A critically damped two-pole filter does not have a sharp knee but does produce a nearly ideal transient response.

> **Example 8.3** In the filter circuit of Figure 8.10a, let $R_1 = R_2 = R_3 = R$ and choose the capacitors such that the Butterworth condition of Problem 3.30 is met when the filter has a corner frequency of 1000 Hz. The largest available capacitor is 0.1 μF.
>
> With all resistors equal to R, Eq. (8.22) reduces to
>
> $$G(s) = -\frac{1}{R^2 C_1 C_2 s^2 + 3RC_2 s + 1}$$

which has poles at

$$s = -\frac{3}{2RC_1} \pm j\frac{\sqrt{4C_1C_2 - 9C_2^2}}{2RC_1C_2}$$

For a two-pole filter, the Butterworth condition is met when the poles are at

$$s = -\left(\frac{1}{\sqrt{2}} \pm j\frac{1}{\sqrt{2}}\right)\omega_c$$

Equating the real parts of these two pole expressions and solving for C_1 gives

$$C_1 = \frac{3}{\sqrt{2}\,R\omega_c}$$

Equating the imaginary parts gives

$$\frac{\omega_c}{\sqrt{2}} = \frac{\sqrt{4C_1C_2 - 9C_2^2}}{2RC_1C_2}$$

After cross-multiplying and squaring, this becomes

$$4C_1 - 9C_2 = 2R^2C_1^2C_2\,\omega_c^2$$

Solving for C_2 and substituting the above expression for C_1 gives

$$C_2 = \frac{2C_1}{9} = \frac{\sqrt{2}}{3R\omega_c}$$

With $f_c = 1000$ Hz and $C_1 = 0.1$ μF, we need an R of

$$R = \frac{3}{\sqrt{2}\,C_1\omega_c} = \frac{3}{1.4(10^{-7}\text{ F})(6.28 \times 10^3\text{ s}^{-1})} = 3400\ \Omega$$

and a $C_2 = 2C_1/9 = 0.022$ μF.

These values satisfy the corner frequency condition of Eq. (8.21).

8.3.7 General Feedback Elements

The feedback element in an operational amplifier design can be more complicated than a simple resistor and capacitor network. Consider the general four-terminal network in Figure 8.11a: If the network is linear, its effect can be described by a complex transfer function **F,** yielding an output voltage

$$\mathbf{V}_{\text{out}} = \mathbf{F}\mathbf{V}_{\text{in}} \tag{8.23}$$

More generally, the output of this network might be described as a function of the input voltage:

$$\mathbf{V}_{\text{out}} = \mathbf{F}(\mathbf{V}_{\text{in}}) \tag{8.24}$$

If such a network is used to implement a feedback loop as Figure 8.11b, the Rule 2 requirement of equal voltages at the input terminals immediately gives the overall amplifier relations:

$$\mathbf{V}_2 = \frac{\mathbf{V}_1}{\mathbf{F}} \tag{8.25}$$

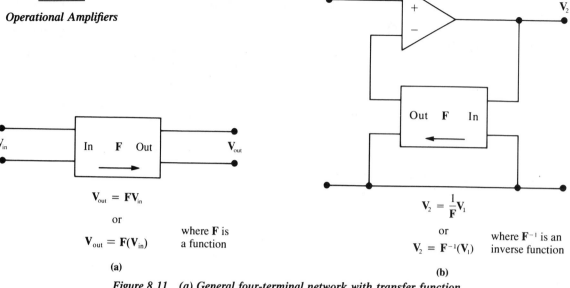

$$\mathbf{V}_{out} = \mathbf{F}\mathbf{V}_{in}$$

or

$$\mathbf{V}_{out} = \mathbf{F}(\mathbf{V}_{in})$$ where **F** is a function

(a)

$$\mathbf{V}_2 = \frac{1}{\mathbf{F}}\mathbf{V}_1$$

or

$$\mathbf{V}_2 = \mathbf{F}^{-1}(\mathbf{V}_1)$$ where $\mathbf{F}^{-1}$ is an inverse function

(b)

Figure 8.11   *(a) General four-terminal network with transfer function F or capable of generating an output signal equal to $\mathbf{F}(\mathbf{V}_{in})$. (b) The same network in a feedback loop generates the inverse transfer function or the inverse functional effect.*

for the linear network, or

$$\mathbf{V}_2 = \mathbf{F}^{-1}(\mathbf{V}_1) \tag{8.26}$$

for the more general functional relationship. In this last expression, $\mathbf{F}^{-1}$ represents the inverse function. A common application of this relationship exploits the exponential current-to-voltage relationship of a diode (or transistor) to obtain a logarithmic response for the overall amplifier.

Another interesting feedback element is the analog multiplier as defined by Figure 8.12a. The multiplier circuit itself can be thought of as another op amp with a feedback resistor whose value is determined by a second input voltage; a JFET operated in the variable resistance region is such a feedback element, although practical multiplier circuits are fairly complicated and are best purchased as packaged units. Multiplication circuits with the ability to handle input voltages of either sign (four-quadrant multipliers) are available as integrated circuits and have a number of direct uses as multipliers, the most obvious of which are in variable gain amplifiers or signal amplitude modulators. But when used in a feedback loop around an operational amplifier, other useful functional forms result. The circuit of Figure 8.12b gives an output that is the ratio of two signals, whereas the circuit of Figure 8.12c yields the analog square root of the input voltage.

In practice, the use of analog circuits involving a multiplier may present problems because of the large voltage range of the output signals. Input signals can, of course, be scaled so that the output resulting from the multiplication of the

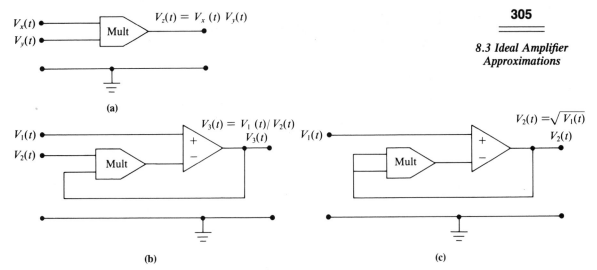

Figure 8.12 *(a) Five-terminal network that performs the
multiplication operation on two voltage signals. As part of a feedback
loop it can result in (b) the division or (c) the square root operation.*

largest expected values (or the division of the largest numerator by a smallest
denominator) does not exceed some fixed maximum. Unfortunately, this can also
mean that the average signal is much smaller than ideal and more susceptible to
noise effects.

8.3.8 Differential Amplifiers

The circuit in Figure 8.13a functions as a precision voltage difference (differential) amplifier. Since the current into the positive input terminal is zero, the lower
pair of resistors forms a voltage divider producing

$$V_3 = \frac{R_2 V_2}{R_1 + R_2} \tag{8.27}$$

This is also the voltage at the negative terminal, and since no current goes into
this terminal, the current in the upper pair of resistors must be given by

$$\frac{V_1 - V_3}{R_1} = \frac{V_3 - V_{out}}{R_2} \tag{8.28}$$

Using (8.27) to eliminate V_3 and solving for V_{out} yields the differential amplifier
expression

$$V_{out} = \frac{R_2}{R_1}(V_2 - V_1) \tag{8.29}$$

The application of this amplifier is sometimes limited because of the relatively
low input impedance of $2R_1$ between two input points. A simple modification

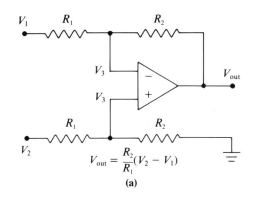

$$V_{out} = \frac{R_2}{R_1}(V_2 - V_1)$$

(a)

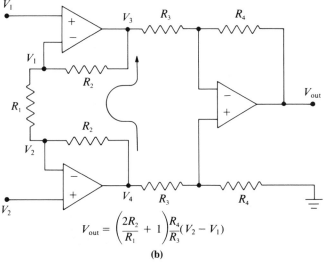

$$V_{out} = \left(\frac{2R_2}{R_1} + 1\right)\frac{R_4}{R_3}(V_2 - V_1)$$

(b)

Figure 8.13   (a) The basic differential amplifier connection. (b) An improved design using three operational amplifiers.

resulting in a large input impedance is to add two noninverting, unity-gain amplifiers, one driving each input. Such a differential amplifier, with a high input impedance and using three operational amplifiers, is an extremely versatile device and is known as an instrumentation amplifier. By adding a few more resistors we obtain the "classic" instrumentation amplifier design of Figure 8.13b. The gain of this amplifier can be calculated in two parts: from $V_4 - V_3$ to the output this amplifier follows the previous equation, with R_3 and R_4 replacing R_1 and R_2; the gain $(V_4 - V_3)/(V_2 - V_1)$ of the two noninverting amplifiers can be obtained from the current relationship

$$\frac{V_2 - V_1}{R_1} = \frac{V_4 - V_3}{R_1 + 2R_2} \tag{8.30}$$

Combining the expressions to eliminate $V_4 - V_3$ yields the overall gain

$$V_{out} = \left(\frac{2R_2}{R_1} + 1\right)\frac{R_4}{R_3}(V_2 - V_1) \tag{8.31}$$

In addition to having a high input impedance, if $R_1 < R_2$ this instrumentation amplifier also has a common-mode rejection that is significantly better than the single op amp design. The bridge circuit of Figure 10.12 is a typical application of an instrumentation amplifier. It can be seen that the operation of the bridge is unaffected by the attachment of an instrumentation amplifier.

8.3.9 Perfect Diodes

An operational amplifier can be combined with a physical diode to yield a circuit with a response very close to that of the ideal diode. In the half-wave rectifier example of Figure 8.14a, diode D_1 is the device whose characteristics are being improved; diode D_2 is present only to provide a current feedback path around the amplifier when D_1 is not conducting.

Figure 8.14   (a) The "perfect" diode circuit overcomes the forward voltage drop of the conducting diode. (b) The "perfect" full-wave rectifier.

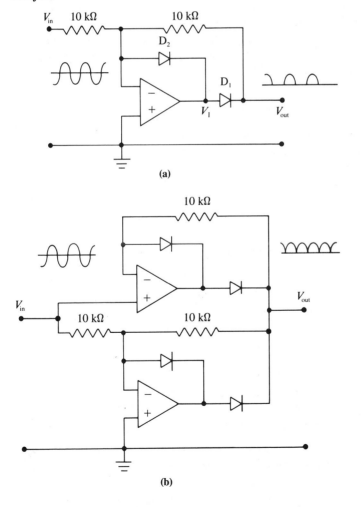

First, consider the operation of this circuit when the input signal is positive: The op amp output V_1 will be driven negative until diode D_2 turns on to provide a feedback path around the amplifier. Since the negative input terminal of the op amp is always at 0 V and the forward resistance of the diode is small, V_1 will never go much below $-V_{PN}$. With V_1 negative, D_1 is nonconducting, there is no current in the feedback resistor, and V_{out} must therefore be at 0 V, just like the negative input terminal.

When the input signal becomes negative, the situation changes completely. The op amp output V_1 will now become positive, jumping suddenly from $-V_{PN}$ to $+V_{PN}$ and continuing up from there as needed. This positive voltage turns D_2 off and opens the unity-gain feedback path through D_1. With this feedback path active, the normal rules apply, making

$$V_{out} = -V_{in} \qquad \text{when } V_{in} < 0 \qquad (8.32)$$

The forward voltage drop V_{PN} is still present across D_1, but the effect causes an offset in the signal V_1 rather than V_{out}.

When the input is a sinusoid, the circuit of Figure 8.14b acts as a full-wave rectifier. Since the forward voltage drop of the diodes has been effectively eliminated, this circuit can be used to provide accurate rectification of signals, even when the amplitude of the input signal is less than V_{PN}. The lower half of this circuit is identical to the half-wave rectifier just discussed, and the upper half is designed to function as a unity-gain, noninverting amplifier during the positive part of the input signal.

8.3.10   The Negative Impedance Converter (NIC)

Another use of operational amplifiers is to make two-terminal circuit elements that would otherwise be unavailable or impractical. The circuit in Figure 8.15a is a negative impedance converter or NIC. As usual, the input impedance to this circuit is defined by

$$\mathbf{Z}_{in} = \frac{\mathbf{V}_1}{\mathbf{I}_1} \qquad (8.33)$$

and assuming an ideal amplifier, it is easily found to be

$$\mathbf{Z}_{in} = -\mathbf{Z} \qquad (8.34)$$

If $\mathbf{Z}$ is a resistance R, we get an input impedance of $-R$, meaning that the op amp is adding energy to the circuit. A simple NIC application is shown in Figure 8.15b. By adjusting R_N, it is possible to reduce the series resistance of the *LCR* circuit, thereby increasing the Q of this bandpass filter.

> **Example 8.4**   The circuit in Figure 8.15b has $R_s = 300\ \Omega$, $L = 30$ mH, and $C = 0.1\ \mu$F. Assuming a 741 op amp connected to plus and minus 15-V power supplies, choose the resistors R and R_N in the NIC so that the Q of the overall circuit is maximized.
>
> To maximize the Q we need to reduce the series resistance to zero, and this can be done if the NIC has an effective impedance of $-R_s$. Equation (8.34) thus speci-

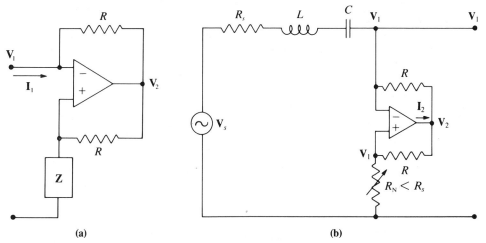

(a)   (b)

Figure 8.15   (a) The NIC exhibits a negative input impedance. (b)
This simple example uses the NIC to increase the Q of an LCR circuit.

fies that $R_N = 300\ \Omega$. With the NIC replaced by an equivalent resistance of $-R_s$, the transfer function $\mathbf{V}_1/\mathbf{V}_s$ of this circuit is

$$\mathbf{G}(j\omega) = \frac{\mathbf{V}_1}{\mathbf{V}_s} = -\frac{j\omega R_s C}{1 - \omega^2 LC}$$

This predicts an infinite Q, since the output signal $\mathbf{V}_1$ goes to infinity at resonance. Obviously, some other effect must limit the amplitude of $\mathbf{V}_1$.

Because of the peculiar operation of the NIC, when the voltage $\mathbf{V}_1$ is positive, the current $\mathbf{I}_1$ is opposite to the direction indicated in Figure 8.15a. This means that when $\mathbf{V}_1$ increases, $\mathbf{V}_2$ increases even more, and the op amp must drive current into both of the R resistors. If $R_N = R_s$, then the voltage and current limitations of the op amp must limit the signal amplitude at resonance. If we wish to maximize the output signal $\mathbf{V}_1$, we must investigate these limits.

In order for the op amp to satisfy the ideal amplifier rules, the voltage divider formed by R and R_N must produce

$$\mathbf{V}_1 = \frac{R_N}{(R + R_N)}\,\mathbf{V}_2$$

It is also easy to see that the op amp's output current must be

$$\mathbf{I}_2 = \frac{2(\mathbf{V}_2 - \mathbf{V}_1)}{R} = \frac{2\mathbf{V}_1}{R_N}$$

Since the current and voltages in these expressions are in phase, we can take the magnitude of both sides and obtain the same expressions with real current and voltage signals. Combining the equations to eliminate V_1 gives

$$\mathbf{I}_2\,(R + R_N) = 2\mathbf{V}_2$$

which defines the relationship between the voltage and current out of the op amp. If we select an $\mathbf{I}_2$ and a $\mathbf{V}_2$ that correspond to the maximum power output of the op amp, this equation yields a value for R that makes full use of that power.

The output amplitude V_2 can certainly not exceed the power supply voltage of 15 V, and the specification sheet for a 741 indicates a short circuit output current of 25 mA; we certainly cannot exceed that. Putting these values into the equation and solving give $R = 900\ \Omega$. With this choice, the V_1 equation predicts a maximum output signal of 4.1 V.

Since the amplitude at resonance is limited, a high Q can be observed only when a relatively small input signal V_s is applied.

8.3.11 *The Gyrator*

Another important circuit of this same type is the gyrator, which can be used to mimic a variety of reactive circuit elements. The general form of this circuit is shown in Figure 8.16a. If ideal amplifiers are assumed, the standard rules permit the definition of currents and voltages as shown in the figure. Using these definitions, the repeated application of Ohm's law and a little algebra allows us to obtain the ratio V_{in}/I_{in}. Starting at the bottom, the current I_{45} is defined by V_{in} across Z_5; this current across Z_4 defines V_{34}; and so on up the impedance chain. At the top the final current, I_{in} will be defined in terms of V_{in}, and the ratio reduces to

$$Z_{in} = \frac{V_{in}}{I_{in}} = \frac{Z_1 Z_3 Z_5}{Z_2 Z_4} \tag{8.35}$$

One of the main justifications for this rather complicated circuit is its ability to provide a circuit element, other than a physical inductor, whose impedance increases with frequency.

Figure 8.16 (a) The gyrator circuit is also used for its effective impedance. (b) A simple filter circuit using a gyrator.

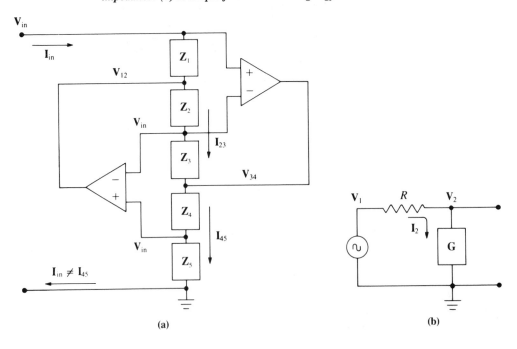

(a)   (b)

If the Z_2 element is a capacitor C_G, and the other elements are resistances R_G, the equivalent impedance of the gyrator is like an inductor and given by

$$Z_{in} = C_G R_G^2 s \qquad (8.36)$$

This arrangement allows a capacitor (and two op amps) to be used as an inductor in any low-frequency application where one side of the inductor can be grounded. If the Z_1 and Z_3 elements are both capacitors, the circuit behaves like a new reactive element—the double capacitor with impedance given by

$$Z_{in} = \frac{1}{C_G^2 R_G s^2} \qquad (8.37)$$

Either of these gyrator forms can be used to form a simple one- or two-pole filter as in Figure 8.16b.

Example 8.5 A tank circuit with a resonance at 60 Hz is needed, and the largest capacitor available is 0.1 μF. (a) What size inductor is required? (b) If another 0.1-μF capacitor is used, what value of R is needed to make a gyrator inductance of this size?

The tank circuit's resonant frequency is $\omega_r = 1/\sqrt{LC}$. At 60 Hz and using a 0.1-μF capacitor, this gives

$$L = \frac{1}{C\omega_r^2} = \frac{1}{(10^{-7}\ \mathrm{F})(377)^2} = 70\ \mathrm{H}$$

From Eq. (8.36), the equivalent inductance of a gyrator is given by

$$L = C_G R_G^2$$

With $L = 70$ H and $C_G = 0.1$ μF, we need an R_G of

$$R_G = \sqrt{\frac{L}{C_G}} = \sqrt{\frac{70\ \mathrm{H}}{10^{-7}\ \mathrm{F}}} = 26{,}500\ \Omega$$

This result shows that a relatively bulky inductor can be replaced by a gyrator using components of reasonable size.

8.3.12 The Wien Bridge Oscillator

Thus far in this text we have viewed oscillators only as the undesirable result of an amplifier or filter gone astray. In fact, oscillators are an important and necessary part of virtually every large electronic system. Fundamentally, they serve as signal generators for other parts of a circuit. Square wave signals are easily obtained with the aid of integrated circuit chips designed for the purpose (the 555 timer is the 741 of oscillators), and we will treat this subject in a later chapter. However, occasionally the need may arise for a sinusoidal oscillator. There are two ways to obtain such a signal with linear circuits: Pass a square wave through a multipole low-pass filter to extract just its fundamental; or build an amplifier with poles exactly on the imaginary axis and apply just enough input signal to start the oscillation. The Wien bridge oscillator is an example of the latter method.

The Wien bridge circuit in Figure 8.17a is a simplified example of a oscillator circuit and may not generate a sine wave output as drawn; a practical circuit generally requires additional components as discussed at the end of this section. However, the simplified version allows us to see the operating principle of the oscillator.

We can treat this and other oscillators using the methods already derived for filters and amplifiers. If we break the oscillator circuit at AB as indicated in the figure and apply an input sinusoidal signal $\mathbf{V}_0$, the standard methods (Problem 13) give the transfer function

$$\mathbf{G(s)} = \frac{\mathbf{V}_2}{\mathbf{V}_0} = \frac{RR_2C\mathbf{s} + (1 + RC\mathbf{s})^2 R_2}{RR_2C\mathbf{s} - R_1(1 + RC\mathbf{s})^2} \tag{8.38}$$

Setting the denominator equal to zero gives

$$R_1 R^2 C^2 \mathbf{s}^2 - (R_2 - 2R_1)RC\mathbf{s} + R_1 = 0 \tag{8.39}$$

which has roots at

$$\mathbf{s} = \frac{R_2 - 2R_1}{2R_1RC} \pm \frac{\sqrt{R_2^2 - 4R_1R_2}}{2R_1RC} \tag{8.40}$$

If $R_2 = 2R_1$, the real part of these roots is zero, thus locating the poles of $\mathbf{H}$ on the imaginary axis as required for constant-amplitude sinusoidal oscillation. With this resistor ratio the pole expression reduces to

$$\mathbf{s} = 0 \pm \frac{j}{RC} \tag{8.41}$$

indicating that the angular frequency of the oscillation is $1/RC$.

Figure 8.17b shows the pole location as a function of the R_1/R_2 resistor ratio. In order to obtain a constant-amplitude sinusoidal oscillation, these poles must lie

Figure 8.17 *(a) The Wien bridge sine wave oscillator. (b) The pole locations for various resistor ratios.*

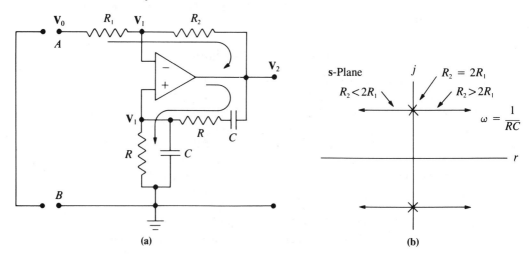

(a)   (b)

exactly on the imaginary axis. Since this requirement can never be obtained exactly with static elements, a self-correcting feature must generally be added to the circuit. This is normally accomplished by using a field effect transistor as a variable resistor to replace part of R_1. By rectifying and filtering V_2, a signal can be derived that will drive the gate of the FET in such a way that its conductance increases when the average amplitude of V_2 increases. An alternative method is to build the basic oscillator with the poles slightly to the left of the imaginary axis, then apply a small driving signal at V_0 as needed to maintain a constant amplitude. This latter method has the advantage that the oscillator can then be "locked" to the phase of the driving signal.

8.3.13 Voltage-to-Current Converters

The operational amplifier can also be used to make a precision voltage-to-current converter. Analysis of the circuit of Figure 8.18a yields the relationship

$$I_L = \frac{V_{in}}{R} \tag{8.42}$$

A similar circuit configuration reverses the ground and input signal connections, but the configuration shown has the advantage of a high input impedance. A circuit such as this is useful for driving a floating load (one with neither terminal grounded), such as a magnetic field coil or a DC motor. Although the resistance of a field coil will change with the temperature of the winding, the voltage-to-current converter can still produce a magnetic field that is strictly a function of the input voltage. Likewise for a DC motor, this circuit can be used to produce a particular torque independent of the rotational speed of the motor.

Circuits that can drive a grounded load are somewhat more difficult to implement. If current is needed in only one direction, the circuit of Figure 8.18b will

Figure 8.18 *(a) A voltage-to-current converter suitable for driving a floating load such as a DC motor. (b) This circuit uses a transistor to drive a grounded load.*

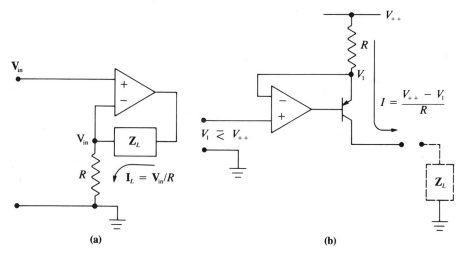

serve. Since the op amp adjusts the transistor base voltage until the transistor's emitter is at V_1, this circuit must yield an output current

$$I = \frac{V_{++} - V_1}{R} \qquad (8.43)$$

which, while not as simple as the previous result, is still independent of the load impedance and the transistor's characteristics.

8.3.14 Power Drivers

In some applications it may be necessary to supply more current to the load than is available from the output of the operational amplifier. (The 741 can supply at most 25 mA.) The addition of a single bipolar or field effect transistor within the feedback loop as shown by the examples in Figure 8.19 can provide the additional current with no loss of circuit precision. The enhancement power MOSFETs are particularly suited to this application; their worst feature is a large turn-on voltage, which means that the gate must be held 5 V to 10 V above the source.

The common drain MOSFET arrangement of Figure 8.19b is the most obvious extension of the common collector bipolar circuit shown in Figure 8.19a. However, since the drain current is a function of the gate-to-source voltage (the turn-on value for V_{GS} is around 4 V) and the source voltage follows the input voltage, the op amp may be unable to supply a sufficiently high gate voltage. The common source circuit of Figure 8.19c overcomes this problem, but it multiplies the operational amplifier's open-loop gain by a factor of $-g_m R$, which may be sufficient to turn even a docile 741 into a potential oscillator—especially for a low closed-loop voltage gain circuit such as the one shown. Note that the voltage inversion

Figure 8.19   (a) A transistor increases the current drive ability of an op amp. (b) An enhancement-type power MOSFET serves the same function. (c) A common source MOSFET allows a larger output voltage swing. Note that the negative feedback connection goes to the positive input terminal.

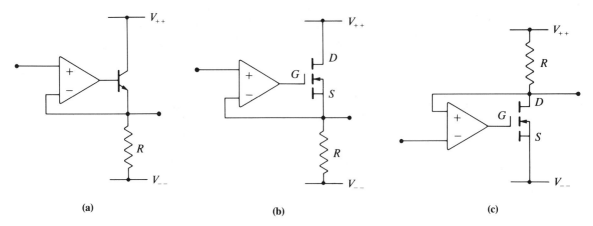

of the common source amplifier reverses the meaning of the plus and minus labels on the operational amplifier's input terminals: The feedback is still negative.

The single-transistor output amplifiers shown above suffer from an unsymmetric drive capability as discussed in Chapter 7. To provide a symmetric output drive, a push-pull amplifier can be substituted for the single transistor. For some applications, a class C design like Figure 8.20a may be sufficient. This circuit relies on the operational amplifier to provide both the bias current and the *PN* junction voltage needed to turn on the appropriate transistor; consequently, there will be an approximately 1-V region around zero when neither transistor is conducting. If a certain amount of crossover distortion can be tolerated, this circuit has the definite virtue of simplicity.

For more exacting work, the class B design of Figure 8.20b can be implemented. In this design the base operating voltage is set by the *PN* voltage drop across the diodes in the base circuit. When the op amp output is at zero, both transistors are on the verge of conduction due to the current and the *PN* voltage drops in the base biasing circuit. The 25-Ω emitter resistors allow for some variation in transistor and diode parameters. The choice of biasing resistors limits the output current and protects the transistors as discussed in Example 8.6. The circuit of Figure 8.20a is not similarly protected.

All of the power drivers just discussed will modify the open-loop gain of the amplifier to some extent. Any additional transistor adds at least one additional delay element or pole in the overall amplifier. If medium-power transistors with a good high-frequency response are used, then the additional poles and correspond-

Figure 8.20 *(a) A class C push-pull transistor stage provides a symmetric current drive. (b) A class B version of the push-pull gives less crossover distortion.*

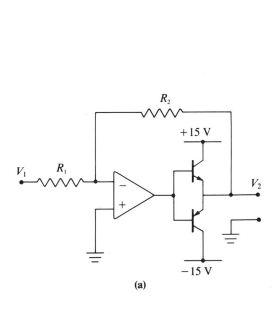

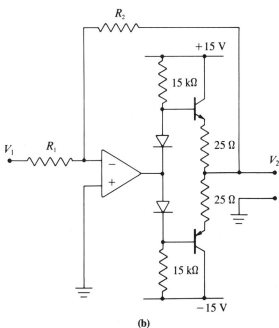

(a) (b)

ing corners in the open-loop tranfer function will also occur harmlessly at high frequency. However, since many high-power bipolar transistors also tend to have a rather poor frequency response, the additional poles introduced by these devices may become troublesome.

Example 8.6 In the power amplifier of Figure 8.20b, assume that the transistors have $h_{FE} = 80$. Determine the worst-case power dissipation in each of the transistors when the output is shorted to ground.

With the output shorted, the worst-case design occurs when the op amp output has turned a transistor full on. Under this condition and with a current I_C in the transistor, the power dissipated by that transistor is

$$P = [(15 \text{ V}) - (25 \text{ }\Omega)I_C]I_C$$

Let us consider the worst-case condition for the upper *(NPN)* transistor. When the op amp output is sufficiently positive, the lower transistor is turned off and the upper one is full on. With the upper diode thus back-biased, the current in the upper 15-kΩ resistor is directed into the base of the upper transistor, and we can write the voltage expression

$$(15 \text{ V}) - (15 \text{ k}\Omega)I_B - V_{PN} - h_{FE}I_B(25 \text{ }\Omega) = 0$$

Setting $h_{FE} = 80$, $V_{PN} = 0.6$ V, and solving for I_B gives

$$I_B = 0.85 \text{ mA}$$

or
$$I_C = h_{FE}I_B = 68 \text{ mA}$$

From the above equation for the power P, we find that the worst-case power dissipation in the transistor is 0.9 W.

8.4 *Analysis Using Finite Open-Loop Gain*

Although the infinite gain approximation used above can lead to a quick understanding and utilization of operational amplifiers, a more complete description is necessary if we are to understand the limitations and more subtle afflictions of these powerful devices. Real operational amplifiers have a large but finite input impedance and a small but nonzero output impedance. They also exhibit internal, self-generated voltage and current asymmetries that result in a nonzero output voltage in the absence of an input signal. Since these problems are all suppressed by the use of an infinite open-loop gain during the analysis of a negative feedback circuit, if we are to look more closely at these problems, the first step must be to introduce an analysis method using a finite open-loop gain.

When dealing with problems encountered at low frequencies, it is reasonable to use an open-loop gain that is a large but finite constant. However, for the problems associated with high frequencies, we must use an open-loop gain transfer function similar to Eq. (8.3) in order to model more correctly the frequency dependence of the operational amplifier's gain and phase shift. Higher-performance amplifiers will require more complicated transfer functions with two or more poles, but can still be modeled effectively by the general form of a multipole low-pass filter.

Although a complete model of an operational amplifier can be quite complicated and algebraically cumbersome, we will be able to show the most common problems by introducing individual features as needed. Since we are in most cases dealing with quite small corrections to the ideal amplifier performance, it is acceptable to consider these nonideal features individually while continuing to treat the other parameters as ideal. Thus we will consider output impedance, input impedance, and voltage and current offsets as separate rather than combined effects.

8.4.1 Output Impedance

A real operational amplifier will generally have a small, resistive output impedance derived from the output impedance of its push-pull output stage. With the frequency and phase dependence of the open-loop amplifier described by $\mathbf{A}(j\omega)$, we can model the open-loop output impedance by an additional series resistor R_0 as shown in Figure 8.21a. The unity-gain amplifier in this figure results in the simplest analysis; more general designs with constant closed-loop gain G generally result in the same expressions we shall derive, but have A_0 replaced by A_0/G for noninverting designs and by $A_0/(1 + |G|)$ for inverting designs.

When connected with negative feedback, the effective output impedance of a circuit will in general be much smaller than R_0. The effective output impedance is obtained from the standard definition

$$\mathbf{Z}_{\text{out}} = \frac{\mathbf{V}(\text{open})}{\mathbf{I}(\text{short})} \qquad (8.44)$$

Using the open-loop transfer function for the operational amplifier,

$$\mathbf{V}_1 = \mathbf{A}(j\omega)(\mathbf{V}_{\text{in}} - \mathbf{V}_{\text{out}}) \qquad (8.45)$$

where $\mathbf{V}_1$ is defined as indicated in Figure 8.21a. Note that since $\mathbf{V}_1$ depends on $\mathbf{V}_{\text{out}}$, it will change with the load applied across the output terminals. We are still assuming no current into the input terminals, and with the circuit unloaded as shown there is no current in R_0, making $\mathbf{V}_{\text{out}}$ the same as $\mathbf{V}_1$. If we identify both $\mathbf{V}_{\text{out}}$ and $\mathbf{V}_1$ in this unloaded case by the notation $\mathbf{V}(\text{open})$, Eq. (8.45) can be solved to give

$$\mathbf{V}(\text{open}) = \left(\frac{\mathbf{A}}{\mathbf{A} + 1}\right) \mathbf{V}_{\text{in}} \qquad (8.46)$$

When a shorting wire is connected to the output as indicated in Figure 8.21a, $\mathbf{V}_{\text{out}}$ becomes exactly zero and the current $\mathbf{I}(\text{short})$ is given by

$$\mathbf{I}(\text{short}) = \frac{\mathbf{V}_1}{R_0} \qquad (8.47)$$

Since the negative input terminal is now at ground because of the shorting wire, Eq. (8.45) is modified to have $\mathbf{V}_{\text{out}} = 0$. In terms of the unchanged $\mathbf{V}_{\text{in}}$, the short circuit current is

$$\mathbf{I}(\text{short}) = \frac{\mathbf{A}(j\omega)\mathbf{V}_{\text{in}}}{R_0} \qquad (8.48)$$

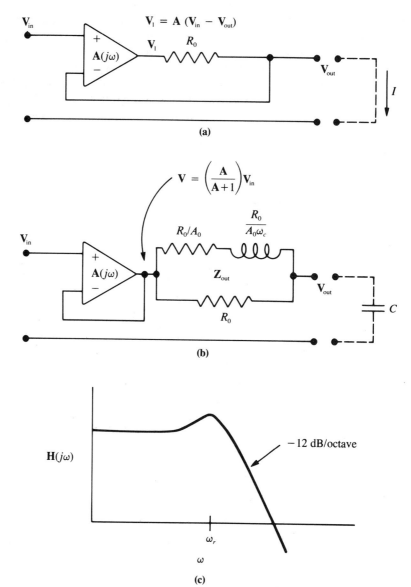

Figure 8.21   (a) Real, current-limiting operational amplifier partially modeled by an ideal amplifier and an output resistor. (b) An equivalent circuit for a 741-type op amp. (c) The overall transfer function when the amplifier drives a capacitive load.

Substituting Eqs. (8.46) and (8.48) into Eq. (8.44), the output impedance expression becomes

$$\mathbf{Z}_{out} = \frac{R_0}{1 + \mathbf{A}(j\omega)} \qquad (8.49)$$

If $\mathbf{A}$ is assumed to be a large frequency-independent constant A_0, the effective output impedance is just a resistor approximately equal to R_0/A_0. (As noted above, for noninverting closed-loop gains greater than unity, this expression changes to R_0G/A_0.)

Now that we know the output impedance of the closed-loop amplifier, it is possible to redraw the circuit in terms of a more ideal op amp, with zero output impedance, and a series $\mathbf{Z}_{out}$ located outside the feedback loop as shown in Figure 8.21b. Consideration of this circuit will show that $\mathbf{V}$(open) is still given by Eq. (8.46), and that the short circuit current,

$$\mathbf{I}(short) = \frac{\mathbf{V}(open)}{\mathbf{Z}_{out}} \tag{8.50}$$

will reduce to Eq. (8.48) if $\mathbf{Z}_{out}$ is given by Eq. (8.49).

The inclusion of the actual frequency dependence of $\mathbf{A}$ will produce a frequency-dependent $\mathbf{Z}_{out}$. The open-loop transfer function of Eq. (8.3) approximates all dominate-pole amplifiers such as the 741 and gives

$$\mathbf{Z}_{out} = \frac{R_0\omega_c + R_0 s}{\omega_c(1 + A_0) + s} \tag{8.51}$$

which can in turn be modeled by three passive components as in Figure 8.21b.

Since the $\mathbf{Z}_{out}$ just derived has an inductive part, its effect can cause even the normally well-behaved 741 operational amplifier to give unanticipated results if it is used to drive a capacitive load. If the load is a capacitor as shown in the figure, we effectively have a signal generator driving a variation of the *LCR* resonant circuit. With the simplifying assumption that $|\mathbf{A}| \gg 1$, we will have $\mathbf{V}_{open} = \mathbf{V}_{in}$, and the transfer function of the overall amplifier will be the transfer function of the *LCR* circuit. Neglecting the small R_0/A_0 series resistance, we have

$$\mathbf{H}(j\omega) = \frac{R_0 + j\omega L}{j\omega L + R_0(1 - \omega^2 LC)} \tag{8.52}$$

where

$$L = \frac{R_0}{A_0\omega_c}$$

For the 741 op amp this effective inductance evaluates to about 17 μH. This transfer function shows a resonance at $\omega_r = 1/\sqrt{LC}$, which can result in a slight peaking of the amplifier's overall transfer function as shown in Figure 8.21c.

Except for being unity gain, there is nothing special about this example circuit. Similar results can be obtained for other circuit configurations, but the method becomes unwieldy for more complicated open-loop transfer functions.

8.4.2 Current Limiting and Slew Rate

Since the voltage $|\mathbf{V}_1|$ in Figure 8.21a is always limited by some outside influence (ultimately the power supply), the presence of the resistor R_0 must limit the current that the amplifier can deliver into a load. For an integrated circuit, the output current is usually limited by some internal circuit element as discussed in Section 7.12, and even when push-pull power transistors are added they will usually be protected by a current limiting feature as discussed in Section 8.3.14. For a large load the output signal will be voltage-limited (the power supply), but as the load

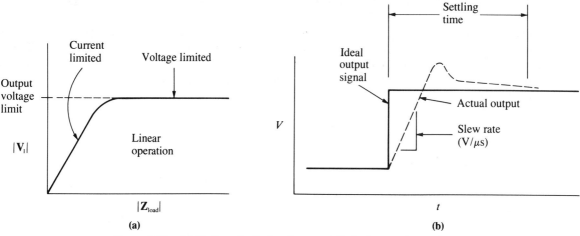

Figure 8.22 *(a) Voltage-limited and current-limited operating regions for an operational amplifier. The region of linear operation is below the curve. (b) Definition of slew rate and settling time for an operational amplifier.*

is reduced the device will become current-limiting and the amplitude of the output signal will shrink. This effect is illustrated in Figure 8.22a.

In the negative feedback amplifier, current limiting is a nonlinear property that voids the normal two-rule analysis. In this operating region, the input terminals are no longer at the same voltage, and the output impedance calculated in Section 8.4.1 becomes meaningless. Similar effects occur when the output of the amplifier is driven into voltage-limited operation, but these effects are more easily recognized on a typical oscilloscope trace.

When an operational amplifier is driven into a current-limiting condition either at the output as just discussed or at some internal point, it goes into saturation and becomes a constant current source. If such a current source drives a capacitive load, it generates a voltage across the capacitor that increases at a constant rate (slew rate) until the amplifier comes out of saturation. The typical result at the output of a real amplifier is as shown in Figure 8.22b. Instead of the desired step function, the actual output has a finite slope, normally overshoots the final value as indicated, and then approaches the final voltage either exponentially as shown or with some damped ringing. Note that the slew rate and overshoot are nonlinear effects that occur in addition to the linear transient responses we have discussed previously; only the final return to zero is related to the transient response function $h(t)$ of the amplifier.

The *maximum slew rate* is an important performance parameter for an operational amplifier and is the result of an internal current limit. Typical slew rates range from about 0.5 V/μs for the 741 to more than 100 V/μs for high-performance devices. In precision applications such as digital-to-analog conversions, where step functions are common, the settling time after amplifier saturation is an equally important parameter. The definition of this time will vary with manufacturer, but it always specifies the time between the edge of the applied step function and the point where the amplifier output settles to within some stated percentage of the target value. Settling time to 0.1% can vary from 100 ns to 10 μs, depending on the choice of operational amplifier.

8.4.3   Input Impedance

In the preceding discussion we kept the earlier ideal amplifier feature of infinite input impedance. Similarly, for this section we will take the output impedance to be zero but consider the effects of a less than infinite resistance at the input. A simple model is to assume an internal resistor R_T connecting the positive and negative input terminals as in Figure 8.23.

The inverting amplifier configuration of Figure 8.23a has had the input resistor removed so that the input impedance can be calculated directly at the amplifier's input terminals. As usual, the input impedance is defined by

$$\mathbf{Z}_{in} = \frac{\mathbf{V}_1}{\mathbf{I}_1} \tag{8.53}$$

where

$$\mathbf{I}_1 = \frac{\mathbf{V}_1}{R_T} + \mathbf{I}_2 \tag{8.54}$$

Figure 8.23   Models for calculating the input impedance of (a) the inverting amplifier and (b) the noninverting amplifier.

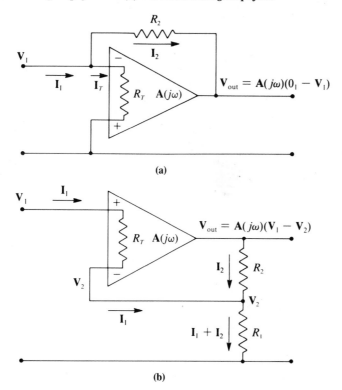

(a)

(b)

The current through the feedback resistor can be defined in terms of the output voltage,

$$\mathbf{I}_2 = \frac{\mathbf{V}_1 - \mathbf{V}_{out}}{R_2} \tag{8.55}$$

and the output voltage is related to $\mathbf{V}_1$ by the open-loop gain $\mathbf{A}$:

$$\mathbf{V}_{out} = \mathbf{A}(j\omega)(0 - \mathbf{V}_1) \tag{8.56}$$

Combining these expressions to eliminate $\mathbf{I}_2$ and $\mathbf{V}_{out}$ gives

$$\mathbf{Z}_{in} = \frac{R_T R_2}{R_2 + R_T(1 + \mathbf{A})} \tag{8.57}$$

For sufficiently large $\mathbf{A}$ this will reduce to

$$\mathbf{Z}_{in} = \frac{R_2}{\mathbf{A}(j\omega)} \tag{8.58}$$

which is just the Miller effect applied to the high-gain op amp. Although it is true that this input impedance is frequency-dependent and will appear inductive at higher frequencies, the important point here is that at low to moderate frequencies, this closed-loop input impedance is small and almost independent of the large R_T of the operational amplifier. Note that although $\mathbf{Z}_{in}$ is small, very little of the current $\mathbf{I}_1$ goes into the negative terminal. The input current $\mathbf{I}_1$ divides between $\mathbf{I}_2$ and $\mathbf{I}_T$ according to the ratio

$$\frac{\mathbf{I}_1}{\mathbf{I}_T} = \frac{R_T}{R_2}(\mathbf{A} + 1) \tag{8.59}$$

As a result, the negative terminal of an inverting amplifier exhibits a very small voltage signal, resulting in its designation as a "virtual ground."

In marked contrast to the low input impedance just found, the same operational amplifier in the noninverting configuration of Figure 8.23b will exhibit a closed-loop input impedance that is much larger than the open-loop value R_T. This calculation is simplified by recognizing that $\mathbf{I}_1$ is much less than $\mathbf{I}_2$, since R_T is much greater than R_1 or R_2. This observation combined with the open-loop gain equation

$$\mathbf{V}_{out} = \mathbf{A}(\mathbf{V}_1 - \mathbf{V}_2) \tag{8.60}$$

yields the result

$$\mathbf{Z}_{in} = R_T \frac{\mathbf{A} + \mathbf{G}}{\mathbf{G}} \tag{8.61}$$

where $\mathbf{G}$ is the closed-loop gain of this amplifier.

8.4.4 The Notch Filter

Not all circuits can be analyzed by the approximate methods based on Rules 1 and 2. One example is the filter circuit of Figure 8.24b, which can be understood only

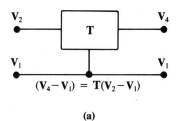

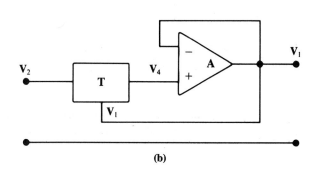

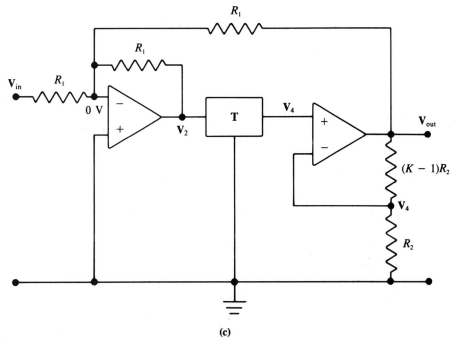

*Figure 8.24 (a) Block diagram for a twin-T notch filter.
(b) Noninverting circuit to produce a narrower notch filter.
(c) Inverting circuit producing a variable-width notch filter.*

in terms of a finite open-loop gain. The filter element in this circuit is a four-terminal network whose transfer function $\mathbf{T}(j\omega)$ is defined by

$$\mathbf{V}_4 - \mathbf{V}_1 = \mathbf{T}(\mathbf{V}_2 - \mathbf{V}_1) \qquad (8.62)$$

as indicated in Figure 8.24a.

If we take the filter element to be a passive twin-T network as described in Section 3.11.2, then the op amp circuit of Figure 8.24b will produce a notch filter that is much narrower than the passive twin-T itself. However, this effect is not seen using the basic two-rule analysis, which would require that $\mathbf{V}_1 = \mathbf{V}_4$, making Eq. (8.62)

$$\mathbf{T}(\mathbf{V}_2 - \mathbf{V}_1) = 0 \tag{8.63}$$

and producing the uninteresting result

$$\mathbf{V}_1 = \mathbf{V}_2 \tag{8.64}$$

Although this result is true at most frequencies, the infinite $\mathbf{A}$ approximation overlooks the important frequency region where the product $|\mathbf{AT}|$ is actually less than 1.

The magnitude of the twin-T transfer function $\mathbf{T}(j\omega)$ shown in Figure 8.25a has a deep notch at a single frequency but a rather wide rejection band. The width of this notch is normally defined at the half-power points where $\mathbf{T}^2 = \frac{1}{2}$, and for a twin-T with perfectly matched components as described in Section 3.11.2, the width is

$$\Delta\omega = \omega_2 - \omega_1 = 4\omega_0 \tag{8.65}$$

Assuming a large and constant open-loop gain $\mathbf{A}$, we can write

$$\mathbf{V}_1 = \mathbf{A}(\mathbf{V}_4 - \mathbf{V}_1) \tag{8.66}$$

With the help of Eq. (8.62), this becomes

$$\mathbf{V}_1 = \mathbf{AT}(\mathbf{V}_2 - \mathbf{V}_1) \tag{8.67}$$

Solving for $\mathbf{V}_1$ and dividing by $\mathbf{V}_2$ gives the overall transfer function:

$$\mathbf{H}(j\omega) = \frac{\mathbf{V}_1}{\mathbf{V}_2} = \frac{\mathbf{AT}}{1 + \mathbf{AT}} \tag{8.68}$$

This expression reduces to (8.64) whenever the product $|\mathbf{AT}| \gg 1$, but this will not be the case at ω_0, since $\mathbf{T}$ can be very small at this frequency. At ω_0 the expression reduces to $\mathbf{AT}$, which can still be small if $\mathbf{T}$ approaches zero closely.

The overall filter transfer function $\mathbf{H}$ can be found by substituting the twin-T transfer function of Eq. (3.77) into (8.68). For $|\mathbf{A}| \gg 1$, the expression reduces to

$$\mathbf{H}(s) = \frac{\mathbf{A}(s^2 + \omega_0^2)}{\mathbf{A}s^2 + 4\omega_0 s + \mathbf{A}\omega_0^2} \tag{8.69}$$

This overall function has zeros at $\pm j\omega_0$ just like $\mathbf{T}$; but unlike $\mathbf{T}$, which has two negative real poles, $\mathbf{H}$ has complex poles at

$$s \simeq -\frac{2\omega_0}{\mathbf{A}} \pm j\omega_0 \tag{8.70}$$

These poles are located adjacent to the zeros as shown in Figure 8.25b and produce the narrow rejection band shown in Figure 8.25a. The graphical method of Section 3.7.2 is a convenient way to obtain $|\mathbf{H}(j\omega)|$ in the vicinity of ω_0. Using

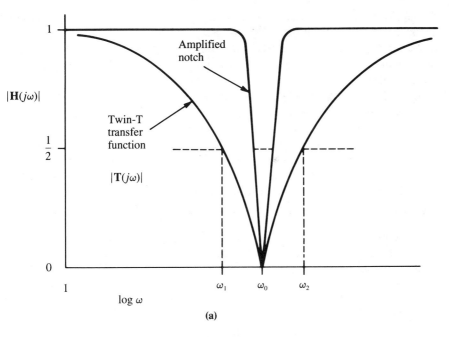

(a)

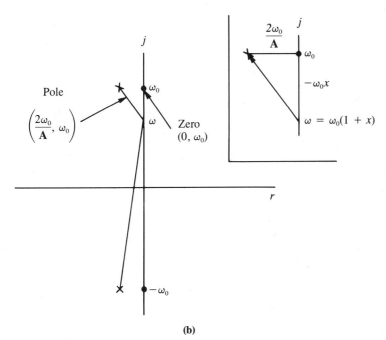

(b)

*Figure 8.25   (a) The transfer function of a well-matched twin-T filter
and of the amplified notch filter. (b) The poles and zeros of the
amplified notch filter*

the graphical technique indicated in the inset of Figure 8.25b and the parameterization $\omega = \omega_0(1 + x)$, it is possible to show that the half-power ($\mathbf{H}^2 = \frac{1}{2}$) width of the notch is $4\omega_0/|\mathbf{A}|$. The lower half-plane poles and zeros can be neglected in this argument: Since they are far away from the variable ω position and very close together, their distances are nearly equal for all positive ω.

Because of the location of the complex poles of $\mathbf{H}$, this notch filter has a transient response that rings at the rejection frequency ω_0. If $\mathbf{A}$ is very large, this ringing can persist for many cycles, producing an output signal at the very frequency the filter was intended to reject. A more flexible circuit with variable gain is shown in Figure 8.24c. This inverting circuit can be analyzed with the two basic rules and, except for a leading minus sign, yields an overall transfer function like Eq. (8.68) but with $\mathbf{A}$ replaced by K. The width and transient response of this filter can now be easily varied by changing the resistor $(K - 1)R_2$.

> **Example 8.7** A 741 op amp is used to build a 60-Hz notch filter of Figure 8.24b. Find the half-power width of this filter and the time constant of the decaying oscillation that follows an input transient.
>
> With $\omega_0 = 2\pi f = 377$ rad/s and $|\mathbf{A}| = 10^5$, the width of this notch is $\Delta\omega = 4\omega_0/|\mathbf{A}| = 1.5 \times 10^{-2}$ rad/s or 2.4×10^{-3} Hz. But this very narrow rejection band comes with a serious problem. From Eq. (8.70), the pole locations are
>
> $$\mathbf{s} = \sigma \pm j\omega_0 = (-7.5 \times 10^{-3} \pm j377)\text{rad/s}$$
>
> These poles will yield a damped sinusoidal transient response function of the form
>
> $$h(t) = \mathbf{C}e^{-\sigma t}\cos(\omega_0 t)$$
>
> where $\mathbf{C}$ is a complex constant whose magnitude is proportional to σ. The time constant of this exponential decay is found by determining the time when $\sigma t = 1$. Solving for t gives $t = 1/\sigma = 133$ s. For a sufficiently large transient, such as might occur when the power is switched on, this filter could thus show a significant ω_0 signal for several minutes. The primary advantage of the circuit shown in Figure 8.24c is that a compromise can be made between the notch width and the ringing time constant. It also allows the width to be narrowed after the circuit has stabilized following turn-on.

8.4.5 *Voltage and Current Offsets*

Modern operational amplifiers are generally DC-coupled throughout. Consequently it is not surprising that a real amplifier can exhibit a nonzero output even when the inputs are grounded or otherwise connected to give no input signal. This manifestation can be modeled by the introduction of three new parameters, V_{os}, I_{B1}, and I_{B2}, as shown in Figure 8.26. The offset voltage is shown explicitly by a voltage source that is actually part of the amplifier; the current offsets are assumed to be driven by current sinks within the amplifier symbol. All of these parameters are constant DC quantities. The voltage offset is the result of slight differences in the two transistors making up the differential input stage, and is generally much smaller for a bipolar transistor than for an FET input stage. Conversely, the bias currents are much less for an FET input stage.

The voltage offset can be significantly reduced by using an external adjustable-bias resistor. Many operational amplifiers provide connections specifically for this

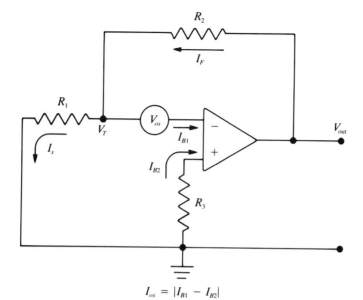

Figure 8.26   *Model for the voltage and current offsets associated with an operational amplifier used in the inverting configuration.*

purpose. With V_{os} thus reduced, we can assume it to be zero and consider only the effects of the remaining offset currents. Since I_{B1} and I_{B2} represent base currents into two identical bipolar transistors, they are both of the same sign (*NPN* transistors as drawn), and their difference I_{os} can reasonably be expected to be much less than either. Indeed, the specification sheets for the 741 operational amplifier show that the quantity $I_{os} = |I_{B2} - I_{B1}|$ is an order of magnitude smaller than either of its constituent currents, and therein lies the reason for the introduction of the resistor R_3 in the circuit of Figure 8.26. This resistor has no first-order effect on the circuit but does reduce the output voltage offset due to leakage currents.

By setting V_{os} in Figure 8.26 to zero, we can determine the output voltage developed by the offset currents alone. The current sum equation at V_T is

$$I_F = I_s + I_{B1}$$

or
$$\frac{V_{out} - V_T}{R_2} = \frac{V_T}{R_1} + I_{B1} \qquad (8.71)$$

Since $V_- = V_+$ and $V_{os} = 0$, we also have

$$V_T = -R_3 I_{B2} \qquad (8.72)$$

Eliminating V_T in the first equation and rearranging terms yields the following expression for the output voltage:

$$V_{out} = R_2 I_{B1} - \left(\frac{R_2 R_3}{R_1} + R_3\right) I_{B2} \qquad (8.73)$$

If we adjust R_3 so that the term in parentheses is just equal to R_2, then this expression becomes simply

$$|V_{\text{out}}| = R_2|I_{B1} - I_{B2}| = R_2 I_{os} \tag{8.74}$$

The value of R_3 that will bring about this order-of-magnitude reduction in the output voltage generated by the bias currents is found by requiring that the coefficient of I_{B2} in Eq. (8.73) be equal to R_2. Solving for R_3 gives

$$R_3 = \frac{R_1 R_2}{R_1 + R_2} \tag{8.75}$$

the parallel combination of R_1 and R_2.

8.5   *Feedback-Induced Oscillations*

An operational amplifier is designed to have an open-loop transfer function that is well behaved with poles only on or near the negative real axis. However, as was demonstrated in Chapter 3, the presence of a feedback loop generates a new closed-loop transfer function whose poles may not be placed so conveniently. Thus the negative feedback amplifier may have a propensity to ring or oscillate even though the operational amplifier itself is not at fault. Although a knowledge of the s-plane locations of the poles and zeros of the closed-loop amplifier is sufficient to predict the behavior of the amplifier, when dealing with a real amplifier rather than a model, this knowledge is not always easy to obtain.

An alternative approach to the same problem concentrates on the phase shift introduced by the amplifier and its feedback loop relative to the signal applied to the amplifier. A typical amplifier will show a phase lag that increases with signal frequency, ultimately reaching an asymptotic value given by

$$\theta(\omega \rightarrow \infty) = -90°(\text{number of poles} - \text{number of zeros}) \tag{8.76}$$

Since a 180° phase shift amounts to a signal inversion, this has the unfortunate effect of converting most negative feedback connections to positive feedback at some high frequency. The smallest perturbation at this frequency may then be iteratively amplified by the feedback loop so that the intended amplifier becomes an oscillator. The only cure for this basic dilemma is to ensure that the amplifier does not amplify at this frequency; that is, the magnitude of its transfer function must fall below unity before the phase shift reaches 180°.

Two methods will be presented for the analysis of amplifiers with feedback: the pole-zero analysis of Section 8.6 and the Nyquist plot method of Section 8.7. The pole-zero approach has been stressed throughout this text. It has the attractive feature of predicting the transient response of the final amplifier, but it has the disadvantage of requiring prior knowledge of all of the basic transfer functions in the problem, including a model for the open-loop amplifier. In contrast, the Nyquist method leads to an experimental procedure that can be used to investigate and perhaps improve an amplifier's performance even when nothing is known about the transfer functions.

8.6 Pole-Zero Analysis

If done without significant approximation, the pole-zero analysis of anything but the most basic amplifier will require the repeated solution of higher-order polynomial equations. Although this is not a particularly difficult problem if a computerized root-finding program is used, it would be tedious to attempt an exact but interesting problem in the format of this text. Rather, we will introduce an approximate form for the open-loop transfer function that is algebraically convenient but still able to give qualitatively correct high-frequency results.

Just as at low frequencies it was helpful to replace the open-loop gain $A(j\omega)$ with an arbitrarily large constant, so at high frequencies it is convenient to replace $A(j\omega)$ with an asymptotic form appropriate to the open-loop gain at a particular frequency. The straight-line approximations used to represent a low-pass transfer function on a Bode plot will serve our purpose and are of the general form

$$A(j\omega) = \left(\frac{\omega_1}{j\omega}\right)^n = \left(\frac{s_1}{s}\right)^n \qquad n = 1, 2, 3, \ldots \qquad (8.77)$$

where $\omega_1 \equiv s_1$ are constants describing a "unity-gain" frequency.

In Section 3.12.2 we derived the general expression for the closed-loop gain of an amplifier employing negative voltage feedback,

$$\text{Noninverting} \qquad H(s) = \frac{A}{1 + AF_N} \qquad (8.78)$$

where A describes the open-loop gain of the amplifier and F_N describes the noninverting amplifier's feedback circuit. In general, both A and F_N will be functions of s. An operational amplifier in the inverting configuration employs current feedback, resulting in the slightly different expression

$$\text{Inverting} \qquad H(s) = -\frac{A}{1 + (A + 1)F_I} \qquad (8.79)$$

where F_I describes the inverting amplifier's feedback circuit.

In the earlier, infinite A analysis we found overall closed-loop transfer functions $G(s)$ for a variety of amplifiers. In that approximation, Eqs. (8.78) and (8.79) reduce to

$$H(s) = G(s) = \frac{1}{F_N(s)}$$

and

$$H(s) = G(s) = -\frac{1}{F_I(s)} \qquad (8.80)$$

We can thus use the infinite A closed-loop transfer function, obtained by the simple two-rule method, as a convenient way to define the feedback parameters $F(s)$ for use in the exact expressions of Eqs. (8.78) and (8.79). By replacing the resistances with complex impedances, Eqs. (8.6) and (8.8) can be used to show that, in general, $\mathbf{F}_N = \mathbf{Z}_1/(\mathbf{Z}_1 + \mathbf{Z}_2)$ and $\mathbf{F}_I = \mathbf{Z}_1/\mathbf{Z}_2$.

In the following pole-zero discussion, $\mathbf{A(s)}$ will be replaced with one of the straight-line forms from Eq. (8.77). Open-loop gain approximations of this type have the obvious problem of being infinite at $\omega = 0$, but they can still provide qualitatively correct results at high frequencies.

8.6.1 Zero-Slope Closed-Loop Gain

The basic amplifiers shown in Figures 8.3 and 8.4 do not make use of complex impedances; they therefore have a closed-loop gain $\mathbf{G}$ that is constant with frequency. The magnitude of this function intersects the open-loop gain of the amplifier at a frequency ω_I, as illustrated in Figure 8.4b. This figure indicates a constant slope of -6 dB/octave for the opening-loop amplifier. As we shall see, the magnitude of the slope change between $\mathbf{G}$ and $\mathbf{A}$ at the point of intersection is one indication of an amplifier's stability; in this case its magnitude is 6 dB/octave.

The high-frequency model for this open-loop transfer function is just

$$\mathbf{A(s)} = \frac{s_1}{s} \tag{8.81}$$

where $s_1 \equiv \omega_1$ corresponds to the unity-gain frequency. With this assumption, the closed-loop transfer function for the noninverting amplifier becomes

$$\mathbf{H(s)} = \frac{s_1/s}{1 + s_1 F_N/s}$$

$$= \frac{s_1}{s + s_1 F_N} \tag{8.82}$$

where F_N is real and positive. This expression has only one real negative root,

$$\mathbf{s} = -s_1 F_N \tag{8.83}$$

and the transient response of the amplifier is a well-behaved decaying exponential.

However, higher-performance operational amplifiers often have poles that result in additional low-frequency corners as in Figure 8.27. It is then possible for the closed-loop gain $\mathbf{G}$ to intersect the open-loop gain at a point where the slope is -12 dB/octave as shown in the figure. The open-loop gain in the intersection region can be described by

$$\mathbf{A} = \left(\frac{s_1}{s}\right)^2 \tag{8.84}$$

and the corresponding closed-loop function becomes

$$\mathbf{H(s)} = \frac{(s_1/s)^2}{1 + (s_1/s)^2 F_N}$$

$$= \frac{s_1^2}{s^2 + s_1^2 F_N} \tag{8.85}$$

The poles in this case are imaginary,

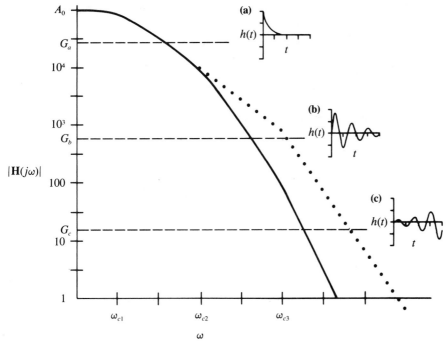

*Figure 8.27 A typical open-loop transfer function for a high-
performance operational amplifier, showing the relationship between
intersection angle and impulse response. The dotted line results from
"pole cancellation."*

$$\mathbf{s} = \pm j\sqrt{F_N}\, s_1 \tag{8.86}$$

and predict an undamped oscillatory transient response. A more exact analysis
would show a smooth transition between an exponential response for an intersection near ω_{c2} and this result at ω_{c3}. Intersections in the -12 dB/octave slope
region would actually result in damped oscillation, with less damping as the intersection frequency nears ω_{c3}.

If we carry this one step further and consider an intersection in a -18 dB/
octave region, the resulting transfer function is

$$\mathbf{H}(\mathbf{s}) = \frac{(s_1/\mathbf{s})^3}{1 + \left(\dfrac{s_1}{\mathbf{s}}\right)^3 F_N}$$

$$= \frac{s_1{}^3}{\mathbf{s}^3 + s_1{}^3 F_N} \tag{8.87}$$

This function has three poles, given by

$$\mathbf{s} = -s_1 \sqrt[3]{F_N}$$

$$\tag{8.88}$$

and
$$\mathbf{s} = s_1 \sqrt[3]{F_N}\left(\cos\frac{\pi}{3} \pm j\sin\frac{\pi}{3}\right)$$

Since the last pair of roots is in the right half-plane, this amplifier will oscillate with increasing amplitude until limited by the power supply. Similar results can be obtained for the inverting amplifier configuration.

Note that the impulse responses in Figure 8.27 become worse as the intersection angle becomes larger at small closed-loop gains. This effect is general, but the intersection angle can be reduced in two ways: (1) Modify the feedback circuit to produce a closed-loop transfer function that rolls off like a low-pass filter; (2) cancel a pole in the open-loop function by placing a zero at the same point in the s-plane. Pole cancellation is achieved by the introduction of an additional reactive element at an internal point in the circuit of the operational amplifier. High-performance devices have a "lead compensation" terminal that is intended for this purpose. The dotted line in Figure 8.27 shows the open-loop response after the middle pole has been canceled.

8.6.2 Positive-Slope Closed-Loop Gain

The high-frequency approximation to $\mathbf{A(s)}$ can also be used to good effect with the differentiation circuit of Figure 8.7. Because this circuit uses a reactive element, its feedback parameter $\mathbf{F}_I$ will be a function of $\mathbf{s}$. From the Bode plot shown in that figure, it is apparent that a closed-loop transfer function that lies lower on the graph will extend to higher frequencies and should therefore be a more faithful differentiator. Remember that the slope is fixed by the power of ω, but the intercept on the ordinate is determined by the product RC. Thus, the choice of RC seems to be clearly predicted by the required operation frequency.

However, if this circuit is constructed using a 741 operational amplifier with an arbitrary choice for C and an R that yields the desired product, the resulting differentiation of an input square wave will generally show substantial ringing, and the use of a less docile operational amplifier can magnify this unwanted effect substantially. As might be expected, the previous discussion of $\mathbf{G}$ and $\mathbf{A}$ intersection angles points the way to a design solution. One can either put a second capacitor in parallel with the feedback resistor as in Figure 8.28a or place a second resistor in series with the input capacitor as in Figure 8.28b. The resulting closed-loop transfer function is shown in Figure 8.28c. Unfortunately, the infinite open-loop gain analysis gives no clues to the best values of either the original R and C or this third component.

By using the exact closed-loop gain expression of Eq. (8.79), together with an approximation to $\mathbf{A}$ at high frequencies, we can find a solution to this design problem. Considering first the basic differentiation circuit of Figure 8.7a, the impedance ratio is seen to be

$$\mathbf{F}_I = \frac{1}{RC\mathbf{s}} \tag{8.89}$$

and the closed-loop transfer function is easily found from Eq. (8.79):

$$\mathbf{H(s)} = -\frac{ARC\mathbf{s}}{\mathbf{A} + 1 + RC\mathbf{s}} \tag{8.90}$$

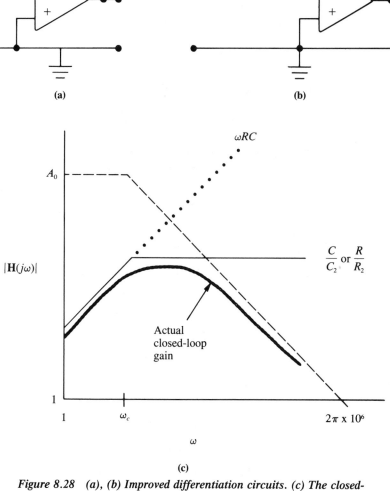

(c)

Figure 8.28   (a), (b) Improved differentiation circuits. (c) The closed-loop gain resulting from either of these circuits.

Replacing **A** at high frequency with s_1/s yields

$$\mathbf{H(s)} = -\frac{RCs_1\mathbf{s}}{RC\mathbf{s}^2 + \mathbf{s} + s_1} \qquad (8.91)$$

The poles of this function are given by

$$\mathbf{s} = \frac{-1 \pm \sqrt{1 - 4RCs_1}}{2RC} \qquad (8.92)$$

and the condition for critical damping is

$$1 - 4RCs_1 = 0 \tag{8.93}$$

A choice of RC meeting this condition produces a 741 circuit with a good transient response.

Now let us consider the improved differentiation circuit of Figure 8.28a. The feedback impedance in this circuit is given by the parallel combination of R and C_2,

$$\mathbf{Z}_F = \frac{R}{1 + RC_2\mathbf{s}} \tag{8.94}$$

making the new impedance ratio

$$\mathbf{F}_I = \frac{1 + RC_2\mathbf{s}}{RC\mathbf{s}} \tag{8.95}$$

Using this ratio in Eq. (8.79) together with $\mathbf{A} = s_1/\mathbf{s}$ gives

$$\mathbf{H(s)} = -\frac{RCs_1\mathbf{s}}{R(C + C_2)\mathbf{s}^2 + (1 + RC_2s_1)\mathbf{s} + s_1} \tag{8.96}$$

The poles of this transfer function are found to be

$$\mathbf{s} = \frac{-(1 + RC_2s_1) \pm \sqrt{(1 + RC_2s_1)^2 - 4R(C + C_2)s_1}}{2R(C + C_2)} \tag{8.97}$$

and the condition for critical damping is itself a quadratic in C_2:

$$(Rs_1)^2 C_2^2 - 2Rs_1C_2 + 1 - 4RCs_1 = 0 \tag{8.98}$$

Solving for C_2 yields

$$C_2 = \frac{1}{Rs_1} \pm 2\sqrt{\frac{C}{Rs_1}} \tag{8.99}$$

For a 741 operational amplifier the unity gain frequency is given by

$$s_1 = 2\pi \times 10^6 \text{ rad/s}$$

Using this value and assuming reasonable values for R and C, we find that the second term in Eq. (8.99) is much larger than the first. The design value for the extra capacitor is therefore given by

$$C_2 \approx 2\sqrt{\frac{C}{Rs_1}} \tag{8.100}$$

This result can be obtained with much less effort by replacing $(\mathbf{A} + 1)$ with $\mathbf{A}$ in Eq. (8.79). This approximation is of course valid whenever $\mathbf{A} \gg 1$ and removes the distinction between this inverting amplifier expression and Eq. (8.78) for the noninverting amplifier. This further approximation is reasonable at frequencies substantially less than s_1, the open-loop unity-gain point.

8.7 Nyquist Plot

When certain reasonable assumptions are made about the open-loop transfer function of an operational amplifier, the Nyquist plot can be used to investigate the stability of the closed-loop amplifier. An advantage of the Nyquist plot is that it can be derived from experimental measurements and does not require the determination of poles and zeros. Since these measurements are made with the feedback loop open, they can be obtained even when the closed-loop amplifier exhibits unstable and perhaps self-destructive behavior. Even though the Nyquist plot can be produced experimentally, the following discussion of transfer functions, poles, and zeros is needed for its interpretation.

We will consider here only noninverting, voltage feedback amplifiers; a similar method suitable for inverting amplifiers is indicated in Problem 8.31. The transfer function of the noninverting feedback amplifier is given by Eq. (8.78), and is restated here in terms of $j\omega$,

$$\mathbf{H}(j\omega) = \frac{\mathbf{A}(j\omega)}{1 + \mathbf{A}(j\omega)\mathbf{F}(j\omega)} \tag{8.101}$$

where $\mathbf{A}$ and $\mathbf{F}$ are in general complex. If $|\mathbf{H}|$ is infinite at some frequency ω_0, then the amplifier will magnify the slightest input and oscillate at frequency ω_0. If we assume that the open-loop amplifier $\mathbf{A}$ is well behaved with no infinities of its own, then an infinite $\mathbf{H}$ implies that

$$1 + \mathbf{AF} = 0 \tag{8.102}$$

Although this function would be satisfied at the poles of $\mathbf{H}$, we have restricted $\mathbf{s}$ to the imaginary axis, which is unlikely to pass through a pole. The Nyquist plot displays $1 + \mathbf{AF}$ (or more commonly $\mathbf{AF}$ or $-\mathbf{AF}$) as a function of frequency: Two examples are shown in Figures 8.29a and 8.29b.

Since we already know how to determine the stability of an amplifier from the location of its s-plane poles, we need to express $1 + \mathbf{AF}$ in terms of these poles. Starting with the closed-loop transfer function

$$\mathbf{H}(\mathbf{s}) = \frac{\mathbf{A}(\mathbf{s})}{1 + \mathbf{A}(\mathbf{s})\mathbf{F}(\mathbf{s})} = \frac{\mathbf{P}(\mathbf{s})}{\mathbf{D}(\mathbf{s})} \tag{8.103}$$

we can extract

$$1 + \mathbf{A}(\mathbf{s})\mathbf{F}(\mathbf{s}) = \mathbf{A}(\mathbf{s})\frac{\mathbf{D}(\mathbf{s})}{\mathbf{P}(\mathbf{s})} \tag{8.104}$$

where polynomials $\mathbf{D}(\mathbf{s})$ and $\mathbf{P}(\mathbf{s})$ can be expressed in terms of their roots: the poles and zeros of $\mathbf{H}(\mathbf{s})$.

The Nyquist expression of Eq. (8.102) is in terms of $j\omega$, so we must evaluate the various polynomials only for $\mathbf{s} = j\omega$. The restriction of $\mathbf{s}$ to the imaginary axis is indicated by the arrows on the s-plane plots in Figures 8.29a and 8.29b. We have chosen to plot the poles and zeros of the function $\mathbf{P/AD}$ so that the roots of $\mathbf{D}$ will have their normal labels as the poles $\mathbf{H}$ (see Eq. 8.103). The ratio $\mathbf{P/A}$ reduces to a polynomial in $\mathbf{s}$; its roots are shown as zeros in these figures.

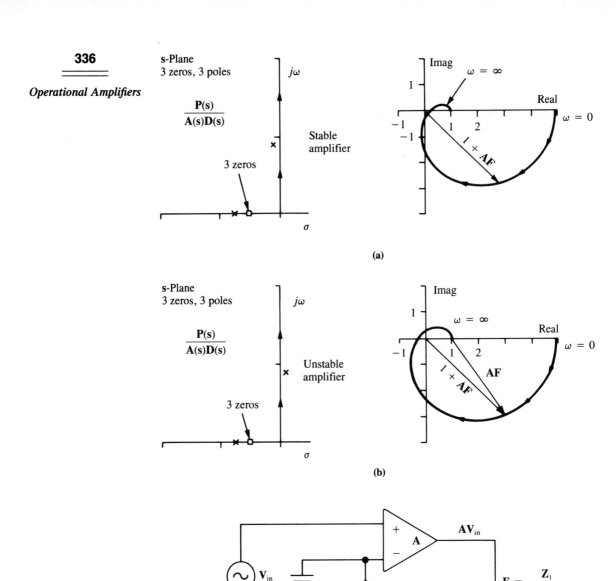

Figure 8.29 *(a) Modified Nyquist diagram for a stable closed-loop amplifier and (b) for an unstable one. (c) Experimental arrangement for measuring the gain-feedback product of a voltage feedback amplifier.*

In Figures 8.29a and 8.29b we show a specific example: D has three roots that are plotted as one real pole and a complex conjugate pair, whereas P/A contributes three real and identical zeros. Figure 8.29a shows a stable amplifier with all poles in the left half of the s-plane: note that the $j\omega$ path passes to the right of the the complex pole on the s-plane, and also to the right of the origin on the Nyquist plot. Figure 8.29b shows an unstable amplifier with the $j\omega$ path passing to the left of the pole and the origin. Thus, the clockwise or counterclockwise sense with which the $j\omega$ path passes the origin of the Nyquist plot can be used to predict the stability of the closed-loop amplifier. (Replace the origin with -1 or $+1$ if **AF** or $-$**AF** is plotted.)

Example 8.8 If an open-loop amplifier has a transfer function $\mathbf{A} = A_0/(1 + s/s_c)^3$, use the Nyquist plot to determine the maximum value of A_0 for which a unity-gain, noninverting feedback amplifier will be stable.

With $\mathbf{F} = 1$ for a unity-gain amplifier, the polar vector $1 + \mathbf{AF}$ in Figures 8.29a and 8.29b can be written as

$$\mathbf{R} = 1 + \mathbf{A} = 1 + \frac{A_0}{(1 + j\omega/s_c)^3}$$

where $\mathbf{R}$ now defines a vector from the origin to the Nyquist curve at a particular value of ω. The complex algebra is simplified by defining a new complex variable $\mathbf{r}$, given by

$$\mathbf{r} = 1 + j\frac{\omega}{s_c} = \sqrt{1 + \left(\frac{\omega}{s_c}\right)^2}\, e^{j\ \arctan(\omega/s_c)}$$

In terms of this variable, the expression for $\mathbf{R}$ reduces to

$$\mathbf{R} = 1 + A_0\mathbf{r}^{-3}$$

As ω increases from zero, it is clear that $\mathbf{R}$ will be real only when $\mathbf{r}^{-3}$ is real; the first time occurs when $\omega = 0$, giving an initial positive $\mathbf{R}$ value of $1 + A_0$, and the second time occurs when the phase angle of $\mathbf{r}^{-3}$ is $-\pi$. From the definition of $\mathbf{r}$, this second occurrence yields the equation

$$3\ \arctan\left(\frac{\omega}{s_c}\right) = \pi$$

which will be satisfied when

$$\omega = \sqrt{3}s_c$$

At this frequency, $\mathbf{R}$ becomes

$$\mathbf{R} = 1 + \left(\frac{A_0}{8}\right)e^{-j\pi} = 1 - \frac{A_0}{8}$$

and the Nyquist criterion says that the amplifier will be stable if

$$1 - \frac{A_0}{8} > 0$$

The transient response of the resulting unity-gain amplifier will thus be damped only if $A_0 < 8$.

This result can also be obtained using the pole-zero method but will require the solution of a cubic. The procedure is first to substitute $\mathbf{A}(s)$ into Eq. 8.78, then locate the poles of $\mathbf{H}(s)$ as a function of A_0. It will be found that at $A_0 = 8$, a complex conjugate pole pair lies directly on the imaginary axis.

The most important use of the Nyquist approach is to experimentally determine and modify the behavior of an amplifier before the feedback loop is closed. For noninverting amplifiers, the technique is indicated in Figure 8.29c: Disconnect the feedback at the negative terminal, ground the negative terminal, and measure the voltage $\mathbf{V}_0$ across the broken loop as shown. The complex voltage ratio $\mathbf{V}_0/\mathbf{V}_{in}$ corresponds to the gain-feedback product $\mathbf{AF}$. Since $\mathbf{A}$ is normally quite large, the input signal must be very small if the amplifier is to stay out of saturation. One word of caution: In the terminology of the Nyquist analysis, the product $\mathbf{AF}$ is itself usually referred to as the open-loop gain. (It is, after all, the gain of the complete amplifier with the feedback loop open.)

This experimental method can be adapted to a wide variety of feedback applications and is particularly well suited to situations where unknown quantities are present in the feedback loop. As an example, consider the problem of maintaining a constant pressure in a container. An amplifier might change the pressure by controlling the speed of a pump, and a mechanical transducer could be used to measure the pressure and generate a feedback to the amplifier. If the power is large enough, a trial-and-error attempt to use the closed-loop system might produce destructive oscillations in the system. Use of the Nyquist method can predict the behavior before risking the oscillations of a closed loop.

Problems

1. If the open-loop gain curve in Figure 8.2 describes the amplifier shown in Figure P8.1, write an expression for $\mathbf{V}_{out}$ when $\mathbf{V}_{in} = 10 \cos(1000t)$ mV.

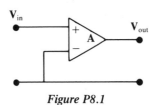

Figure P8.1

2. Use the ideal amplifier approximations to derive expressions for the transfer function of each of the four circuits shown in Figure P8.2.

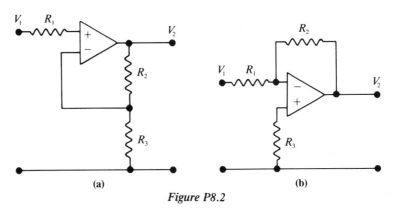

(a)   (b)

Figure P8.2

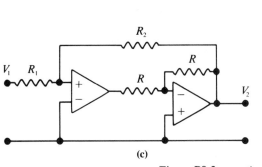
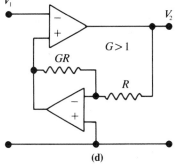

(c) (d)

Figure P8.2 continued

3. (a) Given a signal generator with an output impedance of 600 Ω, design an amplifier with a gain V_{out}/V_s(open) = 10. **(b)** Using the same signal generator, design an amplifier with a gain V_{out}/V_s(open) = −10.

4. The variable resistor in the circuit of Figure P8.4 provides a DC offset to the output voltage; a circuit of this type is often used to provide a zero output for some specific but nonzero input signal. What range of input voltage V_0 can be zeroed out by this circuit?

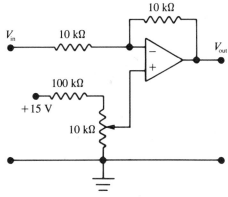

Figure P8.4

5. A two-input current summing amplifier can be used to shift the DC level of an AC signal. For the circuit shown in Figure P8.5, determine the average value of the output signal if the input is $v_{in} = 2\sin(6000t)$ V.

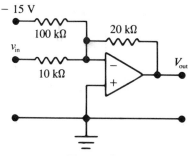

Figure P8.5

6. Assuming ideal operational amplifiers, determine $\mathbf{G}(j\omega)$ for each of the circuits shown in Figure P8.6. Then determine the single-term approximations to the transfer function at various frequencies, sketch the resulting straight-line approximation to $|\mathbf{G}|$, and label all magnitudes, corners, and slopes.

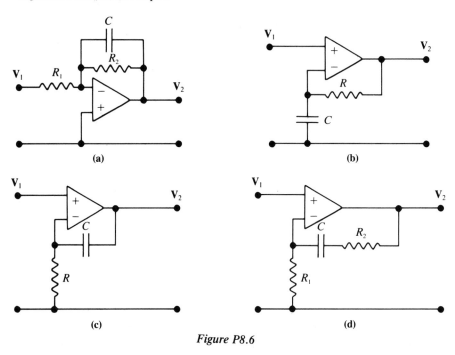

Figure P8.6

7. Assuming ideal operational amplifiers, find the relationship between V_1, V_2, and V_3 for the circuit shown in Figure P8.7.

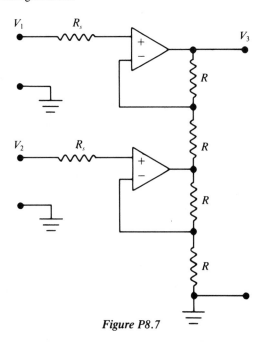

Figure P8.7

8. The transfer function for the two-pole, inverting, low-pass filter of Figure 8.10a is given by Eq. (8.22). For the special case when $R_1 = R_2 = R_3 = R$, determine the relationship between C_1 and C_2 that will result in a critically damped transient response for this amplifier.

9. Write the transfer function for the *LCR* plus NIC circuit in Figure 8.15b when $R_N = 0.9R_s$, and determine the Q of this circuit when $L = 400$ mH, $C = 0.1$ µF, and $R_s = 1000$ Ω.

10. Derive $\mathbf{Z}_{in}$ for the gyrator in Figure 8.16a. (*Hint:* Start at the bottom and work upward while keeping $\mathbf{V}_{in}$ as the independent variable.)

11. (*a*) Write the transfer function $\mathbf{H}(s) = \mathbf{V}_2/\mathbf{V}_1$ for the filter circuit of Figure 8.16b when $\mathbf{Z}_1 = \mathbf{Z}_3 = \mathbf{Z}_5 = R$ and $\mathbf{Z}_2 = \mathbf{Z}_4 = 1/j\omega C$. (*b*) Write expressions for the asymptotic forms of $\mathbf{H}(j\omega)$ and determine the corner frequency in terms of R and C. (*c*) Sketch $|\mathbf{H}(j\omega)|$ and label any slopes in dB/octave.

12. Using Figure 8.23a and Eq. (8.59), determine the voltage on the negative input terminal when $I_1 = 1$ mA, $R_2 = 10$ kΩ, and $\mathbf{A} = 1000$.

13. Derive the transfer function of the Wien bridge oscillator circuit when the input is open as in Figure 8.17. (*Hint:* First replace the series and parallel *RC* impedances with general impedances $\mathbf{Z}_1$ and $\mathbf{Z}_2$.)

14. A Wien bridge oscillator is carefully designed with R_1 and R_2 chosen to make the first term in Eq. (8.40) just slightly positive. However, when the circuit is connected to a signal generator with a 600-Ω output impedance, it seems to exhibit damped oscillation. Explain why this would happen.

15. Derive the transfer function for the notch filter circuit of Figure 8.24c.

16. The circuit shown in Figure P8.16 is a two-pole, high-pass filter.

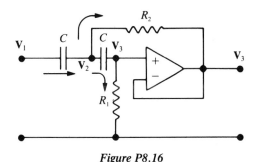

Figure P8.16

(*a*) Show that its transfer function is

$$\mathbf{H}(s) = \frac{R_1 R_2 C^2 \, s^2}{R_1 R_2 C^2 s^2 + 2R_2 Cs + 1}$$

(*b*) From the asymptotic forms $\mathbf{H}_L$ and $\mathbf{H}_H$, determine an expression for the corner frequency of this filter.

(*c*) Derive the relationship between R_1 and R_2 that will result in critical damping.

(*d*) Write an inequality between R_1 and R_2 that will result in an underdamped transient response.

17. The circuit shown in Figure P8.17 is a two-pole, bandpass filter.

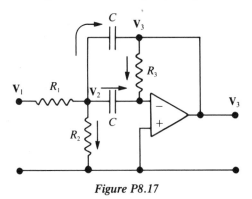

Figure P8.17

(a) Show that the transfer function for this circuit is

$$H(s) = \frac{-R_2R_3C \text{ s}}{R_1R_2R_3C^2s^2 + 2R_1R_2Cs + R_1 + R_2}$$

(b) If $R_1 = R_3 = 100$ kΩ, $R_2 = 1$ kΩ, and $C = 0.1$ μF, determine ω_0, Q, and $|H(j\omega_0)|$. (*Hint:* This general transfer function was the subject of Problem 3.31.) [*Ans.:* 1000 rad/s, 5, -0.5]

18. A physical current source has an output impedance of 1 kΩ and drives a current-to-voltage converter as shown in Figure P8.18. If the amplifier has open-loop gain like that shown in Figure 8.2, and the input signal is a sinusoid of frequency 10^4 Hz, what is the percent error in the amplifier's output voltage compared to the ideal approximation?

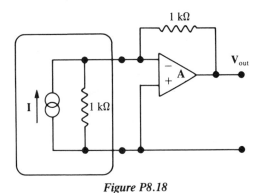

Figure P8.18

19. If the positive input of the current-to-voltage converter in Figure 8.18c is grounded, and the transistor has a *PN* voltage drop of 0.6 V, what is the voltage at the base of the transistor? Explain.

20. The half-wave rectifier of Figure 8.14a is constructed using an operational amplifier with open-loop gain $A = 1000$ at the frequency of the input signal, and the diodes used are ideal except for a forward *PN* voltage drop of 0.5 V. Determine the exact value of the output voltage when $V_{in} = -1$ V. [*Ans.:* 0.997 V]

21. A differentiation circuit like Figure 8.28a is constructed with $R = 10 \text{ k}\Omega$ and $C = 0.1 \ \mu\text{F}$.

   (a) Determine the value of C_2 that will yield critical damping.
   (b) Using Eq. (8.96) for the transfer function of this circuit, determine the two corner frequencies of the critically damped, closed-loop differentiation amplifier.
   (c) Assuming that the asymptotic forms of $\mathbf{H(s)}$ actually describe the circuit's behavior, what is the maximum-frequency sine wave that can be differentiated correctly?

22. Following the procedure outlined in Section 8.6.2 for the circuit of Figure 8.28a, determine an expression for the value of R_2 needed to critically damp the differentiation circuit of Figure 8.28b.

23. Derive an expression for the input impedance of the circuit shown in Figure P8.23b in terms of the open-loop gain $\mathbf{A}$.

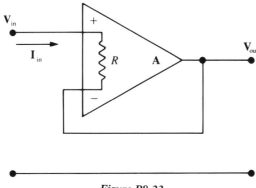

Figure P8.23

24. A current source like that shown in Figure 8.18b is constructed using a resistor $R = 10 \ \Omega$ and a transistor with $h_{ie} = 1000 \ \Omega$ and $h_{fe} = 30$. Assume that the input voltage signal has a DC component that will correctly bias the transistor into operation and a sinusoidal AC component v_{in} at a frequency where the operational amplifier's open-loop gain is 100.

   (a) Calculate the amplitude of the AC current into the load if the amplitude of v_{in} is 1 V.
   (b) What is the amplitude of the AC base current that the operational amplifier must supply?

25. Assume that the noninverting, unity-gain amplifier of Figure 8.21 uses an operational amplifier with an open-loop gain $\mathbf{A}(j\omega)$ exactly like that of Figure 8.2. If $R_0 = 100 \ \Omega$ and $C = 1 \ \mu\text{F}$, evaluate Eq. (8.52) and make a reasonably accurate sketch of $|\mathbf{H}(j\omega)|$ and the open-loop gain $|\mathbf{A}(j\omega)|$ on the same graph. [*Hint:* The general shape of the $|\mathbf{H}|$ curve is determined by $|\mathbf{H}_L|$, $|\mathbf{H}_H|$, and $|\mathbf{H}(\omega_r)|$.]

26. (a) Derive an expression for the closed-loop gain $\mathbf{H(s)}$ for the circuit shown in Figure P8.26 in terms of $\mathbf{A}$. (b) Assuming that the open-loop gain is given by $\mathbf{A(s)} = (s_1/s)^3$, eliminate $\mathbf{A}$ in $\mathbf{H(s)}$ and find the poles of the closed-loop amplifier. (*Hint:* You will need to find the cube root of a number like $-B^3$. One root is obviously $-B$; the other two can be found by first expressing $-B^3$ as $B^3 e^{\pm j\theta}$ with an appropriate choice for θ.)

27. On the modified bridge circuit shown in Figure P8.27, the feedback resistor is parameterized in terms of its fractional change x. Show that, unlike the basic bridge of Problem 1.21, the output voltage is now a linear function of x.

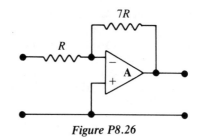

Figure P8.26

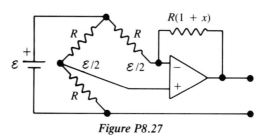

Figure P8.27

28. *(a)* In terms of **A** and *C*, find an expression for the input impedance of the circuit shown in Figure P8.28. *(b)* If $C = 20$ pF and $\omega_1 = 10^8$ rad/s, and you use a high-frequency approximation for **A** that falls off at 6 dB/octave, sketch the input impedance in terms of passive linear elements and evaluate their numerical size when $\omega = 10^7$ rad/s.

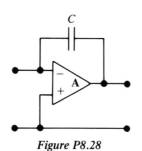

Figure P8.28

29. The circuit shown in Figure P8.29 will not work. Explain why not, with reference to Figures 7.15a and 7.24.

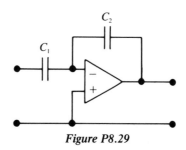

Figure P8.29

30. In the two-section circuit shown in Figure P8.30, the transfer function **F** is unavoidably that of a passive single-pole, low-pass filter with a corner frequency of 1000 Hz. Design an op amp circuit for the first section that has closed-loop gain **G** such that the overall gain **H** = **GF** = 4 up to the natural limits imposed by the open-loop gain of the op amp.

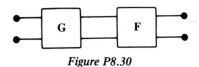

Figure P8.30

31. If the basic inverting amplifier circuit of Figure 8.4a is modified by shorting the input and opening the feedback loop as in Figure P8.31, it is possible to obtain an experimental Nyquist plot in order to determine the stability of the closed-loop amplifier. For the modified circuit given here, show that

$$\frac{V_1 - V_2}{V_1/(F_I + 1)} = (A + 1)F_I + 1$$

where $F_I = R_1/R_2$, or more generally, Z_1/Z_2. When the modified circuit is driven by an input signal V_1 and the difference signal $V_1 - V_2$ is measured, it is possible to determine the left side of this equation experimentally and thus plot the denominator of the inverting amplifier transfer function given in Eq. (8.79). The analysis then follows from the discussion of the noninverting amplifier in Section 8.7.

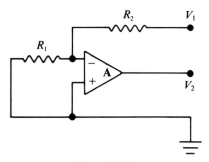

Figure P8.31

9

Digital Circuits

9.1   Introduction

In this chapter we suspend our discussion of analog signals, which can assume a continuous range of values within specified limits, and introduce digital signals, which can typically assume only two discrete values. Signals in the analog domain can be associated with a wide range of continuous physical phenomena such as temperature, pressure, distance, resistance, and brightness, whereas those in the digital domain are appropriate to phenomena that can be represented by integer (or at least rational) numbers: population, radiation counts, lightning flashes, traffic on a highway—indeed, anything that can be counted in some space-time interval. In the next chapter we will discuss circuits that can transform signals from one of these domains to the other.

The techniques employed in the design and analysis of digital circuits are quite different from those developed in the analog chapters. Although the analog methods will still apply to a specific digital circuit if it is investigated in detail, the complexity of most digital applications precludes this sort of analysis except in a few special situations. With digital circuits, we are generally more concerned with the signal (information) flow through the circuit than with the detailed characteristics of the signal at specific points on the circuit.

9.2   The Transistor as a Switch

In digital circuits as in analog ones, the active elements are generally either bipolar or field effect transistors. The difference here is that each transistor is permitted only two rest states, and these states normally correspond to two specific output voltages. Figure 9.1a shows a simple digital circuit that uses a mechanical toggle switch as the active element. This switch has only two states, open and closed, and in the circuit these two states generate two output voltages: V_{CC} when the switch is open and zero when it is closed.

The toggle switch can be replaced by a transistor as shown in Figures 9.1b and 9.1c. When driven by an input signal of sufficiently large amplitude, the transistors in each of these circuits will behave much like the mechanical switch in the previous example: They are either conducting no current (open) and yielding $V_{out} = V_{cc}$, or showing no resistance (closed) and conducting current as needed to make $V_{out} \approx 0$. The transition between the two states occurs as the input signal crosses a "threshold" voltage, which is determined by the design of the transistor switching circuit.

The transistor circuits shown in Figure 9.1 have the characteristic that a high input voltage will turn the transistor on, yielding a low output voltage; circuits of

Figure 9.1 (a) A mechanical switch used to provide a two-state output voltage. A similar result obtained with (b) a bipolar transistor inverting amplifier or (c) a MOS field effect transistor. (d) The schematic circuit symbol for the digital inverter.

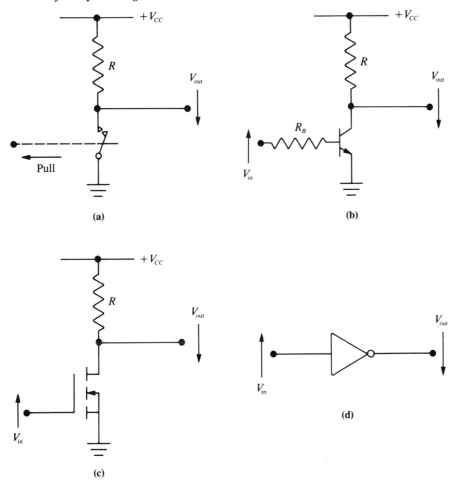

this type are known as inverters and can be represented on a schematic by the symbol shown in Figure 9.1d. Do not confuse the logical inverter with the inverting analog amplifier, which actually changes the sign of the signal voltage between input and output.

9.3   Number Systems

The two states of a digital signal can be given various names: high and low, true and false, or 1 and 0. The 1 and 0 notation leads naturally to the binary (base 2) number system, in which any number can be represented using only these two digits. Binary numbers generally look something like 10010101 (149).

Two other number systems, octal (base 8) and hexadecimal (base 16), are closely related to binary and provide a more condensed number notation. The familiar and deeply ingrained decimal (base 10) system has no natural affinity for digital electronics; indeed, after one has mastered the binary-based number systems, it is easy to wish that there had been a religious taboo against using the thumbs for counting. Most of us (and our hand calculators) are proficient only in base 10 arithmetic, and we generally feel more comfortable in this system. An unavoidable consequence is the need to occasionally convert numbers back and forth between decimal and one of these binary-related systems.

9.3.1   Binary Numbers

The ordinary decimal system has ten digits, 0–9, and in this system the number 139 has the explicit meaning

$$139_{10} = 100 + 30 + 9$$
$$= 1 \times 10^2 + 3 \times 10^1 + 9 \times 10^0 \tag{9.1}$$

where 139_{10} is read 139 to the base 10. As we will see, this represents a different number from 139_8 or 139_{16}. For the number 139_{10}, the 9 is said to be in the units position, the 3 in the tens position, and the 1 in the hundreds position, the names corresponding to the value of a one in that position. The higher the value of the position, the more significant a digit in that position becomes. The power of 10 that determines the position value is known as the order of the digit: Higher-order digits are to the left, lower-order to the right.

In the binary system each position is worth a power of two—2^0, 2^1, 2^2, 2^3—rather than a power of ten; the power again determines the order of the digit position. Our original number 139_{10} could equally well be decomposed into

$$139_{10} = 128 + 8 + 2 + 1$$
$$= 2^7 + 2^3 + 2^1 + 2^0 \tag{9.2}$$

In base 2 notation this would be written

$$139_{10} = 10001011_2 \tag{9.3}$$

where, as usual, zeros identify positions that make no contribution to the number. Each digit position in a binary number is known as a bit and can take on only two values, 0 or 1. The highest-order bit is on the left of the number and is known as the most significant bit; similarly, the lowest-order bit is the least significant bit.

The conversion from decimal to binary can be reduced to a simple arithmetical procedure: repeated division by 2 and recording of the remainder at each step. In the following example, the division process starts at the bottom and the final binary number is read from the remainders with the most significant bit at the top; compare this result with Eq. (9.3).

<u>Decimal-to-Binary Conversion</u>

$$0 \ r = 1$$

$$2 \overline{\big|\ 1\ } \ r = 0$$

$$2 \overline{\big|\ 2\ } \ r = 0$$

$$2 \overline{\big|\ 4\ } \ r = 0$$

$$2 \overline{\big|\ 8\ } \ r = 1$$

$$2 \overline{\big|\ 17\ } \ r = 0$$

$$2 \overline{\big|\ 34\ } \ r = 1$$

$$2 \overline{\big|\ 69\ } \ r = 1$$

$$2 \overline{\big|\ 139\ }$$

If the decimal values for the powers of 2 are available, as in Table 9.1, the method of Eq. (9.2) is often quicker

Table 9.1 Powers of 2

Power	Value	Power	Value	Power	Value
0	1	8	256	16	65,536
1	2	9	512	17	131,072
2	4	10	1,024	18	262,144
3	8	11	2,048	19	524,288
4	16	12	4,096	20	1,048,576
5	32	13	8,192	21	2,097,152
6	64	14	16,384	22	4,194,304
7	128	15	32,768	23	8,388,608

To convert a binary number back to decimal, it is only necessary to add up the decimal value of each 1 in the binary number. Since each 1 is worth some power

Table 9.2 Binary Counting

Decimal	Binary	Decimal	Binary
0	0000	8	1000
1	0001	9	1001
2	0010	10	1010
3	0011	11	1011
4	0100	12	1100
5	0101	13	1101
6	0110	14	1110
7	0111	15	1111

of 2 depending on its position, the power-of-2 table is again handy. Binary representations tend to be overly long and somewhat monotonous when written out on paper, but the binary numbers from 0 to 15 are fundamental and appear with some frequency. The binary equivalents for these numbers are given in Table 9.2.

Single-digit arithmetic in the binary system is particularly simple because the addition and multiplication tables are of minimum length: $0 + 0 = 0$, $0 + 1 = 1$, $1 + 1 = 10$; and $0 \times 0 = 0$, $0 \times 1 = 0$, $1 \times 1 = 1$. Note that only one of these operations produces a carry into the higher-order bit position.

9.3.2 Octal and Hexadecimal Numbers

Although simple, the binary representation of large numbers is quite cumbersome and a poor match for human perception. Numbers represented in the base 8 and base 16 systems scan more easily and still allow quick and easy conversions to and from binary. The eight octal digits are represented by 0–7, but to get the sixteen hexadecimal digits we must include six letters as shown in Table 9.3.

The ease of conversion between base 8, base 16, and base 2 numbers stems from the fact that there is a one-to-one correspondence between the octal and hexadecimal digits and the numbers represented by groups of three or four binary bits. As can be seen from Table 9.2, the numbers generated by a grouping of three binary bits each correspond to one of the octal digits 0 to 7, whereas the four-bit grouping has a similar relationship to the hexadecimal digits 0 to F (15_{10}).

In octal the place values are given by powers of 8, and in hexadecimal they are given by powers of 16. Thus, in octal our original number 139_{10} is

$$139_{10} = 2 \times 8^2 + 1 \times 8^1 + 3 \times 8^0 = 213_8 \qquad (9.4)$$

Table 9.3 Hexadecimal Digits

Decimal	Hexadecimal	Binary
0–9	0–9	
10	A	1010
11	B	1011
12	C	1100
13	D	1101
14	E	1110
15	F	1111

whereas in hexadecimal it would be written as

$$139_{10} = 8 \times 16^1 + 11 \times 16^0 = 8B_{16} \qquad (9.5)$$

As indicated, the octal or hexadecimal representation of this number can also be easily read from the binary form of Eq. (9.3) by grouping the binary bits in threes or fours starting from the least significant bit on the right of the number:

$$10\ 001\ 011 = 213_8$$
$$\text{or} \qquad 1000\ 1011 = 8B_{16} \qquad (9.6)$$

Conversely, an octal or a hexadecimal number is easily converted to its binary representation by replacing each octal or hexadecimal digit with its binary equivalent. Conversion between octal and hexadecimal is also easy if the binary representation is used as an intermediate step.

9.3.3 Nibbles, Bytes, and Words

In digital computer terminology a binary number made up of an arbitrary number of bits is known as a word. In older computers the standard word could always be divided into an integral number of octal digits, but because of the standards imposed by inexpensive integrated circuits, most modern designs have word lengths that are integral multiples of 8 bits. In fact, the use of 8- and 4-bit groupings is now so common that 4 bits (one hexadecimal digit) is known as a ''nibble'' and 8 bits (two digits) is known as a ''byte.'' Thus a byte of information can be any hexadecimal number between 00 and FF (0 and 255_{10}), and several bytes of information make up a complete data word. The most common bit groupings are shown in Figure 9.2a.

If a word has n bits, it can be used to represent exactly 2^n different numbers: generally 0 to $2^n - 1$. If we use such a word for counting, the number that follows $2^n - 1$ is again 0, meaning that the number held in the n-bit word is the actual count modulo 2^n. This converts the normal number line into a number circle, as shown by the 8-bit example of Figure 9.2b.

By giving up half of the positive integers it is possible to represent both positive and negative integers in a fixed-length binary word. The method indicated in Figure 9.2b is known as 2's complement: To obtain the 2's complement representation of -63, complement ($0 \rightarrow 1$ and $1 \rightarrow 0$) every bit in the binary representation of the number $+63$ and add 1 to the result. The steps are

$$
\begin{array}{rl}
& 63_{10} = 0011\ 1111 \\
\text{Complement} & \overline{63}_{10} = 1100\ 0000 \\
\text{Add } 0000\ 0001 & -63_{10} = 1100\ 0001
\end{array} \qquad (9.7)
$$

As can be seen from the figure, the possible 8-bit numbers now run from -128 to 127. Note that all of the negative numbers have a 1 in the most significant bit position.

An alternative representation of negative binary numbers is known as ''signed magnitude.'' This method is much like the normal pencil-and-paper approach; it sets the leftmost bit to represent a minus sign, leaving the remaining bits to define a positive integer. The electronic advantages of the less obvious 2's complement

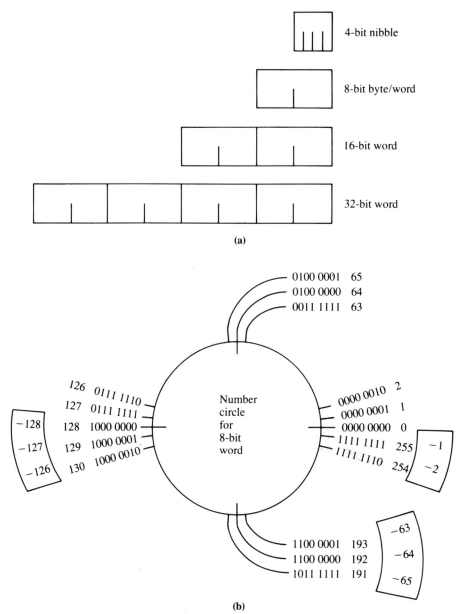

Figure 9.2 (a) Naming conventions for some common bit groupings.
(b) The number circle for an 8-bit binary integer. The boxed negative
numbers correspond to the 2's complement convention.

notation become evident if a simple binary addition is attempted in the two
schemes:

	2's Complement	Signed Magnitude	
Addition	$+63_{10} = 0011\ 1111$	$0011\ 1111 = +63_{10}$	
	$-63_{10} = 1100\ 0001$	$1011\ 1111 = -63_{10}$	(9.8)
	$00_{10} = 0000\ 0000$	$1111\ 1110 = -126_{10}$	

Obviously, the signed magnitude representation requires a different rule for addition if negative numbers are involved.

9.4 Boolean Algebra

Algebraic variables that can assume only two states are known as logical variables since the two states can equally well be defined as false (0) and true (1). Between two such variables A and B, the Boolean algebra defines two operations: AND, denoted by $A \cdot B$; and OR, denoted by $A + B$. If Q is another logical variable, the equation

$$\text{AND} \qquad Q = A \cdot B \qquad\qquad (9.9)$$

expresses the logical statement: "Q is true if and only if A is true AND B is true." As in normal algebra, this "multiplication" operator is often dropped, producing the equivalent expression

$$Q = AB \qquad\qquad (9.10)$$

Similarly, the OR expression,

$$\text{OR} \qquad Q = A + B \qquad\qquad (9.11)$$

is read: "Q is true if A is true OR if B is true."

The characteristics of the AND and OR operations can be easily remembered from the multiplication and addition operator notations if the logical states are taken to be 0 and 1. The operations follow the normal rules of arithmetic multiplication and addition, with the exception that $1 + 1 = 1$ rather than 2. A convenient way to display the results of a Boolean algebra statement is with a truth table, as shown in Figure 9.3.

In addition to these operators between two variables, the NOT or negation operation can be applied to a single variable (or to any part of an expression). A negated variable is indicated by a bar over the variable name. Since we are dealing with variables having only two states (true and false), a quantity that is NOT true must be false and vice versa. Thus, the expression

$$\text{NOT} \qquad Q = \overline{A} \qquad\qquad (9.12)$$

means "whichever state A is in, Q is in the other one."

In many digital applications, logical variables are defined with names rather than with single letters. Since a significant amount of design work is done on computers with logical variables entered from terminal keyboards, the bar symbol for the negation operation is not always convenient, and this operation is often indicated instead by the letter N preceding the name of the variable. Hence, the quantities $\overline{\text{READY}}$ and NREADY would mean the same thing. When choosing names for logical variables, it is thus a good idea to avoid names beginning with N.

Table 9.4   Properties of Boolean Operations

1. $A \cdot 0 = 0$
2. $A + 0 = A$
3. $A \cdot 1 = A$
4. $A + 1 = 1$
5. $A \cdot A = A$
6. $A + A = A$
7. $A \cdot \overline{A} = 0$
8. $A + \overline{A} = 1$
9. $\overline{\overline{A}} = A$
10. $A \cdot B = B \cdot A$
11. $A + B = B + A$
12. $A \cdot (B + C) = A \cdot B + A \cdot C$
13. $A \cdot (B \cdot C) = (A \cdot B) \cdot C$
14. $A + (B + C) = (A + B) + C$
15. $\overline{(A \cdot B)} = \overline{A} + \overline{B}$
16. $\overline{(A + B)} = \overline{A} \cdot \overline{B}$

9.4.1   Theorems

The operations of Boolean algebra have many properties in common with normal algebraic operations, especially if allowance is made for the result $1 + 1 = 1$. The Boolean operations obey the same commutative, distributive, and associative rules of normal algebra, but because the variables have only two states, additional simplifications are possible. The postulates and several of the more fundamental theorems are summarized in Table 9.4. The last two listings make up De Morgan's theorem and will be referenced specifically several times in the following discussion. The other listings should be reasonably self-evident.

9.5   Logic Gates

An electronic circuit that is capable of combining digital signals according to the rules of Boolean algebra is known as a logic gate. The term "gate" comes from the ability of these circuits to control the flow of information from one point to another. In the electronic domain of digital circuits, the true and false logic states are assigned to specific voltage or current conditions. The usual practice is to use two voltages and identify the higher (more positive) voltage with the true state. This electronic representation is known as positive logic. The reverse assignment of the true condition to the lower voltage is known as negative logic. There is rarely a compelling reason to use the less obvious negative logic assignment, but when reading someone else's schematic, it is best to keep the possibility in mind. Positive logic will be used throughout this text, allowing a direct correspondence between true/false, 1/0, and high/low voltage notation.

There are several different designs for digital logic circuits, depending on the type of transistor (bipolar or FET) and the circuit configuration. Since each general method produces circuits with similar characteristics, logic circuits are grouped into families, each with its own set of detailed operating rules. These operating rules for a particular logic family are defined in the manufacturer's data book, which also describes each logic element of the family.

The various families include: resistor-transistor logic (RTL), diode-transistor logic (DTL), transistor-transistor logic (TTL), *N*-channel metal oxide silicon (NMOS), complementary metal oxide silicon (CMOS), and the high-speed emitter-coupled logic (ECL). The family names are fairly descriptive if "T" is taken to mean bipolar transistor and "MOS" is understood to mean MOSFET. The MOS families (including variations not named here) feature very low power consumption and dominate the large-scale integrated circuit technology. However, the most popular logic family for normal or small-scale integrated circuit units is TTL, particularly the 74LS00 (low-power, Schottky diode-clamped) series. The development that follows is not specific to any particular logic family but closely parallels the devices of the various TTL 74xx00 series.

9.5.1 The Basic Gates

The schematic symbols of the four basic gates and the logic inverter (the NOT function) are shown in Figure 9.3. The open circle on the NAND, NOR, and inverter is used to indicate the NOT or negation function. It should be interpreted as meaning that the signal is negated as it crosses the circle. A NAND (NOT AND) gate can thus be taken to be an AND gate followed by an inverter. Although all four of the gates shown here have two inputs, three- and four-input gates are common, and most logic families provide some gates with at least eight inputs.

Also shown in this figure are the logic truth tables for each of these devices. Tables of this type define the output state (*Q* here) for every possible input signal combination and can be applied to individual gates or to entire circuits.

Each logic family has its own intrinsically preferred gates and gate combinations, but all make repeated use of either the NAND or the NOR functions. In a sense, these two gate types are more fundamental than the others, because any gate function can be developed from combinations of one or both of these types.

In some logic designs, NOT circles are used on the inputs to gates as shown in Figure 9.4a. This should not be a source of confusion. As shown in Figure 9.4b, the expanded logic representation can again be easily derived by treating each NOT circle as a distinct inverter. The gate shown is an alternative representation of the OR gate and will be derived in the following discussion.

9.5.2 Implementation with Diodes

Although logic functions are now always implemented using integrated circuit chips with many active transistor elements, the basic AND and OR logic functions

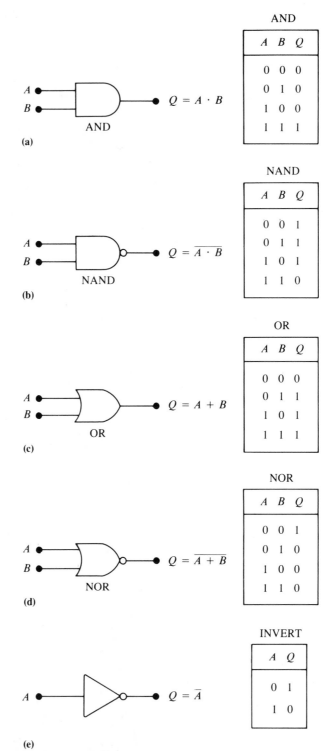

Figure 9.3 *Symbols and truth tables for the four basic two input gates: (a) AND, (b) NAND, (c) OR, (d) NOR, and (e) the inverter.*

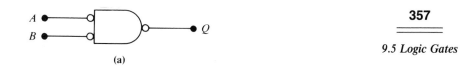

(a)

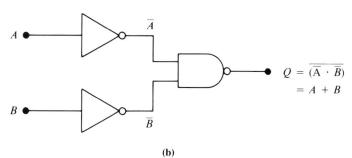

$$Q = \overline{(\overline{A} \cdot \overline{B})}$$
$$= A + B$$

(b)

Figure 9.4 (a) A logic symbol with NOT circles on the inputs, and (b) an equivalent circuit.

can be generated using diodes and resistors as shown in Figure 9.5. For clarity, the truth tables shown in this figure are given in terms of voltages instead of true and false logic states. If 4.5 V to 5.5 V is taken as true and 0 V to 0.5 V is taken as false, these circuits correspond to the AND and OR functions previously defined.

Figure 9.5 (a) A diode-resistor AND logic function and its truth table, and (b) the OR logic function and its truth table.

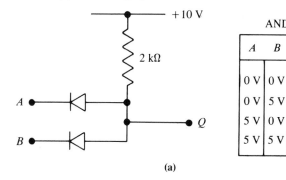

AND

A	B	Q
0 V	0 V	0.5 V
0 V	5 V	0.5 V
5 V	0 V	0.5 V
5 V	5 V	5.5 V

(a)

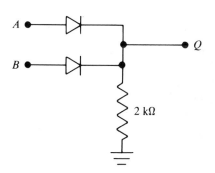

OR

A	B	Q
0 V	0 V	0 V
0 V	5 V	4.5 V
5 V	0 V	4.5 V
5 V	5 V	4.5 V

(b)

These are not practical logic circuits for large systems, because the diode voltage drops cause the output states at Q to be slightly different than the input voltage states. If the output of one logic element is used to feed another, this difficulty will be compounded. In practical systems this problem is overcome by including one or more transistors in each gate. The simplest design would connect one of the inverting amplifiers shown in Figure 9.1 to the Q output of each of these diode circuits. Examples of the actual circuits used for some common logic families are shown in Figures 9.20–9.22.

9.6   Combinational Logic

Logic gates can be used in combination to generate any desired functional relationship between logic variables. In the course of this discussion, we will develop a few of the more useful circuits from the basic gates and then use these combinations as building blocks in still more complicated circuits.

9.6.1   The AND-OR Gate

The typical AND-OR gate, shown in Figure 9.6a, is constructed from two 2-input AND gates and a 2-input OR gate. Its operation is described by the equation

$$Q = AB + CD \tag{9.13}$$

which should be read "Q is true if A AND B are true OR if C AND D are true." Some logic families provide a version of this gate known as an AND-OR-INVERT or AOI gate. As is evident from the name of this gate, the OR gate is replaced with a NOR. The algebraic expression for the AOI gate is

$$Q = \overline{(AB + CD)} \tag{9.14}$$

or, equivalently,

$$\overline{Q} = AB + CD \tag{9.15}$$

As we will see later, AND-OR gates with different numbers of input variables enable us to develop a standard design technique for all combinational logic problems. An example of a more complicated AND-OR gate is shown in Figure 9.6b. The algebraic statement for the operation of this gate is

$$Q = AB + CD + \overline{A}C\overline{D} \tag{9.16}$$

It is convenient to describe a basic AND, NAND, OR, or NOR gate function as being "satisfied" when the inputs are such that a change in any one will change the output. Thus, the 3-input AND gate in this example is satisfied when A and D are false and C is true; the output is then true but will become false if any of the three input variables changes. A satisfied AND or NOR gate has a true output, whereas a satisfied NAND or OR has a false output.

We can shorten the notation of Eq. (9.16) by identifying the four logic variables A, B, C, and D with the 4-bit number $ABCD$. Using this notation, the

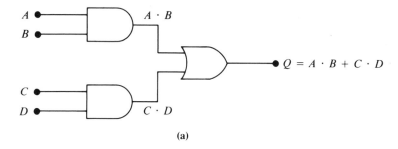

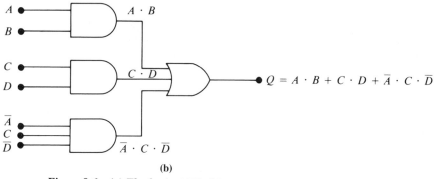

Figure 9.6 *(a) The basic AND-OR gate. (b) A typical logic function.*

numbers that satisfy the 3-input AND gate of Figure 9.6b are 0010 and 0110, or simply 2 and 6. The state of B has no effect on this gate (it is not connected) and can also be indicated by writing $0x10$, where "x" means "don't care." Thus, of the sixteen numbers that can be generated by the nibble, *ABCD,* only two satisfy this 3-input AND gate.

9.6.2 *Combinational Logic Design Using Truth Tables*

There are many ways to approach logic design. With some experience it is often possible to think through the logic of the problem and proceed immediately to a circuit sketch that will produce the desired result. At some level of logical complexity a more formal technique becomes necessary. The procedure described here is not always the quickest, especially when many logic variables are involved, but it does provide a formal approach to any combinational problem. The procedure can be divided into four steps:

1. Devise a truth table of the independent input variables and the resulting output quantities.

2. Write a Boolean algebra statement that describes the truth table.

3. Reduce the Boolean algebra.

4. Mechanize the Boolean statement using the appropriate logic gates.

With the exception of step 3, the reduction of the algebra, these steps are individually quite simple and routine, but they do become increasingly tedious as the number of input variables increases.

Some algebraic reduction of the initial Boolean statement is generally necessary if the number of gates is to be minimized. For this step we have the identities of Table 9.4, which include the normal algebraic associative, distributive, and commutative rules. The De Morgan's theorem statements of entries 15 and 16 deserve special attention because of the dramatic effects these rules have on the logic statement. Later in this chapter we will outline a graphical method (Karnaugh maps) for implementing this minimization step directly from a special form of the truth table.

Unfortunately, minimizing the number of logic functions (gates) is not always the best way to minimize the complexity of the overall circuit. Circuit construction is almost always simpler, faster, and more reliable if the individual package (IC chip) count is minimized—especially in view of the increasing complexity of the logic functions now available on an IC chip. A formal approach to package minimization is impractical, but early in a new design it is wise to make a careful survey of available IC packages.

As a simple example of the four-step design procedure, consider the truth table that defines the OR gate of Figure 9.3c. We can obtain a Boolean statement for Q by reading the lines of this table that yield a true result: Q is true if A is false AND B is true, OR if A is true AND B is false, OR if A is true AND B is true. This statement converts directly to

$$Q = \overline{A}B + A\overline{B} + AB \qquad (9.17)$$

Although different from $A + B$, this statement must also describe the entire truth table since it includes all of the combinations of input variable states that result in a true Q. Since Q is only a two-state variable, all other input state combinations must yield a false Q. If the truth table had more than a single output result, each such result would require a separate equation.

Clearly, this reading of the truth table has not produced the simplest algebraic expression possible, but with a little algebra we can reduce it to the OR function. Since Q is true when the expression AB is true, we do not change Eq. (9.17) by adding a second AB term. With this done, the resulting equation can be rearranged to read

$$Q = B(\overline{A} + A) + A(\overline{B} + B) \qquad (9.18)$$

Using the property given on line 8 of Table 9.4, this reduces to

$$Q = B + A \qquad (9.19)$$

the expected OR function, which can be mechanized with a single gate.

An alternative approach is to write an expression like Eq. (9.17) for the false condition of Q. This alternative reading of the truth table is: Q is false when A is false AND B is false. Again, this single statement represents the entire table, since all other input combinations result in a true Q. The algebraic statement is

$$\overline{Q} = \overline{A}\,\overline{B} \qquad (9.20)$$

This can be inverted to give a new expression for Q,

$$Q = \overline{\overline{Q}} = \overline{(\overline{A}\overline{B})} \qquad (9.21)$$

which can be implemented using two inverters and a NAND gate as shown in Figure 9.4b. The algebraic connection between this equation and the original OR function is given by the De Morgan's theorem statement of line 15 in Table 9.4.

A single output quantity from any truth table can thus be represented either by a "true" equation like (9.16) or a "false" equation like (9.19). The choice of representation in a particular case is determined only by convenience, since both are algebraically connected by De Morgan's theorem. When the number of terms is unequal as in this example (three in the true equation versus one in the false), the equation with the fewer initial terms is a reasonable first choice. When the number of terms in the two representations is nearly equal, the best choice may depend on the availability of noninverting AND and OR gate types versus inverting NAND and NOR types.

9.6.3 Exclusive-OR Gate

The exclusive-OR gate, sometimes abbreviated as EOR or XOR, is a 2-input gate that is very useful in combinational logic problems. Its only distinction from the conventional OR (inclusive-OR) is a false output when both inputs are true (it excludes this state). The schematic symbol and truth table for this gate are shown in Figure 9.7a. From the truth table we can write either the true statement

$$Q = \overline{A}B + A\overline{B} \qquad (9.22)$$

or the false statement

$$\overline{Q} = \overline{A}\overline{B} + AB \qquad (9.23)$$

Using the exclusive-OR operator notation $\oplus$, Eq. (9.22) can be written

$$\boxed{\begin{array}{l} \text{EOR} \\ \text{XOR} \end{array} \quad Q = A \oplus B} \qquad (9.24)$$

and mechanized using five basic gates as shown in Figure 9.7b.

A minimized design for this same function requires only three gates and can be obtained by applying De Morgan's theorem twice to the inverse of Eq. (9.23). This inverse equation is

$$Q = \overline{\overline{Q}} = \overline{(\overline{A}\overline{B} + AB)} \qquad (9.25)$$

The first application of De Morgan's theorem (line 16 in Table 9.4) gives

$$Q = \overline{(\overline{A}\overline{B})}(\overline{AB}) \qquad (9.26)$$

and a second application (line 15 in the table) to the first term only yields

$$Q = (A + B)\overline{(AB)} \qquad (9.27)$$

The implementation of this equation is shown in Figure 9.7c. Note that it requires fewer gates than Figure 9.7b.

9.6.4 Timing Diagrams

Up to this point we have implicitly assumed that logic signals are quiescent at a constant voltage level corresponding to true or false. However, signals of interest

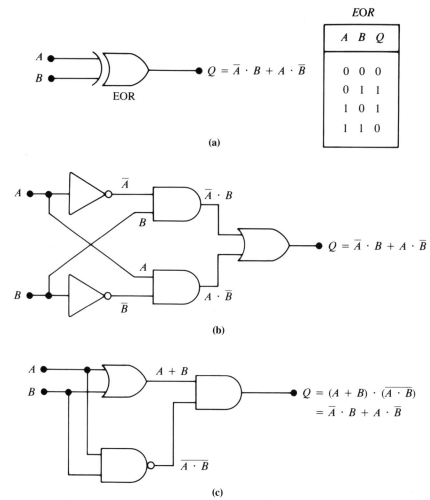

A	B	Q
0	0	0
0	1	1
1	0	1
1	1	0

(a)

(b)

(c)

Figure 9.7   (a) The schematic symbol for the exclusive-OR gate (EOR or XOR) and its truth table. (b) A mechanization directly from the truth table and (c) after applying De Morgan's theorem.

can be expected to change with time, going from one logic state to the other. Whenever the changing inputs to a gate produce a change in its output signal, many voltages and currents within the gate must make the transition from one quiescent state to the other. As a result of stray capacitance between signal wires and grounds and the finite current drive capability of any real circuit element, a signal will show a transition time t_t as it moves between states. Similar delays within the logic element will result in a propagation delay t_{pd} between input signals and output responses. Both effects are shown in Figure 9.8.

Rather than consider the detailed shape of input and output signals (an analog electronics problem), it is common to define the two times t_{pd} (pulse delay) and t_t (transition) as indicated in the figure. For modern logic families, both of these

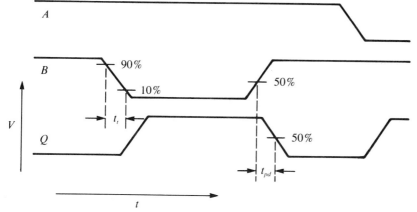

Figure 9.8 *Input and output signal to a NAND gate showing the transition time of input and output signals and the propagation delay through the gate.*

times are in the range of a few nanoseconds to a few tens of nanoseconds, and for many applications they can safely be ignored. On the other hand, these same short times are often the limiting factor in a high-speed digital design. If the interconnecting wires are longer than a few feet, it may also be necessary to consider the propagation time down the wire and signal reflections at the wire ends as discussed in Section 11.6.1.

Whether or not transition and propagation times are important to a specific problem, a timing diagram that shows the variation of the input and output logic signals with time can be very useful. Such a diagram is simply an idealization of the display that would be seen on a multichannel oscilloscope connected to the various signal points of a circuit.

The EOR gate circuit of Figure 9.7b is an example where a timing diagram points up a circuit flaw. Assume that the input terminals of this gate are driven by the signals A and B as shown on the timing diagram of Figure 9.9a. The time dependence of the signals driving this gate has been chosen to display the various lines of the truth table for the EOR gate; the sequence followed by AB is 00, 10, 11, 01, 00, 11.

To emphasize the point, assume that the rightmost positive transitions of these input signals occur at exactly the same time. The gate's output Q faithfully follows the lines of the truth table until the time of the simultaneous transitions: $AB = 00$ to $AB = 11$. In the truth table this transition corresponds to a step from the last line to the first line, and the output Q should be 0 (false) both before and after this step. Indeed, the signal in the figure is constant, except for the small spike (glitch) at the time of the transitions. Figure 9.9b shows the region of the glitch on an expanded time scale and displays the signal at several additional points of the circuit. When the propagation delays, t_1 and $t_2 + t_3$, through the two inverters of this circuit are included as shown, both AND gates will be satisfied for a short time after the simultaneous transitions of A and B. The AND-OR design of Figure 9.7c does not exhibit this problem.

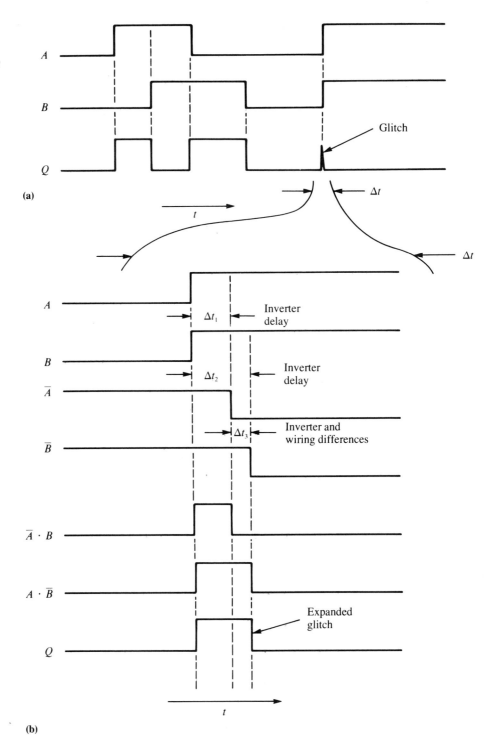

Figure 9.9 (a) A timing diagram for the EOR circuit of Figure 9.7b. (b) An expanded view of the glitch shows it to be caused by a signal race condition.

9.6.5 Signal Race

Even when the individual logic elements are not at fault, the output signal from a combinational logic circuit may still be sensitive to small variations in the relative timing of the input signals. The condition whereby two or more signals change almost simultaneously is known as a signal race, and in general the condition can be expected to produce glitches in the output signals. Such glitches are sometimes a steady and reproducible feature of a circuit, but at other times they may come and go more or less randomly due to small changes in the operating characteristics of various parts of the circuit. It is also worth noting that real-life glitches are often too narrow to be noticed on an oscilloscope display but are still wide enough to invoke a seemingly erratic response by downstream digital circuits.

The effects of glitches can be eliminated by using a synchronous timing technique whereby all logic state changes are initiated by a common timing pulse—a clock pulse—from an oscillator. Glitches in the combinational logical results are then allowed to come and go before the result is interrogated at a later clock pulse; such synchronous or clocked systems, especially the slower ones, have relatively little need for timing diagrams. However, many smaller digital circuits (also the "edges" or interface areas of large ones) do not make use of a clock signal, relying instead on a technique known as asynchronous timing, whereby signals change state at their own pace in various parts of the circuit; in these systems a timing diagram drawn with careful attention to signal race is essential.

9.6.6 Half and Full Adders

The binary arithmetic adder provides a more advanced application of the four-step design procedure outlined above. To show the method, we are going to develop the circuit known as a full adder from basic gates, although in practice this would rarely be done. A circuit capable of adding two binary numbers would almost certainly be constructed from packages that each contain several full adders.

Consider the problem of adding two 2-bit binary numbers, X_1X_0 and Y_1Y_0, as shown in Figure 9.10a. As always, addition starts with the units (2^0) column; in this case it can be expressed as $X_0 + Y_0 = C_1Z_0$, where C_1 is the carry bit into the 2^1 column. Since each of these bits can take on only two values, 0 and 1, the resulting 2-bit C_1Z_0 number can range only from 00 to 10. The truth table for all combinations of X_0 and Y_0 is shown in Figure 9.10b.

The next step is to convert this truth table to Boolean algebra statements. The carry bit C_1 is easy: The truth table reads C_1 is true only if X_0 AND Y_0 are both true. This translates to

$$C_1 = X_0Y_0 \tag{9.28}$$

The column for Z_0 in the truth table has two true entries, leading to

$$Z_0 = \overline{X}_0Y_0 + X_0\overline{Y}_0 \tag{9.29}$$

which is just the exclusive-OR gate. The mechanization of these two equations is shown in Figure 9.11a. Taken together, these circuits are known as the half adder. The two-input half adder is of limited usefulness and is not generally available as a logic element.

Carry bits	$C_2 C_1$
Two-bit number	$X_1 X_0$
Two-bit number	$Y_1 Y_0$
Three-bit sum	$Z_2 Z_1 Z_0$

(a)

X_0	Y_0	C_1	Z_0
0	0	0	0
0	1	0	1
1	0	0	1
1	1	1	0

(b)

C_1	X_1	Y_1	C_2	Z_1
0	0	0	0	0
0	0	1	0	1
0	1	0	0	1
0	1	1	1	0
1	0	0	0	1
1	0	1	1	0
1	1	0	1	0
1	1	1	1	1

(c)

Figure 9.10 *(a) The binary addition of two 2-bit numbers. Truth tables for sum and carry bits from (b) the 2^0 column and (c) the 2^1 column.*

Because there is no carry bit into the units column, the circuit just developed is not able to handle the addition of any other column. As can be seen from Figure 9.10a, the sum of the 2^1 column involves three bits: one bit from each of the two numbers to be added and a carry-in bit from the units column. The circuit that can add this column of three input variables can process any column of the general two-number binary addition.

The three input variables result in a truth table that has eight lines (000 to 111) and two derived output quantities C_2 and Z_1, as shown in Figure 9.10c. Consider

Figure 9.11 *A mechanization of the half adder using an EOR and an AND gate.*

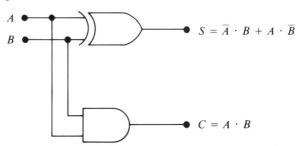

$$S = \bar{A} \cdot B + A \cdot \bar{B}$$

$$C = A \cdot B$$

first the carry-out bit C_2. Examination of the truth table shows that this bit should be true whenever a majority of the three input bits C_1, X_1, and Y_1 are true. The Boolean statement for this "majority detector" can be written directly from the truth table:

$$C_2 = \overline{C_1}X_1Y_1 + C_1\overline{X_1}Y_1 + C_1X_1\overline{Y_1} + C_1X_1Y_1 \tag{9.30}$$

This can be reduced by adding two more terms like the last one (this operation won't change the Q result: $1 + 1 = 1$, $0 + 0 = 0$), then associating one of these $C_1X_1Y_1$ terms with each of the other three terms. The equation then reduces to

$$C_2 = X_1Y_1 + C_1Y_1 + C_1X_1 \tag{9.31}$$

This derived expression for the carry-out bit clearly requires fewer and simpler gates than the original. (It is also logically "obvious" since it will be true whenever any two of the input signals are true.) Reductions of this type can be handled algebraically as in this example or by use of the Karnaugh maps discussed in the next section. The mechanization of this reduced expression is shown in Figure 9.12a.

From the truth table of Figure 9.10c, the expression for the sum bit Z_1 is

$$Z_1 = \overline{C_1}\,\overline{X_1}Y_1 + \overline{C_1}X_1\overline{Y_1} + C_1\overline{X_1}\,\overline{Y_1} + C_1X_1Y_1 \tag{9.32}$$

Factoring out the C_1 and $\overline{C_1}$ terms gives

$$Z_1 = \overline{C_1}(\overline{X_1}Y_1 + X_1\overline{Y_1}) + C_1(\overline{X_1}\,\overline{Y_1} + X_1Y_1) \tag{9.33}$$

The first term involves an exclusive-OR, and from De Morgan's theorem (or Eqs. 9.22 and 9.23), the second term can also be related to an exclusive-OR. With the second term rewritten, the equation becomes

$$Z_1 = \overline{C_1}(\overline{X_1}Y_1 + X_1\overline{Y_1}) + C_1(\overline{\overline{X_1}Y_1 + X_1\overline{Y_1}}) \tag{9.34}$$

But the overall expression is also an exclusive-OR: If we define $U = X_1 \oplus Y_1$, then Eq. (9.34) reduces to $Z_1 = C_1 \oplus U$. In terms of C_1, X_1, and Y_1, the final expression is

$$Z_1 = C_1 \oplus (X_1 \oplus Y_1) \tag{9.35}$$

This can be mechanized by two exclusive-OR gates as shown in Figure 9.12b, with the full adder as indicated schematically in Figure 9.12c. This device is able to add three bits of information and return the column sum bit and a carry-out bit to next higher-order column.

Assuming that the full adder is available as a logic element, it is easy to sketch a circuit capable of adding two numbers of any size; the example in Figure 9.13 will add two 3-bit numbers to yield a 4-bit sum. Note that the carry-in bit into the low-order adder is fixed at zero and this element is only performing the half-adder function.

The circuit developed here is the simplest type of binary addition circuit. It is also the slowest because each column of the sum must wait on the resulting carry bit from its adjoining lower-order column. This is known as serial carry propagation. The time required for the high-order bit in the sum to stabilize is proportional to the number of bits in the adder. More complicated designs have a "look-ahead" feature which, by using additional gates, can determine all sums and car-

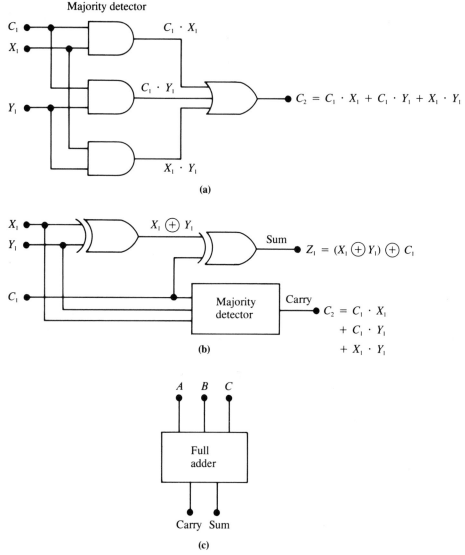

Figure 9.12 *(a) A mechanization of the majority detector. (b) The full adder mechanization. (c) A typical schematic representation of the full adder*

ries simultaneously. The typical modern IC package for this function features a 4-bit plus carry-in adder with "look-ahead" logic over the four bits; the package will yield a simultaneous 4-bit sum and a carry-out bit which can be serially propagated to the next 4-bit package.

9.6.7 Miniterms and Karnaugh Maps

For problems with two, three, or four input variables, the logic minimization problem can be reduced to a graphical procedure known as Karnaugh mapping.

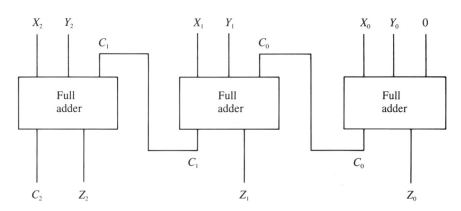

Input
numbers : $X_2X_1X_0$ and $Y_2Y_1Y_0$

Output
number : $C_2Z_2Z_1Z_0$

Figure 9.13   A circuit capable of adding two 3-bit numbers.

Although the minimization problem can be solved by algebraic techniques as used above, Karnaugh mapping formalizes the methods and encourages a notational shorthand that can itself be useful.

A typical three-input truth table might yield a logic equation like

$$Q = ABC + \overline{A}BC + A\overline{B}\,\overline{C} \tag{9.36}$$

where each of the products is known as a miniterm. Other product terms might contain one or two variables, but a miniterm product must use all of the independent variables. If we preserve the *ABC* ordering in each miniterm (the order is arbitrary, but it must be consistent throughout the problem) and replace each true variable with a 1 and each negated variable with a 0, then each miniterm becomes a binary number. In Eq. (9.36) the miniterms would read 111, 011, and 100. A miniterm can thus be described by a number in any convenient base (decimal is normal): 0 to 7 for three-variable miniterms, 0 to 15 for four, and so on. Using this notation, Eq. (9.36) can be described by the three numbers 7, 3, and 4, and could be replaced by the shorthand statement

$$Q = Q(3, 4, 7) \tag{9.37}$$

It must be emphasized that this expression is meaningful only for a well-defined variable ordering within the numbers, and that each term in the AND-OR expression must have the same number of variables.

In a similar manner, the three-variable expression of Eq. (9.30) could be written

$$C_2 = C_2(3, 5, 6, 7) \tag{9.38}$$

where the variable ordering is the same as that used in the equation. Note that these numbers can be read directly from the state of the input variables in the truth table of Figure 9.10c; for a well-organized truth table they are in fact nothing more than the line numbers (the top line is line number zero) of the true results. In some applications, this type of statement may be used as a shorthand notation in place of the truth table itself.

The Karnaugh mapping procedure makes use of some two-dinensional (more than four variables requires more dimensions) arrays as shown in Figure 9.14. Each box of the array is defined by its row and column number and corresponds to a particular miniterm possibility. A consequence of the peculiar row and column numbering of these arrays is that adjacent boxes, left-right or up-down, differ

Figure 9.14 Karnaugh diagrams for (a) two-, (b) three-, and (c) four-element miniterms.

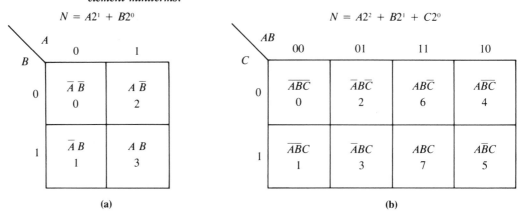

$$N = A2^1 + B2^0$$

$$N = A2^2 + B2^1 + C2^0$$

(a)

(b)

$$N = A2^3 + B2^2 + C2^1 + D2^0$$

(c)

only by the state of a single variable. Thus, when grouped together, two adjacent boxes always span both possible states of one input variable. The 2×4 array of Figure 9.14b maintains this feature even from the rightmost column directly to the leftmost column; in fact, this array should be conceived as being on a cylinder with the right and left edges being the same line. This property is repeated for the left and right edges of the 4×4 array of Figure 9.14c, and also for the top and bottom edges; the result is a toroidal (doughnut) shape.

Using the row and column numbering, each box of the arrays in Figure 9.14 has been labeled by a decimal number corresponding to the state of the variables in the box. For example, the lower left box in Figure 9.14c is in column $\overline{A}\overline{B}$ (00_2) and in row $C\overline{D}$ (10_2); hence its number is $\overline{A}\overline{B}C\overline{D}$ (0010_2) or 2_{10}. To fill a Karnaugh map from a Boolean statement it is only necessary to write a 1 or 0 corresponding to the presence or absence of that particular miniterm in the statement. It is even easier directly from the truth table [or from equations like (9.35) and (9.36)], since the presence of a 1 or 0 on the Karnaugh map just corresponds to a true or false resultant line in the truth table. An example is shown in Figure 9.15a, where the majority detector truth table, expressed also by Eq. (9.38), has been used to fill the Karnaugh map.

Consider the two true terms on this map, boxes 6 and 7, which have been encircled by the vertical dashed loop. Taken separately these terms would each require a three-input AND gate for implementation. But both terms have X_1 and Y_1 true, and the combination spans both values for the other variable C_1. Thus the combination of two terms, enclosed by the dashed loop, can be expressed by the simpler *two*-term statement derived from

$$Q_1 = C_1 X_1 Y_1 + \overline{C}_1 X_1 Y_1$$
$$= (C_1 + \overline{C}_1) X_1 Y_1 \qquad (9.39)$$
$$= X_1 Y_1$$

Note that this product is no longer a miniterm, since it has only two variables. If convenient, the numerical form can still be retained by expressing this product as $x11$, where x means "don't care." The other two encircled groups behave similarly as shown in the figure. The 1 in box 7 is included in all three terms; this is equivalent to the algebraic trick used in the development of Eq. (9.31), whereby this single state was written three times. All of the true input states are now included in at least one of the encircled groups, and the minimized logic statement of Eq. (9.31) is equal to the sum of these three reduced terms.

A more general 4×4 Karnaugh map is shown in Figure 9.15b. The ones and zeros on this map don't correspond to any truth table we have discussed, but have been chosen as good examples of the various combinations possible in the reduction process. The x in box 14 stands for a "don't care" miniterm, not to be confused with a "don't care" bit. Note that the loops on these maps always contain 2^n boxes or miniterms: 1, 2, 4, 8, or 16 boxes are the only possibilities for the maps discussed here. The number of variables in the logical AND corresponding to a combination term is reduced by one each time the number of boxes in the combination is doubled. On an $m \times m$ map, the description of a single box requires m variables, two boxes needs $m - 1$ variables, four boxes needs $m - 2$, or generally 2^n boxes requires $m - n$ variables. Obviously, the simplest terms come from the largest enclosures.

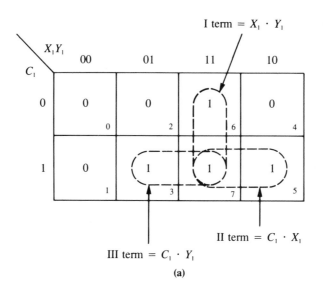

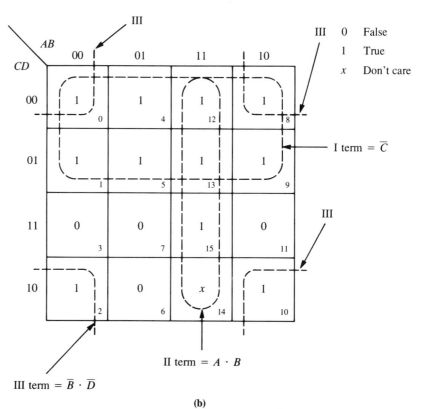

Figure 9.15   (a) The Karnaugh map of the truth table for the majority detector and its reduction to three logical products. (b) Typical Karnaugh mapping of four-element terms showing reduction to three logical products I, II, and III.

Two of the combinations shown are of particular interest. Combination II assumes that the don't care (x) miniterm in box 14 is a 1 and combines it with two terms already defined by combination I. This trick allows the miniterm of box 15 to be included in a larger enclosure; if this "don't care" state had been a zero or simply ignored, a four-term product would have been required to describe the single miniterm in box 15. Combination III is possible because of the commonality of the top and bottom edges and of the left and right edges. The outline around this combination of four boxes loops from top to bottom and from left to right, and being four boxes, is describable by a two-variable term.

The reduced expression for the output Q from this 4×4 map is

$$Q = \overline{C} + AB + \overline{B}\overline{D} \tag{9.40}$$

Although this may yield a minimum algebraic description, it may still require gates that are not readily available. Since either NANDs or NORs alone can be used to generate any other gate function, one or the other of these gates is usually the most common gate in a particular family: for example, NAND in TTL and CMOS, and NOR in ECL. The mechanization of Eq. (9.40) shown in Figure 9.16a uses AND and OR gates. Although this is the minimum logic statement and the easiest circuit to understand, it may not be readily obtainable in practice.

De Morgan's theorem will often yield alternative representations that are more compatible with the available gates. Inverting both sides of Eq. (9.40) yields

$$\overline{Q} = \overline{(\overline{C} + AB + \overline{B}\overline{D})} \tag{9.41}$$

Figure 9.16 *(a) Mechanization of the 4 × 4 Karnaugh map of Figure 9.15b. (b) A different mechanization using mostly NAND gates.*

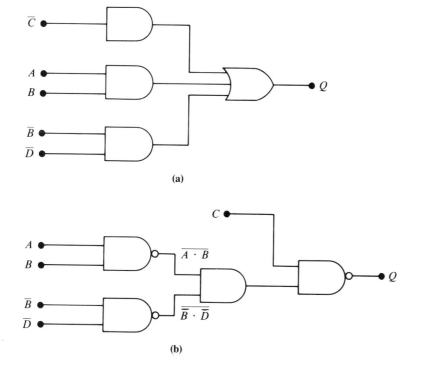

(a)

(b)

and use of De Morgan's theorem will produce

$$\overline{Q} = C(\overline{AB})(\overline{BD}) \tag{9.42}$$

Inverting both sides again gives

$$Q = \overline{[C(\overline{AB})(\overline{\overline{BD}})]} \tag{9.43}$$

which is shown mechanized in Figure 9.16b. This version still uses one AND gate, but that is easily replaced by a NAND and an inverter.

9.7 *Multiplexers and Decoders*

Two important classes of combinational logic devices are known as multiplexers and decoders. Applications for these devices appear frequently in systems where many lines of information are being gated or passed from one part of the circuit to another. Packaged devices performing these functions are available in a number of different logic configurations; only a single example of each type will be discussed here.

The term "multiplex" is used to describe various situations where multiple data signals share a common propagation path. In digital applications it always means time multiplexing, where different signals travel along the same wire but at different times. The 4-line to 1-line multiplexer circuit shown in Figure 9.17 is

Figure 9.17 A 4-line to 1-line multiplexer controlled by the address lines A_0 and A_1.

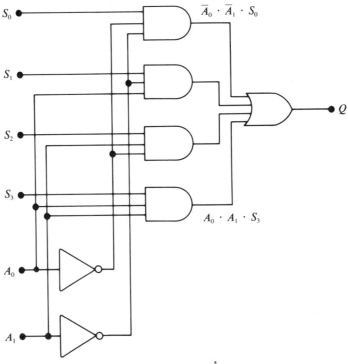

4-line to 1-line multiplexer

designed to permit any one of the four input signals, S_0 to S_3, to appear as the output Q: the particular signal that appears is determined by the condition of the address lines A_0 and A_1. The notation in the figure is such that the number A_1A_0 defined by the address lines is the same as the subscript n on the signal S_n which is passed to the output. In addition to the input lines shown in this figure, most packaged devices also have an "enable/disable" input. When the device is disabled by this input, the output is locked in some particular state or condition, and none of the input signals will appear at Q.

If several signals are to be multiplexed onto one line, we clearly need a way to demultiplex the signals back onto several different lines. This demultiplex function is one of the capabilities of the decoder shown in Figure 9.18. If the output

Figure 9.18 A binary-to-octal decoder: 3-line to 8-line decoder.

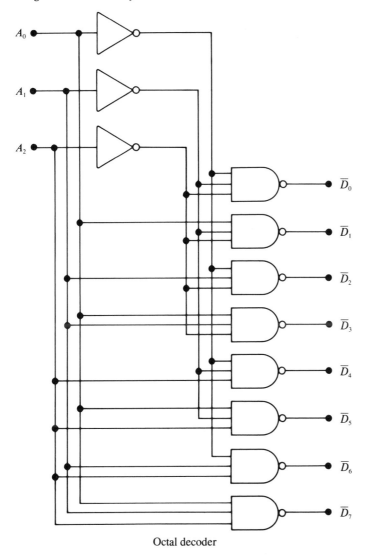

Octal decoder

of the multiplexer Q is connected to the decoder input A_2, and the signals A_0 and A_1 are connected as labeled on both figures, the multiplexer inputs S_n will appear on the corresponding first four of the output lines $\overline{D}_n$ (n = 0, 1, 2, 3). The negated notation of the decoder's output lines is typical of packaged devices but is of no significance in this particular application.

Another use of this decoder is to convert a 3-bit binary number to an output on one of eight lines, hence the name octal decoder. In this application the input number $A_2A_1A_0$ defines a number or address n that selects one of the output lines $\overline{D}_n$ to be low while all other outputs are high. This negated notation is often used on manufacturers' specification sheets to indicate that the selected or active line is driven low by the device: Since such lines rest in the high state, they are known as active low. Other common configurations for decoders are 4-line to 16-line (hexadecimal decoder) and 2-line to 4-line. Like the multiplexers, packaged decoders generally have an additional enable/disable input not shown in this figure. When the octal decoder shown is disabled, all outputs will be high independent of the state of the three address lines. In a typical application, the decoder would be routinely disabled while the address lines are changing, thus eliminating one source of glitches on the output lines.

9.8   Schmitt Trigger

Each input to a logic gate is sensitive to the low or high voltage that is present on that input. Since the input voltages are continuous, time-varying signals, the gate must have an intrinsic threshold voltage that defines the transition point between high and low input voltages (something close to the 50% point in Figure 9.8). Some ambiguity must exist when the input signal is very close to this threshold voltage, since a noisy input signal can be first on one side of the threshold, then the other, then back again. Such repeated crossings of the threshold can show up on the output as rapid full or partial transitions back and forth between logic states. Normally this does not occur, because the typical logic signal passes quickly through the threshold voltage region and noise signals do not have time to affect the output. However, if the input signal changes relatively slowly from one logic state to another, normal gates will sometimes generate outputs with many false transitions near every intended one.

A type of logic input known as the Schmitt trigger is used to reduce the problems caused by input signals with long rise and fall times. These devices use two voltage thresholds, a higher threshold to switch the circuit during a low-to-high transition and a lower threshold to switch the circuit during a high-to-low transition. An example of a slow input signal and the output from a Schmitt-triggered gate is shown in Figure 9.19b. Such a trigger scheme is immune to noise on the input signal as long as the peak-to-peak amplitude of the noise is less than the difference between these threshold voltages. As shown in this figure, the digital output pulse from a Schmitt-triggered gate is generally not centered on the input pulse but is delayed somewhat by the combined effect of the slope of the pulse edges and the voltage difference between the two thresholds.

The symbol for a two-input NAND gate with the Schmitt trigger feature is shown in Figure 9.19a. The small hysteresis curve inside the gate symbol indi-

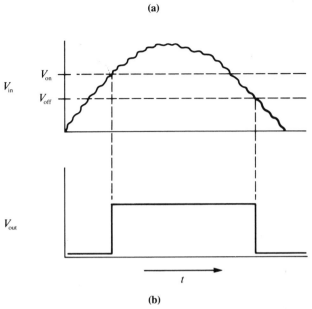

(a)

(b)

Figure 9.19 *(a) The schematic symbol for a Schmitt trigger input and (b) its response to a slowly changing input signal.*

cates the two-threshold Schmitt trigger function. This type of input is limited mostly to inverters and simple gate functions and is used to "condition" slow or noisy signals before passing them to more critical parts of the logic.

9.9 Logic Families

Several different logic families are in current use, and each has its own particular set of operating rules and requirements. No attempt will be made here to list the rules for each family; that is done in great detail in the "data books" available from the various manufacturers. However, it is instructive to look at the input and output characteristics of a typical gate circuit from each of the three common families: TTL, CMOS, and ECL.

9.9.1 Transistor-Transistor (TTL)

In the TTL family, the basic package comprises four NAND gates and is divided into different series depending on the construction, speed, and power consumption of the package. The numbering convention for the basic NAND package in the various series is 7400 for the original package, 5400 for a military version that operates over a wider temperature range, 74L00 for a low-power (and slower) version, 74H00 for a high-speed (and higher-power) version, and 74LS00 for a newer series with low-power Schottky diode clamped operation. Several additional series are also available. The LS series combines the speed of the 7400 with the low power consumption of the 74L00 and is widely used in modern TTL applications. There is a large degree of uniformity within the family, which bridges the series types: In general, the numerical part of the designation defines a logically identical device in the different series, although there are some exceptions in flip-flop triggering between the older series and the LS series.

For the general packaged LS device the numbering is 74LSxxx, where xxx is the device number. The number xxx is more or less chronological with the initial introduction of the device; the result is that larger numbers tend to identify more complex devices. If TTL devices from various series are to be mixed in a single circuit, the most important consideration is the current loading and driving ability of each series. In general, a single gate can drive 10 or more inputs in the same series; but devices from a low-power series (L or LS) can drive only a couple of 7400 series inputs. Conversely a single 7400 series output can drive approximately 40 low-power inputs. The devices of the TTL family normally operate using a single power supply V_{CC} of $+5$ V; the absolute maximum is $+7.5$ V.

The circuit for a 7400 series NAND gate is shown in Figure 9.20a. The design features a split emitter input transistor and a push-pull output stage. This push-pull connection of two like-polarity transistors is known as the totem pole configuration. The base drive to these output transistors is derived from the collector and emitter of Q_2, and since these two points have voltage swings that are $180°$ out of phase, one output transistor is on while the other is off. The operation of

Figure 9.20 *(a) Typical circuit for the TTL 7400 two-input NAND gate, and (b) the open collector output variation.*

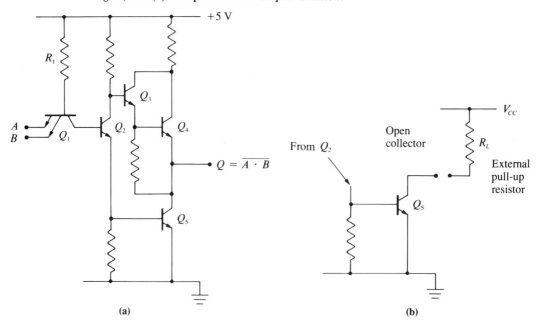

the input transistor Q_1 is such that if either emitter is grounded, Q_1 will conduct, pulling its collector toward ground and turning off Q_2. If both A and B inputs are high (or disconnected), Q_1 will not conduct, and Q_2 will be turned on by the current through R_1 and the base-collector junction of Q_1. Although there are several different input designs for the various series of TTL logic, all behave like this one: The driving circuit must be able to *sink* a current to ground rather than supply (source) a current from the positive voltage. The signal propagation time across a typical LS series gate is around 30 ns.

A variation on the normal totem pole output is shown in Figure 9.20b. This circuit is known as "open collector" and requires an external "pull-up" resistor for operation. It has two primary uses: V_{++} at the top of the pull-up resistor can be as much as 30 V for specially designated gates; and several open collector outputs can be connected directly to a common pull-up resistor to give the wired AND function discussed in Section 9.10.1.

9.9.2 Complementary Metal Oxide Semiconductor (CMOS)

While neither as fast nor as rugged as the TTL family, the various CMOS logic families are extremely flexible and require relatively small amounts of power compared to the TTL family. The lowest-numbered CMOS package contains two 3-input NOR gates and an inverter. It comes in three different series: CD4000A (or MC14000) for the original and best avoided A series; CD4000B (or MC14000B) for the much more predictable buffered B series; and CD4000UB for a faster, and harder to use, unbuffered version of the B series devices. There are also two buffered series, 74C00 and 74HC00, that are pin-for-pin compatible with the TTL logic family. Buffered CMOS gates are typically ten times slower than 74LS00 logic, but the 74HC00 series rivals the LS speed.

A simplified CMOS circuit for an unbuffered 2-input NAND gate is shown in Figure 9.21a. The two power supply voltages $+V_{DD}$ and $-V_{SS}$ can have a wide range of values: The only general requirement is that $V_{DD} - V_{SS}$ be between $+3$ V and $+15$ V. One of the problems with any type of MOS device is that the very-high-impedance MOSFET inputs can be easily damaged by low-current but high-voltage static electricity build-up. The internal diodes on the inputs provide protection against this problem when the chip is wired into a circuit, but careless handling in a dry winter environment can still damage these devices.

Note that on CMOS circuits there is almost no quiescent current within the device. One of the MOSFET's Q_3 or Q_4 is always off, leaving no static current path from $+V_{DD}$ to $-V_{SS}$. As long as the output Q drives only other high-impedance MOSFETs, the DC current out of the device will also be very small. Most of the power consumed by a CMOS device is therefore dynamic and used to charge the stray capacitance of connecting wires and MOSFET inputs.

The basic unbuffered circuit has a relatively low AC current gain between input and output signals. As a result it provides relatively poor isolation between various parts of a logic circuit. Several stages of the amplification circuit shown in Figure 9.21b are included at the output of the buffered devices to increase the AC gain and circumvent this problem. Unfortunately, these buffer stages can increase the signal propagation time across the device to approximately 100 ns.

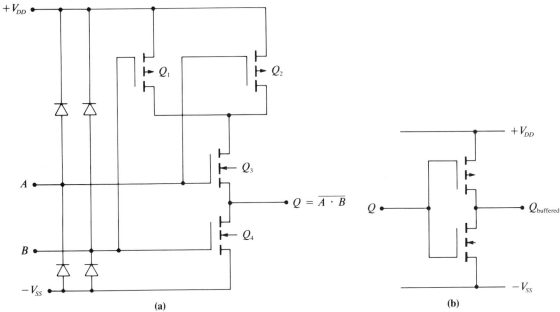

Figure 9.21 *(a) Typical CMOS construction of a NAND gate, and (b) the buffer amplifier that is used as an output stage on some CMOS families.*

9.9.3 Emitter-Coupled Logic (ECL)

The highest-speed standard logic family is ECL. It does not enjoy the general use of either of the previous logic families because of general high-frequency design problems having nothing to do with the logic itself. In the 100- to 500-MHz radio frequency range at which this logic excels, careful attention must be paid to wiring inductance and radiative coupling between various parts of the circuit. As a result, normal breadboarding techniques have no chance of working, and the logic is not as easily learned and applied.

The standard ECL logic series are the 10,000 and the 100,000 or 100K types. The basic gate circuits are the NOR and OR/NOR types, so circuits developed with ECL logic have a different look from the more common NAND-oriented TTL or CMOS designs.

A typical circuit in the ECL family is the OR/NOR gate shown in Figure 9.22. As shown, ECL logic is normally operated with a single -5 V power supply. The heart of this circuit is the differential amplifier formed by Q_1/Q_2 and Q_3, which acts as a current switch for the constant current source I. The input transistors Q_1 and Q_2 are in parallel, and a current into either will switch the current I away from Q_3 and into Q_1/Q_2, thereby turning Q_5 on and Q_4 off. Note that the outputs are current switches to ground; they do not develop an output voltage signal until connected to another gate input or through a resistor to $-V_{EE}$.

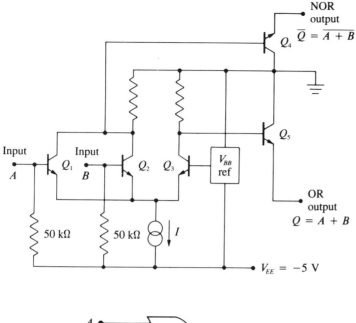

Figure 9.22   Typical ECL construction of an OR/NOR gate.

9.10   *The Data Bus*

In electronics, the term bus is used to mean a common wire connected at various points in the circuit, as for example the ground bus or the power bus. The "data bus" is an extension of this idea to the wires that carry digital information. The term normally refers to a group of parallel wires that connect physically separated parts of the logic circuit with each individual wire carrying a different logic signal. In addition to being received by several gate inputs, a typical wire in a data bus may also be driven by several different gate outputs. This latter feature requires that gate outputs be wired directly together, an option not allowed with normal gates.

A data bus line may be time multiplexed to serve different functions at different times: At any instant only one gate may drive information onto the line, but several gates may be receiving it. Some of the received information may cause a different output gate to gain control of the line at a later time. Such operation can produce a situation where on a given wire information may flow in different directions at different times. The term bidirectional is applied to such wires.

9.10.1   *Wired ANDs and ORs*

As can be seen in Figures 9.20 and 9.21, the most common TTL and CMOS devices use a two-transistor push-pull output (totem pole), which causes the output to always appear as a low-impedance path to some fixed voltage (power supply or ground). In ordinary TTL circuits, the output from one such device should never be wired directly to the output of another; in CMOS circuits it is permissible (to give increased output current) if the driving circuitry independently guarantees that all such connected outputs will try to go to the same state.

However, certain types of logic gates can have their outputs wired directly together. In older diode transistor logic (DTL) families and in the "open collector" types within the TTL family, gate outputs are driven by a single transistor, as shown in Figure 9.20b. Clearly, if the outputs from two such gates are wired together, the output will be high (true) only if neither transistor is turned on. This is known as the wired-AND, since a true result on the joined wire requires both constituent wires to be true. The ECL logic family normally uses a single-transistor output circuit, as shown in Figure 9.22, which can be wired directly to that of another ECL device, but since the output transistors are pulling toward the high state, the result is a wired-OR function.

An example of the data bus application is shown in Figure 9.23a, where three open-collector NAND gates drive a common pull-up resistor R. The gates function as a wired-AND according to the logic equation

$$\overline{Q} = A \cdot G_1 + B \cdot G_2 + C \cdot G_3 \tag{9.44}$$

For this circuit to function as a time-multiplexed bus, only one of the control signals G_1, G_2, and G_3 may be true at any time; the data signal A, B, or C corresponding to the true control line will appear inverted on the bus. The importance of this circuit is not that it is a better multiplexer, but that each of the gates together with its data and control inputs can be physically isolated from the others; only the data bus line must pass from one physical point to another. With the 4-line to 1-line multiplexer of Figure 9.17, all data and control lines must be brought to a common place in order to drive the multiplexing gate.

9.10.2 Three-State Outputs

While open-collector gates can be used in a data bus application, the open-collector drive with a pull-up resistor loses the speed advantage of the symmetrical push-pull output. A better circuit is the 3-state output, which is available in both the TTL and CMOS families. These 3-state devices have a typical two-transistor push-pull output; but unlike the normal push-pull, in which one transistor is always on, 3-state outputs are driven by circuits that are able to turn off both output transistors at the same time. This results in a high-impedance (hi-Z) third state which is neither a high nor a low voltage.

A bus circuit using 3-state logic buffers is shown in Figure 9.23b. The high-impedance state of each gate is controlled by an enabling input, typically an active low input as shown here. When the enabling input is high, the output of the gate will be high-impedance and effectively disconnected from the common bus line; when the enable input is low, the gate will pass the data signal onto the bus line. The connected outputs follow the equation

$$Q = A \cdot \overline{G}_1 + B \cdot \overline{G}_2 + C \cdot \overline{G}_3, \tag{9.45}$$

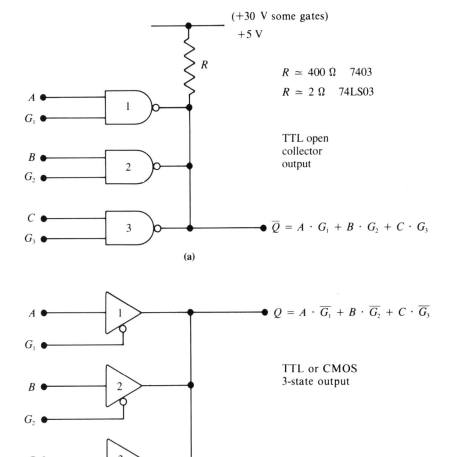

Figure 9.23 (a) Three open-collector NAND gates driving a common bus line. (b) Several 3-state logic noninverting buffers driving a common bus line.

with the additional characteristic that if G_1, G_2, and G_3 are all true, the combined output Q will be in the high-impedance state. The use of three-state logic is widespread, and this type of output is available on a variety of packaged logic elements including flip-flops, various registers, and multiplexers.

9.11 Two-State Storage Elements

In an analog circuit, information can be stored as a *quantity* of charge on a capacitor, with a larger number typically corresponding to a larger charge. However, this information storage method is limited by the gradual leakage of the stored charge into the circuit. As a result, analog voltage storage times are limited. It is much more common, electronically and otherwise, to store information in discrete form. The advantages of discrete storage are fairly obvious when long storage times are required: Compare the slowly aging and subtly unique qualities of a Renaissance painting with the exactly reproducible content of a written work of the same period. Today most information can be quickly converted to discrete form and, by using conversion codes for the various symbols, ultimately stored as numbers.

Since all numbers can be converted to binary form, the problem of discrete storage reduces to the need to store a large number of two-state variables. Various means of performing this function exist; one of the most visible is the ordinary wall light switch, which is a two-state mechanical device capable of retaining (remembering) its current condition. At the present time, there are four commonly used methods for the electronic storage of two-state quantities: magnetic domain orientation, the presence or absence (not the quantity) of charge on a capacitor, the presence or absence of an electrical connection, and the DC current path through the latches and flip-flops of a digital circuit. We will discuss only latches and flip-flops in this chapter; descriptions of the other types of storage devices can be found in Chapter 11.

9.12   Latches and Unclocked Flip-flops

By interconnecting the inputs and outputs of two or more gates, it is possible to generate a circuit that remembers its present condition. Traditionally, the term flip-flop was used to describe a bistable device with symmetrical input and output characteristics, whereas the term latch was reserved for a similar but unsymmetric device that required a different operation for latching and unlatching, in the manner of a mechanical latch. We will retain this definition in the present discussion, although the RS, $\overline{R}\,\overline{S}$, and the statically clocked D flip-flops discussed below are generally listed as latches in manufacturers' literature.

9.12.1   Latches

Whatever the definition, a common feature of all latches is that they have two inputs, data and enable/disable. Some versions have a single output Q, whereas others provide both a Q and $\overline{Q}$ output. The built-up example of Figure 9.24a is known as a "ones catching" latch. When the control input C is false, the output Q follows the input D, but when the control input goes true, the output latches true as soon as D goes true and then stays there independent of further changes in D. The output can only return to false while the control input C is false. The timing diagram of Figure 9.24b shows the action of this latch.

One of the most useful packaged versions of the latch is known as the "transparent" latch or "D-type" latch. This two-input device generates an output Q that follows the data input D whenever the latch is enabled but freezes the state of the output at the instant the latch is disabled. The behavior of a transparent latch is shown by the timing diagram of Figure 9.24c, which again assumes that a true

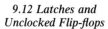

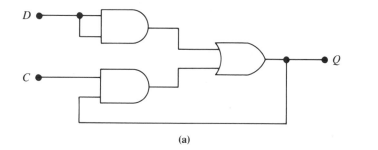

(a)

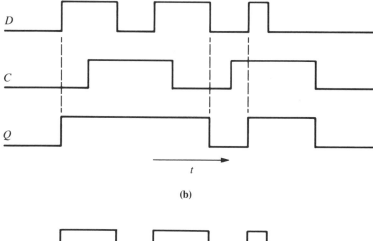

(b)

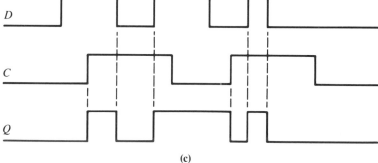

(c)

*Figure 9.24   (a) An AND-OR gate used as a "ones catching" latch
and (b) its timing diagram. (c) The timing diagram of a transparent
latch subject to the same input signals.*

signal C enables the latch. The operation of this latch is the same as that of the
statically triggered D flip-flop discussed below and shown in Figure 9.27.

9.12.2   RS and $\overline{RS}$ Flip-flops

The RS flip-flop (RSFF) is the result of cross-connecting two NOR gates as shown
in Figure 9.25a; its logic symbol and truth table are given in Figure 9.25b. Since
the output of a NOR gate is driven false when either input is true, Q will be false

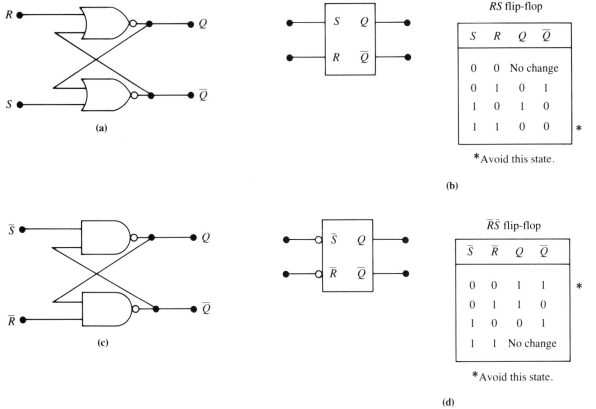

Figure 9.25 (a) The RS flip-flop constructed from NOR gates, and (b) its circuit symbol and truth table. (c) A variation, the $\overline{RS}$ flip-flop, constructed from NAND gates, and (d) its truth table.

whenever R is true, independent of the state of $\overline{Q}$; likewise, $\overline{Q}$ will be false whenever S is true. Thus, ones in the S and R columns directly produce zeros in the repective $\overline{Q}$ and Q columns: The RS inputs are "active ones."

The ideal flip-flop has only two rest states, set and reset, defined by $Q\overline{Q} = 10$ or $Q\overline{Q} = 01$, respectively. The RS flip-flop does not meet this ideal because, as indicated on the bottom line of the truth table, it is possible to drive both Q and $\overline{Q}$ to the same state.

Except in special cases, the bottom state of the RSFF truth table should be avoided. For notational convenience, this state is often labeled as "undefined," but the problem is not with the state itself but rather with the uncertain condition of the $Q\overline{Q}$ outputs if the R and S inputs change simultaneously from true to false. Such a transition directly from the lower line of the truth table ($RS = 11$) to the top line ($RS = 00$) must still lead to the $Q\overline{Q}$ outputs defined on one of the two middle lines of the table; the only alternative is $RS = 00$ and $Q\overline{Q} = 00$, and this is not a stable condition. The state actually assumed by a flip-flop following such a transition is unpredictable and will depend on the individual properties of a particular sample.

A very similar flip-flop can be constructed using two NAND gates as shown in Figure 9.25c. Since a NAND gate's output will be true if either input is false, this flip-flop differs from the RS type by having "active zeros." This effect is seen in the truth table of Figure 9.25d, where zeros on the $\overline{R}$ and $\overline{S}$ inputs determine the location of ones in the respective Q and $\overline{Q}$ columns. The flip-flop designation of $\overline{RS}$ and the inverting circles on the schematic symbol shown in Figure 9.25d are used to emphasize this active zero feature. For this flip-flop the forbidden state is the top line ($\overline{RS}$ = 00), since the outputs are uncertain following a simultaneous transition of the inputs from $\overline{RS}$ = 00 to $\overline{RS}$ = 11.

9.13   Clocked Flip-flops

In addition to various kinds of data inputs such as R and S, a clocked flip-flop has an additional input that allows output state changes to be synchronized to a clock pulse. The same clock pulse often drives a number of flip-flops, thereby generating simultaneous state changes in all. Although there are several different types of data inputs, with names like RS, D, JK, and gated JK, an equally important distinction between the various flip-flops is in the operation of the clock input. Several different types of clocking actions, or triggers, are available, and a clear distinction must be made between the static, or level-sensitive, triggers and the dynamic ones. Circuits that manufacturers designate as flip-flops have a dynamic clock input of either the master-slave or edge-triggered type; of these, the edge-triggered are the simplest to use.

Flip-flops are generally classified by the action of the data inputs, and we will show how the various static-triggered types can be fabricated from the basic gates. Although the dynamic trigger will be introduced after the discussion of static-triggered D and JK flip-flops, packaged examples of these types are usually dynamically triggered.

9.13.1   Clocked RS Flip-flop

The statically clocked RS flip-flop is the simplest example, and although its truth table shows a serious problem, we will use this device to fabricate several of the other flip-flop types. The clock input for the clocked RSFF is formed by the addition of two gates as shown in Figure 9.26a. The schematic diagram of this flip-flop and the truth table for both its static and clocked operation are given in Figures 9.26c and 9.26d.

The first five lines of this truth table give the static input and output states for this flip-flop. When the clock input is false, as shown on the first line, the flip-flop outputs $Q\overline{Q}$ are in one of the two stable states 01 and 10 and are independent of changes in R and S inputs; when the clock is true, the RS inputs have an effect just like those on the unclocked RSFF. The last four lines show the state of the $Q\overline{Q}$ outputs after a complete clock pulse. The outputs on the last line are truly undefined after the clock pulse falls; the state actually assumed by the flip-flop will vary from sample to sample.

Clocked operation of this flip-flop has one other problem: The RS inputs immediately influence the output while C is true. Thus, if the RS inputs change while

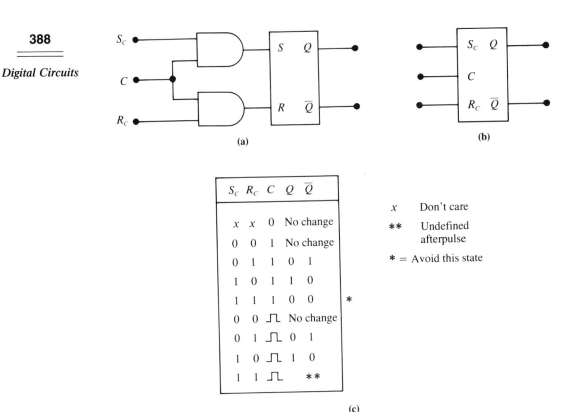

(a) **(b)**

S_C	R_C	C	Q	$\overline{Q}$	
x	x	0	No change		
0	0	1	No change		
0	1	1	0	1	
1	0	1	1	0	
1	1	1	0	0	*
0	0	⊓	No change		
0	1	⊓	0	1	
1	0	⊓	1	0	
1	1	⊓	**		

(c)

x	Don't care
**	Undefined afterpulse
* =	Avoid this state

Figure 9.26 *(a) The clocked RS flip-flop can be constructed from an RS flip-flop and two additional gates. (b) The schematic symbol for the static clocked RSFF and (c) its truth table.*

C is true, $Q\overline{Q}$ will change accordingly until C goes false. The RS inputs to the last four lines of the truth table thus refer to the final condition of RS just before C goes false.

9.13.2 D Flip-flop

One way to avoid the problems associated with the unpredictable state on the RSFF's truth table is to reduce the input options to the flip-flop. The D-type flip-flop can be developed from a clocked RSFF and an inverter as in Figure 9.27a. The inverter guarantees that the input state $RS = 11$ can never be attained and leaves the DFF with only two inputs, D for the data and C for the clock pulse, as shown in Figure 9.27b. The truth table in Figure 9.27c is simplified by the single data input, and it can be seen that a clock pulse simply saves the current state of the data line in the Q output.

The statically clocked D flip-flop is also known as a transparent latch. For this operation the clock input C is treated as a control input. While C is high the output Q follows the input D, then freezes at its current state when C becomes false.

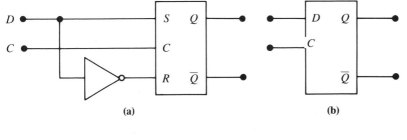

(a)   (b)

D	C	Q	$\overline{Q}$
x	0	No change	
0	⊓	0	1
1	⊓	1	0

   x   Don't care

(c)

Figure 9.27 (a) The statically triggered D flip-flop (transparent latch) mechanized with a clocked RS. (b) The schematic symbol and (c) its truth table.

9.13.3 JK Flip-flop

The basic JKFF in Figure 9.28a uses two AND gates to cross-connect the Q and $\overline{Q}$ outputs back to the JK inputs. The cross-connection again serves to eliminate the undesirable last line on the clocked RSFF truth table but also preserves flexibility by having two input lines, J and K. The last line of the truth table in Figure 9.28c indicates that the JKFF "toggles" on every clock pulse while $JK = 11$, meaning that whatever the flip-flop's initial state, set or reset, after the clock pulse it will be in the other state. As we shall see later, this feature of the JKFF is very useful in counting circuits.

The statically triggered JKFF, indicated by the schematic symbol of Figure 9.28b, still has some serious problems with the last line of its truth table. If both J and K inputs are true, then the inputs to the clocked RSFF are just $R_c = Q$ and $S_c = \overline{Q}$. Now look back at the circuits (Figures 9.26a and 9.25a) that make up the clocked RSFF. While C is true, the additional output to input wiring on the JKFF causes the Q and $\overline{Q}$ outputs from the NOR gates in Figure 9.25a to be effectively and illogically connected to their own R and S inputs. The result is that if the clock pulse stays true too long—longer than the propagation time of the input signal through the gates to the $Q\overline{Q}$ outputs and back to the RS inputs—the final condition of the $Q\overline{Q}$ outputs will be determined by the intrinsic characteristics of the gates and not by the initial state of the circuit: a useless result. This form of the JKFF can be used only with rigidly defined short clock pulses, a very limiting requirement.

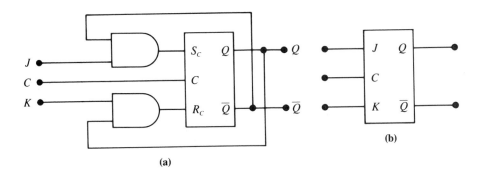

(a)

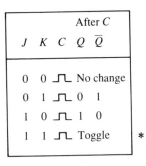

		After C		
J	K	C	Q	$\overline{Q}$
0	0	⊓ No change		
0	1	⊓ 0	1	
1	0	⊓ 1	0	
1	1	⊓ Toggle		*

*If C pulse
is too long,
this state
is undefined.

(c)

Figure 9.28 *(a) The basic JK flip-flop constructed from an RS flip-flop and gates. (b) Its schematic symbol and (c) truth table.*

9.14   *Dynamically Clocked Flip-flops*

The various clocking problems of flip-flop designs discussed above result from the clock input being sensitive to signal levels rather than edges—a static rather than a dynamic clock input. In the days when flip-flops were constructed from discrete transistors (definitely not the "good ol' days"), a series capacitor and a diode were often used to make the clock input sensitive to the edge (rising or falling, not both) of a pulse but not to the pulse level. But the necessary capacitors are not easily incorporated onto the IC chip, and this direct method cannot be used.

One way of simulating a dynamic clock input is to use two flip-flops in tandem, one driving the other in a master/slave arrangement. This design produces a clock input that is sensitive to complete pulses—pulse triggering. A second approach uses special circuits that trigger the flip-flop as the clock input passes through a particular voltage threshold midway between the true and false levels—edge triggering. Because of the importance of dynamic triggering, we will discuss both of these methods in some detail.

9.14.1 Master/Slave or Pulse Triggering

The operating principles of the master/slave arrangement can be seen from the JKFF design in Figure 9.29a. This implementation uses two RSFFs. The gates, A_1 and A_2, which drive the RS inputs of the master flip-flop, both produce false outputs while the clock is false. Since $RS = 00$ is a "no change" condition in the RSFF truth table of Figure 9.25b, the outputs from these gates cause the master flip-flop to be frozen while the clock is false; similarly, the inverted clock signal into gates A_3 and A_4 causes the slave flip-flop to be frozen while the clock

Figure 9.29 (a) An implementation of the master/slave flip-flop and (b) its truth table when JK inputs are stable during clock pulse. (c) The "ones catching" property of this master/slave design.

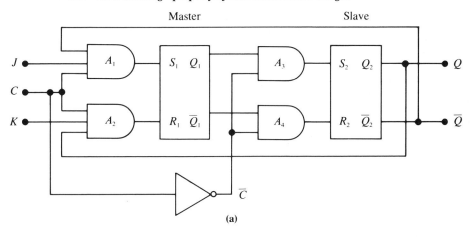

(a)

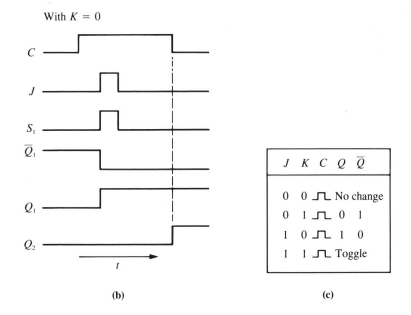

With $K = 0$

(b)

J	K	C	Q	$\overline{Q}$
0	0	⊓	\multicolumn{2}{l}{No change}
0	1	⊓	0	1
1	0	⊓	1	0
1	1	⊓	\multicolumn{2}{l}{Toggle}

(c)

is true. The overall effect is that the data inputs are written onto the master flip-flop while the clock is true and transferred to the slave when the clock becomes false.

The master/slave design guarantees that the $Q\overline{Q}$ outputs of the slave flip-flop can never be connected to its own *RS* inputs. This eliminates the most serious problem of the statically triggered JKFF of Figure 9.28a: the loss of a useful output if the clock pulse is longer than the internal propagation time of the circuit. The circuit is still a pulse-triggered design, and this is indicated by the pulse symbol in the clock column of the truth table in Figure 9.29b.

The master/slave designs can indeed be used with clock pulses of any length, but this design freedom leads directly to a new problem. Consider the situation where a master/slave JKFF is reset ($Q\overline{Q} = 01$), and both inputs are false ($JK = 00$) as shown on the timing diagram of Figure 9.29c. When the clock goes true, nothing happens, since $JK = 00$ holds the outputs of gates A_1 and A_2 false. Now, while the clock is true, suppose that *J* goes true for a short time and then becomes false again. The combination $\overline{Q}JC$ into gate A_1 will satisfy this gate, setting the master RSFF to the $Q\overline{Q} = 10$ state. When the clock finally falls, the transient true condition of the *J* input, as remembered by the master, will be transferred to the slave. A similar state-changing result occurs when the flip-flop is initially set and a transient pulse appears at the *K* input.

The effect just described is known as "ones catching" and is common to many master/slave designs. If the data inputs are held stable from just before to just after the clock pulse, the master/slave flip-flops will function according to their truth tables.

9.14.2   Edge Triggering

A true dynamic clock input should trigger the flip-flop's state change as the rising or falling edge of the clock signal passes through a particular threshold voltage. Ideally, the operation should be insensitive to the slope of the edge or the time (beyond some necessary minimum) the clock spends in the high or low state before the transition. Various circuits have been devised that accomplish this result at the integrated circuit level; we will describe only one technique.

Any of the previous clocked flip-flops, pulse or level sensitive, can be converted to edge-triggered operation by incorporating a pulse-generating circuit into the clock input. A simplified circuit capable of converting a falling edge to a pulse is shown in Figure 9.30. The gate marked *D* is assumed to have a propagation time that is longer than the inverter by an amount Δt_D. As a result of this delay, the AND gate is satisfied for a time Δt_D, resulting in a short pulse. A similar design in which the inverter has the longer propagation time will yield a pulse on the rising edge of an input signal.

A simplified version of the integrated circuit of a falling edge-triggered JKFF is shown is shown in Figure 9.31a. The operation of the circuit is fairly complicated and depends on the assumption that the NAND gates labeled D_1 and D_2 have a longer propagation time than the other gates. As an example of the operation of this circuit, assume that initially $Q\overline{Q} = 01$ and $JK = 11$. According to the truth table of Figure 9.31b, these conditions will allow the next falling clock edge to toggle the flip-flop to $Q\overline{Q} = 10$. This clock edge is indicated on the

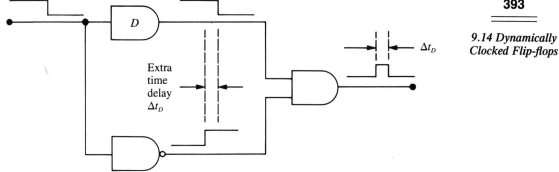

Figure 9.30 *A slow or delayed gate can be used to convert a level
change into a short pulse.*

schematic by the 1/0 notation on the C input, and similar before/after notation
shows the propagation of this signal through the gates A_1, A_2, A_3, A_4, A_5. Even
though the C input going false starts this action and will ultimately drive gate D_1
true, the long propagation time through D_1 allows $\overline{Q}$ to go false first, thus keeping
gate A_5 from ever being satisfied.

Figure 9.31 *(a) Schematic diagram of a TTL edge-triggered JK flip-
flop; gates D_1 and D_2 are delayed gates. (b) The truth table for this
device.*

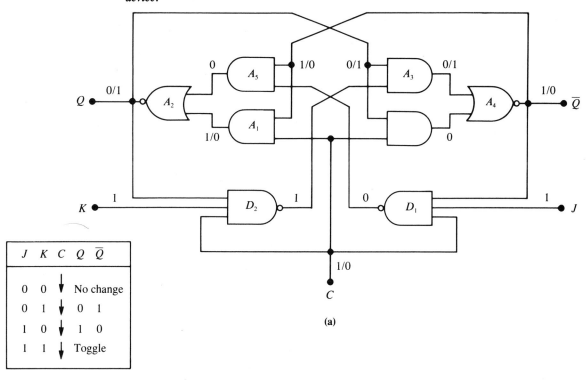

J	K	C	Q	$\overline{Q}$
0	0	↓	No change	
0	1	↓	0	1
1	0	↓	1	0
1	1	↓	Toggle	

(b)

(a)

Circuits designated by the manufacturers as *JK* or *D* flip-flops are either of the master/slave, pulse-triggered type or use some type of edge-triggering circuit; both flip-flop designs are indicated on the schematic drawing by a ">" symbol at the clock input. Used alone, this symbol indicates a positive ($0 \rightarrow 1$ transition in positive logic) edge-triggered flip-flop as shown in Figures 9.32b and 9.32d. When a NOT circle is also used at the clock input as shown in Figures 9.32a and 9.32c, the symbol describes either a master/slave, pulse-triggered or a negative ($1 \rightarrow 0$ transition in positive logic) edge-triggered flip-flop. A signal input that is sensitive to the $0 \rightarrow 1$ transition or to a true pulse is known as an "active true" input. The opposite case is similarly known as an "active false" input, and the same designations are sometimes applied to the driving signals.

In addition to the clock and data (*JK* or *D*) inputs we have discussed, most IC flip-flop packages will also include some type of "set" and "reset" (or "mark" and "erase") inputs. These inputs are level coupled and usually connect to the

Figure 9.32   The schematic symbols for (a) a master/slave or negative (falling) edge-triggered JKFF; (b) a positive edge-triggered JKFF; (c) a master/slave or negative edge-triggered DFF; (d) a positive edge-triggered DFF. Note the locations of the small negating circles on the C, S, and R inputs to some of these flip-flops.

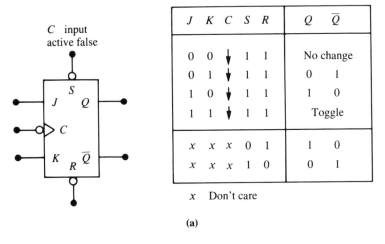

J	K	C	S	R	Q	$\overline{Q}$
0	0	↓	1	1	No change	
0	1	↓	1	1	0	1
1	0	↓	1	1	1	0
1	1	↓	1	1	Toggle	
x	x	x	0	1	1	0
x	x	x	1	0	0	1

x   Don't care

(a)

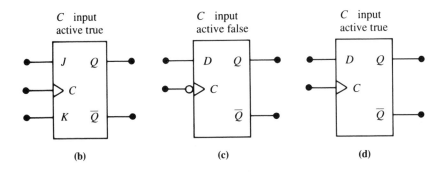

(b)   **(c)**   **(d)**

$Q\overline{Q}$ output much like on the RS and $\overline{RS}$ flip-flops discussed above. The additional inputs allow the flip-flop to be preset to an initial state without using the clocked logic inputs.

9.15   One-shots

The monostable multivibrator, normally called a one-shot, is an extremely useful (and often overused) utility device that is not too different from a flip-flop. The one-shot is essentially an unstable flip-flop that shows an affinity for its reset condition, $Q\overline{Q} = 01$. When set to $Q\overline{Q} = 10$ by an input clock or trigger pulse, the flip-flop will delay for a fixed time and then return to the reset state of its own accord. The delay time is determined by the RC time constant of a resistor and capacitor as indicated on the schematic drawing of Figure 9.33a. Delay times ranging from 20 ns to several seconds are possible, although for many applications the pulse-to-pulse variations in the delay time may be excessive for the longer time constants.

The one-shot symbol shown in Figure 9.33a indicates a device that sets on the falling edge of the input trigger signal, and the corresponding timing diagram is shown in Figure 9.33b. As can be seen from the timing diagram, this one-shot is able to generate a pulse of a particular width following an input pulse.

One shots are often used in pairs, with the output of the first used to trigger the second as shown in Figure 9.34a. This causes the second device to generate a pulse at the end of the first one-shot's delay, as seen on the timing diagram of this figure. When delay or pulse times are critical, or uncertain when the circuit is constructed, a variable resistor may be used to trim the RC time constant. Because of this flexibility, it is all too easy to overuse one-shots in a circuit, with

Figure 9.33   *(a) Schematic symbol for a general-purpose, one-shot multivibrator. (b) The timing diagram for this device.*

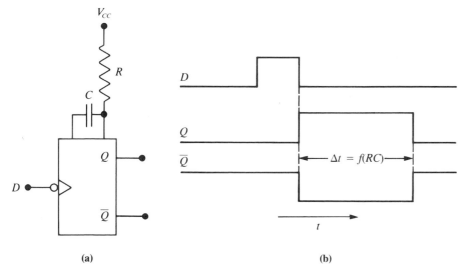

(a)   (b)

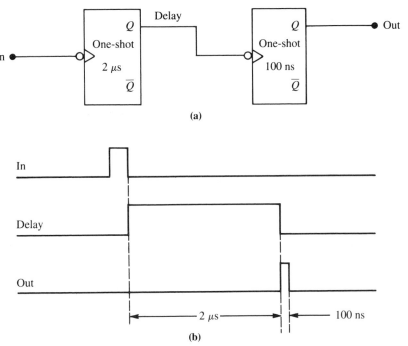

Figure 9.34 (a) Two one-shots connected to generate a delayed pulse
as shown on (b) the timing diagram.

the result that the time relationship between signals becomes excessively interdependent. If the circuit has a clock signal, it is generally better to devise a method that extracts a particular clock pulse than to introduce a one-shot which will generate signal transitions not synchronized with the clock.

9.16   State Diagrams

A state diagram is useful both as a design aid and for circuit documentation. Its relationship to the digital circuit is much like that of a flow chart to a computer program. Figure 9.35 shows a stylized example using two flip-flops to describe the states of an engine and its gas tank. In this example, $A = 1$ means that the tank is empty and $B = 1$ means that the engine is running. The four circles exhaust the possibilities for the states of these two flip-flops, and the labeled connecting lines show the cause of the transitions between these states. The 11 state in the figure is clearly a transient state, since it indicates a running engine with no gas in the tank.

A more detailed example is shown in Figure 9.36. This logic uses two negative edge-triggered *JK* flip-flops with the *JK* inputs wired true, thereby causing each JKFF to toggle on the falling edge of its clock input: *B* changes state on the falling

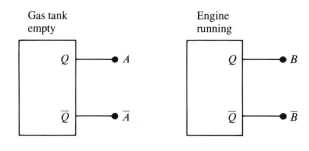

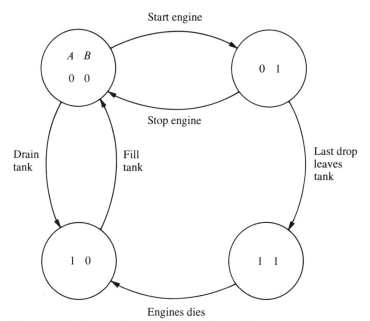

Figure 9.35   A state diagram for an engine and its gas tank. The empty/not-empty condition of the tank and the on/off condition of the engine are each indicated by flip-flops.

edge of T, and A changes state on the falling edge B. The transition lines between the states in Figure 9.36b are labeled only by the clock input transitions. In more complicated circuits driven by a common clock, these labels would instead describe the pretransition state of the JK data inputs to each flip-flop.

One of the main advantages of state diagrams is that they encourage clear thinking about the transition possibilities between states. Unfortunately, since the number of states for n flip-flops is 2^n, this clarity is easily submerged in a sea of circles and lines unless every effort is made to isolate small parts of the total logic system. This turns out not to be so difficult since many of the flip-flops in a logic system can be grouped together into various kinds of registers, leaving only a few to control the sequence of data flow from one register to another. This smaller group generally defines the major states of a system.

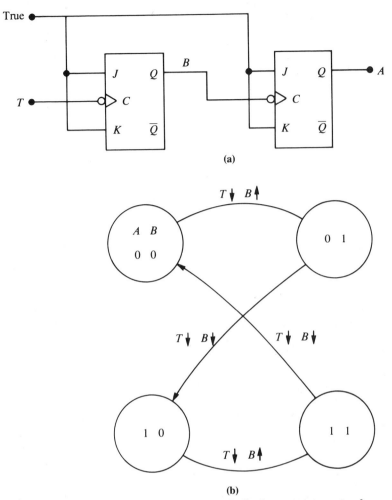

Figure 9.36 *(a) A four-state system. Both flip-flops operate as toggles following the last line of a negative edge-triggered JKFF truth table. (b) The state diagram showing the transitions between the four states.*

9.17   Registers

A register is a group of flip-flops arranged to hold and manipulate a data word using some common circuitry. The examples discussed below are all available on single integrated circuit chips containing at least four flip-flops. Although these packaged registers generally have a multipurpose capability, for clarity only the essential elements of each type are shown here.

9.17.1   Data Registers

The simplest example is the data register of Figure 9.37a. This circuit uses the clocked inputs of *D* flip-flops to load data into the register on the rising edge of a

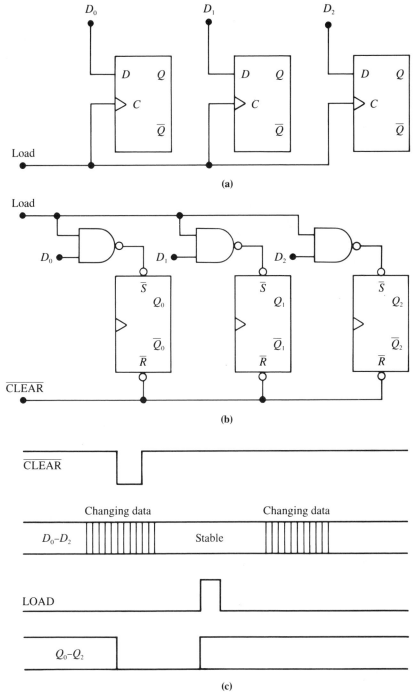

Figure 9.37 *(a) A data register using the clocked inputs to D-type flip-flops. (b) A more complicated data-loading technique leaves the clocked inputs free but requires a clear load pulse sequence as shown on (c) the timing diagram.*

LOAD pulse. Simple registers of this type could be used to hold the two input numbers and the sum for the adder circuit developed previously.

It is often desirable to load data into a register in a manner that leaves the clock inputs free for other purposes such as implementing a shift register as shown below. Most flip-flops and packaged flip-flop arrays (registers) have input circuits similar to those shown in Figure 9.37b. The set/reset ($\overline{S}$ and $\overline{R}$) inputs to these flip-flops are like those of the basic $\overline{R}\,\overline{S}$ flip-flop discussed previously. The data-loading process requires a two-step sequence as shown in Figure 9.37c: First the register must be cleared, then it can be loaded. Inputs of this type are found on most of the register types discussed below, but for clarity they are not shown in these figures.

9.17.2 Shift Registers

A simple shift register example is illustrated in Figure 9.38a. At the falling edge of each CLOCK signal, the value of D moves to C, C moves to B, B moves to A, and A is lost. A register of this type could be used to convert a 3-bit parallel data word (preloaded by the set/reset inputs as discussed above) to a serial-bit stream exiting at output A. Alternatively, it could receive a 3-bit serial-bit stream at input D and save it for parallel use by some other part of the logic.

When the A output from the shift register is connected back to the D input as in Figure 9.38b, the device is known as a circular shift register or ring counter. This register is normally preloaded with a number that is then repeatedly shifted to produce a repetitive signal pattern at output A. The state diagram for this register, shown in Figure 9.38c, displays a disconnected pattern of states; the path actually followed depends on the initial pattern present in the register.

Bidirectional shift registers with all input and output connections are available as packaged units, the most common containing four flip-flops. Extremely large shift registers containing several thousand flip-flops are available in TTL-compatible MOS packages, but these registers have only the serial input and output connections available on external pins.

9.17.3 Counters

Counters can be classified into several different types; binary-coded decimal (BCD) or binary, one direction or up/down, ripple-through or synchronous, and by their clearing and preloading abilities. The most generally useful type is a binary up/down, parallel counter with clear and preset inputs. These are generally available in 4-bit packages that can be used in combination to obtain larger counters. The BCD types are designed to count in decimal, with 4 bits being allocated to each decimal position. They are most useful in counters that must ultimately produce a visual display of a decimal number.

The simplest counter is the binary, ripple-through, up counter of Figure 9.39a. This 3-bit example is built up from pulse or negative edge-triggered JKFFs. When false, the COUNT ENABLE signal forces all of the JK inputs to 00, holding the current condition of each flip-flop; when true, it makes all $JK = 11$, thereby placing each flip-flop into its toggle mode. In this mode, each flip-flop will change state after each negative transition of its clock input. The COUNT input to flip-

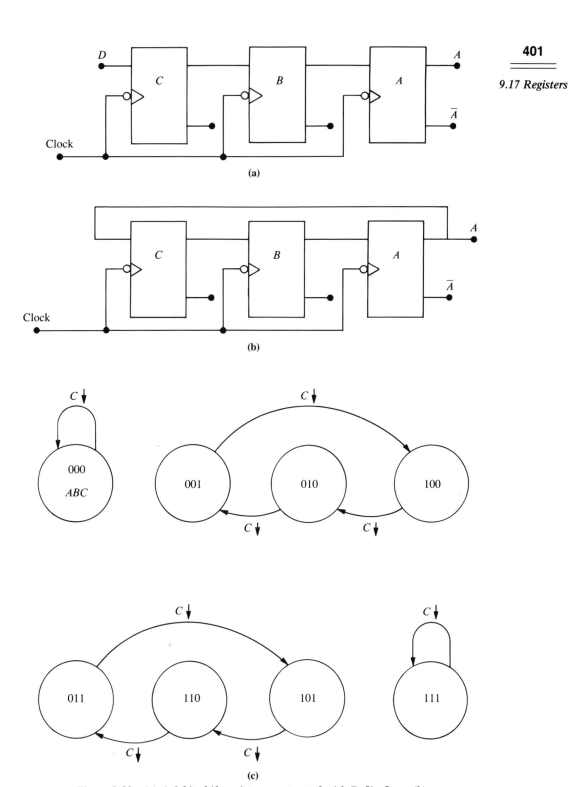

Figure 9.38 (a) A 3-bit shift register constructed with D flip-flops. (b) A 3-bit circular or ring shift resistor, and (c) the state diagram for this device.

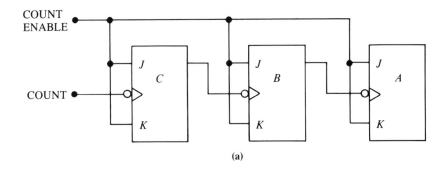

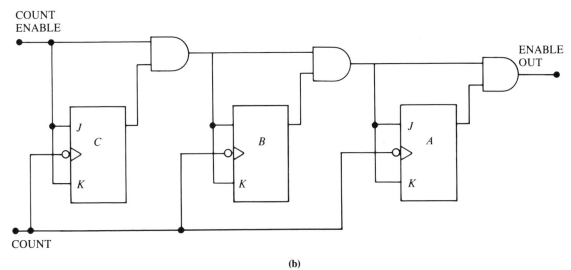

Figure 9.39 (a) A 3-bit ripple counter constructed from JK flip-flops.
(b) A 3-bit synchronous counter.

flop C causes this unit to change state on every falling edge of COUNT; flip-flop B changes state only when C goes from 1 to 0; and A changes state only when B goes from 1 to 0.

Each of the flip-flops used in the counter has a propagation time t_{pd} between the falling edge of its clock input and the state change at its outputs. Because each clock input is driven by the output of the previous stage, a timing diagram would show that the C, B, and A state changes are each delayed from the previous by t_{pd}. During these short times the counter shows a "transient" and incorrect result; if the resulting output is used to drive additional logic elements, these transient states may lead to spurious pulses.

This problem is avoided by using the synchronous clocking scheme of Figure 9.39b. This simplified circuit still has some propagation delay on the information presented to the JK inputs of the various flip-flops. However, the clock input is common to each flip-flop, and the resulting state changes will be synchronized, just t_{pd} behind the falling edge of COUNT. Thus, with a synchronous counter, all outputs signals change state at essentially the same time.

9.17.4 Divide-by-N Counters

A common feature of many digital circuits is a high-frequency clock with a square wave output. If this signal of frequency f drives the clock input of a JKFF wired to toggle on each trigger, the output of the flip-flop will be a square wave of frequency $f/2$. This single flip-flop is a divide-by-2 counter. In a similar manner, an n flip-flop binary counter will yield an output frequency that is f divided by 2^n.

Figure 9.40   (a) A divide-by-3 counter, (b) its state diagram, and (c) its timing diagram.

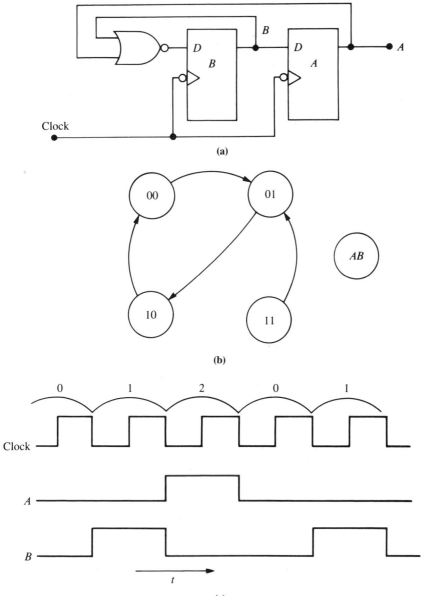

Frequency division by numbers other than 2^n requires a special circuit for each number. A circuit that functions as a divide-by-3 counter is shown in Figure 9.40a. If the CLOCK signal has a frequency f, the output signal at A or B will be repetitive with a frequency $f/3$. The state diagram of Figure 9.40b shows that one of the possible states, $AB = 11$, of these two flip-flops is not in the main counting loop. However, this state changes to the main loop state $AB = 01$ on the first clock pulse—a characteristic known as ''self-starting.'' With a different design it is possible that the counter would stick in the $AB = 11$ state with no path to the main loop; in that case the $AB = 11$ state would be an ''isolated'' state. The timing diagram shows the operation of the counter after it has reached the main counting loop.

Problems

1. Convert the binary number 1001 1110 to hexadecimal and to decimal.

2. Convert the octal number 175_8 to hexadecimal.

3. Add and multiply the two numbers 101_2 and 011_2 directly in binary ($1 + 1 = 10$ and $1 + 1 + 1 = 11$), and show that the results agree with decimal arithmetic.

4. Use the method of division to convert 394 to hexadecimal.

5. Convert the number 146 to binary by repeated subtraction of the largest power of 2 contained in the remaining number.

6. Devise a method similar to that used in the previous problem and convert 785 to hexadecimal by subtracting powers of 16.

7. If we define a new size binary angular degree such that there are 256 degrees in a full circle similar to Figure 9.2b and express all numbers in 16-bit words:
   (a) Show that the angular difference $0 - 254$ is equal to $+2$ when the result is truncated to the least significant 8 binary bits.
   (b) Show that $254 - 0$ gives -2 when the 2s complement sign convention is used to interpret the low order 8-bit result.

8. If the input to the circuit of Figure 9.6b is written as a number $ABCD$, write the nine numbers that will yield a true Q.

9. Design a two-input OR gate using only two-input NAND gates.

10. Using two-input NAND gates, devise a circuit that will mechanize the equation $Q = (A \cdot B \cdot C \cdot D)$.

11. *(a)* Starting with the input signals A, B, and C, use De Morgan's theorem to convert the equation $S = \overline{(A + B)} \cdot \overline{C}$ to one that can be mechanized with a single three-input gate of the AND, NAND, OR, or NOR type. *(b)* Fill a truth table and write a different Boolean algebra expression defining the true state.

12. It is required that Q be true only when the logic signal A equals the logic signal B. Make a truth table of this problem, mechanize the result directly assuming that only A and B signals are available (not $\overline{A}$ or $\overline{B}$), and then use De Morgan's theorem to reduce this to a three-gate mechanization from A and B inputs.

13. Using the 2's complement convention, the 3-bit number ABC can represent the numbers from -3 to 3 as shown in the table in Figure P9.13 (ignore -4). Assuming that A, B, C and $\overline{A}$, $\overline{B}$, $\overline{C}$ are available as inputs, the goal is to devise a circuit that will yield a 2-

bit output *EF* that is the absolute value of the *ABC* number. You have available only two- and three-input AND and OR gates.

    (a) Fill a truth table with the *ABC* and *EF* bits.

    (b) Write a Boolean algebra expression for *E* and for *F*.

    (c) Mechanize these expressions.

Value	A	B	C
0	0	0	0
1	0	0	1
2	0	1	0
3	0	1	1
−4	1	0	0
−3	1	0	1
−2	1	1	0
−1	1	1	1

Figure P9.13

14. If the 3-bit binary number *ABC* represents the digits 0 to 7:

    (a) Make a truth table for *A*, *B*, *C*, and *Q*, where *Q* is true only when an odd number of bits are true in the number. Note that 5 is an odd number but has an even number of bits set true.

    (b) Write a Boolean algebra statement of *Q*.

    (c) Using De Morgan's theorem, convert this equation to one that can be mechanized using only two XOR gates. Draw the resulting circuit.

15. The three-input OR gate follows the Boolean statement $Q = A + B + C$.

    (a) Make a truth table for this gate.

    (b) Write the Boolean statement for $\overline{Q}$ and mechanize the result using 2-input NAND gates and inverters.

16. Suppose that the 2-bit binary number *AB* must be transmitted between devices in a noisy environment. To reduce undetected errors introduced by the transmission, an extra bit *P* is often included to add redundancy to the information. Assume that *P* is set true or false as needed to make an odd number of true bits in the resulting 3-bit number *ABP*. When the number is received, logic circuits are required to generate an error signal *E* whenever the odd number of bits condition is not met.

    (a) Develop a truth table of *E* in terms of *A*, *B*, and *P*.

    (b) Write a Boolean expression for *E* as determined directly from the truth table.

    (c) Using De Morgan's theorem twice, reduce this expression to one EOR and one NEOR operation. (This is very similar to the half-adder problem.)

17. An experimenter observes that the output of a TTL NAND gate is 0 V with both inputs open and does not change when they are both connected to +5 V. Is the gate malfunctioning? Why or why not?

18. The 2-bit numbers *AB* and *CD* are to be multiplied to give the 4-bit number *EFGH*.

    (a) Make a 16 × 8 truth table for *ABCDEFGH*.

    (b) Write shorthand miniterm expressions like Eq. (9.37) for each of the four output bits.

    (c) Using 4 × 4 Karnaugh maps, write a two-term expression for *F*, a five-term expression for *G*, and a single-term expression for *H*.

19. An automated train serves ten stations identified by a 4-bit binary number *ABCD* (0 and 11 to 15 unused). Stations 1, 3, 5, 7, 8, and 10 are major stops and require a longer loading/unloading delay that is to be indicated by a logic output *Q*. Use a Karnaugh map to show that the needed logic can be reduced to a single XOR gate. [*Hint:* You will need to make use of the "don't care" condition.]

20. Make a timing diagram showing the *Q* output of the *RS* flip-flop in Figure 9.25a. As input, let *RS* = *DC* from Figure 9.24b.

21. A mechanical switch will often "bounce" several times as it makes contact so that the connection is alternately low- and high-resistance. When used to drive a digital circuit, this can result in many output pulses per switch closure. Use the single-pole double-throw (SPDT) pushbutton shown in Figure P9.21 together with an $\overline{R}\,\overline{S}$ latch to make a debounced circuit that will yield only one output pulse on each complete push-release cycle of the button.

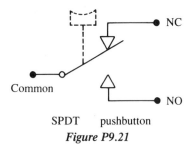

SPDT   pushbutton
Figure P9.21

22. Assuming that the initial state is *AB* = 00, draw the state diagram for the operation of the circuit shown in Figure P9.22.

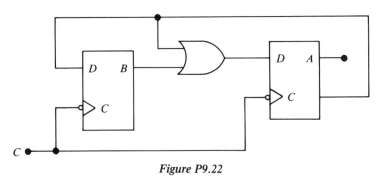

Figure P9.22

23. Using two edge-triggered *JK* flip-flops, design a divide-by-3 counter: The period of one of the *Q* output signals should be three times that of the clock input. Include a state diagram of the operation of your circuit.

24. An oscillator is generating a 5-MHz square wave, 100 ns true and 100 ns false. Using gates and flip-flops as needed, devise a circuit that will have as its output signal every fourth pulse from the oscillator, that is, 100 ns true and 700 ns false, with the true pulse in phase with the true signal from the oscillator.

25. You have three *JK* negative edge-triggered flip-flops that follow the truth table given in Figure 9.32, a 1-MHz square wave clock signal, and a randomly timed 100-ns true event pulse *E*. One of the flip-flops has a two-input NAND connected to its reset input.

    (a) Using these, design a circuit that will follow the timing diagram in Figure P9.25.

    (b) What will your circuit do if a second event pulse occurs while $\overline{FD}$ is true?

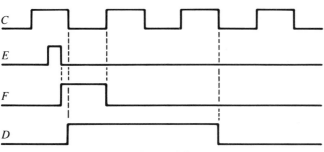

Figure P9.25

26. You have available the outputs C_0 and C_1 of a 2-bit synchronous counter driven by a clock of period *T*. Using three-state logic, devise a circuit that will time multiplex four additional data signals D_0, D_1, D_2, and D_3 onto the same output wire, cycling through the four signals in a time 4*T*.

27. Many laboratory experiments need to measure the time between two signals, in this case the signals START and STOP. The time *T* shown in Figure P9.27 is variable between 0 and 15 μs. You have available various gates and several negative edge-triggered *JK* flip-flops with an additional unclocked input that resets the flip-flop to $Q\overline{Q} = 01$ when false. When false, this CLEAR input overrides all other input signals.

    (a) Assuming that a 1-MHz clock signal is available, devise a register that will contain the number *T* in microseconds after the STOP pulse. The circuit should work repeatedly whenever a START, STOP sequence occurs.

    (b) Add an additional flip-flop that will set whenever *T* exceeds 15 μs and reset otherwise (an overflow bit).

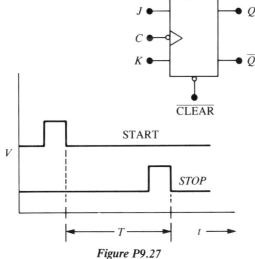

Figure P9.27

10

Data Acquisition and Process Control

10.1 Introduction

The ultimate purpose of most electronic systems is to measure or control some physical quantity, and to this end the system will typically need to both acquire data from its environment and exercise some degree of control over that environment. To effect this interaction, the electronic system must use special devices, known as transducers, which couple other physical processes to the domain of digital and analog electronics. This chapter will be generally descriptive, with primary emphasis on the interface area between analog and digital electronics, but to complete the picture we will also briefly describe a number of different transducers and a few auxiliary circuits.

For several decades, electronic systems have been used for increasingly complex applications in research, medicine, manufacturing, travel, and entertainment. Since the early 1960s, when suitable transistorized versions first became available, computers have been incorporated into the most sophisticated data acquisition and process control systems, adding flexibility and computational power. Today, with the rapidly decreasing size, power consumption, and cost of integrated circuits, it has become practical to include one or more microcomputers in all but the most modest of systems.

10.2 Typical Computer Application

The popular conception of a digital computer focuses on the keyboard and display units, since these comprise the most visible and human-oriented input and output devices to the environment. For our purposes this viewpoint is too limited: Data entry via a keyboard is one of the cruder methods, and the human response to a visual display is slow and sometimes erratic. There are many other ways to couple a computer to its environment, and they are generally faster and more exact than systems that place a human in the chain. In the applications we are discussing, the human functions as a supervisor rather than a serial link in the data flow; in this capacity the human is able to apply judgment skills with minimal disruption of the routine operation of the system.

A typical data acquisition and process control system is shown in Figure 10.1. The figure shows a digital computer at the junction between the input and output data paths, but the figure is made more general if the computer block is under-

Figure 10.1   Schematic for a general electronic system that can acquire data from its environment, convert it for input to a digital computer, then use the output from the computer to control some characteristic of the environment.

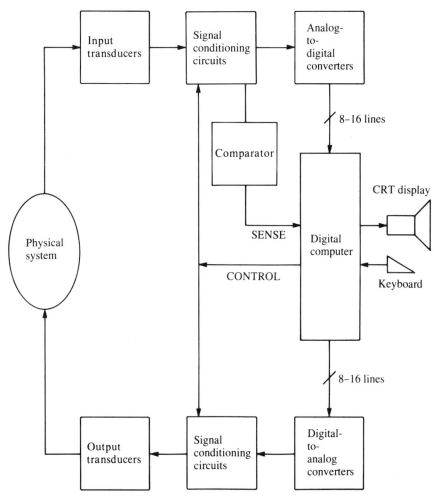

stood to be any digital electronic arrangement that is capable of achieving the desired measurement result.

The flow of information starts with an input transducer that measures some property of the physical sample. The output from this transducer is first conditioned by an amplifier, probably with some low-pass filtering, then converted to digital information by an analog-to-digital converter (ADC). The digital information is then acquired by the computer, processed, and generally displayed in some fashion. As a result of an internal algorithm or external human intervention, the computer may modify the environment of the sample using a digital output signal that is converted to analog by a digital-to-analog converter (DAC). The analog signal is amplified, filtered to remove transients, and otherwise conditioned as needed to drive some appropriate output transducer. The ADC and DAC elements may not always be required, but because of the prevalence of continuous as opposed to discrete physical phenomena, these converters play a major role in most applications.

10.3   Signal Domains

In the electrical domain we have dealt with voltage and current signals that could be described in terms of amplitude, frequency, phase, and time constant. However, the general physical measurement may involve a variety of signals outside the electrical domain: chemical composition, ion concentration, color, temperature, position, velocity, force, light intensity, event count, period, frequency, rates, and so on. These measured quantities may be analog or effectively analog because of large numbers (temperature, chemical composition), or digital (event counting).

Modern measurement technique generally requires that the various nonelectrical signals generated by the response of a physical system be converted into the electrical domain for display and further processing. The processes and instruments used for making such measurements vary widely both in size and complexity, from a simple light-detecting device on an automatic door opener to the sophisticated electromagnetic detection instruments on the Orbiting Astronomical Observatory. However, if the measured system is not inherently electrical, then at the heart of each instrument lies one or more devices capable of converting a signal from its domain of origin into the electrical domain. Such devices are known as input transducers. Similar devices known as output transducers are used to convert information out of the electrical domain, often into domains that can be perceived directly by one of the five human senses.

10.4   Input Transducers

Input transducers may generate electrical signals by varying one or more elements of a circuit. The basic linear circuit elements are voltage and current sources (EMFs), resistance, capacitance, self-inductance, and mutual inductance, but in addition there are the various linearized elements associated with a device such as a Zener diode: V_{PN}, V_Z, R_f, R_r, R_Z. Active devices such as transistors add the "trans" parameters such as h_{fe} and g_m. All of these quantities are subject to variation caused by a changing physical environment and may therefore form the active element of an input transducer.

The most fundamental input transducers respond to temperature, electromagnetic radiation intensity, force, displacement, and chemical concentrations; but if used in combination and with the addition of timing information, these devices can be used to measure any physical or chemical quantity. However, to see how a transducer relates to a circuit, it is better to classify them according to how they operate rather than what they measure.

Common engineering usage splits input transducers into two categories: active or passive, with the word active meaning that the device must be "activated" by an applied EMF known as an "excitation." Although this use of "active" agrees with its application to an active circuit element such as a transistor (which must have external power applied), it results in the peculiar classification of a variable resistor as an active transducer and a variable battery as a passive one.

The transducers in the following tables are classified on the basis of energy flow: those that add energy to the electrical system (EMFs), those that remove energy (resistance), and those that vary the electrical energy by changing some nonresistive parameter within the electrical system (parametric). A few examples of common input transducers are shown in Figure 10.2.

A list of EMF transducers is given in Table 10.1. Although these devices can increase the energy in the electrical domain, many can also function as resistive or parametric devices by the addition of an activating EMF. The photodiodes and phototubes are normally used with an EMF, although both actually convert light energy directly into the motion of free charges.

Table 10.1   EMF Input Transducers

Radio antenna	Detects low-frequency electromagnetic radiation by induced fields in a wire array or loop.
Solid state photovoltaic cell	Detects incident light by a quantum mechanical process that converts radiant energy directly into electrical power.
Solid state photodiode	Detects incident radiation by the charge current resulting from broken valence bonds in a nonconductor.
Vacuum phototube	Detects very low-intensity (down to a single photon) visible and near-visible light using the photoelectric effect. In the photomultiplier tube, an applied electric field accelerates the initial electrons into additional plates to eject more electrons and produce a current amplification.
Piezoelectric crystal	Measures a time-varying applied force or pressure by the generation of an AC EMF across the crystal body.
Variable reluctance	Measures position by the magnetic flux changes produced within a loop or coil.
Thermocouple	Measures the relative temperature between two points over a very wide range. The effect is very small, amounting to only a few millivolts per 100°C.
Electrochemical cell	Measures chemical concentration.

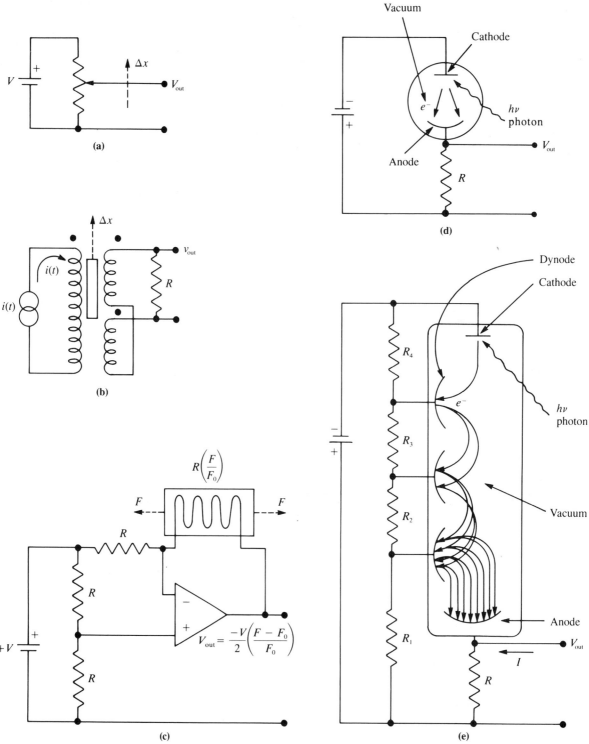

Figure 10.2 Input transducers and their activating circuits: (a)
potentiometer, (b) linear variable differential transformer, (c) strain
gauge, (d) vacuum photodiode, (e) photomultiplier tube.

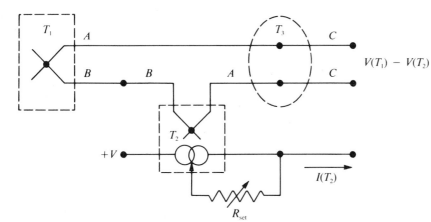

$$V(T_1) - V(T_2)$$

Figure 10.3 *Two thermocouples are needed for a precision temperature measurement. In this arrangement there are three different metals A, B, and C, resulting in four temperature-sensitive junctions between dissimilar metals. In practice, the EMFs from the two junctions at T_3 will cancel, and the temperature T_2 at the reference thermocouple can either be held constant or measured by an absolute-temperature device as indicated in the figure.*

The thermocouple EMF transducer is widely used for temperature measurement, because its various forms span an extremely wide temperature range: $-200°C$ to $2000°C$. Since every junction of dissimilar metals produces the thermocouple effect to some degree, a thermocouple always measure the difference of two temperatures as shown in Figure 10.3. If the reference temperature at the "cold junction" (T_2 in the figure) is reasonably constant, it may be ignored. A high-precision difference measurement can be obtained if the reference temperature is determined with a solid state temperature sensor (see Table 10.3), as shown in the figure.

Some examples of resistive transducers are given in Table 10.2. All of these transducers remove energy from the electrical system at a variable rate. Since

Table 10.2 Resistive Input Transducers

Mechanical switch	Detects motion by opening or closing a contact
Potentiometer	Measures position by the motion of a mechanical contact along a length of resistive material
Resistance temperature detectors	Measures temperature using the varying resistivity of various materials
Strain gauge	Measures force by the changing resistance resulting from the elastic deformation of a metallic or semiconducting material
Photoconductive cell	Measures visible and near-visible light intensity by variations in the conductivity of CdS, CdSe, or CdTe semiconducting materials
Ion concentration	Measures the ionic concentration in a solution or in a gas by changes in conductivity
Ionization chamber	Detects the passage of charged particles by the ionization trail left in a gas

Table 10.3 Parametric Input Transducers

Variable capacitor	Measures changes in the geometry, the electric susceptibility, or
Variable inductor	the magnetic permeability
Variable transformer	Measures position by changes in the mutual inductance between
	a primary coil and one or more secondaries
Solid state	Measures absolute temperature in the $-55°C$ to $150°C$ range
temperature sensor	using the temperature sensitivity of the *PN* junction voltage
Hall effect device	Measures position by the electric potential developed transverse
	to both an applied current and an applied magnetic field

these devices all obey Ohm's law, they can be used either by applying a voltage and measuring the current or applying a current and measuring the voltage.

Some examples of parametric input transducers are listed in Table 10.3. These devices, most of which can be represented on a schematic as an AC EMF, always require an activating power source.

The solid state temperature sensor is a relatively new type of temperature transducer for use near room temperature. The device is an arrangement of several transistors in an integrated circuit, and is important because its output signal depends on the absolute temperature. This feature can be derived from Eq. (6.2). For sufficiently large collector currents, that equation becomes

$$I_C = I_D e^{qV_{BE}/kT} \tag{10.1}$$

where I_D, q, and k are constants and T is the absolute temperature. If two identical transistors are operated such that their collector currents are in a particular fixed ratio $I_{C1}/I_{C2} = r$, then this equation predicts that

$$V_{BE1} - V_{BE2} = T\left[k\frac{\ln(r)}{q}\right] \tag{10.2}$$

showing that the difference in the base-to-emitter voltages of the two transistors is proportional to the absolute temperature. Inexpensive integrated circuit devices are available that use this effect to produce either a temperature-dependent output voltage or a temperature-dependent output current.

10.5 Output Transducers

Transducers used to transfer signals out of the electronic domain are no less varied than input transducers, but they tend to be larger, more visible, and hence more familiar devices. In many cases, a substantial amount of power must be transferred out of the electrical domain in order to produce the desired effect. This power can be transferred to other domains directly as heat or indirectly by generating electric and magnetic fields, which in turn give up their energy to other systems. Some examples are given in Table 10.4.

Electromotive devices that convert electrical energy to motion are widely used. Some common examples are shown in Figure 10.4. Electric motors are particu-

Table 10.4 *Output Transducers*

Radiofrequency	At frequencies above a few 100 kHz, an oscillating current will emit a significant amount of electromagnetic radiation.
Optical	Various materials can be made to emit, or modify, the reflectance or transmission of light as a function of an applied electronic signal.
Resistive	Simple resistive heating converts electronic energy to thermal energy.
Solid state heat pumps	Multijunction assembly of doped semiconductor and metal elements between two thermally conducting plates makes use of energy-level shifts across the junctions to extract energy from an applied current and use it to pump heat from one plate to the other.
Electromagnetic	Broad range of application, including motors, solenoids, loudspeakers, and the magnetic deflection and focusing of charged beams.
Electrostatic	Used for the electrostatic acceleration and deflection of charged beams and to change the dimensions of piezoelectric crystals.
Electrochemical	Applications include electroplating, electrolysis, battery charging, and other reactions whose rates are influenced by an applied voltage or current.

larly versatile; the most convenient for transducer use are the stepping types and the DC types with low rotational inertia. The better DC motors convert current to torque with a high degree of linearity, whereas stepping motors provide precise position control and can be driven with on/off switches.

Figure 10.5a shows the coil-and-permanent magnet arrangement needed for a four-position stepping motor. When the coils are activated in pairs as shown, the electromagnetic fields lock the rotor into a specific position. By activating a different pair of coils, the motor can be made to move to a new position as indicated by the table in this figure. Actual motors have many more magnetic poles on the rotor and more complex pole faces on the electromagnets, but the unipolar types retain the simple four-coil structure shown here and can be activated by a switch circuit like that shown in Figure 10.5b.

One of the most important output transducers for use with a digital computer is the cathode-ray tube or CRT. This device displays information by directing a collimated beam of electrons onto a phosphorescent screen and allowing the computer to control the intensity and/or the position of the beam. The focusing and deflection mechanism of these tubes may be either electromagnetic or electrostatic: The common video monitor uses electrostatic focusing and electromagnetic deflection, whereas laboratory oscilloscopes and random-access XY display devices generally use electrostatic deflection as well. Figure 10.6 shows a tube designed to use electrostatic focusing and deflection.

10.6 *Signal Conditioning Circuits*

Two of the blocks in Figure 10.1 are labeled signal conditioning circuits: one associated with input transducers and the other with output transducers. The circuits represented by these two blocks span all of analog electronics, but typically the input block represents a low-level signal amplifier and a low-pass filter

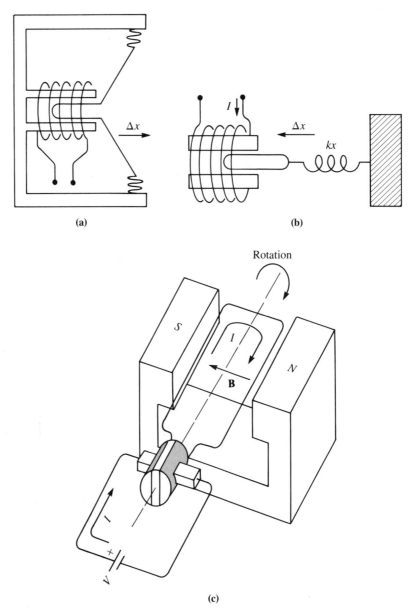

**Figure 10.4   Three common electromotive devices: (a) loudspeaker,
(b) solenoid, (c) DC motor.**

whereas the output block represents a low-pass filter and some type of power amplifier. Since the methods of power amplifier design are outside the scope of this text, this section will concentrate on signal conditioning circuits associated with input transducers.

Drivers for output transducers will often be required to supply substantial amounts of power. For audiofrequency applications such as speaker coils, the commercial audio amplifier is an obvious example of an output driver. However,

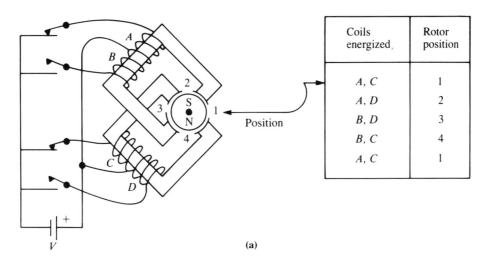

Coils energized	Rotor position
A, C	1
A, D	2
B, D	3
B, C	4
A, C	1

(a)

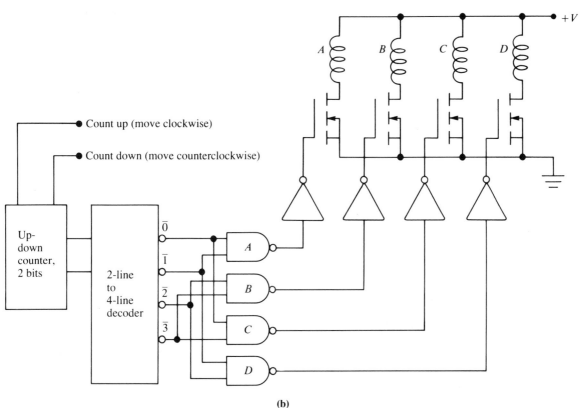

(b)

Figure 10.5   (a) The coil arrangement of a unipolar, four-position stepping motor. By energizing the coils in pairs, the motor can be made to step to the next contiguous line in the table: down for clockwise, up for counterclockwise. (b) This circuit uses power MOSFETs to energize the coils and an up/down counter to remember the present position.

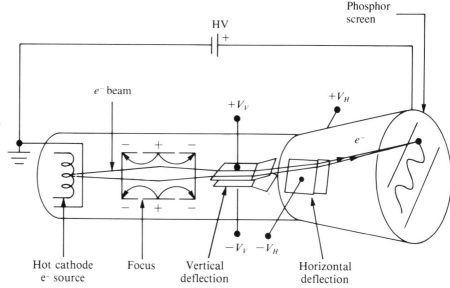

Figure 10.6 *The electron beam of a cathode-ray tube can be focused
and deflected by electrostatic fields. Magnetic focusing and deflection
can also be used.*

at frequency extremes, special-purpose amplifiers are often required. Two simple
but practical high-power drivers are shown in Figure 10.7: The first circuit is a
current driver for heating elements, magnetic field generation, DC motor drives,
and so on; and the second circuit could be used to generate an electric field for
piezoelectric crystals, charged beam deflection, and so on. These power MOSFET

Figure 10.7 *A power MOSFET converts an op amp into a precision
controller for either (a) high-current or (b) high-voltage applications.
For high-current applications, the sensing resistor R_1 must be a high-
power, low-resistance component.*

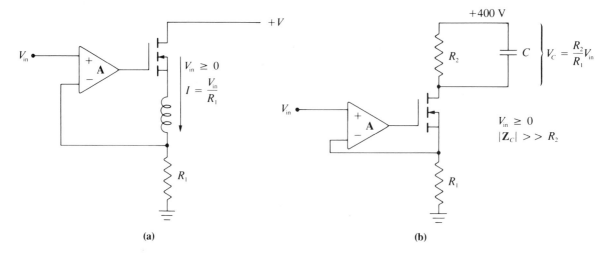

designs can be made to operate from DC to very high frequencies; the main limitation is their unipolar character.

419

10.6 Signal Conditioning Circuits

10.6.1 Debouncing the Mechanical Switch

The simplest form of data entry into a digital system is certainly the mechanical pushbutton or toggle switch, but even these devices often produce signals that are poorly matched to the requirements of the digital system. Based on everyday observation, the mechanical switch would seem to be a simple on/off device ideally suited to the digital application. However, a problem arises from the great disparity between the human reaction time scale (10^{-1} s) and the digital electronic time scale (10^{-8} s). What appears instantaneous to us is really very slow on the electronic time scale, and an unnoticed mechanical contact bounce of a few milliseconds will be seen as several distinct switch closures by a digital system.

Depending on the particular application, mechanical switches may need debouncing circuits to clean up their signal. Some pushbuttons have a type of sliding contact that exhibits very little contact "bounce"; switches of this type can be debounced by an *RC* circuit like that shown in Figure 10.8a. When using this method, it is best to drive a logic gate that has a Schmitt trigger input. On the 74LS14 inverter shown, the falling threshold is 0.6 V and the rising threshold is 1.8 V.

Mechanical switches that have springy closures or snap actions, such as microswitches, present a greater problem, since they may bounce for tens of milliseconds before settling into a new position. The best debouncing circuit for switches of this type is a flip-flop latch of the type shown in Figure 10.8c. Designs of this type require that each switch be a double-throw type. This particular design also requires a break-before-make action, meaning that during the throw there is a time when the common is connected to neither of the other terminals.

10.6.2 Op Amps for Gain, Offset, and Function Modification

The operational amplifier is an important component in most input transducer applications. In addition to simply increasing the amplitude of a signal, filtering it, and decreasing its associated output impedance, an op amp circuit can provide a variable gain control and an offset control. These are particularly useful when it is necessary to calibrate a transducer's output signal. A good example is the room-temperature signal from the circuit of Figure 10.9. Although the basic temperature-sensing device (LM335) is a particularly convenient transducer, being linear with absolute temperature, there are still component variations, both in the transducer and in the other elements of the circuit. Consequently, there are two potentiometers in this circuit: one to control the gain, allowing it to be set to exactly 10 mV/°C; and the other controlling the offset, allowing the output to be set to 0 V at some specific reference temperature (about 25°C in this case).

In some cases the dynamic range of the signal from the input transducer may be too large to process through the system. Often the limiting factor is the analog-to-digital converter (ADC), which must convert the analog signal to a number between 0 and $2^n - 1$, where n is the number of bits in the ADC. If a linear scale is maintained, this limitation will force a choice between overflowing the

ADC for large signals or reducing the overall gain and losing the ability to distinguish between slightly different signals. In some situations the problem can be solved by using a nonlinear amplifier, generally one with a logarithmic gain function such that

$$V_{out} = \log(V_{in}) \tag{10.3}$$

Although such amplifiers can in principle be constructed by placing a diode or bipolar transistor in the feedback loop of an op amp and biasing the *PN* junction to operate in the exponential region of its *IV* curve, this simple design produces a

Figure 10.8 *(a) A sliding-contact pushbutton can often be debounced with an RC delay and a Schmitt trigger logic input as shown by (b) the signals at various points in the circuit.*

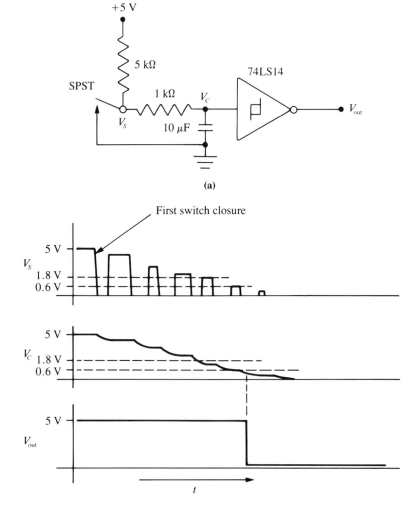

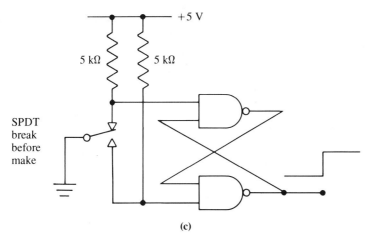

(c)

*Figure 10.8   continued. (c) A double-throw switch can be debounced
using an $\overline{RS}$ flip-flop.*

*Figure 10.9   An operational amplifier adds gain and offset controls to
the temperature-sensitive LM335 device and uses the temperature-
compensated LM329 as a temperature-stable reference. (Adapted from
National Semiconductor Corporation publications.)*

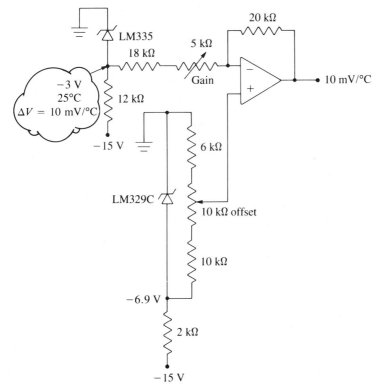

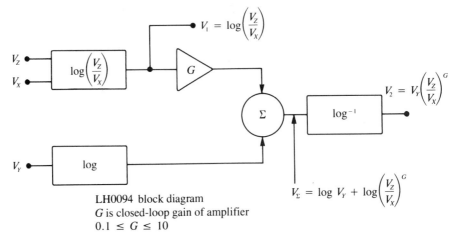

LH0094 block diagram
G is closed-loop gain of amplifier
$0.1 \leq G \leq 10$

*Figure 10.10   The block diagram of the LH0094 general-purpose
function-modification chip. This device accepts only positive input
voltages. (Adapted from National Semiconductor Corporation
publications.)*

highly temperature-dependent circuit. If such an amplifier is needed, it is best to
use an IC designed for the purpose. The multifunction LH0094 device, whose
block diagram is shown in Figure 10.10, provides the log of a voltage ratio at one
of its output pins. To implement Eq. (10.3), one would apply V_{in} to the V_Z input
and a convenient fixed voltage to the V_X input; the output would then appear as
V_1. (Note that the V_1 output from this chip will also vary with temperature but
can be compensated with a thermistor as specified in the *National Semiconductor
Linear Databook*.)

10.6.3   Bridge Circuits

A transducer that operates by producing a small variation in resistance, capacitance, or inductance is usually best measured in a bridge circuit. The DC bridge
of Chapter 1 can be converted to an AC bridge capable of measuring reactances
by replacing the driving DC EMF with an AC EMF operating at a convenient
frequency. Many component arrangements are possible, but the output signal from
an AC bridge will be a sinusoid, and when the bridge is balanced, the sinusoid
will have zero amplitude.

If physical conditions permit, the bridge can be made active by putting one of
the elements in the feedback loop of an operational amplifier as shown in Figure
10.11. This arrangement has the advantage of producing an output variation that
is linear in the variation parameter x. If we define $\mathbf{Z} = 1/j\omega C$ and $\mathbf{Z}_1 = 1/j\omega C(1 + x)$, then the voltage at the positive op amp terminal is

$$\mathbf{v}_+ = \frac{R\mathbf{v}}{R + \mathbf{Z}} \tag{10.4}$$

and the current equation for the feedback loop is

$$\frac{\mathbf{v} - \mathbf{v}_+}{\mathbf{Z}_1} = \frac{\mathbf{v}_+ - \mathbf{v}_{out}}{R} \tag{10.5}$$

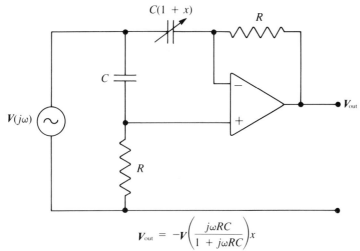

$$V_{out} = -V\left(\frac{j\omega RC}{1 + j\omega RC}\right)x$$

Figure 10.11 *A modified bridge circuit, excited by an AC EMF, can detect small changes in $C(1 + x)$.*

Combining these gives

$$\mathbf{v}_{out} = -\frac{R\mathbf{v}}{\mathbf{Z}_1}\left(\frac{\mathbf{Z} - \mathbf{Z}_1}{R + \mathbf{Z}}\right) \tag{10.6}$$

and after expanding $\mathbf{Z}$ and $\mathbf{Z}_1$, this reduces to

$$\mathbf{v}_{out} = -\mathbf{v}\left(\frac{j\omega RC}{1 + j\omega RC}\right)x \tag{10.7}$$

Note that the term in parentheses is a constant, and the amplitude of the output signal depends only on x and the amplitude of the activating signal.

Transducers such as strain gauges can be obtained prepackaged in the standard bridge arrangement. These devices cannot be incorporated into an active bridge but can be used to drive an instrumentation amplifier as shown in Figure 10.12. For algebraic simplification, a split supply is used to activate the bridge in this example; a single-sided supply is more commonly used and produces only a corresponding offset of the V_{out} signal. For the example shown, the voltage at the positive input to the instrumentation amplifier is given by

$$V_+ = V - R\left[\frac{2V}{R + R(1 + x)}\right] \tag{10.8}$$

which reduces to

$$V_+ = V\left(\frac{x}{2 + x}\right) \tag{10.9}$$

This equation shows a common problem with bridge circuits: Except for very small changes in the component variation parameter x, the output voltage is not proportional to x. Some commercial devices follow the instrumentation amplifier

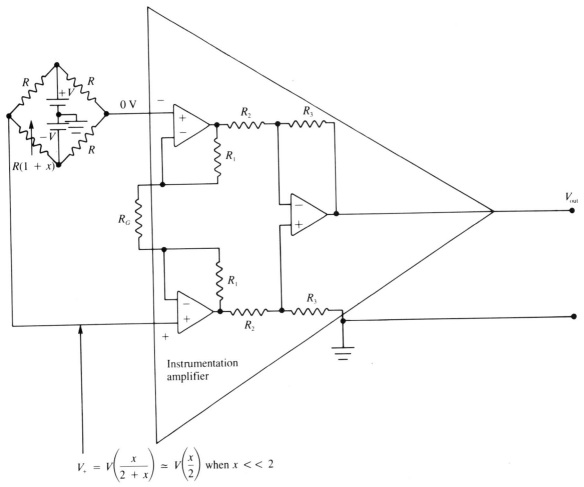

$$V_+ = V\left(\frac{x}{2+x}\right) \simeq V\left(\frac{x}{2}\right) \text{ when } x << 2$$

Figure 10.12 *A high-input-impedance instrumentation amplifier provides the ideal interface to a standard bridge circuit.*

with a nonlinear correction circuit, but if the resulting voltage is going to be digitized and read by a computer, the linearization step is most conveniently performed digitally by the computer.

10.6.4 Analog Switches

The block diagram of Figure 10.1 shows digital control lines going out from the computer to the signal conditioning circuits. Each line of this type can be used to set the position of one switch in the environment outside the computer, usually in some part of the signal conditioning circuits. The switch can be in the form of a mechanical relay or an integrated circuit analog switch. Small mechanical relays that can be driven directly from TTL logic are available in dual inline (DIP) IC packages and are very convenient for some applications. However, mechanical devices are the weak link in most solid state electronic circuits, and for applications requiring either speed, precision timing, or long-term reliability, an analog switch is the preferred device.

For high-power switch applications, such as driving stepping motor coils as shown in Figure 10.5b, the power MOSFET will switch unipolar currents in excess of 10 A; and for signal switching applications, a variety of analog switch IC packages are available that can switch bipolar currents with very little signal degradation.

High-quality analog signal switching circuits can be used in combination with operational amplifiers to perform many useful functions. Figure 10.13a shows a four-switch package used to produce a variable gain amplifier. In this example, the closure of one of the four switches will select a gain of -2, -10, -20, or -100 as needed. If the digital inputs D_0 to D_3 are individually set by control lines from the computer, the computer can select an amplifier gain that best matches the amplitude of the input signal V_{in}. The amplitude of the output signal V_{out} can thus be kept in a reasonable range, even though the input amplitude varies over two decades.

Another switch application is the analog multiplexer shown in Figure 10.13b. By driving the two digital control lines A and A as a 2-bit number 00, 01, 10, or 11, the computer is able to close one of the switches and select one of the four analog input signals V_0 to V_3 as the output signal from the follower.

Although they are fast and reasonably easy to use, analog switches do have some inherent limitations, most notably nonzero "on" resistance and capacitive coupling of the digital switching signal into the analog signal path. The "on" resistance of various types of analog signal switches ranges from a few ohms to several hundred ohms but is reasonably constant for a given switch, varying slightly with the voltage across and the current through the switch. The "off" resistance of analog switches typically exceeds 10^{10} Ω. Any switched amplifier must be given time to settle following the switching transient; the cross-coupling of digital transients into the analog path simply enhances this effect, lengthening the settling time to the final output value. Commercial circuits often use analog switches in pairs, arranged so that the major switch effects cancel.

10.6.5 Sample-and-Hold Amplifiers

The sample-and-hold amplifier, which also makes use of an analog switch, is a critical part of many data acquisition systems. Its purpose is to freeze an analog voltage at the instant a HOLD command is issued and make that analog voltage available for an extended period. Sample-and-hold amplifiers have a variety of uses and may have quite different hold times to suit the particular application. Since the analog voltage is being maintained by charge held on a capacitor, the leakage of charge into the readout amplifier and back through the analog switch will necessarily produce some loss of accuracy with time. For the analog-to-digital conversion application, the hold times will be on the order of 100 μs, just long enough for the analog-to-digital converter to do its job. If necessary, the digital result of such a conversion can be easily held for an extended period in a digital buffer register or a computer memory.

Figure 10.14 shows the relationship between the sample-and-hold (S/H) amplifier, analog-to-digital conversion (ADC), and various types of multiplexers

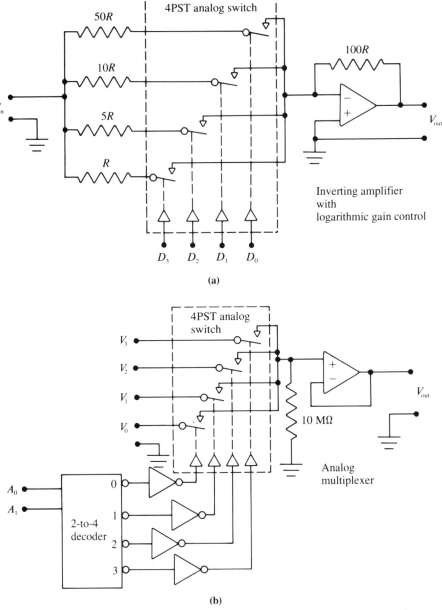

Figure 10.13   (a) An analog switch can be used to make a variable gain amplifier. (b) The analog switch can also be used to multiplex several analog signals onto the same output line.

(MUX). Control lines to the S/H and ADC devices are not shown. The figure shows various ways of converting three analog input signals to digital for acquisition by a single-digital *n*-bit bus. The first figure requires the most components, but since the ADC is the slowest device in the system, it also produces the highest data acquisition rate. A commonly used design is shown in Figure 10.14c, where

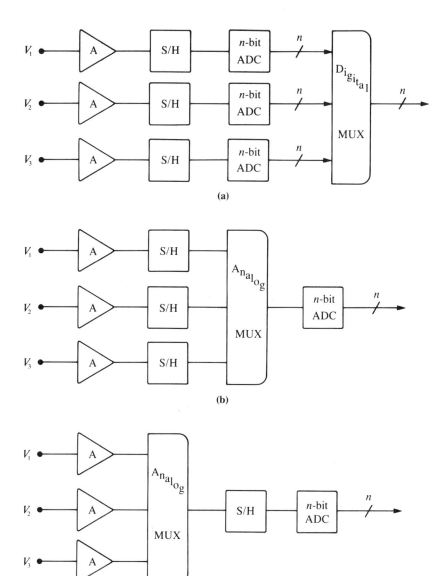

Figure 10.14 *Three schemes for multiplexing several analog signals
down to one digital input path. The notation S/H indicates sample and
hold, ADC means analog-to-digital converter, MUX means
multiplexer, and the $/^n$ symbol across a line indicates n digital signals.*

an analog multiplexer is used to switch various analog signals onto a single S/H
and ADC system. Figure 10.14b shows a compromise that may be useful in cases
where the S/H amplifier's delays are significant compared to those of the ADC.

A typical sample-and-hold amplifier is shown in Figure 10.15a. With the
switch in the position shown in the figure, $V_{out} = V_C = V_S = V_{in}$, neither diode
in the feedback loop of the first amplifier is conducting, no current flows in the

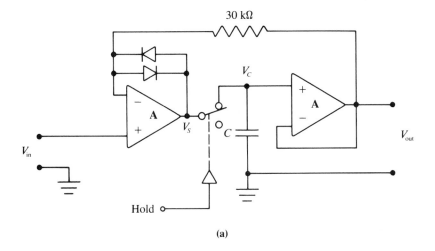

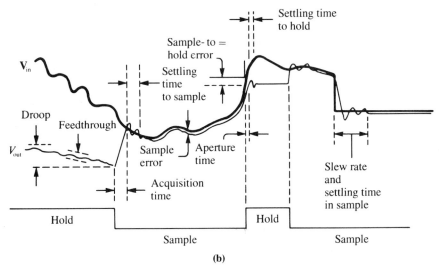

Figure 10.15 *(a) The sample-and-hold amplifier follows V_{in} while
HOLD is false and freezes the value when HOLD goes true. (b) Input
and output signals from a sample-and-hold amplifier showing the
multiple sources of error in the held signal. (Courtesy Analog Devices
Inc.)*

30-kΩ feedback resistor, and the overall amplifier is just a noninverting unity gain
follower. When the switch is opened by the digital HOLD signal, the charge on
the capacitor maintains V_C, which drives the high input impedance follower to
produce a constant V_{out}. The diodes around the first amplifier now come into play
and assure that V_S is always within a *PN* voltage drop of V_{in}, thus keeping the
first amplifier out of saturation. Sample-and-hold amplifiers are available as integrated circuits, with only the charge-holding capacitor C required as an external
component.

There are many sources of error in the sample-and-hold amplifier, and many of these are a function of the magnitude of the charge-holding capacitor. The timing diagram of Figure 10.15b, from Analog Devices, shows the four major errors (droop, feedthrough, sample offset error, and sample-to-hold error) and defines several important times (acquisition time, settling time to sample, aperture time, and settling time to hold). With the exception of "aperture time" these errors and times are obvious from the figure. This "aperture time" is the interval between the rising edge of HOLD and the time when a change in the analog input no longer affects the ultimate held voltage. Both the droop and the sample-to-hold error are reduced by using a larger holding capacitor, but a larger capacitor takes longer to charge and thus results in a longer acquisition time. Because of dynamic effects, the dielectric material in the hold capacitor is also critically important for sampling errors below 1%, with polystyrene, polypropylene, and Teflon being the preferred types.

A wide range of commercial sample-and-hold devices are available, both as integrated circuit devices and fabricated from discrete components. It is possible to make trade-offs between speed and accuracy, but typical times to 0.01% accuracy are 5–10 μs for acquisition, less than 1 μs for settling, and 0.1 μs or less for the aperture time. Droop rates vary from μV/μs to mV/μs, with higher-precision and slower devices having the smallest droop rates.

10.6.6 Gated Charge-to-Voltage Amplifier

The gated charge-to-voltage amplifier shown in Figure 10.16a performs a function that is similar to the sample-and-hold amplifier but is designed specifically as an integrating amplifier to measure the area under a narrow pulse. Because it measures the area directly, without multiple samples, this device is less subject to signal noise and sampling errors than the sample-and-hold amplifier of Figure 10.15. The operation of the gated charge amplifier is straightforward, but its use is complicated by the need to discharge capacitor C before a new sample can be taken. If the initial charge on the capacitor is zero, then the output voltage from the amplifier following a gate signal that extends from T_1 to T_2 is

$$V_{out} = \frac{1}{C} \int_{T_1}^{T_2} I(t)\, dt \qquad (10.10)$$

The signal timing associated with this amplifier is shown in Figure 10.16b.

As shown in the figure, this sampling amplifier is normally used with pulsed signals when the area under the signal is of primary interest. Unlike the example shown in the figure, the gate is normally set long enough to contain the full $I(t)$ pulse including any timing jitter that may exist between the gate and the signal. Under these conditions, the entire signal is integrated, removing much of the high-frequency noise, and making the output insensitive to the details of the signal shape. If the signal pulse rides on a relatively constant but nonzero offset voltage as shown, the effect of the offset can be determined by generating a gate when no signal pulse is present. The resulting output voltage is known as a pedestal and can be subtracted from the data signal at a later point in the system.

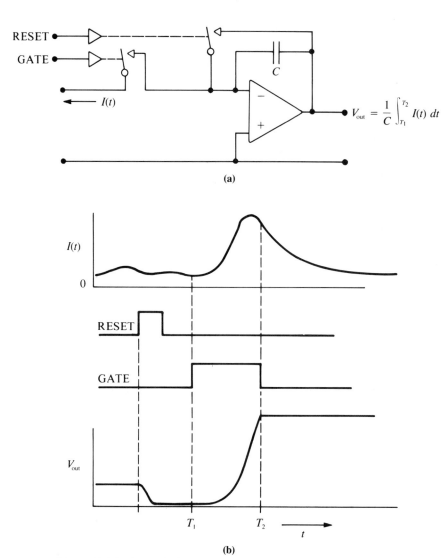

$$V_{out} = \frac{1}{C} \int_{T_1}^{T_2} I(t)\, dt$$

(a)

(b)

Figure 10.16 *A gated charge-to-voltage amplifier integrates a pulsed
signal during the time when GATE is true. These devices are widely
used in nuclear and high-energy physics applications to measure the
energy deposited in a particle detector.*

10.6.7 Comparators

The comparator is used to provide a digital output indicating which of two analog
input voltages is larger: It is a single-bit, analog-to-digital converter. Because the
comparator is a high-gain device much like an operational amplifier, a very small
voltage difference at the input terminals will produce a large change in the output
signal. However, unlike the operational amplifier, the comparator is used without
negative feedback, and consequently the input terminals of the device need not
always be at the same voltage. However, a difference of less than a millivolt at

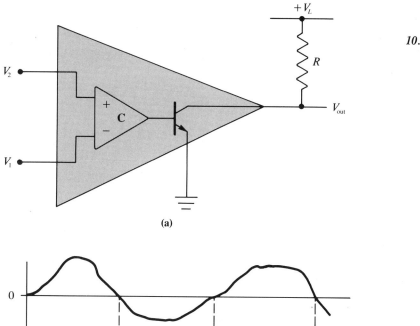

(a)

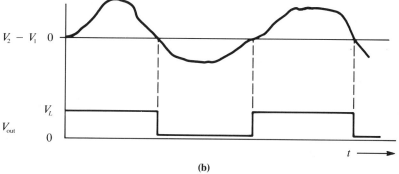

(b)

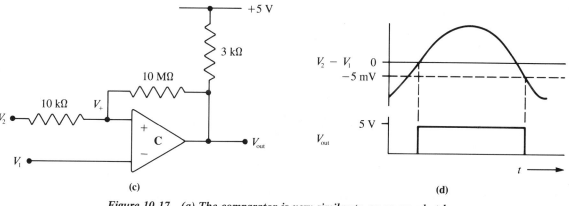

(c)

(d)

*Figure 10.17   (a) The comparator is very similar to an op amp but has
a digital true/false output. Several common designs use an open
collector output that must be pulled up to the logic voltage V_L. (b) The
input and output signals of the comparator. (c) A little positive
feedback adds hysteresis and (d) produces different turn-on and turn-
off thresholds.*

the input terminals is generally sufficient to drive the output to either its high or low limit, and a further increase in the input difference produces no effect other than to speed up the transition.

In addition to having an output that is designed to drive a digital (usually TTL) logic input, the comparator's internal circuits are designed to recover rapidly after the input terminals are driven widely apart in voltage. Since it is basically an amplifier, the op amp schematic symbol is also used for the comparator. To avoid confusion, the symbol C may be inserted inside the comparator symbol as shown in Figure 10.17. The digital output circuit may be a versatile open collector transistor as shown in Figure 10.17a, or a faster totem pole output typical of TTL logic devices.

The operation of an unadorned comparator is shown in Figure 10.17b. Note that the rising and falling transitions both occur as $V_2 - V_1$ passes through zero. If the analog input signal has any noise component (the switching transients generated by the comparator's change of state can themselves produce such noise), then the digital output of a fast comparator will likely show several up and down transitions at each crossover time. This problem can be reduced by introducing a small amount of positive feedback to produce some hysteresis on the input threshold as in the Schmitt trigger. In the example of Figure 10.17c, the 10-MΩ feedback resistor and the 10-kΩ input resistor combine to shift V_+ up by 5 mV whenever the output is high. This effectively shifts the negative going trigger threshold down by 5 mV as shown. The resulting device now has a 5-mV noise margin around each crossover and is more likely to produce a single digital transition.

10.7   Oscillators

Once a digital circuit has expanded beyond a handful of chips, the need invariably arises for some type of repetitive signal to serve as a timing reference for various logic or control functions. This need is served by a constant-frequency square wave oscillator, which can be produced in a variety of ways.

10.7.1   Relaxation Oscillator

The relaxation oscillator is easily implemented using a comparator as shown in Figure 10.18. The operation of this device is quite simple, as can be seen from the timing diagram. Since the comparator's output is either 0 or V_L, the reference voltage V_1 will be either $0.4V_L$ or $0.6V_L$ as a result of the resistor network. The voltage across the capacitor C also varies depending on the comparator's output voltage, but this voltage changes slowly as a result of C charging or discharging through R. Whenever V_2 reaches the reference voltage V_1, the comparator output changes state, the reference voltage switches to its other value, and the current through R changes direction.

The period of the relaxation oscillator can be easily calculated by making the assumption that the charging current through R is a constant given by

$$I = \frac{V_L - 0.5V_L}{R} = \frac{0.5V_L}{R} \tag{10.11}$$

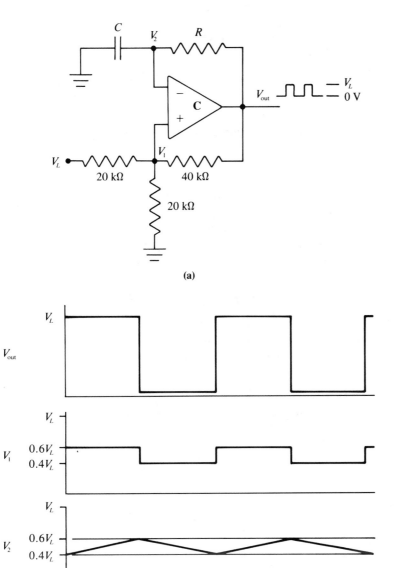

(a)

(b)

Figure 10.18   (a) A comparator-based relaxation oscillator and (b) its signals.

where the time-varying V_2 has been approximated by its mean value of $0.5V_L$. During a half-cycle, the charge added to C is

$$\Delta Q = \frac{IT}{2} = \frac{0.25V_L T}{R} \qquad (10.12)$$

and this charge changes the voltage on C by an amount

$$0.2V_L = \frac{\Delta Q}{C} \tag{10.13}$$

Combining these two equations to eliminate ΔQ and then solving for the period T gives

$$T = 0.8RC \tag{10.14}$$

whereas the exact integral solution using the time-varying current gives

$$T = 0.81RC \tag{10.15}$$

If an exact frequency is needed, a trimming potentiometer can be used to form part of the resistance R. On any type of relaxation oscillator (including the 555 timer below), the capacitor type should be chosen with care if frequency stability is important. In particular, the common ceramic capacitor should be avoided. Comparator-based oscillators are typically limited to frequencies below 1 MHz.

10.7.2 TTL Digital Logic Oscillators

If it is necessary to build a high-frequency oscillator for use in a digital circuit, it is most convenient to build the oscillator from digital logic elements. The *RC* design of Figure 10.19a is a type of relaxation oscillator that makes use of the

Figure 10.19 *(a) This 1-MHz relaxation oscillator uses TTL inverters. (Courtesy Signetics Corporation.) (b) The addition of a crystal yields a more precise frequency. (c) Another version of the crystal oscillator yields an 8-MHz output signal.*

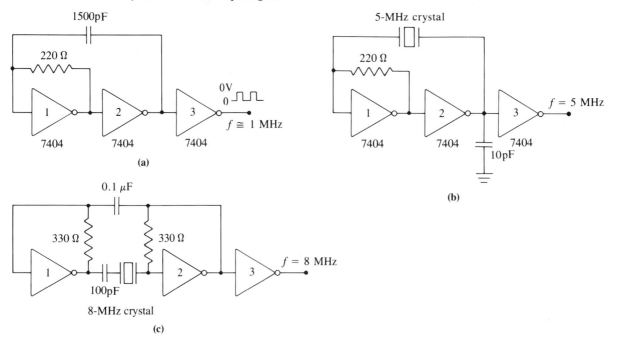

logic thresholds of the first inverter. If the input to the first inverter is initially low, then its output will be at the logic true voltage V_L. The second inverter will hold the right side of the capacitor low, but the voltage on its left side will be rising as a result of the current in the resistor. When the input to the first gate crosses the logic threshold V_T (somewhere between 0.8 V and 2.4 V), the output of the first gate goes low; this drives the second gate high, pulling the first gate's input voltage to $V_T + V_L$. The capacitor now discharges through the resistor until the threshold is crossed again, causing everything to switch and the input voltage to change to $V_T - V_L$. If we assume that $V_L = 5$ V and $V_T = 1.5$ V, then the period of the oscillation can be shown to be about

$$T = 2.4RC \tag{10.16}$$

but this result must be taken as a rough approximation because of the variation in logic thresholds.

The *RC* oscillator just described can be expected to have temperature variations that are on the order of 0.1%/°C. When greater frequency stability is required, a similar design can be used to produce the crystal-controlled oscillators of Figures 10.19b and 10.19c. These designs make use of the piezoelectric effect in thin quartz crystal to produce an electrical oscillation at the natural mechanical frequency of the crystal. The crystal itself can be modeled by a series *LCR* circuit in parallel with a second capacitor, but because the natural frequency depends on the physical dimensions of the crystal, these designs are frequency-stable to better than 0.5 ppm/°C. In both designs the small capacitor may be adjusted as needed to improve the oscillator's operation.

10.7.3   The 555 Timer

At frequencies from 0.0005 Hz to 500 kHz, the inexpensive 555 timer chip can be used for a variety of applications, the simplest of which is a square wave oscillator. As shown in Figure 10.20, the chip requires only three external components to provide a TTL-compatible output square wave signal that is stable to 50 ppm/°C. Although the circuit has been the subject of hundreds of clever design applications, we will limit the present discussion to its use as an oscillator.

The figure shows an internal block diagram of the 555 timer as well as the external wiring needed for the oscillator function. The operation of the chip is much like a relaxation oscillator, but it uses two comparators for improved temperature stability. The three internal resistors *R* set the comparator thresholds at two-thirds and one-third of the power supply voltage, as indicated in the figure. The oscillator frequency is determined by two external resistors R_A and R_B and a capacitor *C*, with the junction of R_B and *C* connected to both comparators as shown. The comparators drive an *RS* flip-flop, which also has an overriding reset input.

Since the first half-cycle of operation following a reset is unique, it is best to start the operational description from that point. If the reset pin is held low for a time, the output and *Q* will be low, the transistor connected to the discharge output will be nonconducting, and the capacitor will ultimately charge up to V_{++}, making *S* true and *R* false. When the reset pin is taken high, the flip-flop will set

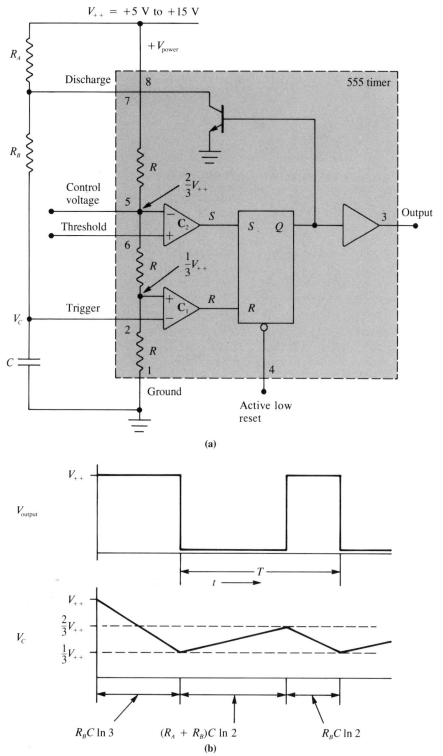

Figure 10.20 *(a) The 555 timer wired as an oscillator and (b) its
signals.*

because of the RS inputs. This turns the transistor on and pulls the discharge pin to ground, thus causing the capacitor to begin a discharge through R_B to ground. When V_C passes the threshold of C_2, S goes false; but this has no effect on the flip flop. Only when V_C drops below the threshold of C_1 and R becomes true does the flip-flop reset, turning off the transistor, and allowing C to recharge. But the capacitor does not charge all the way to V_{++} on these later half-cycles: When V_C passes C_1, R goes false, and when V_C passes C_2, S goes true, setting the flip-flop and ending the charging half-cycle.

The times for all three of these charging intervals can be calculated from the same equation. We start with the differential equation that describes the charge accumulation on the capacitor,

$$dV_C = -I\frac{dt}{C} \tag{10.17}$$

which can also be written as

$$dV_C = \frac{V_C - V}{R_C C}\,dt \tag{10.18}$$

where R_C is the resistor between V_C and some fixed voltage V that will be either V_{++} or zero in the present application. After solving for dt, this can be put into the integral form

$$\int_0^{T_n} dt = R_C C \int_{V_1}^{V_2} \frac{dV_C}{V_C - V} \tag{10.19}$$

where T_n is the time it takes to charge the capacitor from V_1 to V_2. This expression integrates to

$$T_n = R_C C \ln\left(\frac{V_1 - V}{V_2 - V}\right) \tag{10.20}$$

and can be evaluated for different choices of R_C, V, V_1, and V_2.

During the anomalous first discharge of capacitor C, we have $R_C = R_B$, $V = 0$, $V_1 = V_{++}$, and $V_2 = V_{++}/3$, producing a time

$$T_1 = R_B C \ln 3 \tag{10.21}$$

For a charging half-cycle we have $R_C = R_A + R_B$, $V = V_{++}$, $V_1 = V_{++}/3$, and $V_2 = 2V_{++}/3$, and this gives a charging time of

$$T_2 = (R_A + R_B)C \ln 2 \tag{10.22}$$

This is the time during which the 555 output signal is high. Similarly, the typical discharge half-cycle time is

$$T_3 = R_B C \ln 2 \tag{10.23}$$

matching the time when the output signal is low. Combining T_2 and T_3, we find that the oscillation period is

$$T = (R_A + 2R_B)C \ln 2 \tag{10.24}$$

giving a frequency of

$$f = \frac{1.44}{(R_A + 2R_B)C} \tag{10.25}$$

Note that this oscillator spends more time in the high state than in the low state. The ratio of the high to low times is known as the duty cycle and is obviously given by $(R_A + R_B)/R_B$. Since R_A must be nonzero, it is not possible to obtain a 50% duty cycle without the addition of a diode in parallel with R_B.

10.7.4 *Application to Interval Timers*

With increasing system complexity, the need may arise for several repetitive timing signals with periods ranging from a few microseconds to many hours. If each timing signal is obtained from a separate oscillator, the signals will have a random and variable phase relationship: They will be asynchronous. This is an error-prone design technique, since it may lead to glitches that occur only occasionally, at times related to some beat frequency between two or more oscillators. A much better technique is to use one high-frequency oscillator with a short period T and from it derive all longer-period signals. If the longer periods can be of the form $2^n T$, the easy solution is a counter with n flip-flops. Each flip-flop will yield an output that has twice the period of its driving signal.

Figure 10.21 An interval timer fabricated from digital components is based on a clock of known frequency.

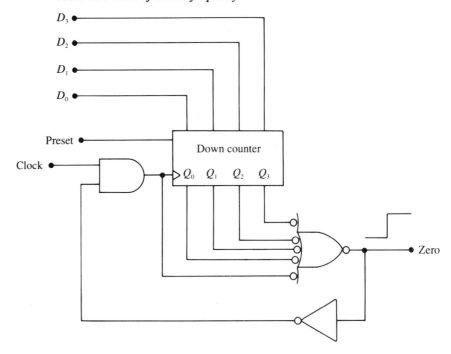

For periods that are of the form NT, where N is an integer other than 2^n, the solution is a divide-by-N counter. For small values of N, this may be a specific circuit designed for just this function; but as N increases, a more viable solution is to use a presetable down-counter. A 4-bit down-counter used as an interval timer is shown in Figure 10.21. In order to produce a square wave output, this circuit would have to be augmented by a register to hold the initial count, a toggling flip-flop on the output, and some control logic to reload the down-counter each time it counts to zero.

Integrated circuit interval counters of this type are available in each of the popular microcomputer lines. There are generally several counters in each package, and they offer a variety of signal output options, including variable-width pulses and square waves. An additional feature useful in the computer environment allows the computer not only to preset the counter but to read its numerical value at any instant.

10.8   Digital-to-Analog Conversion

The process of converting a number held in a digital register to an analog voltage or current is accomplished with a digital-to-analog converter or DAC. An early and still important application of the DAC is as an interface device between a digital computer and an analog output transduce. However, the availability of inexpensive integrated circuit DACs makes this device increasingly popular as a control element in self-contained and mostly analog circuits such as amplifiers and filters.

10.8.1   Current Summing and IC Devices

Modern DACs are almost exclusive switched current devices designed to drive the current-summing junction of an operational amplifier. The DAC itself is often nothing more than a precision resistor network and a number of analog switches. In fact, we have already used the principle of DAC circuit in the variable-gain amplifier of Figure 10.13a. To use this circuit as a kind of DAC we need only hold the input voltage V_{in} constant, thereby causing the analog output of the op amp to vary only with the digital input D_0–D_3.

The 4-bit circuit of Figure 10.22a is only a slightly modified version of the variable-gain amplifier. The most significant change is to the resistor sequence, which is now proportional to 2^1, 2^2, 2^3, 2^4, and so on. The current through each resistor is thus one-half of the previous one, making each current proportional to the value of a bit position in a binary number. The analog switches in this figure are driven by digital input signals as in Figure 10.13, but for clarity these inputs are not shown here. With the switches in the position shown, the current through each resistor is added on the current-summing line (bus), which feeds the negative input to the op amp.

The significance of the 2^n resistor set is easily seen. For discussion purposes we will assume a 4-bit DAC driven by the digital input number $ABCD$, where each letter is a Boolean variable that can take on the values 0 and 1 corresponding to whether the switch is connected to ground or to the current-summing bus. The bus current I is then given by

$$I = (A/2 + B/4 + C/8 + D/16)\frac{V_{ref}}{R} \tag{10.26}$$

and factoring 1/16 out of the sum gives

$$I = (8A + 4B + 2C + D)\frac{V_{ref}}{16R} \tag{10.27}$$

In this form it is clear that the bus current I is proportional to a number for which

Figure 10.22 (a) The current DAC uses the summing input to an op amp to yield a voltage output. LSB and MSB refer to the least and most significant bits of a binary number. (b) This resistor arrangement produces an 8-bit DAC with only six different resistor values.

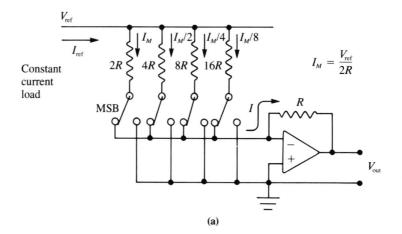

(a)

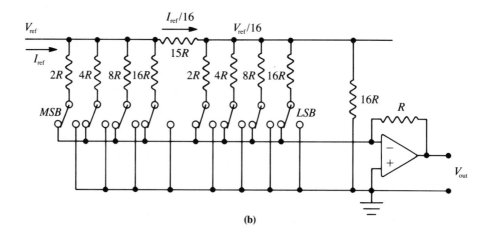

(b)

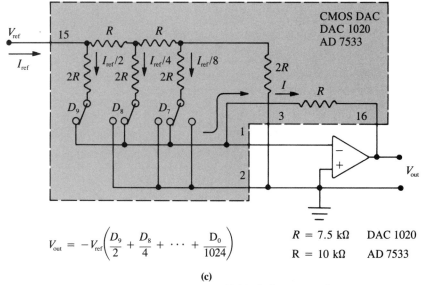

$$V_{out} = -V_{ref}\left(\frac{D_9}{2} + \frac{D_8}{4} + \cdots + \frac{D_0}{1024}\right)$$

$R = 7.5 \text{ k}\Omega$ DAC 1020

$R = 10 \text{ k}\Omega$ AD 7533

(c)

*Figure 10.22 continued. (c) This common 10 bit design uses only two
different resistor values. Only three of the ten switches are shown.*

ABCD is the binary representation. The op amp produces an output voltage that
is obtained by multiplying Eq. (10.27) by $-R$.

The design of Figure 10.22a requires a number of different resistors, each at a
precisely defined value. Because of manufacturing difficulties, commercial products replace this simple circuit with one that requires fewer distinct values. One
possibility is shown in Figure 10.22b. This design is in the form of a sequence of
4-bit hexadecimal digits; each digit is like our original design, but the reference
current into each successive digit is reduced a factor of 16 by a coupling resistor.
This circuit is not often used for hexadecimal (binary) applications but can be
modified for decimal digits by simply changing the coupling resistor so that the
reference current to each digit is reduced by a factor of 10 instead of 16.

One of the more common resistor network designs is shown in Figure 10.22c.
This particular DAC uses CMOS (complementary MOSFET) analog switches,
which can handle currents of either polarity, and is known as a multiplying DAC,
since V_{ref} may be a time-varying signal taking on both positive and negative values. Similar designs use less flexible but more accurate bipolar transistor switches,
which require a monopolar, usually fixed-voltage V_{ref}.

The resistor configuration shown in this figure can be thought of as an extreme
case of the previous design: Each section is now only a single-bit binary digit,
and the reference current to each successive digit is reduced a factor of 2 by the
coupling resistor. The major advantage of this design is that only two resistor
values are needed, independent of the number of binary digits in the complete
DAC. As shown in this figure, a typical integrated circuit DAC contains only a
precision resistor network and analog switches; the op amp is usually an external
component, although its feedback resistor is part of the DAC package.

Both bipolar and CMOS current DACs have very fast switching times (around 1 μs), and the rate of change of the output voltage is usually limited by the settling time of the external operational amplifier. Fast response requires high-performance op amps, and this usually leads to the addition of compensation capacitors, which are not shown in these figures.

An important general feature of all of the circuits shown in Figure 10.22 is that the current I_{ref} out of the voltage reference is a constant independent of the switch positions. This is because the negative terminal of the op amp is a virtual ground, causing the current through each DAC resistor to be constant independent of switch position. A constant I_{ref} improves V_{ref} stability, reduces switching transients, and generates a constant amount of internal heating. The last feature helps to stabilize the resistor ratios by keeping them in thermal equilibrium.

10.8.2 DAC Limitations

Although the output signal from a DAC is analog, it can assume only a discrete set of steady state values. If the DAC is perfect, then the relationship between the input binary number and the analog output is given by the solid dots in Figure 10.23a. Note that the output signal is defined only for these discrete points on the digital input axis. The associated stair-step function is often used to show the DAC's output, as it correctly shows a DAC's analog output when the digital input is driven by a steadily increasing binary number. However, in that case the x axis is really the continuous variable t, and the vertical steps are transients caused by switch and amplifier transition and settling times. This stair-step signal is commonly seen on an oscilloscope display generated to check the behavior of a DAC.

For a perfect set of resistors and ideal analog switches, each of the vertical steps in Figure 10.23a will be the same height, and the dots will lie on a straight line passing through the origin. A variety of problems can occur with the imperfect set of resistors found in real DACs, but the most common shows up as an anomalous step size between adjacent binary numbers such as 01111111 and 10000000. The difference between these two numbers is just 00000001, the least significant bit (LSB), and the current out of the DAC should correspondingly change by its smallest increment. However, the first current is determined by a path through seven resistors and switches, whereas the second uses only a single, entirely different resistor and switch: This is where the resistor match and switch imperfections are most critical.

Figures 10.23b,c show examples of anomalous steps between 011 and 100 that are either too large or too small. Bipolar switches generally have smaller variations in their "on" resistances than CMOS switches and thus are closer to the ideal, at least for monopolar currents.

10.8.3 DAC Applications

One application of the multiplying current DAC is to permit digital control of an analog device such as an amplifier. An example of this application to a variable-gain amplifier is shown in Figure 10.24. The resistor network and analog switches are modeled in this schematic by the variable resistors R_1, R_2, and R_3. The resis-

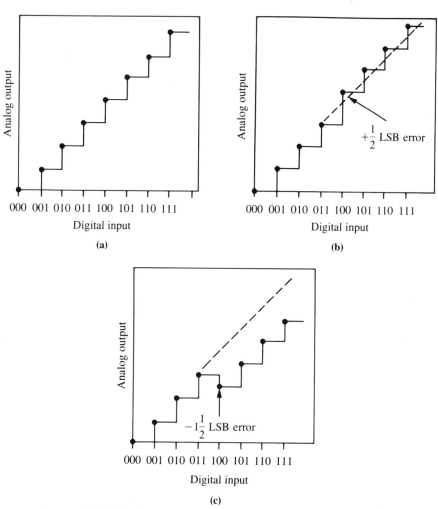

Figure 10.23 *Output signals from DACs showing (a) the ideal result,
and (b) a differential nonlinearity or (c) nonmonotonic behavior, both
caused by imperfectly matched resistors.*

tance R_1 between pins 1 and 15 is in the feedback path of this inverting op amp
design, whereas R_2 and R_3 have no significant effect: There is no current through
R_3, and R_2 is simply a variable-load resistance on the output of the op amp. By
adjusting the feedback resistance digitally, the gain of this amplifier can be
changed.

The analysis of this circuit is simplified by observing that the input impedance
looking into pin 15 is always R, making the input current (I_{ref} in Figure 10.22c)
just V_{out}/R. Applying the switched currents indicated in Figure 10.22c to the design of Figure 10.24, the current sum at pin 1 can be written

$$\frac{V_{in}}{R} + \left(\frac{V_{out}}{R}\right)\left(\frac{D_9}{2} + \frac{D_8}{4} + \cdots + \frac{D_0}{1024}\right) = 0 \qquad (10.28)$$

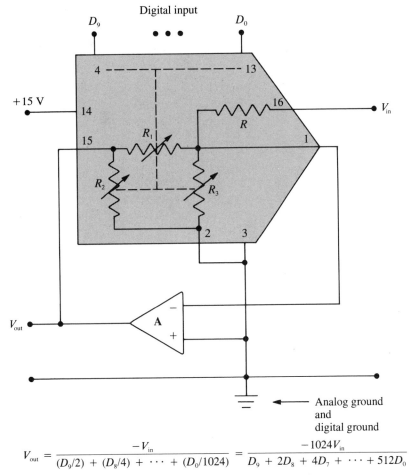

$$V_{out} = \frac{-V_{in}}{(D_9/2) + (D_8/4) + \cdots + (D_0/1024)} = \frac{-1024V_{in}}{D_9 + 2D_8 + 4D_7 + \cdots + 512D_0}$$

DAC 1020 is a multiplying DAC

*Figure 10.24 A variable-gain amplifier has its gain set by the digital
input to the current DAC.*

where each of the digital signals D_0 to D_9 is either 0 or 1. Rearranging terms gives

$$V_{out} = -\frac{V_{in}}{(D_9/2) + (D_8/4) + \cdots + (D_0/1024)} \tag{10.29}$$

or

$$V_{out} = -\frac{1024V_{in}}{D_0 + 2D_1 + 4D_2 + \cdots + 512\,D_9} \tag{10.30}$$

It can be seen that this amplifier can be digitally programmed to yield gains form
$-1024/1023$ to -1024. Although this design is similar to that of Figure 10.13a,
in this case we need only two IC devices and no other components.

A simple graphics output device for a computer is shown in Figure 10.25. In
this design, a computer word placed in the output buffer register by a LOAD pulse

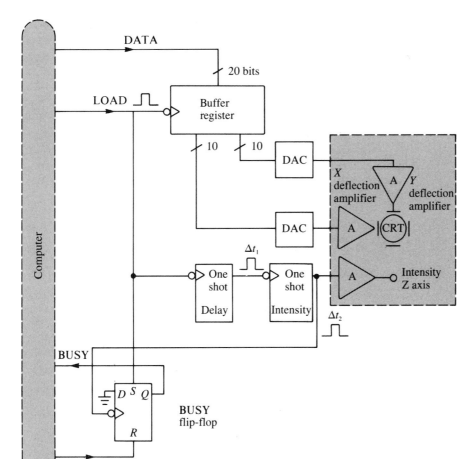

Figure 10.25   Two DACs are a key part of the interface between a digital computer and an XY oscilloscope display.

drives the phosphorescent dot on an XY oscilloscope to any of over a million dots on the CRT face. To produce a useful display, the CRT's electron beam is held off except during the short time Δt_2 (typically 1 μs) of the Z axis pulse. This produces a screen image composed of a number of dots, one for each output word. The Z axis pulse is generated by the output of a one-shot that is triggered an interval Δt_1 after the buffer register is loaded; this Δt_1 delay allows the DACs and the CRT deflection amplifiers time to settle to a final value before the burst of electrons is sent down the tube. The BUSY flip-flop generates a handshake signal back to the computer, causing it to wait for the Z axis pulse before sending the next output word.

The major drawback of this simple display mechanism is the time needed to draw a complete picture or frame. Since the human eye is a light-integrating device, an image will be visible only if it can be held on the CRT screen for an extended period. This can be accomplished using a type of CRT known as a storage tube or by repeating the output sequence at a time interval that is short compared to either the phosphor decay time or the persistence of the human retina. If the phosphor decay time is short, as on a laboratory oscilloscope or a normal television screen, then we are dealing with human physiology, and to avoid excessive flicker the display must be repeated at least 30 times each second. Even if the computer is able to output successive points very quickly, the DAC settling time Δt_1 is still likely to be around 5 μs per point, thus limiting an image to about 6000 points without flicker. By contrast, a computer terminal screen full of text typically excites 30,000 points on the phosphor. The increase in speed is achieved by sequentially accessing every point on the screen, rather than randomly accessing only a few.

10.9   Analog-to-Digital Conversion

The analog-to-digital converter or ADC is used to convert an analog voltage to a digital number. In addition to being a key element in the computerized data acquisition scheme of Figure 10.1, circuits that appear to be analog at their input and output terminals may use internal DACs and ADCs to combine the best features of both the analog and digital domains.

A generalized circuit of the latter type is shown in Figure 10.26. On this circuit the analog signal is first digitized into an n-bit data word, then, depending on the details of the digital circuit, this digital word can be used in a number of different ways: It can be transmitted as digital information over noisy lines without loss of information; it can be stored in a register to produce a long-term sample-and-hold circuit; it can be shifted through n parallel shift registers to delay the output signal; or it can be used to form the address to a read-only memory device that can then return a new n-bit word bearing a preprogrammed functional relationship to the original. After being digitally processed, the number can be converted back to analog with a DAC.

Figure 10.26   A generalized hybrid analog and digital circuit by which input analog data can be transmitted, stored, delayed, or otherwise processed as a digital number before reconversion back to an analog output.

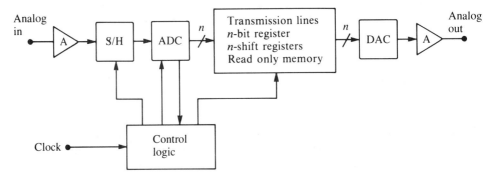

By incorporating the discrete signal techniques of Chapter 14, a device of this general type can also function as a frequency filter. Such digital filters are particularly useful at frequencies below 1 Hz, where a completely analog circuit would require excessively large capacitor values.

10.9.1 Parallel-Encoding ADC

The parallel-encoding or flash ADC provides the fastest operation at the expense of high component count and high cost. As shown in Figure 10.27, this design uses a resistor network to set discrete thresholds for a number of comparators, which all connect directly to the analog input signal. As the analog input voltage varies, all comparators with thresholds below the input will be true and all those

Figure 10.27 A 3-bit, parallel-encoding or flash ADC.

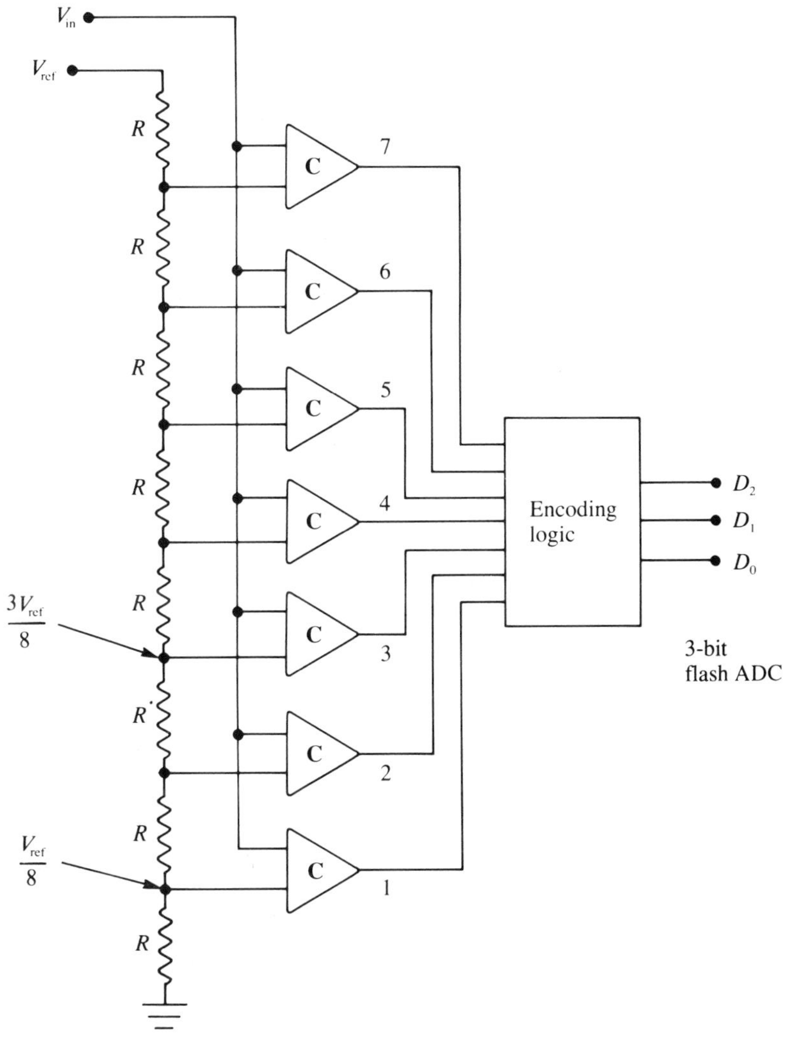

with thresholds above the input will be false. Digital encoding logic is then used to convert this result to a digital number. ADCs of this type can encode signals in less than 100 ns but are not yet available as inexpensive integrated circuits.

10.9.2 Successive-Approximation ADC

The block diagram of Figure 10.28a shows an ADC method that needs only a single comparator. In this diagram, the analog output of a high-speed DAC is compared against the analog input signal. The digital result of this comparison is used to control the contents of a digital buffer that both drives the DAC and provides the digital output word. Independent of the details of the digital control logic, it is clear that whenever the comparator output changes state, the digital buffer must be within one count of the correct digital result. The simplest control logic would clear the buffer register, then count it up until the comparator output switched from false to true, but this method could take a long time: up to $2^n - 1$ comparisons for an n-bit binary result.

The successive-approximation ADC uses a more complicated, but much faster, control logic, which requires only n comparisons for an n-bit binary result. As indicated in Figure 10.28b, the logic first clears all of the register bits, then sets the most significant bit (MSB) and waits for the comparator response. If the comparator is true, the digital number 100 is too small and the MSB is retained; if false, 100 is already too big and the MSB is reset. In either case the testing is then repeated for the next bit in sequence. As can be seen by the assumed path in the figure, only one test is required for each bit.

The successive-approximation ADC is the most commonly used type and is suitable for a wide range of applications. However, when this type of ADC is being used for data acquisition, it is important to remember that the ADC result is only as good as the DAC, which determines the reference voltage for the comparator. In particular, the anomalous DAC step sizes shown in Figures 10.22b and 10.22c between the digital numbers 011 and 100 will lead to an incorrect bias toward one of these two numbers on the ADC's output. If a randomly distributed analog signal is digitized by a successive-approximation ADC, a histogram of the digital numbers will have anomalous peaks and valleys whose sizes are determined by the nonuniformity of the DAC.

10.9.3 Dual-Slope ADC

The problems associated with the DAC can be avoided by using the analog method of charging a capacitor with a constant current; the time required to charge the capacitor from zero to the voltage of the input signal becomes the digital output. When charged by a constant current I_C, the voltage on a capacitor $V_C(t)$ is a linear function

$$V_C(t) = \left(\frac{I}{C}\right)t \tag{10.31}$$

and this characteristic can be used to connect the analog input voltage to the time as determined by a digital counter.

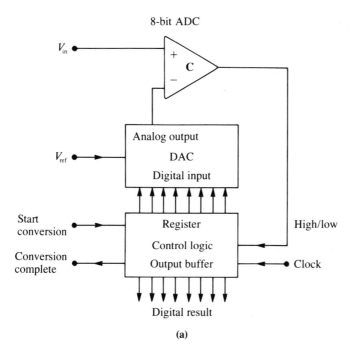

8-bit ADC

(a)

Successive Approximation Testing

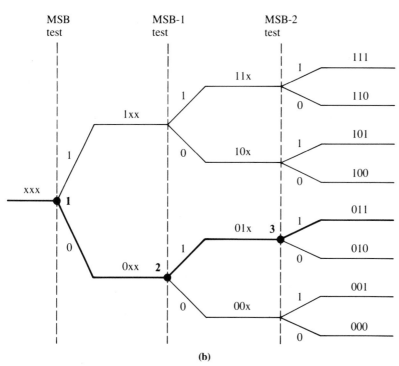

(b)

Figure 10.28 (a) The block diagram of an 8-bit successive-approximation ADC. (b) The bit-testing sequence used in the successive approximation method. The dark path shows the test sequence for an analog input that yields a digital output of 011.

To achieve higher accuracy and temperature stability, the somewhat more complicated dual-slope method is used on commercial devices. A simplified version of a dual-slope ADC and its associated timing diagram are shown in Figure 10.29. For this design we assume a negative analog input signal $-V_{in}$. The conversion is started by the leading edge of the GATE signal. This signal moves switch S_1 to the position shown, blocks the AND gate so there can be no COUNT signals, and triggers the one-shot. The one-shot sets the BUSY flip-flop, which in turn moves

Figure 10.29   (a) The circuit and (b) operation of a dual-slope ADC.

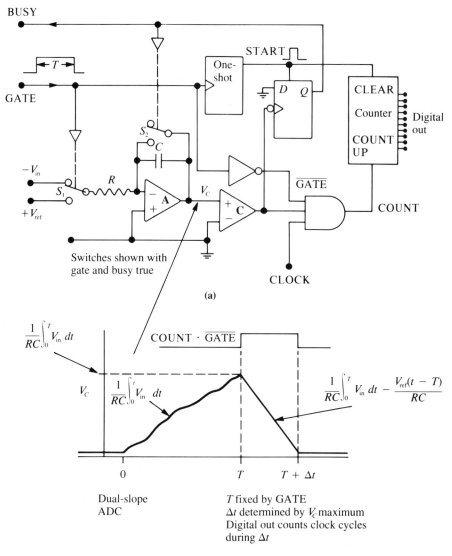

switch S_2 to the position shown, allowing the capacitor C to begin charging from zero and causing V_C to become increasingly positive. The positive V_C immediately causes the comparator output to become true.

The increasing voltage on the capacitor V_C is given by

$$V_C(t) = \frac{1}{RC} \int_0^t V_{in}\, dt \qquad (10.32)$$

which will not have a constant slope if $-V_{in}$ is changing with time. The GATE signal goes false after a time T, moving S_1 to the positive voltage $+V_{ref}$ and satisfying the AND gate, which now passes the CLOCK signals to the counter. At the end of the GATE time T, the voltage on the capacitor had become

$$V_C(t) = \frac{1}{RC} \int_0^T V_{in}\, dt \qquad (10.33)$$

so that after S_1 connects the positive reference, the voltage $V_C(t)$ is given by

$$V_C(t) = \frac{1}{RC} \int_0^T V_{in}\, dt - \frac{(t-T)V_{ref}}{RC} \qquad (10.34)$$

This constant-slope discharge continues until V_C becomes zero, at which time the comparator goes false, stopping the counter and resetting the BUSY flip-flop. Setting Eq. (10.34) to zero and solving for the time $(t - T)$ during which COUNT was driving the counter to its digital result, we get

$$t - T = \frac{1}{V_{ref}} \int_0^T V_{in}\, dt \qquad (10.35)$$

which is proportional to the integral of $V_{in}(t)$ over the gate time T. We see that a significant advantage of the dual-slope method is that the quantity RC has canceled, making this ADC insensitive to temperature variations in these components.

10.10   Time-to-Digital Conversion

It is possible to digitize relatively long time intervals by incrementing a counter with a repetitive signal derived from an oscillator; the operation is much like a stopwatch. However, with TTL logic elements, devices of this type are limited to clock speeds less than 30 MHz, giving a least time interval of about 30 ns. For short intervals requiring better resolution, a circuit such as that shown in Figure 10.30a can be used. Once again, this device operates by charging a capacitor with a constant current, but it differs from previous examples in that the resulting voltage V_C on the capacitor is proportional to the time between the START and STOP pulses. The circuit's operation is indicated by the timing diagram of Figure 10.30b. By using very high-speed amplifiers, special-purpose circuits of this type are able to measure time intervals with picosecond (10^{-12} s) resolution.

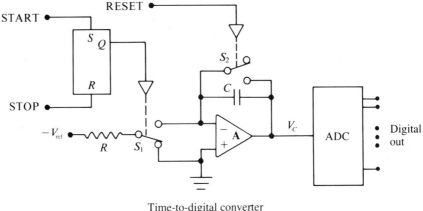

Time-to-digital converter

(a)

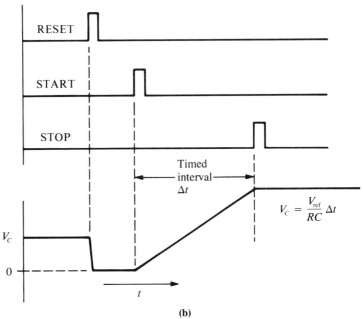

(b)

Figure 10.30   (a) A time-to-digital converter TDC and (b) its timing signals. The circuit is shown with the switches in the signal-holding position after a STOP pulse and with RESET false.

10.11   Voltage-to-Frequency Conversion

When a slowly changing analog signal must be transmitted over a long distance where it is susceptible to significant amplitude loss, it may first be desirable to convert from an amplitude-modulated (AM) signal to a frequency-modulated (FM) signal. There are several oscillator circuits, known as voltage-controlled oscillators (VCOs), whose output frequency can be changed by varying an input voltage, but the frequency range and linearity of such circuits is limited. However, the wideband integrated circuit device (LM331) shown in Figure 10.31 is capable of linear operation from 1 Hz to 100 kHz. The first part of this figure shows a block diagram of the internal workings of this device and indicates the external components needed to produce the basic VCO function.

Figure 10.31 (a) *The circuit and* (b) *operation of the LM331 voltage-to-frequency converter. The switch position is shown for* V_P *false.*

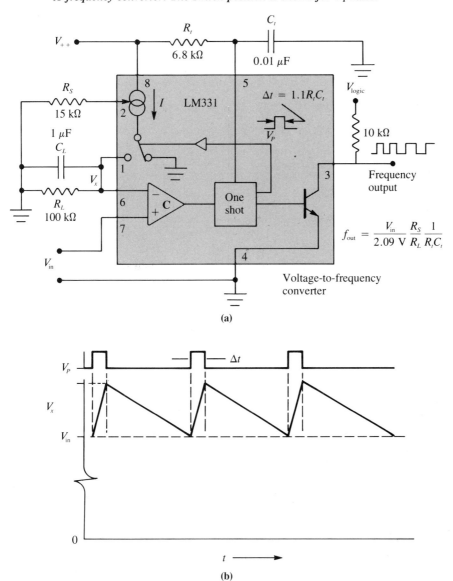

$$f_{out} = \frac{V_{in}}{2.09 \text{ V}} \frac{R_S}{R_L} \frac{1}{R_t C_t}$$

(a)

(b)

Once again, a constant current I (the magnitude of I is programmed by resistor R_S) is used to charge capacitor C_L through an analog switch. Since this charging process occurs only during the time when the one-shot is high, as shown in Figure 10.31b, we will begin the analysis by assuming that the one-shot has just fired, moving the switch to the left: The one-shot stays true for a time $\Delta t = 1.1R_tC_t$ set by additional external components, allowing C_L to be charged during this interval. At the end of this time the switch opens and C_L discharges back toward V_{in} as shown; when it reaches V_{in}, the comparator goes true, starting the cycle over again. This interaction between the comparator and the analog switch is such that V_x is never allowed to fall below V_{in}.

If we assume that the amplitude of the AC component of V_x is small compared with V_{in}, then the current in R_L will have an approximately constant value given by V_{in}/R_L. This means that during the time T of one complete cycle, the charge flow onto C_L is

$$Q_{on} = I \, \Delta t = I1.1R_tC_t \tag{10.36}$$

whereas the charge flow off C_L is

$$Q_{off} = \left(\frac{V_{in}}{R_L}\right) T \tag{10.37}$$

When in equilibrium, the net charge flow over one cycle must be zero, and we can equate these two expressions to give

$$f = \frac{1}{T} = \frac{V_{in}}{IR_L1.1R_tC_t} \tag{10.38}$$

which shows that the output frequency is proportional to the input voltage V_{in}. For the actual device this expression is replaced by

$$f = \frac{V_{in} R_S}{(2.09 \text{ V})R_L \, R_tC_t} \tag{10.39}$$

which includes the effect of the current programming resistor R_S.

Since frequency is easily converted to a binary number with the aid of a counter and a fixed time gate, the voltage-to-frequency converter represents another form of analog-to-digital converter. It is particularly useful in remote applications because the digital frequency requires only two wires for transmission. Temperature measurement over a large and electrically noisy volume is one example: A number of circuits like that shown in Figure 10.32a can be placed at various locations, with each connected to a central data acquisition system by only two wires.

In those applications where it is necessary to convert from a frequency-modulated signal back to an amplitude-modulated signal (perhaps for display on a meter), the same IC device can be used in the circuit of Figure 10.32b to form a frequency-to-voltage converter. This circuit operates from 1 Hz to 10 kHz with less than 0.1% nonlinearity.

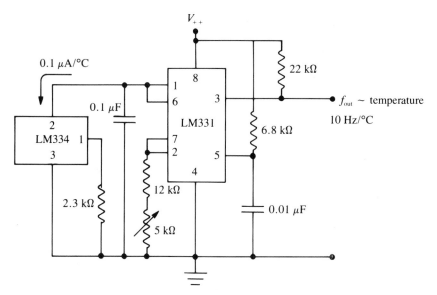

(a) Temperature-to-frequency converter

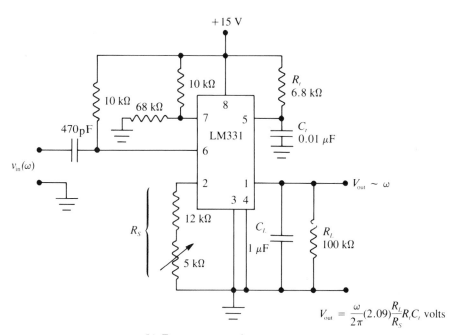

$$V_{out} = \frac{\omega}{2\pi}(2.09)\frac{R_L}{R_S}R_t C_t \text{ volts}$$

(b) Frequency-to-voltage converter

*Figure 10.32   (a) A temperature-to-frequency converter using the
LM331. (b) A broad-band frequency-to-voltage converter using the
same IC. (Adapted from National Semiconductor Corporation
publications.)*

10.12 Phase-Locked Loop

The phase-locked loop (PLL) is an analog device that uses a form of negative feedback to cause a variable-frequency oscillator to adjust itself to either the fundamental or some odd harmonic of an external AC signal. As shown in Figure 10.33a, the device consists of a signal multiplier called a phase detector, a low-pass filter, an amplifier, and a voltage-controlled oscillator (VCO) with a natural frequency ω_1, which is set by external components. When the PLL has settled into a condition where the VCO output v_0 has matched the desired frequency ω in the incoming signal, there exists a definite phase difference θ between the incoming signal v_{in} and the VCO output; this condition is known as phase lock, and the resulting PLL output signal V_{out} has a DC component that is proportional to $\cos \theta$, which is in turn proportional to $\omega - \omega_1$.

The complete description of the operation of a PLL requires a linear feedback analysis (using frequency instead of voltage) that is somewhat more general than we have used previously. Instead of pursuing that development, we will use an alternative description that is sufficient to describe the steady state operation of the PLL.

If the input signal is $V_1 \cos(\omega t)$ and the VCO signal is $V_0 \cos(\omega_0 t + \theta)$ as shown in the figure, then the multiplier in the phase detector produces an output V_P that is

$$V_P = V_1 V_0 \cos(\omega t) \cos(\omega_0 t + \theta) \qquad (10.40)$$

After expanding the cosine product, this becomes

$$V_P = V_1 V_0 [\cos(\omega t + \omega_0 t + \theta) + \cos(\omega t - \omega_0 t - \theta)] \qquad (10.41)$$

Note that V_P is not strictly an AC signal; for example, the product of two identical cosine signals gives $\cos^2$, which is always positive. The low-pass filter removes the higher-frequency first term, leaving an output signal of the form

$$V_{out} \simeq V_1 V_0 \cos[(\omega - \omega_0)t - \theta] \qquad (10.42)$$

When in phase lock, $\omega = \omega_0$, and only the DC part of this signal remains:

$$V_{out} \simeq V_1 \cos(\theta) \qquad (10.43)$$

The amplitude of the VCO has been dropped because it is constant. This result is made useful by the fact that the dynamic operation of the PLL is such that $\cos(\theta)$ is also proportional to the frequency difference $\omega - \omega_1$ between the incoming signal and the natural frequency of the VCO. Using this proportionality, we have

$$V_{out} \sim V_1(\omega - \omega_1) \qquad (10.44)$$

where ω_1 is set by external components. The external connections necessary for the operation of the NE565 device are shown in Figure 10.33b.

The PLL circuit is used as a frequency-to-voltage converter (FM to AM), but its range of application is quite different from the wide-band device of Figure 10.32b. The PLL will only lock onto an input signal that is fairly close to the natural frequency of the VCO. This band of lockable frequencies is known as the

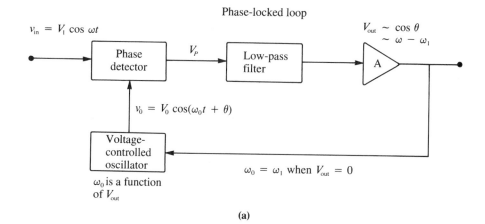

(a)

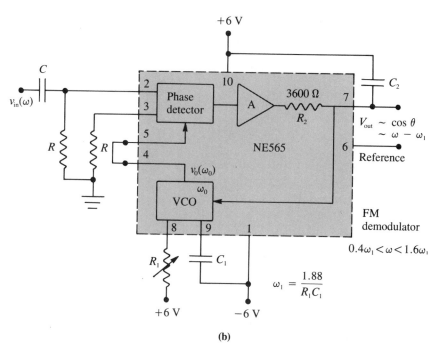

(b)

Figure 10.33   (a) The operation of the phase-locked loop or PLL. (b) The Signetics NE565 IC wired to convert a frequency-modulated FM input signal to a variable-voltage output.

capture range and serves to make the PLL sensitive only to a narrow band of frequencies, a feature that is used in radio and television receiver applications.

The narrow band of the PLL also allows the natural frequency of the VCO to be set so that the PLL ignores the fundamental and locks to an odd harmonic of the input signal. This will result in an output v_0 from the VCO (pin 4 in Figure 10.33b) that is at the harmonic frequency and phase-locked to that component of the input signal.

Problems

1. Assuming that a microcomputer performs all of the logic functions as in Figure 10.1, make a block diagram for an intelligent thermostat for a natural gas home heating system. *(a)* Show and describe the type of input and output transducers necessary to make this system work. *(b)* Describe two features whose complexity might justify the use of a computer.

2. *(a)* Using a microcomputer as in the previous problem, draw a block diagram for a home burglar alarm. *(b)* Explain the operation of at least four different types of input transducers for this system. *(c)* What types of output transducers are possible? What fundamental physical process do they use at the electronic/nonelectronic interface? *(d)* Assuming that false alarms caused by wind storms are a serious problem, what cures might be possible using a microcomputer and perhaps additional transducers and/or controls?

3. Design a circuit to debounce a "make-before-break" microswitch.

4. Design an inverting op amp circuit that has potentiometers for gain and offset.

5. Assuming that the diode shown in Figure P10.5 exactly follows the equation $I = I_0(e^{V/V_T} - 1)$ with $I_0 = 10^{-7}$ A and $V_T = 50$ mV, sketch V_{out} versus V_{in} over the input range -2 V to $+2$ V. Show the scale on both axes.

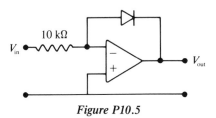

V_{in} — 10 kΩ

Figure P10.5

6. A patterned target of varying reflectivity $r(x)$ is illuminated by fluorescent room light that produces a signal $V_R(t)$ when measured by a photodiode that "sees" the entire target with some average reflectivity r_{avg}. A second identical photodiode is connected to a telescope so that it can scan the target along x and see only a small part at any one time.
(a) Assuming that the voltage signals are proportional to the reflectivity times the intensity of the incident light, devise a circuit whose output will be the target signal independent of room light variations. Consider using either a summing amplifier or the ratio portion of the LH0094 device of Figure 10.10. *(b)* Explain your choice and why the alternative will not give the desired result.

7. If you can measure the output of the active bridge circuit of Figure 10.11 to 1 mV, what is the smallest variation x that can be measured in a 0.1-μF capacitor if $\omega = 100$ rad/s and the amplitude of $\mathbf{V}(j\omega)$ is 1 V?

8. Determine an expression for $V_+ - V_-$ into the instrumentation amplifier when the input bridge is activated as shown in Figure P10.8.

9. Make a 16-line truth table showing all of the gains possible with the amplifier of Figure 10.13a.

10. Using a 4PST analog switch, devise a circuit to decode the multiplexed analog signal of Figure 10.13b back to four distinct analog outputs. Use the same two digital signals A_0 and A_1.

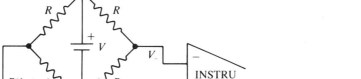

Figure P10.8

11. On the S/H circuit of Figure 10.15a, assume that the first op amp has an output current limit of 30 mA, and the second has an input impedance of 10 MΩ.

(a) If we need to hold the output signal for 1 s with only 0.1% droop, determine the size of *C* and estimate the slew rate (volts/microsecond) during the acquisition interval.

(b) Is the acquisition time a constant for this circuit?

12. Derive Eq. (10.16).

13. If you are using a 0.1-μF capacitor with the 555 timer and wish to generate a 1-kHz output signal, choose the resistors R_A and R_B such that your duty cycle is 70% true.

14. (a) Using an ideal diode, modify the 555 timer external circuit of Figure 10.20 to give a 50% duty cycle. (b) Now assume that the diode is real and add a second one that will compensate the effects.

15. Replace the 15*R* resistor in Figure 10.22b so as to make a binary-coded decimal (BCD) DAC. The BCD digits 0–9 are given by the binary numbers 0000–1001, and the other numbers 1010–1111 are not used.

16. On an 8-bit DAC, between which pair of decimal numbers would you expect to find the largest step error; next largest; etc.?

17. Using the 74193 up/down counter chip, expand the register/control logic block of Figure 10.28a to produce a 4-bit ADC whose digital output follows the analog input as long as an input signal HOLD is false and retains the last digital output while HOLD is true. The operation should be like a digital version of the S/H amplifier.

18. The integrating operation of the dual-slope ADC of Figure 10.29 neatly removes certain high-frequency noise components of the input signal and has variable sensitivity in between these canceled frequencies.

(a) In terms of the notation in the figure, determine the sequence of the exactly canceled frequencies.

(b) Assuming an AC white noise input with unity signal amplitude at all frequencies, determine an expression for the integrated digital noise signal midway between each exactly canceled frequency. Assume the worst-case phase relationship between the sinusoidal components of the noise signal and the sampling period.

(c) Sketch the magnitude of the digital noise signal as a function of frequency.

19. Using two 1020-type DACs and two op amps, design a circuit whose analog output is proportional to the product of two digital numbers.

20. How many comparators are needed to build an 8-bit flash encoder?

21. Using a TDC, devise an experiment and show a complete block diagram of a laboratory system to determine the speed of a bullet.

22. On the circuit of Figure 10.31a, what input voltages V_{in} correspond to output frequencies 100 Hz, 1 kHz, and 10 kHz?

23. *(a)* Using the internal block diagram of Figure 10.31a, explain the operation of the frequency-to-voltage converter of Figure 10.32b. *(b)* For the LM331 device, the internal constant current I is given by 1.9 V/R_S. If R_S is set to 15 kΩ and the frequency changes suddenly from zero to 1 kHz, what is the average V_{out} at equilibrium? *(c)* How many cycles will it take for V_{out} to reach 99% of this value?

24. You have a slowly varying (0.1 Hz or less) input signal that must be recorded on an audio tape deck.
   (a) Using the NE565 PLL chip shown in Figure 10.33b, design a system that will frequency-modulate the input signal around a 1-kHz center frequency.
   (b) Using a second PLL, design a circuit to convert the tape deck's frequency-variable output back to the original input signal.

25. Show how you could use a digital divide-by-3 counter to lock the PLL to three times the frequency of the v_{in}.

11

Computers and Device Interconnection

11.1   Introduction

In the previous chapter we discussed a number of devices and circuits suitable for the data acquisition and process control (DAPC) application shown in Figure 10.1, but omitted the central element: the digital computer. In this chapter we will discuss some general elements of the computer, outline some of the more common device interface standards, and investigate some of the noise problems that become particularly acute when digital and analog circuits are combined. Although we will be discussing computers in a general way, specific references will be to microcomputers made up of microprocessors and other large-scale integrated circuit (LSI) devices.

Although it is possible to devise a DAPC system that does not use a digital computer, the inclusion of this element adds a degree of complexity unattainable with other components. This qualitatively different character stems not only from the computer's ability to remember and execute a complex sequence of operations, but also from its ability to immediately modify the sequence based on external stimuli. It is this latter feature, more than any other, that distinguishes the DAPC or ''real-time'' computer application from the more familiar applications involving computers whose external environment consists mainly of a keyboard and display.

When a dedicated computer is connected to its environment via transducers such that it has both measurement and process control capabilities, it may acquire characteristics that are unfamiliar to most computer users. The instantaneous response to external stimuli, coupled with the ability to evaluate data-dependent equations quickly, can yield a system whose behavior is often unexpected and difficult to explain. To develop and debug such systems effectively, a knowledge of both computer software and hardware is necessary. Without going into software details, the present chapter will concentrate on the interface area between computer software and instrumentation electronics.

11.2   Elements of the Microcomputer

The inexpensive microprocessor has dramatically expanded the range of computer applications, not only in the direction of low-cost, general-purpose DAPC applications, but also to situations where the microprocessor serves a very specialized function, either as one of several computers in an application or as just another IC chip on a circuit board. We will describe here the elements of a general-purpose microcomputer, one that is suitable for the DAPC application in a research or development laboratory. A scaled-down version of the general-purpose laboratory computer outlined here, omitting such things as a terminal and other program development aids, results in a computer that is suitable to the more specialized, single-process or single-circuit application.

In a typical research laboratory arrangement, the computer would be used to control equipment, acquire data from measurement apparatus, perform mathematical operations on the data, and make these available on a display device for operator review. Most useful in this arrangement is a graphics display that can present either raw data or the results of sophisticated calculations in an easily understood form. When such displays operate in "real time" to show the results of ongoing measurement, they enable the experimenter to detect time variations or errors in the data acquisition process that would otherwise pass unnoticed, at least until long after the data acquisition phase is complete.

11.2.1   Microprocessors and Microcomputers

The key element of a microcomputer is the LSI microprocessor chip, whose circuitry can acquire, interpret, and execute a sequence of logical and arithmetic instructions. Although most computer programming is done in assembly or higher-level languages whose mnemonics and organization obscure the basic microprocessor operation, the processor itself deals only with binary numbers: numbers that represent data, numbers that define operations, numbers that represent addresses or locations, even numbers that represent letters.

Stripped of its various language disguises, a computer program is nothing more than a list of binary numbers (8, 16, or 32 bits long), which the processor executes in sequence, interpreting part of the number as an operation code (move data, add, subtract, compare, jump, etc.) and the remainder of the number as the address (routing information for the data flow) of registers, memory locations, or input/output data ports. Each bit of one of these numbers is a digital true/false signal and will appear at various times on a single wire at some point within the computer or on one of its external pins.

The microprocessor is only one of the elements that make up a microcomputer: The complete computer requires power supplies, memory, interface circuits to provide "ports" to external devices, and input/output devices such as keyboards, displays, DACs and ADCs, and magnetic data storage. To be general-purpose and suitable for hardware and software development work, the computer must have some form of human interface, usually a video terminal, and a bulk memory device such as a magnetic disk. These more routine input and output devices are excluded from the following discussion.

11.2.2 Functional Elements of the Computer

The typical computer contains several functional subsystems: the central processing unit (CPU), random access storage (memory), and input/output to external devices (I/O). Among these subsystems, the CPU is the most complicated and may be subdivided into smaller units: instruction decode and CPU control, control of addressing for memory and I/O ports, data transfer control, data and address registers, and an arithmetic logic unit capable of performing binary arithmetic and Boolean logic operations between two data words. Microcomputer designs vary from the extremes of placing simple versions of each of these functions on a single LSI microprocessor chip, to placing different operations on different chips or even different circuit boards. With increasing miniaturization, the trend is to place more and more of the computer's subsystems onto a single chip.

The operation of the CPU can be demonstrated by tracing the steps it performs during the execution of a single instruction. To keep track of its progress, the processor maintains a special register, known as the program counter, which points to (contains the address of) the next instruction to be executed. The instruction itself is just a number, 1 to 4 bytes long, part of which specifies the operation to be performed and the processor registers that will be used. In addition, the number may contain one or two bytes that represent data, or point to other data still located in memory.

To begin an execution cycle, the processor utilizes the program counter to fetch its next instruction from memory. A part of the CPU decodes the instruction, fetches an additional number from memory if needed, then performs the required operation; a typical example would be to add the number from memory to a number held in one of the processor's internal registers, with the result left in the same register. Typically, the processor will also set some internal flip-flops that indicate whether the sum is positive, negative, zero, or has overflowed the limits of the register; these are known as status bits and are usually grouped into a status register.

At some point during the execution of an instruction, the processor must increment or otherwise modify the program counter as needed to point to the next instruction. Much of the power of a computer comes from its ability to execute an instruction known as the conditional branch or jump, which varies the contents of the program counter based on the results of a previous operation. Typically, the action taken depends on the contents of the status register, thereby giving the processor the ability to alter the sequence of instruction execution based on the results of data-dependent calculations.

Whether on the same chip or on different circuit boards, the various functional units of the computer are for the most part connected by one or more multiwire digital buses which pass data, address, and control information between the units as shown on Figure 11.1.

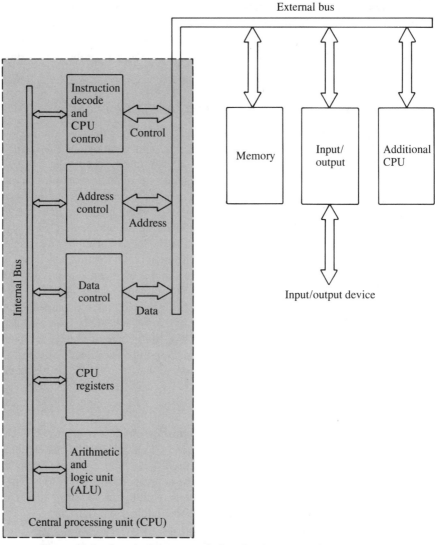

Figure 11.1 *A typical computer design showing two multiwire buses,
an internal bus connecting functional units within the CPU and an
external bus for connecting additional computer subsystems. Often the
subsystems on the external bus are on physically distinct circuit boards
that plug into the bus as shown in Figure 11.2.*

11.2.3 *Mechanical Arrangement*

An extensive and rapidly changing inventory of microcomputers is available to
the modern experimenter, and the choice of a best system is not easy. The mechanical and physical layout of a computer can have a significant influence on its
cost and reliability on the one hand, and on its flexibility and ease of maintenance
on the other. Machines with the lowest cost and highest reliability are generally

those with the fewest mechanical connectors and socketed components, while
those with more connectors and sockets are more easily expanded and maintained.

The computer shown in Figure 11.2a is designed so that all circuits are on
mechanically equivalent boards that plug into a common bus; this is the most
modular arrangement and is easily maintained. Most electronic faults in a com-

Figure 11.2 (a) A computer in which all circuit boards plug into a common bus. (b) A single-board design that features an external or expansion bus.

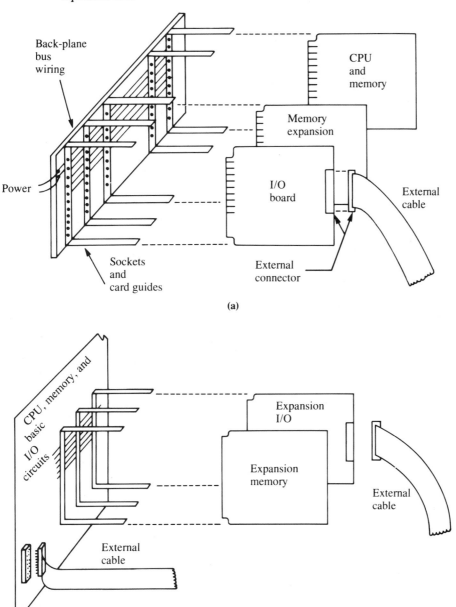

(a)

(b)

puter of this design can be quickly isolated by exchanging suspect circuit boards with spares kept in reserve or taken from another identical computer.

An alternative design is the single-board computer which places the CPU and a substantial portion of the memory and I/O interface circuits onto a single circuit board. In a design that is suitable for laboratory use, this large computer board will be connected electrically to an external expansion bus as shown in Figure 11.2b. This design is generally more compact and less expensive, and it allows the more routine activity of the computer to occur without use of the external bus, thereby reducing the data transfer rates on that bus.

11.2.4 Addressing Devices on the Bus

A full appreciation of the term "address" is vital to understanding of the operation of a computer, and fortunately the term has the same meaning here as on an envelope entrusted to the U.S. Postal Service: It defines—and if all goes well determines—either the destination or the source of information. However, the analogy with the mail system cannot be carried very far: Since the wires of a bus are common to all functional units, each unit will "see" all of the data placed on the bus lines. Rather than being used to route information along different bus paths as in the mail system, the address lines are used within a receiving unit to determine if available information should be processed or ignored.

Normally each data repository on a common bus will have a unique address; thus, each byte of memory information, each register associated with an I/O port, and each register of an associated LSI circuit component will have a unique address as seen from the bus. Typically a chip or plug-in circuit board will span one or more addresses depending on the complexity of the unit.

When the CPU needs to transfer data between itself and a particular location, it implements a sequence of signals as specified by the read or write operation protocol for the bus. Bus protocols vary, but in the simplest write operation, the CPU sets the address and data lines to their appropriate values, then generates a write pulse on the appropriate bus control line. Each unit on the bus matches the address lines against its internally defined addresses, and if a match is found, the write pulse is accepted as a data strobe for the appropriate element. (An example of a data strobing operation into a flip-flop register is shown in Figure 9.37b; the write strobe is labeled LOAD in that figure.) In the simplest read operation protocol, the CPU first sets the address lines and a read-enable signal, hesitates for 50 to 500 ns to allow the transmitting unit time to recognize its address and gate its data onto the bus, then strobes the data from the bus lines into its own registers.

The range of numbers that can be represented by the available address lines (wires) on a bus is known as the address space; for 16 address lines this space runs from 0 to 65535 (normally called 64K), as shown in Figure 11.3a. A range of numbers of this or larger size is used mostly to access information from memory and is thus known as the memory address space. Some processors assign a few of these "memory" addresses to other input/output devices as indicated in the figure, a feature known as memory-mapped I/O. Other microprocessors feature a dual address space, a large one for memory or I/O devices and a smaller one strictly for I/O operations. (The bus protocol must then implement two read and two write operations, one for memory and one for the I/O devices.) Although the

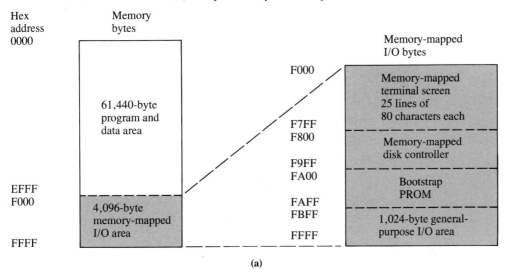

65,536-Byte Memory Address Space

Hex address	Memory bytes			Memory-mapped I/O bytes
0000				
	61,440-byte program and data area		F000	Memory-mapped terminal screen 25 lines of 80 characters each
			F7FF F800	Memory-mapped disk controller
EFFF F000	4,096-byte memory-mapped I/O area		F9FF FA00	Bootstrap PROM
			FAFF FBFF	1,024-byte general-purpose I/O area
FFFF			FFFF	

(a)

256-Byte Port Address space

Hex address	Device	Output/input
00	Keyboard	Data out/data in
01		Status/control
04	Modem	Data out/data in
05	8251 UART	Status/control
06	Serial printer	Data out/data in
07	8251 UART	Status/control
08	8255 PIO port A	Data out/data in
09	port B	Data out/data in
0A	port C	Data out/data in
0B	Control port	Status/control
0E	Video controller	/Register select
0F		/Function select
10	Timer	Read/write counter 0
11	8253	Read/write counter 1
12		Read/write counter 2
13		/Control
18	Tone generator	Data out/data in
19	76489	Status/control
1A–FF	Unused	

(b)

Figure 11.3 (a) The address space of a microprocessor that has 16 address lines spans the number range 0 to 65535, and these must be shared between memory and any circuit board that uses this address space. (b) The input/output address space may be distinct, and if so is usually smaller; in this example, an 8-bit I/O address is divided among various devices.

use of a separate and smaller I/O address space reduces the complexity of the decoding logic needed in the attached devices, the method also reduces the number of computer instructions that can access the I/O data directly.

Generally, a single unit on the bus must respond to many addresses; a memory unit is an extreme case. Figure 11.4a shows the address decoding logic needed to access 64K bits from four different 16K-bit IC memory chips. Since 14 bits of the address are decoded within the memory IC, in order to select 1 bit from the 16K bits stored in the chip, the external logic needs only to select which of the

Figure 11.4   (a) The address decoding logic needed to access data stored in a 64K-word memory unit. The schematic shows only 1 bit of a memory word. (b) The timing diagram for the bus signals associated with a memory read from this unit. When multiple address and data lines are represented by one signal waveform, it is common to show the signal in both the true and false conditions. Sloping transitions indicate timing uncertainty, and the wiggly line indicates the hi-Z or otherwise unknown line condition.

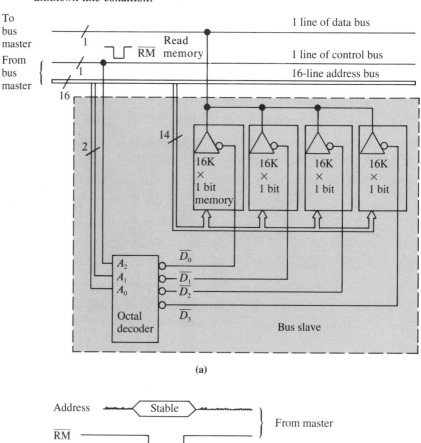

(a)

(b)

four IC chips is allowed to place its data onto the bus line. The design in the figure uses the simplified octal decoder of Figure 9.18. The input to the decoder consists of the two most significant address bits and an active false signal on the read memory control line $\overline{RM}$. When the read signal is active (false), one of the decoder's four output lines D_0 to D_3 will be false, thereby enabling the 3-state output driver on one of the memory ICs. The outputs from the other memory ICs remain in the disconnected hi-Z state. In this example only a single data bit is presented to the bus. However, the design can be easily extended to a full byte of data: Each memory unit in the figure would be replaced by a "stack" of eight chips, all of which receive the same address and chip-enable lines, but with the output from each of the eight ICs connected to a different data line on the bus.

11.2.5 Control of the Bus

The information flow on the computer bus is time-multiplexed to allow different functional units to use the same bus lines at different times. As discussed above, addressing solves the problem of which device should read (receive) the bus lines at any particular time, but some provision must also be made to assure that only one device at a time tries to write (drive) a given bus line. As in Figure 11.4, the most common solution is to drive individual lines with 3-state logic devices (true, false, and hi-Z), with all such drivers remaining in the hi-Z or disconnected state until specifically enabled by a properly addressed control signal as shown in the timing diagram of Figure 11.4b. A special unit, known as a bus master, has the responsibility for controlling the other units, which are correspondingly known as bus slaves. At any instant, the flow of data on a bus is between the controlling "bus master" unit and a slave unit.

In a simple computer, a single master unit—the CPU—will drive the address and most control lines on the bus, thus controlling the flow of information between itself and subsidiary or slave units. A more complex and flexible design may feature several units capable of becoming bus masters (possibly multiple CPUs), and these must arbitrate among themselves to determine which is to have control of the bus for a given interval of time. Even if only a single CPU is used, the use of auxiliary master units can speed the flow of information between units on the bus.

A block diagram of a simple dual master system, consisting of a CPU and a direct memory access (DMA) unit, is shown in Figure 11.5. Since the DMA can be a bus master, it can exchange information directly with the slave memory unit, generally at much higher rate than would be possible by a two-step process such as I/O operation to a CPU register followed by a CPU write operation to memory. As shown in the figure, the direct memory access unit is usually attached to an external device (disk, tape, data acquisition) that requires a high-speed path for the exchange of external data with the computer.

11.2.6 Clock Lines and Timing Diagrams

The digital signal on a typical line of the computer bus shows a very complex time variation and can be understood only in combination with other bus signals. In general, the changes of state of all bused signals are synchronized to one or

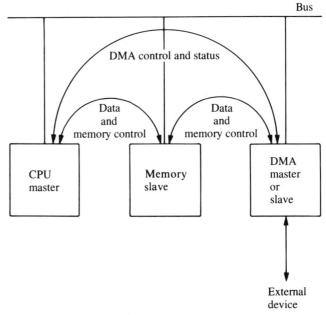

*Figure 11.5 On this bus the CPU and the DMA are bus masters,
whereas the memory unit is a bus slave. The arrows on the bus
indicate the paths along which data, control, and status information
can be exchanged between individual bus units. The dual master
arrangement shown permits the direct exchange of information between
the DMA and memory.*

more clock signals, which are distributed to all functional units on the bus. Read
and write operations between the CPU and memory units are the most common
operations on the computer bus, and representative timing diagrams for the necessary address, data, and control signals for a write operation are shown in Figure 11.6.

Figure 11.6a shows an open-ended bus write operation (similar to the read
operation of Figure 11.4b), where the master unit sets the address and data lines
and then blindly issues two control pulses, assuming that the slave unit will respond as needed within the allowed time. The first pulse of this read control
sequence enables the memory unit addressing logic, thereby establishing a path to
a particular bit within each selected memory chip before the write strobe occurs.
A more elegant but slightly slower technique known as handshaking is shown in
Figure 11.6b; here the master unit sets the address and data lines and raises a
control line as before, but then waits for an acknowledgment response from the
slave before lowering the control line and proceeding. In this design the slave can
extend the delay time as needed in a particular situation simply by delaying its
acknowledgments.

11.2.7 RAM Memory, Static and Dynamic

Until the introduction of large-scale integrated circuits, random-access memory
(RAM) was composed of small magnetic donuts made from ferrite material and

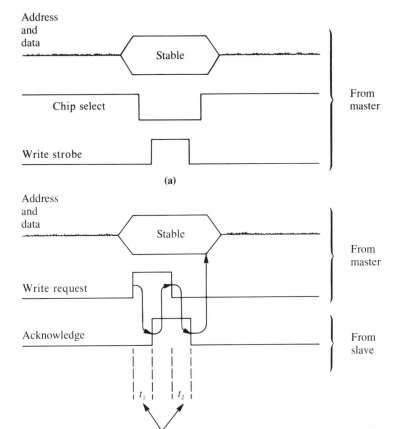

Figure 11.6 *The bus signal timing diagram for (a) a write-to-memory operation, and (b) the same operation with handshaking. The arrows indicate the cause-and-effect relationship between transitions.*

assembled by hand into matrix-like arrays threaded by read and write wires. In all modern designs this magnetic memory—known as core memory—has been replaced by IC devices which are not only much cheaper but also much smaller. The most popular IC memories, the dynamic types, also consume very little power and thus permit quite large memories in compact, even portable, computers. Although not a part of the mnemonic, the term RAM also implies that the memory will support both read and write operations.

Integrated-circuit RAM memory comes in two main types, static and dynamic, operating on quite different principles. A single bit of static memory is simply a digital flip-flop in one of its two stable states and requires only the continuous application of power to maintain its state. Operating in an entirely different manner, a single bit of dynamic memory is a few-picofarad capacitor in either the charged or uncharged condition, and even though the state of the bit storage capacitor is read by a high-impedance MOSFET amplifier as shown in Figure 11.7,

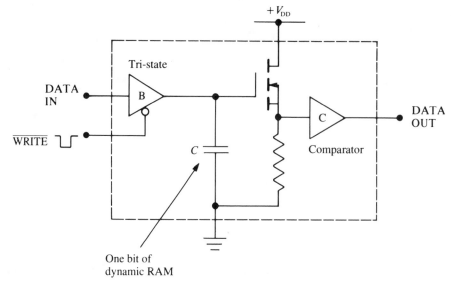

*Figure 11.7   A simplified design for 1 bit of dynamic memory. The
input buffer amplifier B has a tri-state output and is connected to the
capacitor only during a write operation, whereas the high-input
impedance MOSFET and output comparator C sense the charge being
held on the capacitor.*

an initially charged capacitor will slowly discharge and require periodic charge
renewal or refreshing—thus the term dynamic.

Compared to dynamic memory, static memory has several advantages: It is
much simpler to use, about 10 times faster to read or write, and somewhat more
reliable. On the other hand, it is more expensive, consumes much more power,
and requires more physical space. Because of fundamental problems associated
with heat dissipation, there is a definite upper limit to the power consumption of
an integrated circuit chip; and since memory units tend to push this limit, it is
significant that at the time of this writing, the largest available static RAM contains only 16,384 bits versus 262,144 bits for the largest dynamic unit. Typically
then, dynamic RAM is used for most of a computer's memory and static RAM is
used only for special high-speed applications.

Both of the solid state RAM memories just described are volatile, meaning that
stored information is lost when power is removed from the chip. Some computer
designs provide a limited amount of nonvolatile read/write RAM storage by using
special low-power (and slower) dynamic memories powered by rechargeable batteries.

11.2.8   ROM Memory

Another form of nonvolatile random access storage is the read-only memory
(ROM). Here a single memory bit is nothing more than a connection that is either
open or closed. The most common ROM types are known as field (as opposed to
factory)-programmable and with the help of special equipment can be programmed

by the user to any true/false bit pattern. This programming process consists of stepping through all of the bits and setting the necessary ones by "burning" open the fuse-like material associated with that bit.

There are many varieties of field-programmable read-only memory units, and most have descriptive mnemonics: PROM for programmable read-only memory; EPROM for erasable PROMs, which can be reset to all zeros by exposure to ultraviolet light; and EEPROM or E^{2}PROM for electrically erasable PROMs, which can be reset, albeit slowly, by the application of a special voltage. The erasure of a programmed PROM is accomplished by the diffusion of excited conducting molecules back into the opened area of all burned fuses and is thus a slow process taking typically 30 minutes.

The most common uses of ROM memory in a computer are to provide initialization (memory test, disk bootstrap) programs and certain translation tables such as character shapes, but in some applications it may be desirable to commit a large program to ROM to ensure its availability and permanency in a laboratory situation. This last use is most important when nonvolatile memory devices such as disk drives are not available.

11.2.9 I/O Ports

Input and output ports are the pathways by which the CPU communicates with the world outside the computer. Depending on the data transfer rates needed, the pathway presented to the external world may be either 1 bit wide (bit-serial), 8 bits wide (byte-serial), or 16 to 32 bits wide (word-serial). As seen by the computer, a port may appear as simply a special memory address accessible by any normal memory read/write operation, or it may appear as a separate and smaller space of port addresses requiring special port commands to provide access.

The electronics associated with the port must decode as many address lines as are required to identify the port uniquely and must provide the needed conditioning of data lines to match or interface the computer environment to the external environment. On output, for example, the port electronics might latch the transiently available computer data into a register and hold it for as long as the external device requires. To avoid overwriting the register before it is released by the external device, this port electronics would provide the computer with a separately addressed status register showing the current state of the register (busy or done).

Several important LSI circuits are available to support the I/O port function; the two most important types are the parallel I/O chip (PIO) and the universal asynchronous receiver transmitter (UART) serial I/O chip, both available in several different versions. These chips feature data registers and status and control registers, and they are software-programmable into several different operating modes.

11.2.10 Interrupts

In many situations it is desirable for the computer to respond immediately to an external stimulus; this is especially true of a real-time application, where the interval between events is short. One way to keep the computer aware of its environment is to program the CPU to test the state of some input port with the

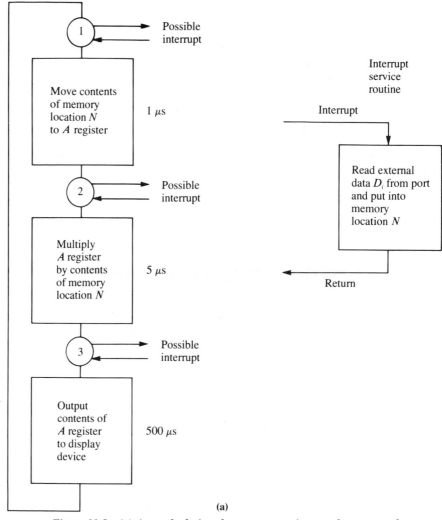

Main program to display $(D_i)^2$

(a)

*Figure 11.8 (a) A poorly designed program to print out the square of
an input number obtained by an interrupt routine. The program will
produce different results depending on exactly when the interrupt
occurs.*

necessary regularity. However, if the computer is performing a variety of other
tasks, insertion of software tests at regular time intervals may be difficult or impractical. A better alternative in these situations is the hardware interrupt, which
can be used to suspend the current sequence of instructions, perform a specific
and usually short I/O task, then return to the original sequence.

The operation of the hardware interrupt is much like a software CALL subroutine instruction. An interrupt is requested by changing the state of an input line to
the CPU, causing the CPU to break the regular program sequence and execute a
CALL instruction to the interrupt's service routine. As with any other subroutine,

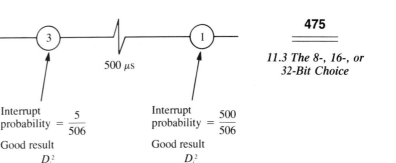

Figure 11.8 continued. (b) An analysis of the assumed timing for the various software steps shows that a bad result (D_iD_{i+1} rather than D_i^2) will occur less than 0.2% of the time.

the CPU executes the instructions in this list until it encounters an order to return to the original program sequence. A certain time delay necessarily exists between an interrupt request and the execution of the CALL instruction, and for a dedicated computer this delay can be as short as the time required to complete or terminate a single instruction, usually only a few microseconds. By comparison, an interrupt on a multiuser computer will typically cause the software operating system to take control and execute many instructions, taking several hundred microseconds, before finally responding to the interrupt.

Great care must be exercised when designing and programming a computer system that uses interrupts. Since interrupt signals are externally derived and generally not synchronized to the internal program execution, the exact conditions surrounding an interrupt may not be exactly reproducible. Consequently, in an imperfect system a running program may produce different results, depending on exactly when it is interrupted or on the data obtained during the interrupt. The result is that the perfect reproducibility of a computer program is lost; different results can be obtained under superficially identical conditions. Thus, when debugging an interruptable program, one must not assume that a program and system are perfect just because they have worked on a number of occasions. An example of such an occasional, and thus very difficult to locate, failure is shown in Figure 11.8.

11.3 The 8-, 16-, or 32-Bit Choice

A microprocessor can be characterized in several different ways, but one of the more important is the width of its internal and external buses. Since the internal buses connect various registers and functional units within the processor, they determine the magnitude of the largest number that can be transferred or processed in a single clock cycle. Likewise, the width of the external data bus determines the amount of information that can be transferred between the processor and memory (or I/O) in a single memory cycle. Even more important is the width of the external address bus, since this determines the number of memory addresses that are accessible by the processor. In general, a wider bus is faster, more convenient for the programmer, and provides many more choices for the overall hardware design, whereas a narrower bus is cheaper and, if adequate for the task, requires simpler external hardware with fewer wires and connections.

The following three sections comprise a short description of three of the more popular microprocessor series.

11.3.1 Intel 8080, Zilog Z80

The first successful integration of a microprocessor into a low-cost but general-purpose computer was the Altair, developed in 1975 by MITS, a short-lived company. This computer used an Intel 8080 microprocessor and introduced the S-100 bus, which was quickly adopted by a number of other microcomputer manufacturers. Although it has many programming limitations, the 8080 became widely used in a number of microcomputer designs.

Some insight into the operation of the 8080 can be obtained by considering its register set, shown shaded in Figure 11.9a. In this design, a typical register has special characteristics and capabilities that distinguish it from other registers. For example, the 8-bit A register is the only one able to execute add, subtract, and compare instructions; whereas the 16-bit HL pair has unique memory-addressing capabilities. Such a register design results in a very unsymmetric instruction set, where, for example, ADD A,B is valid but ADD B,A is not.

From the arrangement of the registers in Figure 11.9a, it is clear that certain pairs of 8-bit registers can be treated as a single 16-bit register; thus BC, DE, and HL can be used to manipulate 16-bit operands. Two additional 16-bit registers round out the 8080 design: a program (or instruction) counter PC, which holds the address of the current instruction; and a stack pointer SP, which holds the address of the current top of a first-in, last-out (FILO) byte list in memory.

The Z80 from Zilog is a newer and more powerful design that is upward-compatible from the 8080, meaning that every 8080 register and instruction is included in the Z80. In addition, the Z80 provides the extra (unshaded) registers shown in Figure 11.9a and many additional 16-bit instructions not found in the 8080.

11.3.2 Intel 8088, 8086

Another growth path from the 8080 is provided by the Intel 8086 series of processors. At the bottom of this series is the 8088, which features a 16-bit internal data bus and an 8-bit external data bus. This split internal/external bus width design, which is most sensibly used in small dedicated applications where the lower cost of the external bus is a major saving, is also available in other 16- and 32-bit microprocessor series. The 8086 is a full 16-bit design both internally and externally and is followed by two more powerful designs: the 80186 and the 80286.

The register set for the 8086/8088 processor is shown in Figure 11.9b. Note that again the 8080 register model has been included as a subset of this design. This has the advantage of a certain amount of software compatibility (though not

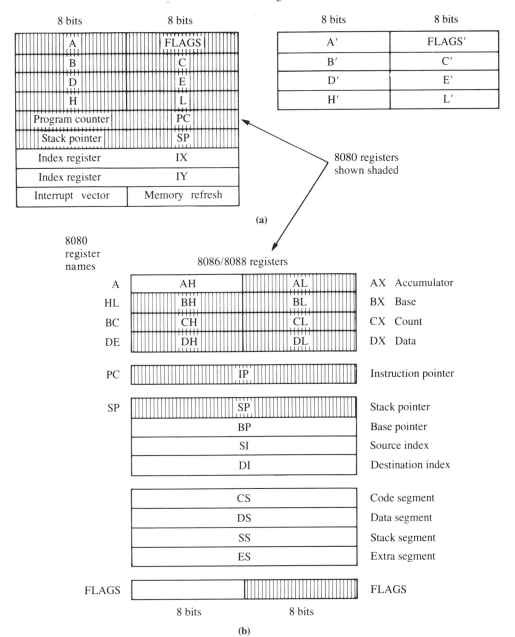

Figure 11.9 (a) The register set for the 8080 (shaded) and the Z80 8-bit microprocessors. (b) The register set for the 8086/8088 16-bit and 16/8-bit microprocessors. The 8080 register subset is shown shaded.

complete like the 8080/Z80), but has the disadvantage of perpetuating the lack of instruction and register symmetry in the original 8080 design.

As memory prices fall and microprocessor power increases, it becomes more important for the processor to address large amounts of memory. The 8080/Z80 designs use 16 address lines and are therefore able to access only 64K bytes. Input/output ports can be used to extend this address space, but the software needed to utilize this expansion is cumbersome. The newer processors all have more address lines to access a much larger address space.

The 8086/8088 has 20 address lines and can thus address over a megabyte of memory. However, the internal registers are only 16 bits wide and are unable to easily manipulate 20-bit addresses. As a result, the 20-bit external address is always obtained by adding an internal 16-bit "segment" register to a 16-bit memory reference as shown in Figure 11.10. As shown in the figure, the two 16-bit numbers are offset before the addition so that their sum will produce a 20-bit result. The least significant 4 bits of the resulting memory address always match the corresponding bits of the memory reference. The resulting processor can access 2^{20} bytes, but since the programmer works with the 16-bit internal registers, he or she must be constantly aware of the 2^{16}-byte limitations of the registers. As can be seen from Figure 11.9b, the 8086/8088 has four such segment registers, which the program can manipulate as needed.

11.3.3   Motorola 68000-68020

In contrast to the upward-compatible route followed by Intel, Motorola chose to adopt an entirely new register and instruction model when it converted from its 8-bit 6800 series of processors to the newer 16- and 32-bit designs. The first entries in this new series were the 68000 and 68010 processors, which feature 32-bit internal registers, 16-bit internal and external data buses, and 24 external address lines providing more than 16 million unique addresses for memory and I/O ports. The 68020 is fully software-compatible with these 16-bit processors but features 32-bit internal and external data buses and a 4000-megabyte address space.

The relative simplicity of the register model shown in Figure 11.11 shows the advantage of a 32-bit design. In this model the registers are nearly equivalent, the major break being the division into data and address registers as shown: These

Figure 11.10   The 8086/8088 processors obtain a 20-bit address by adding a 16-bit reference address (obtained from an instruction or an internal register) to a 16-bit segment register that has been offset 4 bits to the left.

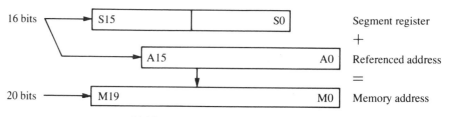

20-bit address generation in 8086/8088

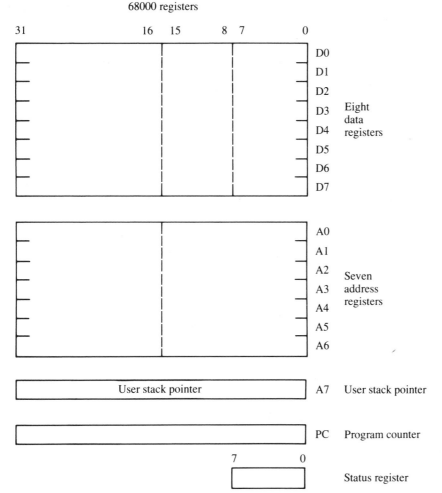

Figure 11.11 *The register set for the Motorola 68000 microprocessor
series. Except for the register of status bits, all registers are 32 bits
long.*

two types of registers can be used interchangeably in some instructions, but not
in others. One of the big advantages of the 68000 design stems from the use of
32-bit address registers. These can easily hold and manipulate the 24 bits needed
for an external address, and allow program access to the entire 16-megabyte address space without undue concern for the artificial 64-kilobyte boundaries imposed by 16-bit registers.

11.4   Auxiliary LSI Devices

The commercial success and utility of a microprocessor series is strongly dependent on the availability of supporting IC chips. These chips range from relatively
simple clock generators to highly complex, floating-point arithmetic units, which

greatly enhance the utility of a microprocessor in scientific and engineering applications. The following is a list of some of the more important devices that attach to and enhance the operation of a microprocessor's external bus:

1. Memory Management—This device accepts data, address, and control lines from the microprocessor and modifies them before passing them on to the external bus. Its purpose is to provide a software-controlled mapping of the internal address space seen by the program (the logical address space) into the external address space of the bus (the physical address).

Figure 11.12 The Intel 8255 programmable peripheral interface features three 8-bit ports, one of which can be split and shared by the A and B ports. These ports can be placed into several different configurations under software control from the CPU. (Adapted from Intel Corporation publications.)

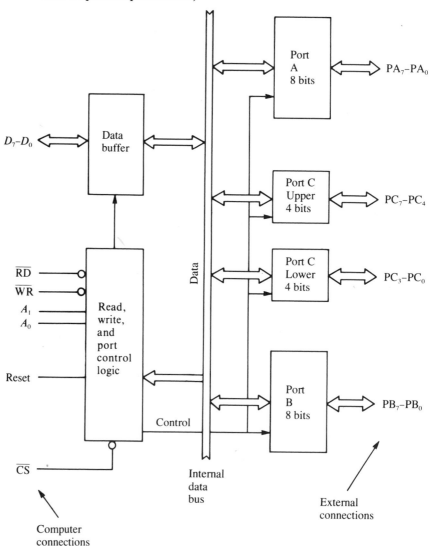

2. *Floating-Point Arithmetic*—This device permits numbers coded in scientific notation to be manipulated quickly and accurately. Its use can increase the execution speed of a calculation intensive program by more than a factor of 10.

3. *CRT Controller*—This device is used to transfer information from memory to a raster-scanning cathode-ray tube such as a television monitor or a terminal display. These devices can be used to display text or graphical information and are most common in computer terminals connected to the computer by a bit-serial communication line. However, the information display is much faster, more flexible, and better for high-data-rate graphics displays when the controller and CRT are incorporated directly into the computer and have direct access to the data bus.

4. *Interrupt Processor*—This device is used to accept several external interrupts, determine which has the highest priority, interrupt the computer, then place the memory address of the appropriate interrupt service routine on the bus data lines to complete the interrupt operation. The device can remember pending interrupts until the CPU is able to service them.

5. *Clocks and Timers*—These devices are used to provide time-of-day information, interval timing between events, or event counting. A timer is normally loaded with an initial count by the CPU, then counted down by an external signal. When the count reaches zero, it can be used to generate an interrupt to the CPU.

6. *Parallel Input and Output Port*—This device (PIO) is used to buffer information between the computer's external bus and an external device. Most units are organized as a series of 8-bit ports, which can be configured, loaded, or read under software control. An example is the Intel 8255 in Figure 11.12.

7. *Universal Asynchronous Receiver and Transmitter*—This device (UART) is used to connect byte-serial information on the computer bus with bit-serial information on a two-wire transmission line. An example is the Intel 8251, shown in Figure 11.13a, which also has a synchronous mode of operation. In the asynchronous or UART mode, the receiving device is able to generate a properly timed data strobe directly from the received data signal.

8. *Direct Memory Access*—As discussed above, this device contains the control logic and registers needed to function as a bus master controlling the flow of information between memory and some external device.

11.5   *Standard Device Interfaces*

The exchange of digital information between the computer bus and an external device can be in increments of bits, bytes, or words, where a word may be 16, 24, or 32 bits wide. Thus, one data-carrying wire and its ground return (2 wires) can transmit information only in bit-serial form; eight data wires and a return (9 wires) can transmit byte-serial, and similarly (17, 25, and 33 wires) for word-serial. Since there are practical limits to the speed at which a single wire can transmit digital information, more wires generally correspond to higher transmission rates. The following sections describe the most commonly used, general-purpose interface standards.

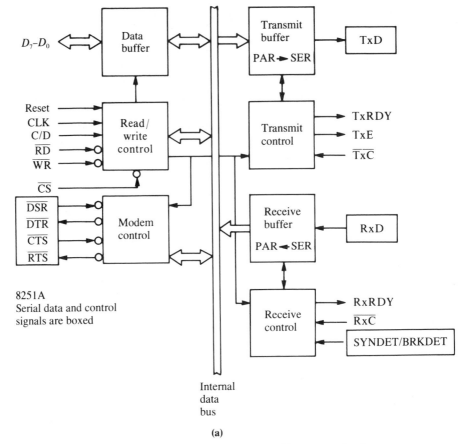

D_7-D_0

(a)

Figure 11.13 (a) The Intel 8251 programmable communication interface, which is used to convert between byte-serial and bit-serial information. The boxes all contain data holding flip-flops or registers that are interconnected by the chip's internal bus. (Courtesy Intel Corporation.)

11.5.1 Bit-Serial: Asynchronous RS232C

The Electronics Industry Association (EIA) RS232C bit-serial data link is a feature of most computers. This interface is asynchronous, allowing the clocking signals in the transmitter and receiver to be uncorrelated, with the receiver deriving its timing from the data signals themselves. This standard is widely used between computers, terminals, printers, plotters, keyboards, digitizers, and any other devices that can tolerate maximum data rates in the kilobyte-per-second range. Because it is asynchronous, the RS232C interface can also be connected to a modem for long-distance communication over telephone lines.

The RS232C standard achieves noise immunity by using very large signal voltages: at the receiver, a true (marking) signal that can be anywhere in the range -1.5 V to -36 V and a false (spacing) signal in the range $+1.5$ V to $+36$ V. Note that these are reversed from what might be expected. Typically, RS232C

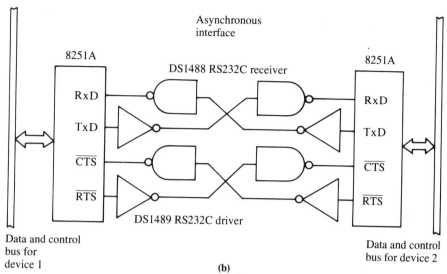

Figure 11.13 continued. *(b) A typical RS232 standard asynchronous
connection between two devices. The ground lines are not shown.
Boxed signal names connect to the serial link; the others typically
connect to the data bus of the CPU.*

lines are driven by integrated circuit devices (such as the DS1488 driver with
matching receiver DS1489) using plus and minus power supplies in the 12-V to
15-V range. These IC drivers and receivers interface RS232C signal levels to
normal TTL levels.

The 8251A shown in Figure 11.13a is a multipurpose device with more input/
output pins than are needed for the typical serial application. Figure 11.13b shows
how drivers and receivers can be used to establish a handshaking RS232C link
between two UARTs, which are connected to the digital buses of two otherwise
isolated devices (typically a terminal and a computer). However, in most serial
applications even the handshaking links CTS to RTS are not connected. The most
important connections to the bus of the controlling device are through the data
lines D_0 to D_7, the chip select line CS, and the read and write strobe lines RD
and WD. The RESET line is needed to initialize the chip, the clock input CLK
determines the speed of serial transmission, and C/D selects paths within the
8251A that determine whether the digital bus will be connected to control/status
registers or to data registers.

The RS232C standard also defines all the signals on a 25-pin connector known
as the DB-25. Table 11.1 lists the most commonly used subset of these signals.
Even this truncated listing is excessive; most cabling uses only three lines—TxD,
RxD, and SG—with data, control, and status information all being exchanged on
these lines. When hardware handshaking is desired, it can be implemented using
the lines RTS and CTS. Some of the communication modems discussed below
provide signals on more of the lines, but their use is usually optional. Printers that
support the RS232C interface often use the DTR line to show a busy, or off-line
condition.

Table 11.1   RS232C Signals

Pin number		Mnemonic	Function	Direction relative to terminal
DTE*	DCE*			
1	1	FG	Frame (chassis) ground	
2	3	TxD**	Transmitted data	From
3	2	RxD**	Received data	To
4	5	RTS†	Request to send	From
5	4	CTS†	Clear to send	To
6	6	DSR	Data set ready	To
7	7	SG**	Signal ground	
8	8	DCD	Data carrier detect	To
20	20	DTR	Data terminal ready	From
22	22	RI	Ring indicator	To

*Data terminal equipment (DTE) includes terminals, printers, and computer ports configured to connect to modems; data communication equipment (DCE) includes modems and computer ports configured to connect to terminals and printers.

**Essential connections.

†Basic handshaking connections.

A single block of asynchronous information consists of a start bit, data bits, an optional parity bit, and one or more stop bits, with the full range of variability being determined by the UARTs used for transmission and reception. The most common 7-data-bit format is shown in Figure 11.14. Note that the 7 data bits and the parity bit are preceded by an always high (false) start bit and followed by one or two always low (true) stop bits. These leading and trailing bits allow the receiving UART to establish and maintain the correct phase relationship between its internal data strobe and the incoming data.

The 7-data-bit format has become a standard because virtually all I/O devices such as terminals and printers expect characters to be represented in the 7-bit American Standard Code for Information Exchange (ASCII). This 7-bit code defines 128 characters, which include upper- and lowercase letters, punctuation sym-

*Figure 11.14   The most commonly used asynchronous transmission
format passes 7 bits of information in a single block. Typically, the
"mark" signal is −15 V and the "space" signal is +15 V.*

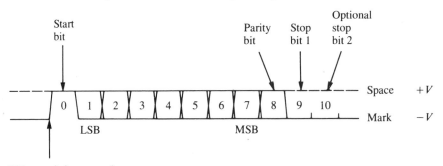

bols, and control codes such as backspace, line feed, carriage return, and so on. The presence of control codes makes it possible to establish software handshaking procedures between devices that have the minimal bidirectional 3-line data connection of TxD, RxD, and SG. The parity bit appended to these 7-bit codes may be chosen true or false to make the total number of true bits in the 8-bit code either even or odd as specified by the protocol in use.

Asynchronous data rates of 300, 1200, 2400, 4800, 9600, and 19200 bits per second (baud) are commonly used, and this corresponds to approximately 30, 120, etc., ASCII characters per second. Much higher data rates are possible with synchronous techniques, especially if low-impedance, twisted-pair or coaxial cable transmission lines are used.

11.5.2 Modems

As the distance between communicating devices increases, the frequency content of the transmitted signals becomes increasingly important. For example, a normal voice-grade telephone line can transmit signals only in the frequency range 700 to 3000 Hz; higher-frequency signals are not only lost but may produce cross-talk onto other telephone lines. The device that converts the sharp-edged (and high-frequency content) digital signal into a narrow-band frequency-modulated signal suitable for long-distance transmission is known as a modem (Figure 11.15). Several modem types are available, each employing different signal-coding techniques and designed for operation at different frequencies.

The most commonly used modems operate at 300 and 1200 baud; 300 baud works with most any telephone link, including the cheaper long-distance lines, and 1200 baud works with most local phone systems and the higher-quality long-distance lines. These devices are normally driven by an ASCII-coded RS232C bit-serial connection to a terminal or computer port. In addition to coding outgoing and incoming data, modems may also be capable of dialing and answering the phone.

Figure 11.15   The modem is used to convert a digital signal into a narrow-frequency-band analog signal suitable for very-long-distance communication.

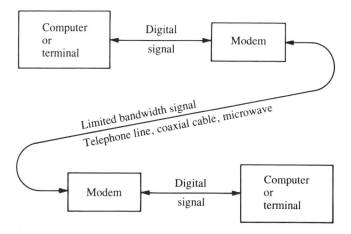

11.5.3   Byte-Serial: IEEE 488

The byte-serial interface is widely used in laboratory applications involving room-scale distances. Although there are several interface standards designed explicitly for magnetic tape, rotating magnetic disk, or printer applications, only the General Purpose Instrumentation Bus (GPIB) standard developed by Hewlett Packard, Inc., offers the flexibility needed for general laboratory instrumentation.

The GPIB has been adopted as an Institute of Electrical and Electronics Engineers standard and is designated as IEEE 488. Interface electronics conforming to this specification are widely available as an I/O option on microcomputers, and

Figure 11.16   The IEEE 488 or general-purpose instrumentation bus (GPIB) features a byte-wide data path and is suitable for the one-room laboratory.

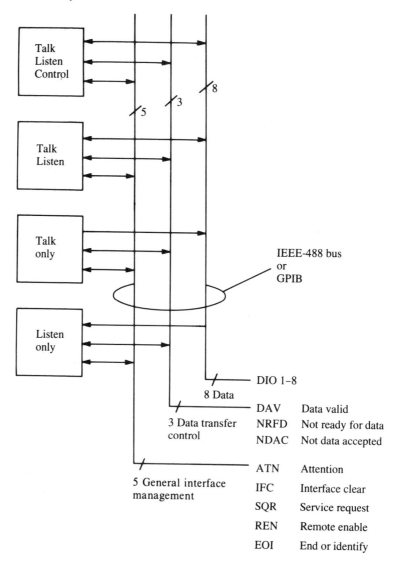

DIO 1–8	
DAV	Data valid
NRFD	Not ready for data
NDAC	Not data accepted
ATN	Attention
IFC	Interface clear
SQR	Service request
REN	Remote enable
EOI	End or identify

several manufacturers make laboratory equipment that can be connected to this bus. As can be seen from Figure 11.16, this bus features 8 data lines and 8 control lines, all with bidirectional information flow. As indicated in the figure, devices can be classified into four categories based on their ability to talk, listen, and control. The GPIB bus is intended specifically for use with a single laboratory arrangement: The specification states that the bus can support at most 15 instruments with no more than 4 m of cable between individual instruments and not more than 20 m of overall cable length. The maximum data rate is listed as 1 megabyte per second in the specification, but most commercial interfaces operate at much lower rates.

11.5.4 Word-Serial: CAMAC

The foremost word-serial interface standard, IEEE 583, is designated CAMAC, meaning computer automated measurement and control. Because it originated in Europe when 24-bit computers were common, it uses a 24-bit word length. The basic standard defines an electromechanical arrangement known as a crate, which includes power, slots for 25 electronics boards plugging into 86 pin connectors, and a backplane bus as described in Figure 11.17. The crate is designed to fit into a standard 19-in. equipment rack. Each crate must have a double-width controller module filling slots 24 and 25; this controller can connect to a computer and to additional crates.

Figure 11.17   The CAMAC bus features 48 data lines, 4 subaddress lines, 5 function select lines, and 10 general control lines. In addition, slots 1 through 24 are each connected to slot 25 by two direct wires used to select and accept service requests from individual modules.

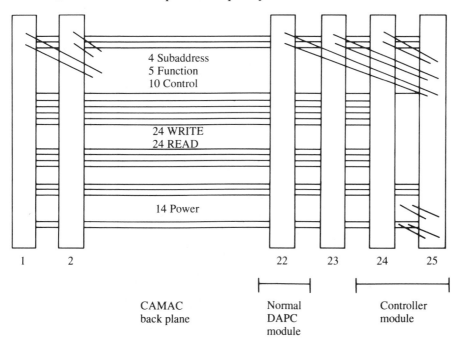

4 Subaddress
5 Function
10 Control

24 WRITE
24 READ

14 Power

1   2   22   23   24   25

CAMAC
back plane

Normal
DAPC
module

Controller
module

Each of the module positions 1 through 24 is selected by a unique control line fanning out from the controller directly to the module. In addition, there are 4 subaddress lines, and 5 function lines that can be used to select specific operations on the selected module. In large-scale applications, external cabling from a single module may fan out to a much larger instrument.

The CAMAC system is widely used for large-scale and high-data-rate applications in medicine, nuclear, and high-energy physics. Although the system is expensive, with a basic system costing about as much as a high-quality laboratory microcomputer, it provides a very flexible and convenient laboratory arrangement. An exceptionally wide range of high-performance plug-in modules are available from several manufacturers. The CAMAC bus with its two 24-bit-wide data paths is capable of 1 megaword per second data readout speed and in most applications is limited by the transfer rate of the associated computer.

11.6 Device Interconnection

In most applications the laboratory computer will be connected to a variety of instrumentation that is not produced by the computer manufacturer. These "foreign" devices may be designed to interface to the computer through the ports of some standard I/O board following some established communication standard as discussed above, or they may be designed to interface directly to the external bus through a special board. Most manufacturers of computers that are suitable for laboratory use supply expansion boards that provide the bus interface logic while leaving a large amount of board space free for the external device interface.

Whatever the interface method, whenever equipment from different sources is mixed, careful attention must be paid to the details of the interface circuitry. Individually simple problems associated with signal voltage levels, input and output impedances, stray capacitance, and noise pickup must all be considered before reliable performance can be expected from the interconnected devices.

In the following sections we will describe some of the common features of interface electronics, starting with various kinds of short- and intermediate-distance transmission lines and ending with a survey of the techniques for making power and ground connections. The analysis of transmission lines must use analog methods, independent of whether the information on the line was originally digital or analog. If the transmitted signal is digital, the receiver will compare the incoming analog signal with a threshold to determine the corresponding digital state.

11.6.1 Low-Impedance Lines

In all previous discussions we implicitly assumed that electronic signals propagate with infinite speed down the wires of a circuit, producing the same instantaneous voltage and current signals at all points of a common wire. In fact, electrical signals propagate down wires (and the surrounding dielectric insulators) at somewhat less than the speed of light, taking approximately 1.5 ns to traverse 1 ft. As the signal frequency and wire length increases, it becomes increasingly necessary to take the propagation time of signals into account.

A low-impedance line consists of a pair of wires terminated with a resistive load R_T corresponding to the characteristic impedance of the line as shown in Figure 11.18a. Such lines are known as transmission lines and are typically driven by a signal source whose output impedance R_S is also close to R_T. When a wire pair is terminated in its characteristic impedance, the signal propagating down the line does not "see" the end of the line and does not reflect. An alternative and

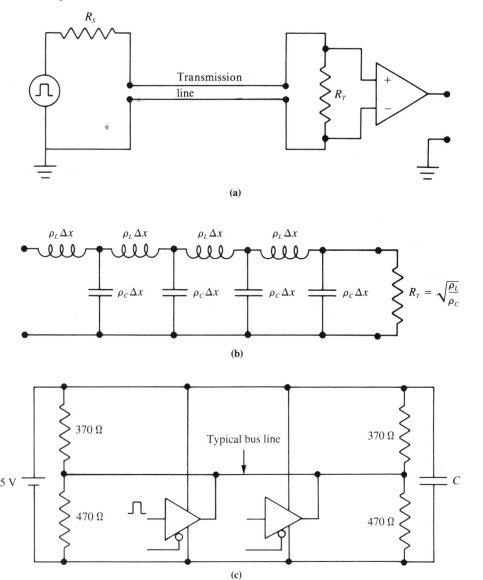

Figure 11.18 *(a) A typical transmission line terminated in its characteristic impedance R_T. (b) The lumped-parameter model of the distributed impedance of the transmission line. (c) A typical data bus line functions as a terminated transmission line.*

equivalent view observes that when the terminating resistance is R_T, it will completely dissipate the signal as it arrives.

The discrete element model for the continuously distributed inductance and capacitance of a transmission line is shown in Figure 11.18b. On this figure ρ_L and ρ_C are respectively inductance and capacitance per unit length, and in the limit as Δx goes to zero the effective impedance looking into this terminated line from the left is a resistance R_T, given by

$$R_T = \sqrt{\frac{\rho_L}{\rho_C}} \qquad (11.1)$$

The derivation of this equation is left as a problem.

The reflected signal from the end of a transmission line will be inverted if the line is terminated in a resistance less than R_T and upright if the terminating resistance is greater than R_T. Larger deviations from R_T will reflect a larger proportion of the incident signal; the extreme cases of infinite and zero impedance terminations are shown in Figures 11.19a and 11.19b. As shown in these figures, it is helpful to think of an image line propagating the reflected signal in the reverse direction, allowing the two signals to add on the actual transmission line.

A signal propagating along a line that is unterminated on either end will be multiply reflected, losing some amplitude at each traversal and reflection, and will be seen by the receiver as a composite but ultimately damped waveform. Two possible shapes are shown in Figure 11.19c for cases when the round-trip signal propagation time T_{round} is approximately the same as the initial signal's pulse width. Although such ringing is always present whenever unterminated lines are used, its significance is a function of the high-frequency response of the signal receiving amplifier.

A receiver designed to pass a signal with either of the initial waveforms shown in Figure 11.19c would also respond to the multiply reflected received signal and generate a false version of the transmitted signal. However, if the pulse width of the initial waveform is much greater than T_{round}, and the receiver is designed to pass only these lower-frequency signals, then the rapid fluctuations arising from the reflections will be converted to a smooth decay by the receiver. As a general rule, if the corner frequency associated with the pass band of the receiver corresponds to a period greater than $10T_{\text{round}}$, then the reflections present on an unterminated line can be ignored.

As an example, consider the transmission of a 1-MHz square wave. If the receiver is to output a signal that is reasonably square, it must have a corner frequency of at least 10 MHz. Thus, we can write

$$100 \text{ ns} = 10T_{\text{round}} \qquad (11.2)$$

and T_{round} is given by

$$T_{\text{round}} = 2L(1.5) \text{ ns} \qquad (11.3)$$

where L is the transmission line length in feet. Solving for L suggests that a cable in excess of 3 ft should be properly terminated.

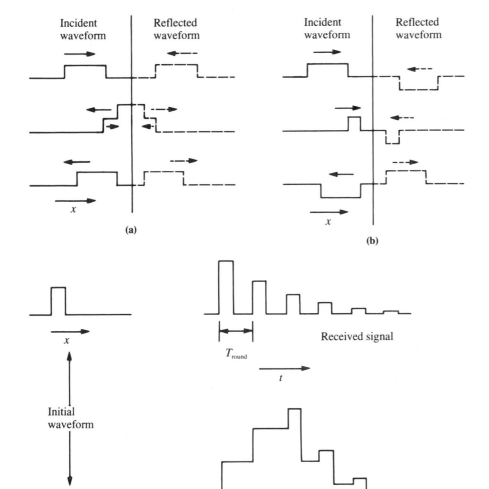

*Figure 11.19   The traveling waveform of a pulse signal as it is
reflected at the end of a transmission line terminated in (a) an infinite
impedance and (b) a zero impedance. (c) Typical composite signals seen
by a high-input-impedance receiver following a pulse placed on the line
by a high-output-impedance driver.*

Low-impedance transmission lines can take several forms: a coaxial cable featuring a central signal-carrying conductor completely surrounded by a conducting
shield; a twisted pair of wires usually driven differentially so that they carry equal
and opposite voltage and current signals; or a data bus wire, usually etched onto
a circuit board and running parallel to a nearby plane conductor held at ground
potential. The various types of coaxial cable have tightly controlled characteristic

impedances in the range of 50 to 100 Ω, whereas the characteristic impedance of twisted pairs or backplane wires is typically in the range of a few hundred ohms. A typical line on the external bus of a computer may be only a few feet long, but if signals above a few megahertz are to be received cleanly, even these short lines must be terminated. A common digital bus termination technique holds an undriven bus line at an intermediate voltage as shown in Figure 11.18c. In this example, each end of the bus line is terminated in an effective resistance of about 200 Ω.

For long lines, lines in noisy environments, or lines carrying low-level analog signals, a twisted pair of wires carrying equal and opposite signals is preferred. In extreme cases these wires may be surrounded by a shielding conductor. A typical arrangement is shown in Figure 11.20. The twisting of the wires in the pair means that induced signals resulting from extraneous electric and magnetic fields will tend to be equal on both wires, even if the field has a high gradient in the vicinity of the line. Thus, induced signals appear in common mode on both wires and can be rejected by a differential amplifier used as a receiver. Because the electric and magnetic fields generated by the two opposite-polarity signals tend to be confined to the immediate vicinity of the wires, this technique also produces less electromagnetic coupling onto adjacent wires, further reducing the cross-talk in a multiwire cable.

11.6.2 *High-Impedance Lines*

Unfortunately, if a significant proportion of the signal power is to be transmitted to the receiver, it is necessary for the output impedance of the transmitter to approximately match the terminating impedance of the line or the input impedance of the receiver, whichever is lower. In some cases, especially when the signal comes directly from a transducer, the output impedance of the signal source is high and cannot be used to drive a low-impedance line. The obvious remedy is to

Figure 11.20 A terminated twisted pair of lines may be used over distances of several thousand feet at megahertz speeds if driven differentially as shown here.

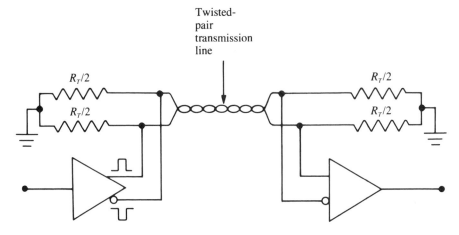

pass the signal through a low-output-impedance amplifier, but sometimes for mechanical reasons a short length of cable (a few feet at most) is required between the source and the amplifier.

A simple connection between a high-impedance source and an amplifier might use a coaxial cable as shown in Figure 11.21a, with the external shield of the cable doubling as a grounded shield and the signal return conductor. An alternative and generally better arrangement would use two signal-carrying wires contained in an external shield; the internal wires would carry the signal and return, with the external shield connected to chassis ground. At the higher frequencies, both of these techniques suffer from the stray capacitance resulting from the conductor geometry of the cable. The simple model of Figure 11.21b can be used to calculate the low-pass filter effects of this stray capacitance. As indicated in the figure, the stray capacitance of a typical cable is on the order of 50 pF per meter of cable length; for a 1-MΩ source impedance, a meter of this cable would result in a corner frequency of slightly more than 3 kHz.

Figure 11.21   (a) A single-conductor coaxial cable is often used to shield a high-impedance signal, but (b) the coaxial cable loads the signal source with its stray capacitance.

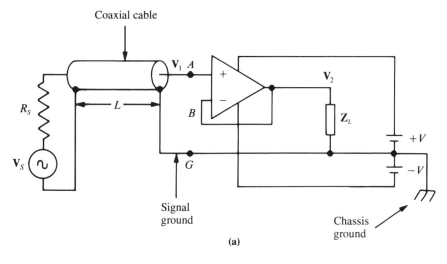

(a)

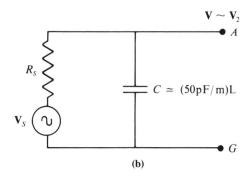

(b)

One solution to the stray capacitance problem is to use a more complicated cable geometry consisting of a central conductor surrounded by guard and shield conductors as shown in Figure 11.22a. Since the follower amplifier has a low output impedance, it can easily drive the guard conductor to the same voltage as the central conductor, effectively eliminating the capacitive loading on the central

Figure 11.22 *(a) A guarded and shielded signal line has a much smaller effective impedance. (b) A simplified model with C representing the stray capacitance between the central conductor and guard conductor. (c) The equivalent input circuit when A is constant, and (d) when it is falling at 6 dB/octave.*

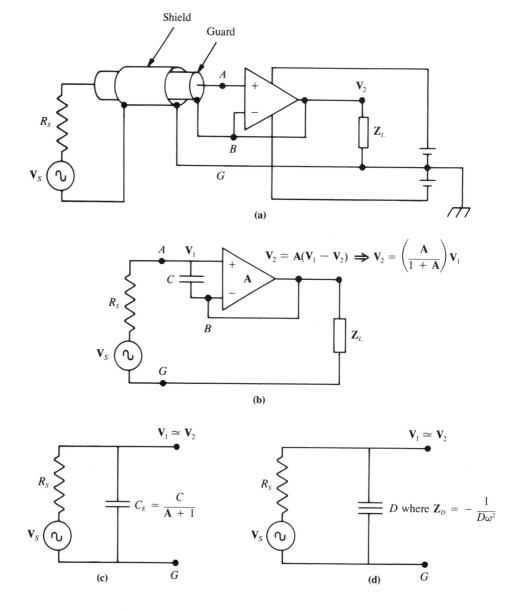

conductor. Using the simplified model of Figure 11.22b, where C is the central wire to guard capacitance, it can be shown that the input impedance looking into terminals AG is

$$\mathbf{Z}_{in} = (\mathbf{A} + 1)\mathbf{Z}_C \qquad (11.4)$$

If we assume that the amplifier has a constant open-loop gain A, it can be seen that the effective capacitance seen by the central conductor is thus reduced to $C/(A + 1)$.

However, the capacitance C provides a feedback path from the amplifier's output terminal to its positive input terminal, raising the possibility of unbounded oscillation. We can investigate this problem using the techniques of Chapter 8. A typical dominate-pole operational amplifier does not have a constant open-loop gain but instead has a transfer function given by

$$\mathbf{A} = \frac{A_0}{1 + j\omega/\omega_c} \qquad (11.5)$$

where ω_c is about 25 rad/s. Thus, at any frequency where stray capacitance is likely to be a problem, $\omega/\omega_c \gg 1$ and the open-loop transfer function is more like $A\omega_c/j\omega$ than a constant. Again the derivation is left as a problem, but if this approximate transfer function is expressed in the more convenient form $\omega_1/j\omega$, where ω_1 is the angular frequency corresponding to unity gain, it is straightforward to show that the effective impedance behaves like an unphysical double capacitor. The impedance of this device is

$$\mathbf{Z}_D = -\frac{1}{D\omega^2} \qquad (11.6)$$

where D is the magnitude of the double capacitance; its value in this case is given by

$$D = \frac{C}{\omega_1} \qquad (11.7)$$

Because of the unusual frequency dependence of this impedance, the effective input circuit shown in Figure 11.22d has a pole on the imaginary axis indicating undamped oscillations at a frequency given by

$$\omega_r = \sqrt{\frac{\omega_1}{CR_s}} \qquad (11.8)$$

In a more complete model, the Q of this resonance would be reduced to a finite value by the input impedance of the amplifier. The Q can be further reduced and turned into a design parameter by substituting a resistor R for the connecting wire between point B and the guard conductor of Figure 11.22a. The transfer function $\mathbf{V}_2/\mathbf{V}_s$ for this modified circuit is

$$\mathbf{H}(s) = \frac{CR\omega_1 s + \omega_1}{CR_s s^2 + CR\omega_1 s + \omega_1} \qquad (11.9)$$

Its derivation is left as a problem.

11.6.3 Grounds and Induced Signals

Some of the most difficult and persistent interface problems are associated with ground connections within and between devices, and are particularly troublesome when analog and digital signals are mixed. It is convenient to classify ground connections as either chassis, digital, or analog. Ideally, these three types of grounds would be maintained separately within the equipment and connected at only one point to avoid current mixing between the grounds. Practically, this division is not always possible because of preexisting connections within devices and IC chips (see Figure 10.24).

The chassis ground is intended to provide a common electrostatic potential at 0 V relative to the earth, and should not be used as the ground return for any signal. Its purpose is to provide shielding and to eliminate the shocking possibility of encountering high voltage at external points of a chassis or cabinet. Chassis ground should be connected to signal ground at no more than one point in each device, but even this limited design goal is often unattainable because of the widespread use of single-conductor coaxial cables in which the connectors are mounted so that the external shield serves as both the signal return and the chassis ground. It is possible to electrically isolate coaxial connectors from their mounting panel, but this is not always done on commercial equipment.

Since chassis ground is usually connected to the local AC power ground (the third power cord wire) within each device for safety reasons, it is sometimes impossible to establish a common chassis potential between devices powered by distinct (as in separate buildings) or faulty AC power sources. In cases where this ground imbalance is large, there is little recourse but to maintain electrical isolation between the offending devices, using transformer or optical coupling for all signals that must be exchanged.

Significant voltage drops can also occur along the interconnecting wires of a circuit, and are especially important on ground return lines that carry large currents. At low frequencies the magnitude of these added signals is determined by the small resistance of the conductors and mechanical connectors, but at higher frequencies the inductance of the wires can also have a significant effect. In any analog design in which both high- and low-level signals are present, it is vitally important to consider the location of all high-current loops.

Figure 11.23a shows a two-stage, high-gain amplifier with good power and ground connections. Note especially that the high-current loop i_L from the positive supply through the load impedance (there is an equivalent loop from the negative supply) is well isolated from the low-level signals producing i_S at the first amplifier. In particular, i_L does not produce voltage drops on either the power lines running to the first stage or on the ground line between points A and B. By contrast, the circuit shown in Figure 11.23b suffers from both of these flaws: The output current i_L produces a voltage drop Δv_1 in the power line to the first amplifier, and more importantly produces a voltage drop Δv_2 that transfers directly to the input terminals of the first amplifier.

Even with a properly designed circuit such as in Figure 11.23a, it is possible to make some serious mistakes when fabricating the actual circuit. The input signal loop indicated by the current path i_S is particularly important. This low-level

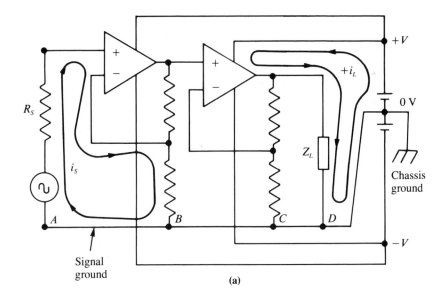

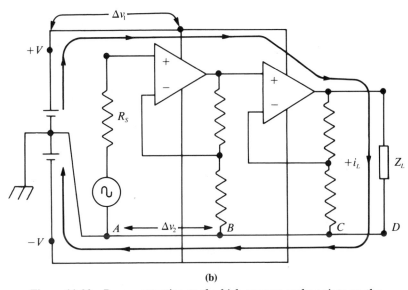

*Figure 11.23 Proper attention to the high-current paths points up the
difference between (a) good and (b) bad grounding techniques.*

signal loop can be influenced by externally generated electric and magnetic fields
acting either independently or as a propagated high-frequency electromagnetic
wave.

Electric fields in the vicinity of the input loop result in charge redistribution on
the stray capacitances of the input loop, allowing time-varying electric fields to
produce AC current in the input loop. Because of the compensating effects of

reduced voltage drops but higher stray capacitance, reducing the size of the input loop has little effect on this problem. Because the induced signal is a current, the problem is most severe for high-impedance circuits; in this case elimination of the fields by positioning or shielding is the only solution.

Time-varying magnetic field lines that link the physical loop followed by i_S will produce an additional EMF in the loop. This can be a problem for both high- and low-impedance circuits but can be reduced by moving the input wires close together, eliminating the enclosed area, and thereby reducing the magnetic flux lines that cut the loop. Twisted pairs used as the signal and return lines are particularly effective, since flux lines tend to cancel in successive twists of the pair.

Above a few megahertz, electromagnetic or radiofrequency (RF) radiated waves can become a problem. Any wire in the input circuit can act as an antenna, producing a signal matching the passing wave. Shielding, physical location, and reduction of the physical size of the input circuit can all be used to reduce this type of pickup. If the offending radiation is produced by high-current rather than high-voltage sources, it may also be possible to reduce the problem at the source by reducing the physical size of the current loop generating the signal.

11.6.4 Bypass Capacitors

It is accepted technique to place capacitors across the power and ground connections of most high-frequency digital and analog IC chips. These bypass capacitors are most effective when physically located very near the chip. This technique is always a good idea on digital circuits, where it serves the dual purpose of filtering the power supply at the chip and providing a local source of charge for the highest-frequency signals. This latter effect is often overlooked but is vitally important: It keeps the high-frequency currents off the power lines, thereby dramatically reducing the physical size of the current loop followed by these signals and minimizing the amount of RF radiation generated by the state changes of the chip. In a mixed digital and analog system, these capacitors can mean the difference between success and failure.

With analog circuits, the use and placement of bypass capacitors must be approached with more caution. Because two power connections and ground are usually involved, there are three possible places to locate two bypass capacitors: Both capacitors may run from power to ground, or one may run from $+V$ to $-V$ and the other from one of these to ground. In many cases the best arrangement will be found by trial and error, but the paths taken by the output currents from the amplifier should always be considered. The circuit of Figure 11.23a is a good example. If a bypass capacitor is placed from the positive terminal of the second amplifier to signal ground, and if the ground connection happens to be physically located between circuit points A and B, then the high-frequency current path for i_L will be altered to produce unwanted voltage drops in the input loop. The best physical connection for the ground side of this bypass capacitor is between points C and D of the circuit. Likewise, the bypass capacitor of the first amplifier is most effective if it connects to signal ground between points A and B of the circuit.

1. If the load register instruction of a CPU specifies the operation, the register, and the memory location:

(a) What is the minimum length of an instruction (in bytes) that can load a byte from any location in a 64K-byte memory?

(b) If it takes 500 ns to obtain a single byte from memory, how long will it take the CPU to fetch the instruction and load the byte? [*Ans.:* 2 μs]

2. Using fundamental NAND, AND, and OR logic elements with two, three, or four inputs, devise a circuit that will be false only when presented with the 8-bit address $F1_{16}$.

3. If a computer like that shown in Fig 11.1 takes 2 μs to execute a read or write memory instruction as described in Problem 1, estimate the time needed to transfer a single byte of information from an external device to memory. Assume that the I/O device is memory-mapped and also takes 500 ns to return a byte.

4. (a) Using only one control signal from the CPU and one response from the memory unit, sketch a read-memory timing diagram that uses handshaking like the write diagram of Fig. 11.6b. Indicate the relative timing of the address and data lines. (b) Assuming that the CPU is fabricated using edge-triggered D-type flip-flops like those shown in Figures 9.23c and 9.23d, draw a 4-bit CPU register and show its connection to the data and control lines of your timing diagram.

5. Assuming that you have write-enable (WENB), data write (WRITE), and refresh (RSFH) signals as shown in Figure P11.5, devise a logic circuit that will allow the dynamic memory of Fig. 11.7 to be loaded with DATA-IN during the WRITE strobe and refreshed by DATA-OUT during the RSFH strobe. Assume that external logic guarantees that WRITE always occurs in the middle of a WENB true pulse, and that RSFH will never occur in or near a WENB pulse. You can make connections only at the three external signal pins of this device.

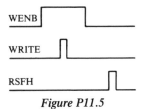

Figure P11.5

6. With MEM used to represent a 16-bit memory reference address, the Intel 8086 instruction MOV AL,MEM will transfer a byte of data to the AL register from a memory byte whose 20-bit address is determined as shown in Fig. 11.10. If the segment register (DS in this case) contains the number $FA10_{16}$, what value of MEM will fetch the byte from the memory location $FD521_{16}$?

7. The following is a Motorola 68000/68010 instruction sequence to transfer 100 bytes of information from one block of memory to another.

	Operation	Operand	Comments
	MOVE.W	#99,D1	Move 16-bit number (99) into register D1.
	LEA	BLOCK1,A6	32-bit address of first byte of BLOCK1 to A6.
	LEA	BLOCK2,A7	32-bit address of first byte of BLOCK2 to A7.
LOOP	MOVE.B	(A6)+,(A7)+	Move byte from A6 address to A7 address and increment both address registers.
	DBF	D1,LOOP	Decrement D1 and loop until D1 = −1. LOOP appears here as a 16-bit relative address. Next instruction follows.

In contrast to the 68000, the 68010 has special electronics that allows it to avoid having to repeatedly fetch the last two instructions and the relative LOOP address from memory. If each 8- or 16-bit memory access takes 500 ns, and other times can be neglected, compare the total execution time of this program for each of these microprocessors. The instruction itself is 16 bits long and includes all register identification, but the other numbers on the above lines will require additional memory accesses. For example, the second line accesses memory three times. [Ans.: 254 μs versus 105.5 μs]

8. Synchronization of the transmitter and receiver is a serious problem with RS-232C serial data transmission and must be accomplished using the data stream itself. Normally this is accomplished by looking for the first rising edge following a long string of "marking" signals. However, one 7-bit character is useful because it has only one rising edge independent of whether the transmitted parity is odd or even. If MARK corresponds to "1," write this 7-bit number, and sketch the RS-232C signal for both even and odd parity.

9. *(a)* Determine an expression for the input impedance to the short section of terminated transmission line shown in Figure P11.9. *(b)* In the limit where the $(\Delta x)^2$ term can be neglected, show that this input impedance reduces to R_T when $R_T = \sqrt{\rho_L/\rho_C}$. *(c)* Explain how this result can be used to show that the input impedance to the transmission line of Figure 11.18b is also R_T.

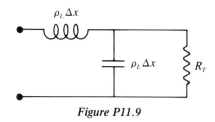

Figure P11.9

10. *(a)* Sketch a received signal like the ones shown in Figure 11.19c when the initial square-pulse waveform persists for a time that is five times longer than T_{round}. Assume that 20% of the signal amplitude is lost at each reflection. *(b)* Repeat for a pulse that is 20 times longer than T_{round}.

11. Redraw Figure 11.21a with a two-conductor shielded cable replacing the single-conductor coaxial cable.

12. Show that the effective capacitance looking into the *AG* terminals of the amplifier in Figure 11.22b is $C/(\mathbf{A} + 1)$ for any $\mathbf{A}$.

13. *(a)* If the open-loop gain $\mathbf{A}$ of the amplifier in Figure 11.22b is approximated by $\omega_1/j\omega$ in a frequency range where $|\mathbf{A}| \gg 1$, show that the input impedance looking into the *AG* terminals is given by Eq. (11.6). *(b)* Show that the equivalent circuit of Figure 11.22d has a resonant frequency given by Eq. (11.8).

14. The circuit of Figure 11.22a can be given a better transient response if the wired connection between *B* and the guard conductor is replaced by a resistor *R*. The circuit shown in Figure P11.14 is the equivalent of this improved design.

   (a) Using the approximations of the previous problem, derive an expression for the input impedance.
   (b) Show that the resulting transfer function $\mathbf{V}_1/\mathbf{V}_S$ is given by Eq. (11.9).

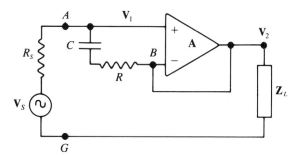

Figure P11.14

15. *(a)* Determine the value of the resistance *R* that will result in critical damping for the circuit of the previous problem. *(b)* If $R_S = 10^6 \ \Omega$, $C = 50$ pF, and $\omega_1 = 2\pi \times 10^6$ rad/s, compare the low-pass corner frequencies of the unguarded equivalent circuit of Figure 11.21b and this guarded and critically damped circuit. Express this result in hertz, not radians per second.

12

Signal Analysis

12.1   Introduction

In this chapter we extend the signal analysis discussion of Chapter 3 to the more powerful techniques of the Fourier and Laplace transforms. We have already seen that a signal can be analyzed into any useful set of constituent or basis signals, but only investigated the Fourier series decomposition into $\sin(n\omega t)$ and $\cos(n\omega t)$ terms. A signal can be more generally described as a sum of phasors, each with a different amplitude, phase, and frequency, but we are not even restricted to frequency analysis; a totally different description is as a sum of narrow pulses, each having a different amplitude or area and occurring at a different time (see Figure 12.7). This pulse description is analogous to the numerical integration techniques of integral calculus and opens a new perspective to signal composition.

Some of the mathematical methods, terminology, and concepts introduced here are so widely used that it is easy to forget that they are all derived from the principle of superposition (see Figure 3.1) and relate only to linear systems. Real devices such as diodes and transistors are not linear, so the methods can be applied only to linear models of these devices. Under well-defined operating conditions for the devices, this approach yields useful results; but it is wise to remember that some approximation is always involved.

Although it is not necessary to have a working relationship with analysis methods for most electronic applications, a general understanding of the material can provide some important new insights and viewpoints. This chapter will present the methods in some detail, although generally without proofs or other mathematical refinements.

12.2 Frequency-Domain Signal Analysis

The Fourier series expansion described in Chapter 3 provides a starting point from which we can develop more powerful methods of signal analysis. In this section we develop three additional signal expansion methods, each of which is a generalization of the preceding one. Taken together they show the relationship between the Fourier series techniques, which are most frequently discussed in introductory courses, and the Laplace transform methods, which are most useful in electronic applications.

12.2.1 Complex Fourier Series

One problem with the Fourier series expansion of Chapter 3 is that it uses real trigonometric functions $\sin(n\omega t)$ and $\cos(n\omega t)$ instead of the more efficient phasor notation $e^{jn\omega t}$. The conversion to complex notation is straightforward and simplifies the series expressions. In analogy with Fourier series Eq. 3.2, a complex $\mathbf{f}(t)$ can be expanded as a sum of phasors according to the expression

$$\mathbf{f}(t) = \sum_{n=-\infty}^{\infty} \mathbf{C}_n e^{jn\omega_0 t} \tag{12.1}$$

The constant term of Eq. (3.2) is included in this sum, since $e^0 = 1$, and the factors of 2 in that earlier equation have been eliminated as a result of summing from $-\infty$ to ∞. As with the real Fourier series, the angular frequency ω_0 is $2\pi/T$, where T is the period of the signal. The complex amplitudes $\mathbf{C}_n$ are given by

$$\mathbf{C}_n = \frac{1}{T} \int_{-T/2}^{T/2} f(t) e^{-jn\omega_0 t}\, dt \tag{12.2}$$

Like all complex numbers, each of these coefficients, $\mathbf{C}_n$, can be expressed in phasor form—in this case $C_n e^{j\theta_n}$—showing that they determine the amplitude and phase of each phasor component in the expansion of $\mathbf{f}(t)$.

If $\mathbf{f}(t)$ is a real function $f(t)$, then $\mathbf{C}_{-n} = \mathbf{C}_n{}^*$, and Eqs. (12.1) and (12.2) can be reduced to the more common forms given by Eqs. (3.2) and (3.3). Thus, the real-variable Fourier series is contained in this complex formulation.

The complex Fourier series expansion is not particularly useful in its own right, but serves as a stepping stone to the more powerful methods of the Fourier transform.

12.2.2 Fourier Transform

As we have seen, it is possible to obtain the Fourier series expansion coefficients for any periodic signal, that is, one that repeats itself at regular intervals and extends from $-\infty$ to $+\infty$. Although this condition is easily met in mathematical exercises, laboratory signals can only approximate it since they are generally zero except during some finite interval when the equipment is turned on. The single

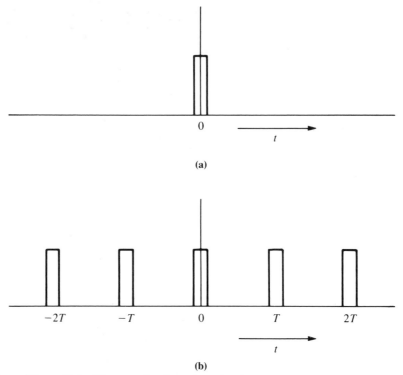

Figure 12.1   The true signal consists of a single pulse centered on t = 0, but it can be approximated with a signal that has the pulse repeating with period T.

pulse shown in Figure 12.1a is an extreme example: It is not periodic and cannot be expanded in a Fourier series.

However, a single pulse can be approximated by a periodic signal as shown in the figure. Clearly, the approximation is improved by increasing the period T, since the fictitious pulses are then further away from the true signal. The periodic approximation can be expanded in a Fourier series, but as T increases, the corresponding angular frequency ω_0 and the frequency difference between consecutive terms in the sum of Eq. (12.1) become smaller. In the limit of infinite T the approximation becomes perfect, and Eq. (12.1) can be converted to an integral over $d\omega$, and the discrete coefficients $\mathbf{C}_n$ in Eq. (12.2) convert to a continuous function $\mathbf{F}(j\omega)$.

There are several equivalent ways to define the Fourier transform and its inverse, depending on where the normalization constant 2π is placed; we will use the asymmetric form, where the constant is placed entirely in the inverse transform. With this choice, the Fourier transform is given by the relatively uncomplicated

$$\mathbf{F}(j\omega) = \mathscr{F}[\mathbf{f}(t)] = \int_{-\infty}^{\infty} \mathbf{f}(t)e^{-j\omega t}\, dt \qquad (12.3)$$

and the inverse transform becomes

$$\mathbf{f}(t) = \mathscr{F}^{-1}[\mathbf{F}(j\omega)] = \frac{1}{2\pi} \int_{-\infty}^{\infty} \mathbf{F}(j\omega)e^{j\omega t} \, d\omega \qquad (12.4)$$

Note the similarities between Eqs. (12.1) and (12.4), and between Eqs. (12.2) and (12.3).

The Fourier transform thus maps a complex signal $\mathbf{f}(t)$ into a complex function $\mathbf{F}(j\omega)$. This function describes the frequency content of the signal $\mathbf{f}(t)$ and plays the same role in the inverse transformation as the discrete coefficients $\mathbf{C}_n$ in the complex Fourier series expansion—or the a_n and b_n coefficients in the normal Fourier series expansion.

12.2.3 The Delta Function

It is clear that the Fourier transform of a periodic signal must yield the same results as the complex Fourier series, but we need a way to relate the continuous function $\mathbf{F}(j\omega)$ to the discrete coefficients $\mathbf{C}_n$.

For this and other purposes the delta function $\delta(x)$ is most useful. This function is operationally defined by the integral Eq. (12.8), but it can also be visualized as the limit of a tall, narrow pulse as shown in Figure 12.1a. If the area A of this pulse is held constant as we approach the limit of zero width and infinite height, the limiting shape will be given by the expressions

$$f(t) = 0 \qquad t \neq 0$$
$$\text{and} \qquad \int_{-\infty}^{\infty} f(t) \, dt = A \text{ volt-sec} \qquad (12.5)$$

The delta function has the same shape as this signal and is given by

$$\delta(t) = \frac{f(t)}{A} \text{ sec}^{-1} \qquad (12.6)$$

Expressed in terms of a general variable x, a delta function located at point x_0 is described by

$$\delta(x - x_0) = 0 \qquad x - x_0 \neq 0$$
$$\text{and} \qquad \int_{-\infty}^{\infty} \delta(x - x_0) \, dx = 1 \qquad (12.7)$$

This function is often used in an expression like $A\delta(x - x_0)$ to describe an infinitely narrow pulse of area A located at a position x_0. When it is necessary to sketch such a function, we will use a vertical arrow whose height is proportional to A, as has been done in Figure 12.4a and 12.4f.

The delta function is defined mathematically by the identity

$$f(x_0) = \int_{-\infty}^{\infty} f(x) \, \delta(x - x_0) \, dx \qquad (12.8)$$

which is useful for evaluating certain types of integrals. As an example and to express the delta function in another of its many guises, consider the Fourier transform of a unit area pulse expressed as a delta function $\delta(t)$:

$$\mathbf{F}(j\omega) = \int_{-\infty}^{\infty} \delta(t) \, e^{-j\omega t} \, dt \qquad (12.9)$$

If the independent variable x in Eq. (12.8) is changed to t and $f(t)$ is identified with $e^{-j\omega t}$, comparison shows that the transform integral evaluates to

$$\mathbf{F}(j\omega) = e^{-j\omega 0} = 1 \qquad (12.10)$$

This surprisingly simple result says that the frequency spectrum of $\delta(t)$ is constant: All frequencies are present in equal amounts.

Now that we have the frequency spectrum $\mathbf{F}(j\omega)$ of the delta function, it can be used in the inverse transform of Eq. (12.4) to yield the original delta function. Applying the inverse Fourier transform to Eq. (12.10) gives

$$f(t) = \delta(t) = \frac{1}{2\pi} \int_{-\infty}^{\infty} e^{j\omega t} \, d\omega \qquad (12.11)$$

This new expression for the delta function shows that an infinitely narrow pulse of unit area can be formed from an integral sum of rotating phasors, all with $1/2\pi$ amplitude.

Suppose that we wish to go to the other extreme and describe a function composed of only a single oscillation at angular frequency ω_0. The requirement of a single-frequency, time-varying signal is equivalent to describing the frequency spectrum with a delta function

$$\mathbf{F}(j\omega) = \delta(\omega - \omega_0) \qquad (12.12)$$

This statement can be verified by taking the inverse Fourier transform:

$$\mathbf{f}(t) = \frac{1}{2\pi} \int_{-\infty}^{\infty} \delta(\omega - \omega_0) \, e^{j\omega t} \, dt \qquad (12.13)$$

$$= \frac{1}{2\pi} e^{j\omega_0 t}$$

The delta function in frequency thus describes a *single* phasor rotating at frequency ω_0. Using this, we see that the Fourier transform of a periodic signal can be expressed by an $\mathbf{F}(j\omega)$ that is a sum of terms like $\mathbf{C}_n \delta(\omega - n\omega_0)$.

12.2.4 Fourier Transform Applications

The Fourier transform and its inverse allow us to view a signal as either a function of time or a function of frequency. When a signal's shape is given as a function of time, it is being described in the time domain; if given by its frequency content, it is being described in the frequency domain. As long as all functions are such that the integrals can be performed (actual signals derived from real apparatus always meet this requirement), there will be a one-to-one relationship between the two descriptions and either can completely determine the signal.

Several of the more basic theorems relating to Fourier transforms are presented without proof in Table 12.1. These theorems allow us to quickly relate a few known transforms to a large number of signals. The scaling theorem describes the effect of stretching or compressing a signal along the time axis; it will be used

Table 12.1   Fourier Transform Theorems

Name	Signal	Transform
	$\mathbf{f}(t)$	$\mathbf{F}(j\omega)$
	$\mathbf{g}(t)$	$\mathbf{G}(j\omega)$
1 Linearity	$A\mathbf{f}(t) + B\mathbf{g}(t)$	$A\mathbf{F}(j\omega) + B\mathbf{G}(j\omega)$
2 Reversal	$\mathbf{f}(-t)$	$\mathbf{F}(-j\omega)$
3 Scaling	$\mathbf{f}(t/a)$	$\lvert a\rvert\mathbf{F}(aj\omega)$
4 Shift	$\mathbf{f}(t - t_0)$	$\mathbf{F}(j\omega)e^{-j\omega t_0}$
5 Phasor modulation	$\mathbf{f}(t)e^{j\omega_0 t}$	$\mathbf{F}(j\omega - j\omega_0)$
6 Cos modulation	$\mathbf{f}(t)\cos(\omega_0 t)$	$\mathbf{F}(j\omega - j\omega_0) + \mathbf{F}(j\omega + j\omega_0)$
7 Convolution	$\displaystyle\int_{-\infty}^{\infty} \mathbf{f}(t')\mathbf{g}(t - t')\, dt'$	$\mathbf{F}(j\omega)\mathbf{G}(j\omega)$

and discussed more fully in some following special cases. The cos modulation theorem describes the effect of modulating the amplitude of a constant-frequency sinusoid with a function $f(t)$ and hence could be used to describe the frequency content of an AM (amplitude-modulated) radio station. We will use it later to determine the frequency content of a sinusoidal signal that is switched on and then off after a finite time. The convolution theorem and its variations are of particular importance in advanced work. We will describe one application of the convolution integral later in this chapter.

Although the Fourier transform of the phasor $e^{j\omega t}$ is a single delta function at frequency ω, the transform of its real (or imaginary) part produces two delta functions: one at ω and one at $-\omega$. This result is apparent from the linearity theorem and the cosine (or sine) expansions:

$$\cos(\omega t) = \frac{1}{2} e^{j\omega t} + \frac{1}{2} e^{-j\omega t}$$
$$\sin(\omega t) = \frac{1}{2j} e^{j\omega t} + \frac{1}{2j} e^{-j\omega t} \tag{12.14}$$

However, since negative frequencies are unphysical and simply a reflection of the positive frequency side of the transform, the added terms that distinguish real $f(t)$ signals from their complex phasor representations are of little consequence in most applications.

The Fourier transform of a general $\mathbf{f}(t)$ will be complex, but there are some important classes of real functions that have simpler transforms. This can be demonstrated by expanding a general $\mathbf{f}(t)$ into a sum of its even part $\mathbf{f}_e(t)$ and its odd part $\mathbf{f}_o(t)$

$$\mathbf{f}(t) = \mathbf{f}_e(t) + \mathbf{f}_o(t) \tag{12.15}$$

The Fourier transform can then be written

$$\mathbf{F}(j\omega) = \int_{-\infty}^{\infty} \mathbf{f}_e(t)e^{-j\omega t}\, dt + \int_{-\infty}^{\infty} \mathbf{f}_o(t)e^{-j\omega t}\, dt \tag{12.16}$$

The exponentials can be expanded using

$$e^{-j\omega t} = \cos(\omega t) - j\sin(\omega t) \tag{12.17}$$

where the first term is even in both t and ω and the second is odd in both variables. Since an integral over dt from $-\infty$ to $+\infty$ is zero if the total integrand is odd in t, Eq. (12.16) reduces to

$$\mathbf{F}(j\omega) = \int_{-\infty}^{\infty} \mathbf{f}_e(t) \cos(\omega t)\, dt - j \int_{-\infty}^{\infty} \mathbf{f}_o(t) \sin(\omega t)\, dt \qquad (12.18)$$

Note that the two terms in this expression are respectively even and odd in the variable ω.

Equation (12.18) can be used to determine several simple relationships between an $\mathbf{f}(t)$ and its corresponding $\mathbf{F}(j\omega)$. For example, if $\mathbf{f}(t)$ is even and real, then $\mathbf{F}(j\omega)$ will also be even and real; if $\mathbf{f}(t)$ is odd and real, then $\mathbf{F}(j\omega)$ will be odd and imaginary; if $\mathbf{f}(t)$ is odd and imaginary, then $\mathbf{F}(j\omega)$ will be odd and real. Note that the odd and even content of a sinusoidal signal is determined by its absolute phase: $\cos(\omega t)$ is even, $\cos(\omega t + \pi/4)$ has odd and even content, and $\cos(\omega t + \pi/2)$ is odd.

If $f(t)$ is real, as it must be for all physical signals, then the most general transform will be of the form

$$\mathbf{F}(j\omega) = E(\omega) + jO(\omega) \qquad (12.19)$$

where $E(\omega)$ and $O(\omega)$ represent real, even and odd functions in the variable ω. If a real signal $f(t)$ is even, as in Figure 12.4, then it will have a transform that is real and easily displayed. A real laboratory signal, defined only at $t > 0$, can always be converted to an even signal by assuming that $f(-t) = f(t)$.

One way around the problem of displaying a complex frequency spectrum is to use the magnitude of Eq. (12.19):

$$\begin{aligned}|\mathbf{F}(j\omega)| &= \sqrt{\mathbf{F}^*\mathbf{F}} \\ &= \sqrt{E^2(\omega) + O^2(\omega)}\end{aligned} \qquad (12.20)$$

Because each term is squared, the magnitude must always be an even function, symmetric around zero. It is therefore common to plot $|\mathbf{F}(j\omega)|$ or $|\mathbf{F}(j\omega)|^2$ only in the positive frequency region. The squared version of this spectrum is called the energy or power spectrum of the signal $\mathbf{f}(t)$.

Since the phase information is lost in the squaring process, different $\mathbf{f}(t)$ can have the same power spectrum, and it is generally not possible to regenerate $\mathbf{f}(t)$ from its power spectrum. Nevertheless, the power spectrum is a convenient way to display the frequency content of a signal.

Two examples for signals that extend in time from $-\infty$ to $+\infty$ are shown in Figures 12.2 and 12.3. For these signals of infinite duration, the Fourier transform is a line spectrum with the same frequency content as the corresponding Fourier series. (To get the same amplitude for both series and transform terms, it is necessary to add the transform's $+\omega$ and $-\omega$ amplitudes.) Note that although both of the example waves shown in the figures have the same fundamental frequency, the triangle wave has much less high-frequency content than the square wave.

The transforms of several important signals are shown in Figure 12.4. Of particular interest is the Gaussian, given by

$$f(t) = Ae^{-(1/2)(t/\sigma)^2} \qquad (12.21)$$

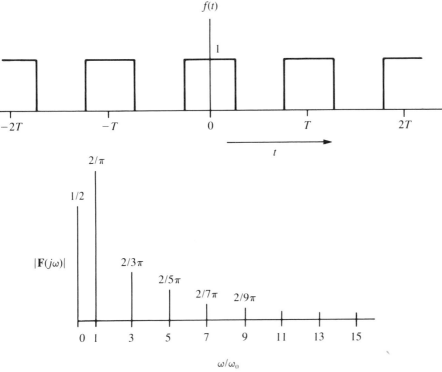

Figure 12.2 *The magnitude of the Fourier transform of an infinitely
long square wave. The square of this function would be the power
spectrum.*

where A is the maximum height of the curve and σ determines the width or standard deviation of the curve. This function is symmetric around $t = 0$; it is therefore real and even, and its transform must also be real and even. In fact, the
transform of Eq. (12.21) is also a Gaussian, given by

$$F(j\omega) = \sqrt{2\pi}A\sigma^2 e^{-(1/2)(\sigma\omega)^2} \tag{12.22}$$

The amplitude of this transform is related to the amplitude and width of the signal
by the scaling theorem in Table 12.1. Note that when σ increases, this $f(t)$ gets
broader while its transform $F(j\omega)$ gets narrower.

This simple relationship is an example of a general "uncertainty" principle
that can be applied to all signals: If the signal is localized in time, it must contain
a broad spectrum of frequencies; and if it contains a narrow spectrum of frequencies, it must be broadly distributed in time. The extremes are characterized by the
delta function and the sinusoid. In quantum mechanics, similar mathematics is
used to describe the Heisenberg uncertainty principle.

Another important Fourier transform pair is the square pulse and the $\sin(\theta)/\theta$
or sinc function. The example pulse shown in Figure 12.4 is positioned as an even

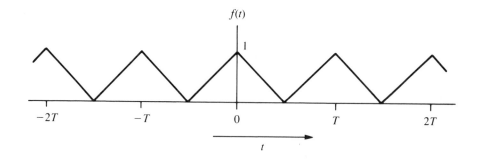

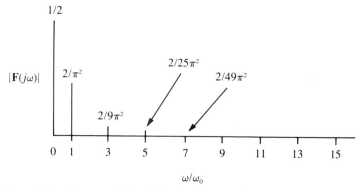

Figure 12.3   The magnitude of the Fourier transform of an infinitely long triangular waveform.

function; its transform is a real and even function also. If the pulse is zero everywhere except

$$f(t) = A \qquad \frac{-a}{2} < t < \frac{a}{2} \qquad (12.23)$$

then its transform is

$$F(j\omega) = Aa \frac{\sin(a\omega/2)}{a\omega/2} \qquad (12.24)$$

Again the widths of these two functions are inversely related; as the pulse gets narrower, the effective width of its transform gets larger.

A common signal is a cosine wave of finite length, easily produced in the laboratory by turning a sinusoidal signal generator on and off. Mathematically, this signal can be expressed as the product of an infinite cosine wave and a square pulse of unit height. The cos modulation theorem from Table 12.1 defines the transform of this product if $f(t)$ is taken to be the square pulse of Eq. (12.23). The resulting transform is the sum of two sinc functions: one centered at ω_0 (the first term), and one centered at $-\omega_0$.

The example in Figure 12.5a shows a cosine function that is turned on for only two cycles. (Note that the signal is positioned on the time axis to make an even function with a real and even transform). Figure 12.5b shows the magnitude of the $+\omega_0$ term in the Fourier transform of the time-limited cosine wave. The com-

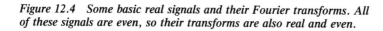

*Figure 12.4   Some basic real signals and their Fourier transforms. All
of these signals are even, so their transforms are also real and even.*

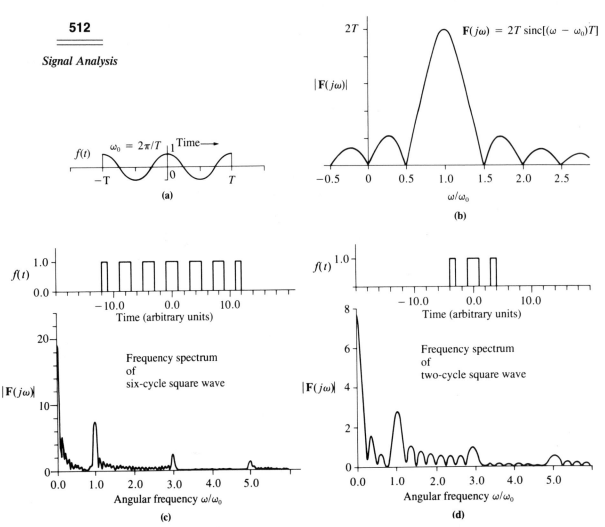

Figure 12.5 *(a) A two-cycle cosine wave, and (b) its frequency spectrum shown as the magnitude of the first term of the Fourier transform obtained from the cos modulation theorem in Table 12.1. (c) The spectrum for a square wave of finite extent. (d) The spectrum for a square wave of the same frequency but generated for only half as many cycles.*

plete transform at positive frequencies will have a contribution from the positive frequency tail of the matching sinc function centered at $-\omega_0$. The amplitude of the sinc function is proportional to the amplitude of the cosine wave.

A square wave of finite extent can be treated in a similar manner. Just as for the cosine wave, this time-varying signal can be formed from the product of an infinite square wave and a square pulse of unit height. Since an infinite square wave has the frequency spectrum given in Figure 12.2, where each line corresponds to an infinite cosine wave, when we limit the extent of the square wave we are also limiting the extent of each of the cosine waves.

The result is that the frequency spectrum of a finite square wave has a sinc function shape replacing each line in the spectrum of an infinite square wave. The amplitude of each function is again proportional to the amplitude of each sinusoidal oscillation, as specified by the magnitude of each frequency line. Since the overall transform would have to describe all of these sinc functions centered at different frequencies, an analytic expression for the overall transform is clearly an infinite sum of sinc functions.

The frequency spectra of Figures 12.5c and 12.5d are derived from the magnitude of the sum of sinc functions just described. In the first example, the sinc functions are narrow with respect to the line spacing of the infinite square wave, and the two contributions to the transform are clearly evident. In the second example (Figure 12.5d), the square wave is cut off after only two cycles, resulting in broad sinc functions that are not so clearly resolved.

In general, if a composite signal can be formed as the product of a signal of infinite time extent and a single pulse (not necessarily square), then the general shape of its Fourier transform is easy to predict. The infinite-duration signal will have a line spectrum, and the transform of the product will replace each line with the transform of the pulsed signal, appropriately shifted in frequency and scaled in amplitude.

12.2.5   The Laplace Transform

The Laplace transform is structurally similar to the Fourier transform and includes the latter as a special case. Its defining expression can be obtained directly from that of Fourier transform by replacing the imaginary frequency variable $j\omega$ in Eq. (12.3) with the complex frequency variable $\mathbf{s} = \sigma + j\omega$. This substitution gives

$$\mathbf{F(s)} = \mathcal{L}[\mathbf{f}(t)] = \int_{-\infty}^{\infty} \mathbf{f}(t)e^{-st}\,dt$$

$$= \int_{-\infty}^{\infty} \mathbf{f}(t)e^{-\sigma t}e^{-j\omega t}\,dt$$

$$(12.25)$$

Note that we use the symbol $\mathcal{L}[\mathbf{f}(t)]$ to indicate the Laplace transform of the function $\mathbf{f}(t)$. Because the exponential term $e^{-\sigma t}$ grows without bound for negative times, it is common to restrict the application of this transform to functions $\mathbf{f}(t)$ that are zero until after $t = 0$; such signals are known as causative signals.

The restriction to causative signals is equivalent to using any signal but integrating only from $t = 0$ to infinity:

$$\mathbf{F}_s(\mathbf{s}) = \int_0^{\infty} \mathbf{f}(t)\,e^{-st}\,dt \qquad (12.26)$$

In this form the integral is known as a single-sided Laplace transform, but since we will restrict our discussion to causative signals, we do not need to make a distinction between these two types of Laplace transforms. Although the restriction to causative signals does not limit our description of physical signals (which are zero before the apparatus is switched on), it does preclude the use of some simple mathematical examples, as for example $A \cos(\omega t)$.

The inverse Laplace transform can also be obtained directly from the inverse Fourier transform by replacing the basis function $e^{j\omega t}$ with e^{st}.

$$\mathbf{f}(t) = \frac{1}{2\pi} \int_{-\infty}^{\infty} \mathbf{F(s)} \, e^{st} \, d\omega \tag{12.27}$$

Note that the integration in this expression continues to be over the variable ω. Fortunately, it is possible to make effective use of the Laplace transform and its inverse without actually performing any of these transformation integrals.

It is still helpful to think of the inverse transform as the analysis of the signal $\mathbf{f}(t)$ in terms of component signals, just as with the Fourier techniques, but in this case the basis signals are not sinusoids, but include all variations on the function

$$\mathbf{u}(t) = e^{st} = e^{\sigma t} e^{j\omega t} \tag{12.28}$$

where both σ and ω can take on negative values. At times greater than zero, the real part of this unit-amplitude function can describe several distinct laboratory signals: (1) When σ is negative, the real part is a sinusoidal oscillation whose amplitude decays as time increases. (2) When σ is positive, the amplitude of the oscillation increases as time increases. (3) The special case when $\sigma = 0$ corresponds to a constant-amplitude oscillation as in the Fourier transform. (4) If $\omega = 0$, the functions display no oscillation, just the decreasing or increasing exponential behavior.

The basis functions of the Laplace transform span the family of transient signals that can be produced by a simple linear system following a disturbance. Consequently, the application of this transform is not just to a signal but to the device that processes the signal.

12.3   Time-Domain Signal Analysis

The Fourier methods are most convenient for describing smoothly varying signals that persist for a long time; such signals have a limited frequency content. However, a single pulse is described by a sinc function in the frequency variable as shown in Figure 12.4b, and the sinc function gets broader as the pulse gets narrower. In the limiting case—as the pulse width is reduced to zero—the signal can be described in terms of a delta function in time, but its frequency spectrum would then be flat and contain an infinite number of sinusoidal components, as shown in Figure 12.4a.

A disturbance whose duration is short compared to the natural time constants of a system is known as an impulse. Even though the impulse does not simplify through frequency analysis, it generally produces a relatively simple response when applied to a linear system.

An example of an impulse applied to a mechanical system would be a clapper against a bell or a pencil thumped against a desk. These examples can be approximately described by linear systems, and the response, which in these cases is mechanical and generally audible, can be well described by one or more decaying oscillations. A tuning fork, when struck gently, will vibrate in a simple manner and can be well described by a single, slowly decaying oscillation. The natural system responses are clearly related to the basis functions of the Laplace transform.

If a system is initially in an equilibrium or resting condition, the application of an impulse will typically result in an output signal that moves away from zero, reaches a peak, and then returns to zero—possibly with some overshoot and oscillation. A typical system output $p(t)$ resulting from the application of a short pulse of unit area at $t = 0$ is shown in Figure 12.6b. Similarly, a driving pulse of area A occurring at time t' would produce the response shown on Figure 12.6c. In this case the output remains at zero until $t = t'$ and then follows the function $Ap(t - t')$. Figure 12.6d shows the situation when a second pulse occurs before the effects of the first have died away. The principle of superposition allows us to obtain the composite output easily: It is simply the sum of the individual responses.

As the shape of a single applied pulse approaches the shape of a delta function (narrower and taller while maintaining unit area), the output signal $p(t)$ will approach a function $h(t)$ known as the impulse response or the transfer function in the time domain. In the language of linear differential equations, the impulse response is the transient solution evaluated for a particular choice of initial conditions. This transfer function has a one-to-one relationship with the transfer function in the frequency domain $\mathbf{H}(j\omega)$, so that either can be used to represent the response of a linear system.

In general, the impulse response will be a linear combination of several of the basis functions given by Eq. (12.28), but in many cases (especially far away from $t = 0$) the response can be adequately approximated using only one (or a complex conjugate pair) of the basis functions.

The delta function and the impulse response can be used to describe the response to an input signal of arbitrary shape. To demonstrate, we first restate the definition of $h(t)$ mathematically. With reference to Figure 12.6a, if a real input voltage pulse is

$$f(t) = A\,\delta(t) \tag{12.29}$$

where A defines the area of the delta function in units of volt-seconds, then the output signal is given by

$$g(t) = Ah(t) \tag{12.30}$$

Note that the basic functions $\delta(t)$ and $h(t)$ both have units of inverse time but can be converted to more familiar voltage signals by the simple ploy of multiplying by an area factor equal to one volt-second.

A general signal $f(t)$ can be described by a linear combination of delta functions if each delta function is located at a different time t' and given an area that is proportional to $f(t')$. Because the delta function occupies an infinitely small width, the linear combination takes the form of an integral and turns out to be a restatement and reinterpretation of Eq. (12.8):

$$f(t) = \int_{-\infty}^{\infty} f(t')\,\delta(t - t')\,dt' \tag{12.31}$$

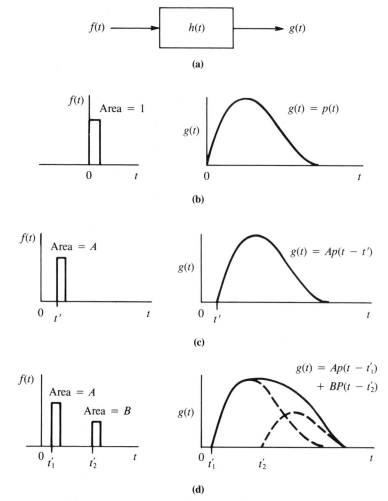

Figure 12.6 *(a) Linear system characterized by impulse response function h(t). (b) Response of a linear system to an impulse of unit area applied at t = 0. (c) Response of the same system to an impulse of area A applied at t = t'. (d) Response of the system to the application of two impulses. In the limit where the impulse becomes proportional to the delta function, the p(t) of these curves would become the impulse response function h(t).*

The resulting picture is like a limiting case of the numerical integration technique shown in Figure 12.7.

When the composite signal $f(t)$ of Eq. (12.31) passes through the linear system, it becomes the total output signal $g(t)$. The process is the same as that shown in Figure 3.1, but the discrete sum over f_n indicated there is now an infinite sum over $f(t') \delta(t - t') dt'$, and the subscript n has become the integration variable t'. A particular component delta function $\delta(t - t')$ has an area $f(t') dt'$ and will

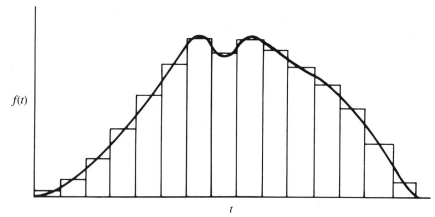

Figure 12.7 Approximation of a general signal by a sum of narrow impulse functions.

therefore generate an impulse response $f(t')\ h(t - t')\ dt'$. The output signal is the integral sum of these responses and is given by

$$g(t) = \int_{-\infty}^{\infty} f(t')\ h(t - t')\ dt'$$ (12.32)

This expression is known as a convolution integral and is the prescription by which $h(t)$ converts an input signal $f(t)$ into an output signal $g(t)$. If desired, the upper limit of infinity on the integral can be replaced by t, since the impulse response $h(t - t')$ is zero whenever $t - t'$ is less than zero.

12.4 Connection Between the Time and Frequency Domains

The complex impedance methods make it reasonably straightforward to obtain the frequency response $\mathbf{H}(j\omega)$, even for complicated circuits. To determine a circuit's effect on a signal, we can Fourier analyze the input signal, multiply each term by $\mathbf{H}(j\omega)$, then recombine the terms to yield the output signal. But if the input signal is an impulse (or other abruptly changing signal), it will require many Fourier terms for an adequate description. In this situation, the alternative provided by the time-domain transfer function $h(t)$ is most helpful. Since this function defines the network's response to an ideal impulse, it will naturally resemble the network's response to a real impulse or even an abrupt step.

To demonstrate the mathematical relationship between $\mathbf{H}(j\omega)$ and $h(t)$, we start with the convolution integral of Eq. (12.32). Using $f*h$ to denote the convolution integral and the symbol $\mathcal{L}$ to denote the Laplace transform, we can compactly write the Laplace transform of Eq. (12.32) as

$$\mathcal{L}[g(t)] = \mathbf{G}(\mathbf{s}) = \mathcal{L}[f*h]$$ (12.33)

An important Laplace transform theorem (which we leave as a problem) allows us to reduce this double integral to the product of two Laplace transforms:

$$\mathscr{L}[g] = \mathscr{L}[f*h] = \mathscr{L}[f]\,\mathscr{L}[h]$$
$$= \mathscr{L}[h]\,\mathscr{L}[f] \tag{12.34}$$

Denoting the transformed expressions by capital letters, this equation becomes

$$\mathbf{G(s)} = \mathbf{H(s)}\,\mathbf{F(s)} \tag{12.35}$$

Along the imaginary axis where $\mathbf{s} = j\omega$, we have the special case

$$\mathbf{G}(j\omega) = \mathbf{H}(j\omega)\,\mathbf{F}(j\omega) \tag{12.36}$$

If we identify $\mathbf{F}(j\omega)$ with the input signal to a four-terminal network and $\mathbf{G}(j\omega)$ with the output signal, then $\mathbf{H}$ must indeed be the frequency-domain transfer function. Thus, we see that

$$\mathbf{H(s)} = \mathscr{L}[h(t)]$$

$$\tag{12.37}$$

and

$$h(t) = \mathscr{L}^{-1}[\mathbf{H(s)}]$$

Given an $\mathbf{H}(j\omega)$ obtained directly from the analysis of an electrical circuit, only two steps are needed to derive an exact expression for the impulse response $h(t)$. The first step is simple: Just replace each occurrence of $j\omega$ in $\mathbf{H}(j\omega)$ with $\mathbf{s}$ to form $\mathbf{H(s)}$. The functional forms are identical, and only the variable changes. A little care is required, since terms like $-\omega^2$ in $\mathbf{H}(j\omega)$ are really $(j\omega)^2$. The second step is to take the inverse Laplace transform of $\mathbf{H(s)}$ to obtain $h(t)$. When necessary, this step can usually be performed by the use of tables of Laplace transforms and their inverses; the entries in Table 12.2 are particularly useful in electronics. For complicated circuits, the exact derivation of $h(t)$ by this method can become fairly difficult.

However, if we only need the general character of the impulse response (Does the output ring like a tuning fork, or is it critically damped?), then the procedure can be quite simple. Since the most persistent parts of $h(t)$ generally have the most practical importance in circuit development, we can often get by with an approximate form of $h(t)$ that is valid only at large t. In these very common situations, the $h(t)$ functions obtained from the simple circuit examples of the following section can also have application to more complicated circuits.

12.4.1 Poles and Zeros in H(s) and the Impulse Response Function h(t)

From the discussion in Section 3.7.1, we know that $\mathbf{H(s)}$ is the ratio of two complex polynomials in $\mathbf{s}$ and is defined, except for a scale factor, by the poles and zeros of this function. Except for a real multiplicative constant, which we take to be 1 in this section, the general form of $\mathbf{H}$ is

$$\mathbf{H(s)} = \frac{(\mathbf{s} - \mathbf{a_1})(\mathbf{s} - \mathbf{a_2}) \cdots (\mathbf{s} - \mathbf{a_n})}{(\mathbf{s} - \mathbf{b_1})(\mathbf{s} - \mathbf{b_2}) \cdots (\mathbf{s} - \mathbf{b_m})} \tag{12.38}$$

$f(t)$	$F(s) = \mathcal{L}[f(t)]$
1 $f(t)$	$F(s) = \int_0^\infty f(t)\, e^{-st}\, dt$
2 $\mathbf{a}f(t) + \mathbf{b}g(t)$	$\mathbf{a}F(s) + \mathbf{b}G(s)$
3 $e^{\mathbf{a}t}f(t)$	$F(s - \mathbf{a})$
4 $f(\mathbf{a}t)$	$F(s/\mathbf{a})/\mathbf{a}$
5 $df(t)/dt$	$sF(s) - f(0^+)$
6 $\delta(t)$	1
7 $U(t) = \int_{-\infty}^t \delta(t')\, dt'$	$1/s$
8 $U(t)e^{\mathbf{a}t}$	$1/(s - \mathbf{a})$
9 $\delta(t) + U(t)\mathbf{a}e^{\mathbf{a}t}$	$s/(s - \mathbf{a})$
10 $U(t)te^{\mathbf{a}t}$	$1/(s - \mathbf{a})^2$
11 $U(t)[1 + \mathbf{a}t]e^{\mathbf{a}t}$	$s/(s - \mathbf{a})^2$
12 $U(t)(Ae^{\mathbf{a}t} + Be^{\mathbf{b}t})$ where $A = -B = 1/(\mathbf{a} - \mathbf{b})$	$\dfrac{1}{(s - \mathbf{a})(s - \mathbf{b})}$
13 $U(t)(Ae^{\mathbf{a}t} + Be^{\mathbf{b}t})$ where $A = (\mathbf{a} - \alpha)/(\mathbf{a} - \mathbf{b})$ $B = (\mathbf{b} - \alpha)/(\mathbf{b} - \mathbf{a})$	$\dfrac{(s - \alpha)}{(s - \mathbf{a})(s - \mathbf{b})}$
14 $U(t)(Ae^{\mathbf{a}t} + Be^{\mathbf{b}t} + Ce^{\mathbf{c}t})$ where $A = \dfrac{1}{(\mathbf{a} - \mathbf{b})(\mathbf{a} - \mathbf{c})}$ $B = \dfrac{1}{(\mathbf{b} - \mathbf{a})(\mathbf{b} - \mathbf{c})}$ $C = \dfrac{1}{(\mathbf{c} - \mathbf{a})(\mathbf{c} - \mathbf{b})}$	$\dfrac{1}{(s - \mathbf{a})(s - \mathbf{b})(s - \mathbf{c})}$
15 $U(t)(Ae^{\mathbf{a}t} + Be^{\mathbf{b}t} + Ce^{\mathbf{c}t})$ where $A = \dfrac{(\mathbf{a} - \alpha)(\mathbf{a} - \beta)}{(\mathbf{a} - \mathbf{b})(\mathbf{a} - \mathbf{c})}$ $B = \dfrac{(\mathbf{b} - \alpha)(\mathbf{b} - \beta)}{(\mathbf{b} - \mathbf{a})(\mathbf{b} - \mathbf{c})}$ $C = \dfrac{(\mathbf{c} - \alpha)(\mathbf{c} - \beta)}{(\mathbf{c} - \mathbf{a})(\mathbf{c} - \mathbf{b})}$	$\dfrac{(s - \alpha)(s - \beta)}{(s - \mathbf{a})(s - \mathbf{b})(s - \mathbf{c})}$ for $\alpha \ne \beta$ and $\mathbf{a} \ne \mathbf{b} \ne \mathbf{c}$

Notes:

1. The time-dependent function, $f(t)$, is assumed to be zero until just after $t = 0$.

2. The notation $f(0^+)$ in line 5 means the limiting value of $f(t)$ as $t \to 0$ from the right, i.e., the value of $f(t)$ just after $t = 0$.

3. $\delta(t)$ is the delta function, and $U(t)$ is its integral, the unit step function.

4. A sometimes useful theorem is $f(t) = sF(s)$
$$t = 0^+ \qquad s = \infty$$

The $\mathbf{a}_k$ represent points on the s-plane where $\mathbf{H}$ is zero and are thus known as the zeros of $\mathbf{H}$. Similarly, the $\mathbf{b}_k$ correspond to points where $\mathbf{H}$ is infinite and are known as the poles of $\mathbf{H}$. Since these few points determine $\mathbf{H}$, they must also provide a complete description of the linear system and must also define $h(t)$. Although the methods for taking the inverse Laplace transform of the general $\mathbf{H}$ are beyond the scope of this discussion, we can make use of Table 12.2 to show some important special cases.

Consider the simple case when $\mathbf{H}$ has no zeros and only one pole,

$$\mathbf{H(s)} = \frac{1}{\mathbf{s} - \sigma_1} \tag{12.39}$$

where σ_1 is a real number. This transfer function describes a single-pole, low-pass filter (see Eq. 3.57), but we have suppressed the constant factor and introduced an explicit minus sign into the denominator. In this form, the function has a single pole on the real axis at $\mathbf{s} = \sigma_1(= -1/RC)$. Line 8 of Table 12.2 gives the inverse as

$$h(t) = U(t)e^{\sigma_1 t} \tag{12.40}$$

where $U(t)$ is a unit step function defined by

$$U(t) = 0 \quad t \leq 0 \tag{12.41}$$
$$\text{and} \qquad U(t) = 1 \quad t > 0$$

The only effect of this step function is to require that the function $h(t)$ remain at zero until after $t = 0$. At times greater than zero, the impulse response is simply an exponential:

$$h(t) = e^{\sigma_1 t} \quad \text{for } t > 0 \tag{12.42}$$

Thus, if a circuit has a frequency domain transfer function that is like Eq. (12.39), then its response to an impulse applied at $t = 0$ will be given by this $h(t)$. If σ_1 is negative (as it is for a low-pass filter), the exponential will decay with time; if σ_1 is positive, the exponential will increase with time; if σ_1 is zero, $h(t)$ will be constant.

The simplest two-pole transfer function is

$$\mathbf{H(s)} = \frac{1}{(\mathbf{s} - \mathbf{a})(\mathbf{s} - \mathbf{b})} \tag{12.43}$$

This transfer function can also describe several circuits; the simplest is shown in Figure 12.8. Depending on circuit details, the poles $\mathbf{a}$ and $\mathbf{b}$ may be real and different, real and identical, or a complex conjugate pair. Each case results in a different functional form for $h(t)$.

When the poles are distinct, the impulse response is given by line 12 in Table 12.2:

$$h(t) = U(t) \frac{1}{\mathbf{a} - \mathbf{b}}(e^{\mathbf{a}t} - e^{\mathbf{b}t}) \tag{12.44}$$

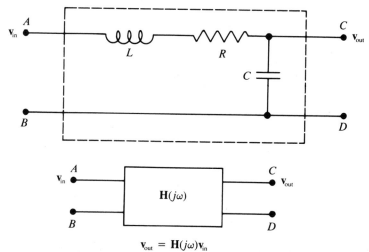

$$\mathbf{V}_{out} = \mathbf{H}(j\omega)\mathbf{V}_{in}$$

Figure 12.8   A variation of the series LCR circuit.

For real poles of the form

$$\mathbf{a} = \sigma_1$$

$$(12.45)$$

and

$$\mathbf{b} = \sigma_2$$

the impulse response reduces to the sum of two real exponentials:

$$h(t) = U(t) \frac{1}{\sigma_1 - \sigma_2} (e^{\sigma_1 t} - e^{\sigma_2 t}) \qquad (12.46)$$

The transfer function $\mathbf{H}(s)$ of a passive network such as that shown in Figure 12.8 will always have poles with negative real parts. With σ_1 and σ_2 negative, this $h(t)$ decays back to zero from a positive value at small t. A system that behaves in this way is said to be overdamped and has a relatively slow return to zero following the application of an impulse.

The two distinct poles can also appear as a complex conjugate pair,

$$\mathbf{a} = \sigma_0 + j\omega_0$$

$$(12.47)$$

and

$$\mathbf{b} = \sigma_0 - j\omega_0$$

where σ_0 and ω_0 are both real. Substituting these poles into Eq. (12.44) gives

$$h(t) = U(t) \frac{1}{2j\omega_0} [e^{(\sigma_0 + j\omega_0)t} - e^{(\sigma_0 - j\omega_0)t}] \qquad (12.48)$$

The exponentials can be expanded with Euler's equation,

$$h(t) = U(t) \frac{e^{\sigma_0 t}}{2j\omega_0} [\cos(\omega_0 t) - \cos(-\omega_0 t) + j\sin(\omega_0 t) - j\sin(-\omega_0 t)] \qquad (12.49)$$

then reduced to the required real expression by canceling the cosine terms,

$$h(t) = \frac{e^{\sigma_0 t}}{\omega_0} \sin(\omega_0 t) \qquad \text{for } t > 0 \tag{12.50}$$

Thus, we see that a complex conjugate pair of poles corresponds to an $h(t)$ that is the product of an exponential and a sinusoidal term. When σ_0 is negative, the system is said to be underdamped, with an impulse response that is an oscillation with a decaying amplitude.

A special case occurs when the poles are real and identical:

$$\mathbf{a} = \mathbf{b} = \sigma_0 \tag{12.51}$$

The impulse response in this case is obtained from line 10 of the table:

$$h(t) = U(t) \, t e^{\sigma_0 t} \tag{12.52}$$

Because σ_0 always lies between σ_1 and σ_2, a negative σ_0 will cause Eq. (12.52) to return to zero more quickly than the response function described by Eq. (12.46). A system with this impulse response is said to be critically damped.

Example 12.1 The differentiation theorem given on line 5 of Table 12.2 can be used to add entries to this table. Use this theorem to determine the inverse transform of line 9 from the inverse transform of line 8.

The theorem on line 5 can be written

$$\mathcal{L}\left[\frac{df(t)}{dt}\right] = \mathbf{s}F(\mathbf{s}) - f(0^+)$$

After taking the inverse transform of both sides and rearranging, we obtain

$$\mathcal{L}^{-1}[\mathbf{s}F(\mathbf{s})] = \frac{df(t)}{dt} + \mathcal{L}^{-1}[f(0^+)] \tag{12.53}$$

From line 8, the inverse transform of $F(\mathbf{s}) = 1/(\mathbf{s} - \mathbf{a})$ is $f(t) = U(t) \, e^{at}$, and at times greater than zero (not including zero), its derivative is

$$\frac{df(t)}{dt} = U(t) \, \mathbf{a} e^{at}$$

The limit of $f(t)$ as $t \to 0$ from positive values is just

$$f(0+) = 1$$

Substitution of these two results into Eq. (12.53) yields

$$\mathcal{L}^{-1}[\mathbf{s}F(\mathbf{s})] = U(t) \, \mathbf{a} e^{at} + \mathcal{L}^{-1}[1]$$

Entry 6 in the table says that the inverse transform of 1 is $\delta(t)$, making the final result

$$\mathcal{L}^{-1}[\mathbf{s}F(\mathbf{s})] = U(t) \, \mathbf{a} e^{at} + \delta(t)$$

as given on entry 9 of the table.

The impulse response is closely related to the Q of the LCR circuit such as that shown in Figure 12.8. For this particular circuit, the frequency-domain transfer function is

$$\mathbf{H}(j\omega) = \frac{1/j\omega C}{j\omega L + R + 1/j\omega C} \qquad (12.54)$$

and conversion to the **s**-plane is made by replacing each $j\omega$ with an **s.** This operation, followed a multiplication of numerator and denominator by sC, puts the transfer function into the standard form

$$\mathbf{H(s)} = \frac{1}{LC\mathbf{s}^2 + RC\mathbf{s} + 1} \qquad (12.55)$$

The denominator of this function is a quadratic in **s** and thus has two roots corresponding to two poles in $\mathbf{H(s)}$.

These roots are easily found by setting the denominator equal to zero,

$$LC\mathbf{s}^2 + RC\mathbf{s} + 1 = 0 \qquad (12.56)$$

then using the quadratic formula to find the roots:

$$\mathbf{a} = \frac{-RC + \sqrt{R^2C^2 - 4LC}}{2LC}$$

$$\mathbf{b} = \frac{-RC - \sqrt{R^2C^2 - 4LC}}{2LC} \qquad (12.57)$$

and

If R, L, and C are limited to their normal positive values, the first term in each of these roots is always real and negative. The sign of the term under the radical depends on the relative sizes of the three circuit elements and determines whether the two roots are either both real or both complex.

If R is sufficiently large, the term under the radical will be positive, yielding two real negative roots (poles of $\mathbf{H}$):

$$\mathbf{a} = \sigma_1 = \frac{-R}{2L} + \sqrt{\left(\frac{R}{2L}\right)^2 - \frac{1}{LC}}$$

$$\mathbf{b} = \sigma_2 = \frac{-R}{2L} - \sqrt{\left(\frac{R}{2L}\right)^2 - \frac{1}{LC}} \qquad (12.58)$$

and

Both of these poles are be located on the negative real axis, and the largest is always be to the right of $-R/2L$. Each of these real poles will contribute a non-oscillatory decaying exponential to the impulse response, but the rightmost pole **a** will dominate the impulse response after a sufficient time. Since this overdamped circuit has no oscillation, all of its stored energy is lost in the first cycle and its Q factor is zero.

If R is sufficiently small, the term under the radical in Eq. (12.57) will be negative, and **a** and **b** will be a complex conjugate pair:

$$\mathbf{a} = \sigma_0 + j\omega_0$$

$$\mathbf{b} = \sigma_0 - j\omega_0$$

(12.59)

and

where

$$\sigma_0 = \frac{-R}{2L}$$

(12.60)

and

$$\omega_0 = \sqrt{\frac{1}{LC} - \left(\frac{R}{2L}\right)^2}$$

(12.61)

In Chapter 2 we found the resonant angular frequency of this circuit,

$$\omega_r = \frac{1}{\sqrt{LC}}$$

and defined the Q factor of this circuit to be

$$Q = \frac{\omega_r L}{R}$$

In terms of these two quantities, the real part of these complex roots is always negative and given by

$$\sigma_0 = \frac{-\omega_r}{2Q}$$

(12.62)

and the angular frequency in the imaginary parts is given by

$$\omega_0 = \omega_r \sqrt{1 - \frac{1}{4Q^2}}$$

(12.63)

From Eq. (12.62) we see that $|\sigma_0|$ becomes smaller as Q increases, causing the oscillating impulse response to decay more and more slowly. A high-Q circuit is thus also highly underdamped, and its impulse response will "ring" with slowly diminishing amplitude like the audible sound from a struck bell. Equation (12.63) shows that the ringing frequency of the impulse response will be very close to the circuit's resonant frequency.

Critical damping is achieved when the expression under the radical in Eq. (12.57) is zero. This requirement produces a useful equation for selecting the relative sizes of circuit components. As a result, critical damping is often a convenient design goal. However, in practice the preferred circuit is usually slightly underdamped, since it will exhibit a somewhat faster return to equilibrium without obvious oscillation.

Figure 12.9 shows the locus of points for the poles of Eq. (12.55) as R is increased from zero. It is significant that a two-pole electrical system with positive values for R, L, and C can never have a pole in the first quadrant and will therefore always have an $h(t)$ that returns to zero.

In order to get poles in the first quadrant, we will have to introduce active circuit elements. Something that behaves like a negative resistance would work

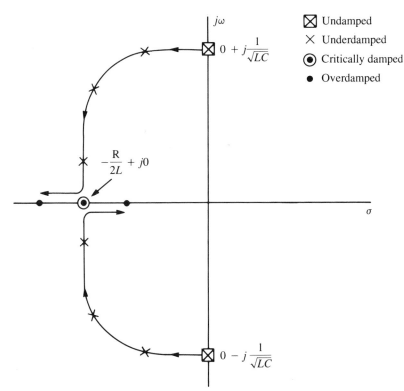

Figure 12.9   The locus of points of the poles of **H(s)** *for the series
LCR circuit as R is increased from zero.*

for the *LCR* circuit discussed above. But since poles in the first quadrant describe
an impulse response signal with an ever-increasing amplitude, they also imply that
the energy stored in the circuit continues to increase even after the driving signal
has disappeared. This effect is possible only if the four-terminal network has its
own internal energy source. Thus, poles in the first (and fourth) quadrant can
occur only when a circuit has some internal energy source (such as an amplifier)
other than the driving signal.

12.4.3   Identifying the Most Persistent Term in h(t)

Complex circuits will produce more complicated **H(s)** functions with many poles
and zeros. However, if we have a method (it can always be done numerically) for
finding the roots of polynomials, we can determine the locations of all poles and
zeros on the **s**-plane. If we identify $f(t)$ with $h(t)$ and **F(s)** with **H(s)**, then a quick
study of lines 8 to 15 of Table 12.2 shows that each pole of **H** defines an exponential term in the expansion of $h(t)$; a real pole always produces a term like Eq.
(12.42), whereas a complex conjugate pair always combines to an oscillatory term
like Eq. (12.50).

It can also be seen from the table that the presence of zeros in **H** does not add
exponential terms but only modifies the complex coefficients multiplying these

terms, affecting the real amplitude of the term and the phase of any oscillatory component. Because of the rapid variation of real exponential terms, the relative size of various amplitude factors is often of secondary importance—the longest-lived exponential will ultimately dominate the impulse response as in Example 12.2. An exception to this general rule occurs when a nearby zero strongly suppresses the effect of a pole, effectively removing both from the transfer function.

The location of single real poles or complex conjugate pairs thus determines the shape of a corresponding term in an $h(t)$ expansion. Typical shapes resulting from various pole placements are shown in Figure 3.13. As can be seen from that figure, the most persistent—and therefore the most important—term in the impulse response of a circuit is determined by the rightmost pole or pole pair on the s-plane. The observation is derived using the Laplace transform, but the application requires only the rightmost roots of a polynomial.

Example 12.2 Given a transfer function with no zeros and three real poles at -1 ms^{-1}, -5 ms^{-1}, and -10 ms^{-1}, make a table showing the contributions of the three exponential terms in $h(t)$ at times 0.1 ms, 0.5 ms, and 1 ms.

Using time units of milliseconds, this transfer function has the form $\mathbf{H(s)} = 1/[(s + 1)(s + 5)(s + 10)]$, and from entry 14 in Table 12.2 we find $h(t) = 0.0278e^{-t} - 0.05e^{-5t} + 0.0222e^{-10t}$. Evaluating at the required times gives:

t (ms)	$0.0278e^{-t}$	$0.05e^{-5t}$	$0.0222e^{-10t}$
0.1	0.025	0.030	0.008
0.5	0.017	0.004	0.0001
1.0	0.010	0.0003	0.000001

This result clearly shows how the rightmost pole, located at -1 ms^{-1}, dominates $h(t)$ for large values of t.

12.5   *Intuitive Techniques*

The pragmatic student may wonder if this mathematical elegance really has much to do with practical electronics, especially since the application of the methods outlined in this chapter is not always easy or obvious. Although it is certainly possible to make extensive use of electronics without giving a thought to transform techniques, an intuitive grasp of the fundamental ideas expressed in this chapter can be extremely important in some situations. The good news is that a lot of the advantages can be realized without performing the detailed mathematics.

The Fourier series allows us to frequency-analyze a periodic signal, and the Fourier transform provides a method for extending this concept to the frequency analysis of nonperiodic signals. Because the general laboratory signal can be analyzed into a sum of $e^{j\omega t}$ terms, and we have a simple method for obtaining the the transfer function $\mathbf{H}(j\omega)$, we can determine the effect of the network on each of these terms. Without even decomposing a signal in detail, the transform methods make it clear that any short or abruptly changing signal will have frequency components extending to high frequencies, whereas a long and slowly changing signal will have only a few low-frequency components.

For abruptly changing signals, where the Fourier methods become unwieldy because of many frequency components, we are able to switch to the methods arising from the Laplace transform, finding in particular that the inverse transform of $\mathbf{H(s)}$ is $h(t)$, the impulse response of the four-terminal network. Thus, when a four-terminal network is driven by an input signal closely approximated by a delta function or a step function (the integral of the delta function), its output can be approximated by $h(t)$ or $\int h(t)\,dt$. The Laplace transforms given in Table 12.2 provide an analytic result for many common circuits, but even this degree of detail is not often necessary. The most important intuitive idea arising from this technique is that $h(t)$ is a sum of exponential terms—one for each pole in $\mathbf{H(s)}$—and the rightmost pole on the complex plane (not canceled by a zero) will identify the longest-lived part of a circuit's response to a transient in the input signal.

The transfer function with its poles and zeros points up the intimate connection between the frequency response $\mathbf{H}(j\omega)$ and the impulse response $h(t)$. An awareness of this fundamental connection can often pave the way to the solution of an electronic problem, sometimes just by making the inherent incompatibility of certain design goals readily apparent.

Problems

1. Use Euler's equation to express the $\mathbf{C}_n$ of Eq. (12.2) in terms of the Fourier series coefficients a_n and b_n of Eqs. (3.3) and (3.4).

2. Suppose that the $\mathbf{C}_n$ coefficients in Eq. (12.1) are real such that $\mathbf{C}_{-n} = \mathbf{C}_n$ with $\mathbf{C}_0 = 0$. **(a)** Is the resulting $\mathbf{y}(t)$ real, imaginary, or complex? **(b)** Is it odd or even? [*Hint:* Expand the exponents with Euler's equation.]

3. If Eq. (12.1) is to describe a $\mathbf{y}(t)$ that is purely imaginary and odd, what is $\mathbf{C}_0$, and what is the relationship between $\mathbf{C}_n$ and $\mathbf{C}_{-n}$? Are the $\mathbf{C}$'s real or imaginary?

4. On the interval $-T/2$ to $T/2$, a real function $y(t)$ is everywhere zero except for a narrow pulse of width $2d$ and unit height centered at time $T/8$. If the function is periodic with period T: **(a)** Sketch three cycles of the function on the interval $-3T/2$ to $3T/2$ and label the time axis in terms of T. **(b)** With the ssumption that $d/T \ll 1$, evaluate the integral of Eq. (12.2) to show that $\mathbf{C}_n = (2d/T)e^{-jn\pi/4}$. **(c)** If $2\pi d/T \ll 1$, sketch the phasors that represent $\mathbf{C}_0$, $\mathbf{C}_1$, and $\mathbf{C}_2$. Indicate the magnitude of each phasor.

5. Equation (12.9) is the Fourier transform of a delta function centered at $t = 0$.
   (a) Write a similar equation for a delta function centered on $t = T$ and evaluate this integral using the delta function definition of Eq. (12.8).
   (b) Sketch the locus of end points of the phasor $\mathbf{F}(j\omega)$ as ω increases from zero.

6. Using the Fourier transform and its inverse in a method similar to that used to derive Eq. (12.11), show that

$$\delta(\omega) = \frac{1}{2\pi}\int_{-\infty}^{\infty} e^{-j\omega t}\,dt$$

7. Show that $\mathbf{Y}(j\omega) = \sum_{n=0}^{3} e^{-j\omega nT}$ is zero when $\omega = \pi/2T$. [*Hint:* Do it graphically on the complex plane.]

8. Two sinusoidal signals of slightly different frequencies combine to form a signal $y(t)$ that exhibits a beat frequency. **(a)** Draw $2e^{j\omega t}$, $4e^{j(\omega + \Delta\omega)t}$, and their sum on the complex plane, then determine an expression for the times when $|y(t)|$ is a minimum. **(b)** If $\Delta\omega = 8\omega/25$, what will be the value of the real part of $y(t)$ at the first of these times? **(c)** At the second?

9. Use Euler's equation to prove Eqs. (12.14).

10. By substitution of $f(t/a)$ into Eq. (12.3), prove the scaling theorem of Table 12.1.

11. By substitution of $f(t - t_0)$ into Eq. (12.3), prove the shift theorem of Table 12.1.

12. The unit step function $U_0(t)$ is defined as

$$U_0(t) = 0 \qquad \text{when } t \le 0$$

and
$$U_0(t) = 1 \qquad \text{when } t > 0$$

Sketch the odd and even parts of this function.

13. Use a trigonometric identity to find the odd and even parts of $y(t) = A \cos(\omega t + \theta)$.

14. Find the Fourier transform of $f(t) = A$. [*Hint:* Use the result of the previous problem.]

15. Find the Fourier transform of $f(t) = A \sin(\omega_0 t)$. [*Hint:* Expand the sin to exponentials, then use the result of Problem 14.]

16. Using the exponent expansion of the cosine function, find the Fourier transform of $f(t) = A \cos^2(\omega_0 t)$.

17. By counting squares on a graph, integrate $f(t)$ in Figure 12.2 between the limits $t = 0$ and $t = 2T$ and plot the integrated signal.

18. **(a)** Make a sketch of the real part of the signal

$$y(t) = 100e^{-8t}e^{j\pi t} + 2e^{-0.1t}e^{j0.5\pi t}$$

on the interval $0 \le t \le 6$. **(b)** At times $t > 1$, what angular frequency would you expect to observe on an oscilloscope display of this signal?

19. Assume that the impulse response of a system is $h(t) = te^{-t}$. If the input signal to the system is a square pulse given by $f(t) = 1$ for $0 < t < 2$ and $f(t) = 0$ elsewhere, we can obtain the approximate output by decomposing this input into a number of narrower pulses and following the procedure of Figure 12.6. Note that this pulse has area 2. **(a)** On graph paper, plot the composite response $g(t)$ when the input signal is approximated by one delta function at $t = 0$ and by two delta functions at $t = 0$ and $t = 2$. **(b)** Using the two-pulse approximation, determine the approximate functional form of $g(t)$ at times $t \gg 2$. [*Ans.:* $2te^{-t}$]

20. Derive Eq. (12.34) by direct substitution of the $g(t)$ convolution integral into the Laplace transform definition of Eq. (12.25). [*Hint:* Reverse the order of integration of t and t' and multiply the expression by 1 in the form $e^{-st'}e^{st'}$.]

21. **(a)** If $\sigma_0 = A\omega_0$, rewrite $h(t)$ from Eq. (12.50) in terms of A and T, the period of the oscillation. **(b)** On the time interval $t = 0$ to $t = 2T$, plot $h(t)$ when $A = -0.1$ and when $A = -3$.

22. A rearrangement of the *LCR* circuit, with R as the vertical element, will give $\mathbf{H(s)} = s/[(s - \mathbf{a})(s - \mathbf{b})]$, where $\mathbf{a}$ and $\mathbf{b}$ are given by Eqs. (12.47).
    (a) Use line 13 of Table 12.2 to show that for this circuit,

$$h(t) = \frac{e^{\sigma_0 t}}{\omega_0} [\sigma_0 \sin(\omega_0 t) + \sigma_0 \cos(\omega_0 t)]$$

(b) Following the method of Problem 21, plot this $h(t)$ when $A = -0.1$ and when $A = -3$.

23. Suppose that a three-pole $\mathbf{H(s)}$ is given by

$$\mathbf{H(s)} = \frac{1}{(s - \mathbf{a})(s - \mathbf{b})(s - 0.1\sigma_0)}$$

where $\mathbf{a}$ and $\mathbf{b}$ are given by Eq. (12.47). If $\sigma_0 = -2 \text{ s}^{-1}$ and $\omega_0 = 0.1 \text{ s}^{-1}$, write an approximate expression for $h(t)$ when $t > 5$ s.

24. A series *LCR* circuit like Figure 12.8 has $Q = 1000$ and a resonant angular frequency of 100 rad/sec. If it is excited by a unit impulse, what will be the amplitude of its oscillation at $t = 0$, $t = 5$, and $t = 10$ seconds?

25. Following the technique used in the discussion of Fourier transforms, show that $\mathcal{L}^{-1}[1] = \delta(t)$.

26. Using the inverse Laplace transforms from Table 12.2, find an expression for the impulse response $h(t)$ of a network whose frequency domain transfer function is

$$\mathbf{H}(j\omega) = \frac{1}{(1 + j\omega RC)(1 + 3j\omega RC)}$$

27. Using the inverse Laplace transforms from Table 12.2, find an expression for the impulse response $h(t)$ of the network shown in Figure P12.27.

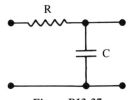

Figure P12.27

13

Noise and Statistics

13.1   Introduction

In experimental applications of electronics, the desired information signal is usually superimposed on a significant admixture of unwanted contributions loosely known as "noise." Although the term is frequently used to identify unwanted signals induced by changing electromagnetic fields, our concern in this chapter is only with random noise of a statistical nature such that it can be reduced by repeated measurement.

To introduce the treatment of random noise, we will briefly review some statistics, relate it to the standard models for random noise sources in electronic devices, then develop some of the mathematical techniques that allow us to separate signals from noise in experimental arrangements. The analysis in this chapter is done with continuous analog signals, but in the following chapter we will extend it to the domain of discrete signals as seen by a data acquisition computer.

13.2   Statistical Description of Signals with Random Noise Content

The following presentation is neither a complete nor a rigorous treatment of statistical analysis. The intention is to connect the terminology of more formal treatments to the practices of everyday signal analysis.

13.2.1 The Measurement Process

Let us begin by assuming a measurement process that yields a signal $x(t)$ over some time interval 0 to T. We will also assume that this process can be repeated as often as desired, with each measurement producing a different signal $x_k(t)$, where k runs from 1 to the number of measurements N, and t is over the same time interval 0 to T for each measurement.

The signal $x_k(t)$ resulting from a single measurement can be thought of as the sum of two different signals: an exactly repeatable "deterministic" signal $u(t)$, such as a sine wave, which is fixed by knowledge of amplitude, frequency, and phase; and a noise component, composed of a purely random signal which we identify with the underlined notation $\underline{x}_k(t)$. As shown in Figure 13.1, the complete signal is then given by

$$x_k(t) = u(t) + \underline{x}_k(t) \tag{13.1}$$

where the lack of a subscript on $u(t)$ indicates that this signal is exactly reproducible in repeated measurements. The goal of a measurement is to separate these two components by a combination of a priori theoretical knowledge and repeated measurement of $x_k(t)$.

The collection of $x_k(t)$ functions is known as an ensemble of functions and can be represented by the notation $\{x_k(t)\}$. Thus each $x_k(t)$ is a continuous function, but when the ensemble is evaluated at a particular time t_1, it yields a set of discrete values $\{x_k(t_1)\}$ as indicated in Figure 13.2.

Figure 13.1   A typical measured signal $x_k(t)$, which is assumed to be the sum of a deterministic part $u(t)$ and a random process $\underline{x}_k(t)$.

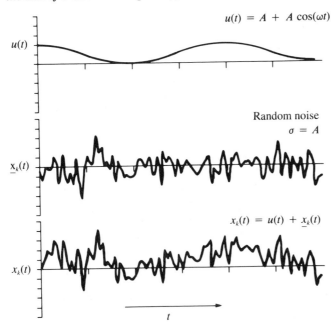

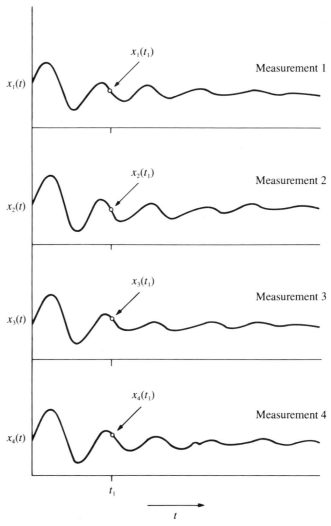

Figure 13.2   A number of repeated measurements of the same process produce an ensemble of functions $\{x_k(t)\}$ and a set of discrete values $\{x_k(t_1)\}$ at the common time t_1.

With the understanding that we are now discussing a set of values determined at a specific point t_1 in each measurement, we will temporarily simplify the notation by writing $x_k(t_1)$ as simply x_k. This set of numbers will have a mean given by the average of x_k over the N measurements

$$\bar{x}(t_1) = \bar{x} = \frac{1}{N} \sum_{k=1}^{N} x_k \qquad (13.2)$$

This average value, which we expect to approach $u(t_1)$ in the limit of infinite N, is an ensemble average at time t_1. In a typical experiment involving a finite number of repeated measurements, the mean is used as an ''estimator'' of $u(t_1)$.

The actual measurements x_k will be distributed about their mean value, and for large N the smooth curve outlining the histogram of these values determines the frequency (of occurrence) distribution function of the x_k measurements. The shape of this curve can be characterized by the ensemble average of quantities like $(x_k - \bar{x})^n$, which are known as moments of the distribution. The simplest nonzero moment has $n = 2$, is known as the variance σ^2 (the standard deviation is σ), and is defined by

$$\sigma_x^2(t_1) = \frac{1}{N} \sum_{k=1}^{N} (x_k - \bar{x})^2 \tag{13.3}$$

The variance can be put into a more useful calculational form by expanding the squared factor to give

$$\sigma_x^2 = \frac{1}{N} \left[\sum x_k^2 + \sum \bar{x}^2 - 2\bar{x} \sum x_k \right]$$
$$= \frac{1}{N} \sum x_k^2 + \bar{x}^2 - 2\bar{x}^2$$

$$\sigma_x^2 = \overline{x^2} - \bar{x}^2 \tag{13.4}$$

If the distribution of $\{x_k\}$ happens to be Gaussian (Eq. 13.32), as is usually the case, its shape is completely determined by the mean and variance and no other ensemble averages are needed.

Note a particularly useful result from Eq. (13.4): If the mean is zero, then

$$\sigma_x^2 = \overline{x^2} \tag{13.5}$$

Since the noise sources in electronic systems are assumed to be AC with mean values of zero, this relationship gives the variance whenever the mean square value of a noise source is given.

We further note that the variance in the random signal $\{\underline{x}_k\}$ is equal to the variance in the measured signal $\{x_k\}$. This can be shown by starting with Eq. (13.3) and working in the limit of infinite N. We observe that $\bar{x}$ can be replaced by u and x_k by Eq. (13.1) to give

$$\sigma_x^2(t_1) = \frac{1}{N} \sum_{k=1}^{N} \underline{x_k}^2 \tag{13.6}$$

which is the variance in $\{\underline{x}_k\}$.

The square root of the right side of Eq. (13.5) is often identified as x_{RMS}, the square root of the mean of the squared values, and is known as the root-mean-square value of $\{x_k\}$. On a schematic, a noise EMF is often labeled as simply e (or i), but its real meaning is e_{RMS}, and by Eq. (13.5) can be interpreted as the standard deviation of the random signal.

In the following discussion we will refer to a general signal like $x_k(t)$ as a signal with random content, and to a signal like $\underline{x}_k(t)$ (which must be AC with mean value zero) as a random signal.

13.2.2   The Ergodic Assumption and Time Averages

Our random signal $x_k(t)$ is said to result from a "stationary" random process if, in the limit of an infinite number of measurements, ensemble averages like $\bar{x}$ and σ_x^2 turn out to be constants independent of the evaluation time t_1. Further, a stationary random process is said to be ergodic if these ensemble averages equal the corresponding time averages over one infinitely long measurement interval. The value of the ergodic assumption is that it allows us to replace the ensemble average over many measurements with the time average over a single measurement. To illustrate, we again consider the mean and variance of a typical process.

If $y[x(t)]$ is an arbitrary function of the measured variable x, then its time average over the interval $T_2 - T_1$ is defined by the integral

$$\langle y(x) \rangle = \frac{1}{T_2 - T_1} \int_{T_1}^{T_2} y[x(t)] \, dt \tag{13.7}$$

The time average over an infinite interval can be calculated by the limiting process

$$\langle y(x) \rangle = \lim_{T \to \infty} \frac{1}{2T} \int_{-T}^{T} y[x(t)] \, dt \tag{13.8}$$

At the appropriate limits, an ergodic process measured by $\underline{x}$ will produce equivalent time and ensemble averages

$$\langle y(\underline{x}) \rangle = \bar{y}(\underline{x}) \tag{13.9}$$

Specifically, if $\underline{x}(t)$ is the random contribution to one measurement on the time interval 0 to T, then its time-averaged mean is given by

$$\langle \underline{x} \rangle = \lim_{T \to \infty} \frac{1}{T} \int_0^T \underline{x}(t) \, dt \tag{13.10}$$

and its time-averaged variance by

$$\langle \sigma_{\underline{x}}^2 \rangle = \lim_{T \to \infty} \frac{1}{T} \int_0^T (\underline{x}(t) - \langle \underline{x} \rangle)^2 \, dt \tag{13.11}$$

These must equal the corresponding ensemble averages for the mean and variance,

$$\langle \underline{x} \rangle = \bar{x} \tag{13.12}$$

and

$$\langle \sigma_{\underline{x}}^2 \rangle = \sigma_x^2 \tag{13.13}$$

where the right-hand sides are defined by Eqs. (13.2) and (13.3) in the limit of infinite N.

Note that these results apply only to the random part $\underline{x}_k(t)$ of the measured signal $x_k(t)$ in Eq. (13.1). Before Eqs. (13.10) and (13.11) can be used to estimate the mean and variance of the random component of a signal, some method must first be found to remove the time-varying deterministic part $u(t)$. Of course, the determination of $u(t)$ is usually a primary goal of the measurement process.

From this point, we assume that each noise signal results from an ergodic random process whose ensemble averages can each be determined by a time average. Fortunately, this assumption is appropriate to the vast majority of physical situations.

To make full use of the Fourier and Laplace transform methods, we will need to work with time averages over infinite time intervals. This presents a formal problem for a measured signal $x(t)$ that is known only on the interval 0 to T_m. If the signal is assumed to be zero outside this interval, the time average over an infinite interval will clearly yield zero. One way around this problem is to replace the measured $x(t)$ with an extended signal $x_a(t)$ that exactly duplicates itself on successive T_m intervals,

$$x_a(t) = x(t - nT_m) \qquad \text{for } nT_m \leq t \leq (n + 1)T_m \qquad (13.14)$$

The time average of this function over an infinite interval,

$$\langle y(x) \rangle = \lim_{T \to \infty} \frac{1}{T} \int_{-T}^{T} y[x_a(t)] \, dt \qquad (13.15)$$

is equal to

$$\langle y(x) \rangle = \frac{1}{T_m} \int_{0}^{T_m} y[x(t)] \, dt \qquad (13.16)$$

and to

$$\langle y(x) \rangle = \frac{1}{T_m} \int_{-\infty}^{\infty} y[x(t)] \, dt \qquad (13.17)$$

if $y[x(t)]$ is taken to be zero outside the measurement interval 0 to T_m.

These expressions can be viewed as estimates of the time average, the best that can be done with a signal of finite extent. The equivalent process for an ensemble average is to estimate a quantity such as the mean from only a finite number of measurements.

13.3 *Probability Density Distributions*

By definition, the value of a random variable is unpredictable from moment to moment and must be described in terms of averages such as the mean and variance. Alternatively, a random variable x can be described in terms of a probability distribution $p(x)$ over its possible values.

We first define a probability $P(x)$ by again assuming that $x_k(t)$ is a random variable that yields the ensemble $\{x_k\}$ on repeated measurements indexed by k and evaluated at a specific time t_1:

$$P(x) = \text{probability that } x_k \leq x \qquad (13.18)$$

This expression for $P(x)$ specifies the *fraction of measurements* that yield $x_k \leq x$. Clearly, $P(x)$ is a monotonically increasing function of x that varies from 0 at $x = -\infty$ to 1 at $x = \infty$. If the random variable is ergodic, we can also write

$$P(\underline{x}) = \text{probability that } \underline{x}(t) \le \underline{x} \qquad (13.19)$$

which states that $P(\underline{x})$ is also the *fraction of the time* that $\underline{x}(t) \le \underline{x}$. For an ergodic process, $P(\underline{x})$ can thus be interpreted as either a probability over measurements or a probability over time.

The function $P(\underline{x})$ is known as the cumulative probability distribution, and its differential element $dP(\underline{x})$ defines the probability that $\underline{x}_k$ lies in the interval $\underline{x} < \underline{x}_k \le \underline{x} + d\underline{x}$. This differential element can be written as

$$dP(\underline{x}) = p(\underline{x}) \, d\underline{x} \qquad (13.20)$$

where $p(\underline{x})$ is just the derivative of $P(\underline{x})$. The quantity $p(\underline{x})$ is known as the probability density function, and its integral over all $\underline{x}$ is 1:

$$\int_{-\infty}^{\infty} p(\underline{x}) \, d\underline{x} = P(\underline{x}) \Big|_{-\infty}^{\infty} = 1 - 0 = 1 \qquad (13.21)$$

Expressed in terms of the $P(\underline{x})$ distribution, the ensemble average of any well-behaved function $y(\underline{x})$ is

$$\bar{y}(\underline{x}) = \int_{-\infty}^{\infty} y(\underline{x}) \, dP(\underline{x}) \qquad (13.22)$$

and in terms of $p(\underline{x})$ is

$$\bar{y}(\underline{x}) = \int_{-\infty}^{\infty} y(\underline{x}) \, p(\underline{x}) \, d\underline{x} \qquad (13.23)$$

Since the $\underline{x}$ variable in these expressions is derived from $\{\underline{x}_k(t_1)\}$, these averages are determined at a specific time t_1. Note that the exact evaluation of the integrals requires knowledge of the full probability distribution from $-\infty < \underline{x} < \infty$; such information is available only as an assumption or as the result of an infinite number of measurements.

13.3.1 Poisson Distribution

The Poisson distribution, which is associated with the process of counting randomly occurring events, is easily derived from a few fundamental assumptions. Suppose that we repeatedly count a random process over the time interval T and find that the average number of events (counts) during this interval is $\langle n \rangle$. Now, if we subdivide the measurement time T into K smaller time intervals Δt, we can reasonably expect that each interval Δt will contain $\langle n \rangle / K$ counts. For large enough K this ratio becomes much less than 1 and can be taken as the probability p of one event occurring in the interval Δt. Thus we have

$$p = \frac{\langle n \rangle}{K} \qquad (13.24)$$

with $p \ll 1$.

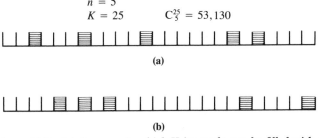

$n = 5$
$K = 25 \qquad C_5^{25} = 53,130$

(a)

(b)

*Figure 13.3   (a) One way in which K intervals can be filled with n
events and (b) a second way.*

The smallness of p allows us to neglect complications resulting from two or
more events occurring in one time interval: If two events are independent, the
probability of both occurring is the product of the two separate probabilities. Consequently, the chance of two events occurring in the same interval is p^2, and with
$p \ll 1$ can be safely neglected.

Thus, if we count n events in a measurement time T, then subdivide this time
into K very small intervals, we must have exactly one event in each of n intervals
Δt, with no events in the remaining $K - n$ intervals. The probability of having
no events in a particular interval Δt is $1 - p$. Again multiplying independent
probabilities, we find that the chance $P_1(n)$ of any specific arrangement of n filled
and $K - n$ empty intervals as shown in Figure 13.3a is given by

$$P_1(n) = p^n(1 - p)^{K - n} \qquad (13.25)$$

But we will still count n events during the time T no matter which of the K
intervals is filled, so we must multiply this probability by a factor indicating the
number of different ways the intervals can be filled. This is a standard combina-
torial problem of K things taken n at a time, and the result is known to be

$$C_n^K = \frac{K!}{(K - n)!n!} \qquad (13.26)$$

Combining these results, we find that the probability of counting n events in a
measurement time T is given by

$$P(n) = C_n^K P_1(n) = \frac{K!}{(K - n)!n!} p^n(1 - p)^{K - n} \qquad (13.27)$$

which is known as the binomial distribution. Replacing p with the expression in
Eq. (13.24) gives

$$P(n) = \frac{K!}{(K - n)!n!} \frac{\langle n \rangle^n}{K^n} \left(1 - \frac{\langle n \rangle}{K}\right)^{K - n} \qquad (13.28)$$

which can be simplified by taking the limit as $K \to \infty$. A workable procedure is
to first show that

$$\lim_{K \to \infty} \frac{K!}{(K - n)!\, K^n} = 1 \qquad (13.29)$$

and

$$\lim_{K \to \infty} \left(1 - \frac{\langle n \rangle}{K}\right)^{K - n} = e^{-\langle n \rangle} \qquad (13.30)$$

Given these results (which can be shown directly by writing out a few terms and allowing K to dominate all sums), the distribution can be reduced to

$$p(n) = \frac{e^{-\langle n \rangle}\langle n \rangle^n}{n!} \qquad (13.31)$$

which is the most useful form of the Poisson distribution. This distribution is shown in Figure 13.4 for several values of $\langle n \rangle$.

Figure 13.4   The Poisson distribution for various values of $\langle n \rangle$. The distribution is defined only at integer values of n, as indicated by the solid dots; the lines in this figure are only to guide the eye. Note that large $\langle n \rangle$ values give distributions very similar to the Gaussian distribution shown below.

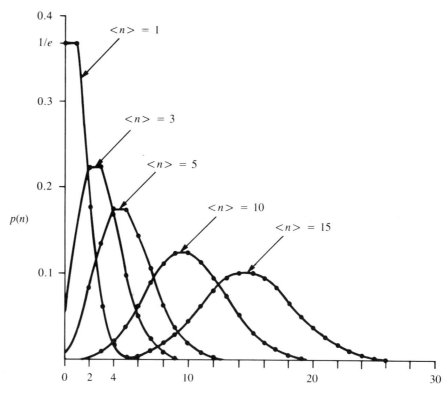

13.3.2 Gaussian Distribution

The general form of the Gaussian distribution of probability density is

$$p(x) = \frac{1}{\sqrt{2\pi\sigma^2}} e^{-(x-\langle x \rangle)^2/2\sigma^2} \tag{13.32}$$

where $\langle x \rangle$ is the mean value of the random variable x, and σ^2 is the variance. The square root of the variance is known as the standard deviation σ and identifies the values of $x - \langle x \rangle$ where the distribution has fallen to $e^{-1/2}$ of its maximum value. The standard deviation is thus a measure of the width of the Gaussian, and in this regard the following integrals are useful:

$$\int_{\langle x \rangle - \sigma}^{\langle x \rangle + \sigma} p(x)\, dx = 0.682$$

$$\tag{13.33}$$

and
$$\int_{\langle x \rangle - 2\sigma}^{\langle x \rangle + 2\sigma} p(x)\, dx = 0.954$$

These results indicate the fraction of measurements that can be expected to fall within one and two standard deviations of the mean. The Gaussian distribution plotted with mean value zero and showing the range of these two integrals is shown in Figure 13.5.

The Gaussian distribution arises naturally from a variety of random processes, including the large $\langle n \rangle$ and n limit of the Poisson distribution derived above. This important result will now be derived. For notational convenience we replace $\langle n \rangle$ with N and define a new variable $z = n - N$, which measures n with respect to the mean value N. In terms of these variables, the Poisson distribution becomes

$$p(z + N) = \frac{N^{N+z}\, e^{-N}}{(N + z)!} \tag{13.34}$$

and this can be expanded to

$$p(z + N) = \frac{N^N e^{-N}}{N!} \frac{N^z}{(N + 1)(N + 2) \cdots (N + z)} \tag{13.35}$$

where the denominator of the second fraction has z factors. Taking the natural logarithm of both sides gives

$$\ln[p(z + N)] = \ln[p(N)] + \ln\left(\frac{N}{N + 1}\right)$$

$$+ \ln\left(\frac{N}{N + 2}\right) + \cdots + \ln\left(\frac{N}{N + z}\right) \tag{13.36}$$

and for large N this can be approximated by

$$\ln[p(z + N)] = \ln[p(N)] + \ln\left(1 - \frac{1}{N}\right)$$

$$+ \ln\left(1 - \frac{2}{N}\right) + \cdots + \ln\left(1 - \frac{z}{N}\right) \tag{13.30}$$

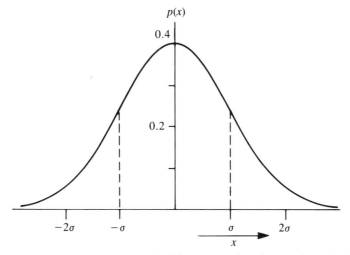

Figure 13.5 The Gaussian probability density distribution plotted with mean value of zero.

For large N, each of the last z terms can be further simplified using the small $|x|$ approximation $\ln(1 + x) \simeq x$. This gives

$$\ln[p(z + N)] = \ln[p(N)] - \sum_{k=1}^{z} \frac{k}{N} \tag{13.37}$$

The sum evaluates to $z(z + 1)/2N$, and for $|z| \gg 1$ (many counts away from the mean), we have

$$\ln[p(z + N)] = \ln[p(N)] - \frac{z^2}{2N} \tag{13.38}$$

Taking the antilog of both sides and expanding $p(N)$ with Eq. (13.34) gives

$$p(z + N) = \frac{N^N e^{-N}}{N!} e^{-z^2/2N} \tag{13.39}$$

which with the help of Stirling's approximation, $N! \simeq \sqrt{2\pi N}(N/e)^N$, takes on the Gaussian form

$$p(z + N) = \frac{1}{\sqrt{2\pi N}} e^{-z^2/2N} \tag{13.40}$$

Making the replacements $z = n - N$ and $N = \langle n \rangle$ puts this expression back into Poisson notation,

$$p(n) = \frac{1}{\sqrt{2\pi\langle n \rangle}} e^{-(n-\langle n \rangle)^2/2\langle n \rangle} \tag{13.41}$$

and shows that when a measurement yields $\langle n \rangle$ counts, the variance is given by

$$\sigma^2 = \langle n \rangle \qquad (13.42)$$

This important result is used frequently in experimental work and is loosely expressed as "the error in N is the square root of N."

13.3.3   Sum of Random Variables

In Section 1.8 we described the standard method for propagating the errors in measured quantities X and Y into a derived quantity Z. We now have the tools to approach this problem again from a statistical perspective. If Z is determined according to some functional relationship

$$Z = Z(X, Y) \qquad (13.43)$$

then a random fluctuation in Z away from its expectation value $\langle Z \rangle = Z(\langle X \rangle, \langle Y \rangle)$ can be approximated by the first two terms of its Taylor series expansion around that expectation value:

$$\Delta Z_r = \frac{\partial Z}{\partial X}\bigg|_{\langle X \rangle, \langle Y \rangle} \Delta X_r + \frac{\partial Z}{\partial Y}\bigg|_{\langle X \rangle, \langle Y \rangle} \Delta Y_r \qquad (13.44)$$

In this equation, the partial derivatives evaluate to constants, whereas ΔX_r and ΔY_r represent the random fluctuations in the measured variables X and Y. Changing notation, we write the three terms as

$$\underline{z} = \underline{x} + \underline{y} \qquad (13.45)$$

where each quantity is a random variable with mean value zero. The general probability density distribution for $\underline{z}$ will be of the form

$$p(\underline{z}) = q(\underline{x}, \underline{y}) \qquad (13.46)$$

but if the fluctuations in the random variables $\underline{x}$ and $\underline{y}$ are independent (uncorrelated), the function q will separate to give

$$p(\underline{z}) = p(\underline{x})\, p(\underline{y}) \qquad (13.47)$$

If we further assume that $p(\underline{x})$ and $p(\underline{y})$ are Gaussians, with, for example, $p(\underline{x})$ given by Eq. (13.32) with mean equal to zero,

$$p(\underline{x}) = \frac{1}{\sqrt{2\pi\sigma_x^2}}\, e^{-\underline{x}^2/2\sigma_x^2} \qquad (13.48)$$

then $p(\underline{z})$ will also be a Gaussian given by

$$p(\underline{z}) = \frac{1}{\sqrt{2\pi\sigma_{\underline{z}}^2}}\, e^{-\underline{z}^2/2\sigma_{\underline{z}}^2} \qquad (13.49)$$

An expression for the variance in $p(\underline{z})$ can be obtained using Eq. (13.4), which reduces to $\sigma_{\underline{z}}^2 = \langle \underline{z}^2 \rangle$ when $\langle \underline{z} \rangle = 0$. Using Eq. (13.45), we can then write the variance of $\underline{z}$ in terms of the $\underline{x}$ and $\underline{y}$ variables:

$$\sigma_{\underline{z}}^2 = \langle \underline{z}^2 \rangle = \langle \underline{x}^2 \rangle + 2\langle \underline{x}\underline{y} \rangle + \langle \underline{y}^2 \rangle \qquad (13.50)$$

Since the mean values $\langle \underline{x} \rangle$ and $\underline{y} \rangle$ are also zero, the first and last terms are just the variances $\sigma_{\underline{x}}^2$ and $\sigma_{\underline{y}}^2$, respectively. The term $\langle \underline{xy} \rangle$ is known as the covariance,

$$\sigma_{\underline{xy}} = \langle \underline{xy} \rangle \tag{13.51}$$

and we can show that it is zero for independent variables $\underline{x}$ and $\underline{y}$. In terms of the variances and covariances, Eq. (13.50) can be written

$$\sigma_{\underline{z}}^2 = \sigma_{\underline{x}}^2 + 2\sigma_{\underline{xy}} + \sigma_{\underline{y}}^2 \tag{13.52}$$

By analogy with Eq. (13.23), the covariance term $\langle \underline{xy} \rangle$ can be written as a double integral over the two-variable probability distribution $q(\underline{x}, \underline{y})$,

$$\langle \underline{xy} \rangle = \sigma_{\underline{xy}} = \int_{-\infty}^{\infty} \int_{-\infty}^{\infty} q(\underline{x}, \underline{y}) \, \underline{xy} \, d\underline{x} \, d\underline{y} \tag{13.53}$$

where the function $y(x)$ in (13.23) has been replaced by the product xy. For independent random variables, $q(\underline{x}, \underline{y}) = p(\underline{x}) \, p(\underline{y})$, and the integral becomes

$$\langle \underline{xy} \rangle = \sigma_{\underline{xy}} = \int_{-\infty}^{\infty} \int_{-\infty}^{\infty} p(\underline{x}) \, p(\underline{y}) \, \underline{xy} \, d\underline{x} \, d\underline{y} \tag{13.54}$$

Taking advantage of the independence of the measured variables, we rearrange this integration to yield

$$\langle \underline{xy} \rangle = \int_{-\infty}^{\infty} p(\underline{x}) \, \underline{x} \, d\underline{x} \int_{-\infty}^{\infty} p(\underline{y}) \, \underline{y} \, d\underline{y} \tag{13.55}$$

which reduces to

$$\langle \underline{xy} \rangle = \langle \underline{x} \rangle \langle \underline{y} \rangle = 0 \tag{13.56}$$

With the covariance equal to zero, Eq. (13.50) reduces to

$$\sigma_{\underline{z}}^2 = \sigma_{\underline{x}}^2 + \sigma_{\underline{y}}^2 \tag{13.57}$$

Remembering that our definitions of $\underline{x}$ and $\underline{y}$ include the partial derivative factors from Eq. (13.44), this result is the same as Eq. (1.45).

If the random variables $\underline{x}$ and $\underline{y}$ are correlated such that the distribution function of Eq. (13.46) does not separate into the product of Eq. (13.47), then $\langle \underline{xy} \rangle$ will not be zero. It is possible to extend the propagation-of-error method to include such cross-terms in the measured errors, but we will not pursue that development here.

It is also common to describe the magnitude of the covariance of two variables in terms of a correlation coefficient, defined as

$$\rho_{xy} = \frac{\sigma_{xy}}{\sigma_x \sigma_y} \tag{13.58}$$

This coefficient will always be in the range $-1 < \rho_{xy} < +1$, with both extremes indicating complete correlation.

Statistical noise sources are a fundamental part of any electronic device and can arise from several different physical processes. We will briefly discuss three types: thermal noise caused by the Brownian motion of electrons, shot noise caused by the counting statistics of electron flow, and excess or flicker noise caused by a variety of low-frequency phenomena. Each of these noise sources can be represented by a Gaussian distribution with a mean value of zero, but in real systems multiple sources of noise may not be uncorrelated as we will assume in the following discussion.

13.4.1 Thermal or Johnson Noise

Johnson noise is a result of the randomized thermal motion of electrons in a conducting medium. Because the electrons carry charge, these random motions correspond to random currents resulting in voltage fluctuations across the body of the conducting medium. The mean velocity along any direction of motion is zero, hence the mean noise voltage and current must also be zero. However, the mean squared velocity is not zero, and this yields a nonzero value for the mean squared noise voltage and current. Indeed, from kinetic theory we know that the average kinetic energy of a point mass free to move in three dimensions is $\frac{3}{2}kT$, where k is Boltzmann's constant ($k = 1.38 \times 10^{-23}$ joule/deg) and T is the absolute temperature; this provides us with the temperature dependence of Johnson noise.

In 1928, Nyquist derived an expression for the noise signal generated by the thermal motion of charges in a resistance R. If we identify $d\langle e_n^2 \rangle$ with that part of the noise EMF lying within the frequency interval df, his result can be written

$$d\langle e_n^2 \rangle = 4kRT\, df \tag{13.59}$$

If this signal passes through an amplifier or other network with transfer function $\mathbf{A}(jf/2\pi)$, the observed output signal will be

$$\langle e_n^2 \rangle = 4kRT \int_0^\infty A^2\, df \tag{13.60}$$

Assuming an idealized A^2 that is 1 on the interval $f_1 < f < f_2$, and zero otherwise, the expression clearly reduces to

$$\langle e_n^2 \rangle = 4kRTB \tag{13.61}$$

where $B = f_2 - f_1$ is the frequency bandwidth in hertz. This is the most frequently used expression for the mean square voltage associated with Johnson noise in a resistance R. If the bandwidth covers several decades in frequency, then B is approximately f_2, and this expression is sufficiently accurate. However, when the bandwidth B is narrow, the detailed shape of A^2 becomes important, and Eq. (13.60) may need to be evaluated for the specific case.

We can also express a Johnson noise source as a current EMF. Since the noise power present in a frequency interval df can be expressed as either $\langle e_n^2 \rangle / R$ or $\langle i_n^2 \rangle R$, Eq. (13.60) could also be written

$$\langle i_n^2 \rangle = \frac{4kT}{R} \int_0^\infty A^2 \, df \qquad (13.62)$$

and Eq. (13.61) converts to an equivalent expression for a current source,

$$\langle i_n^2 \rangle = \frac{4kTB}{R} \qquad (13.63)$$

Equations (13.61) and (13.63) have units of volts^2 and amperes^2, respectively, but in practice the frequency bandwidth B is often set to unity, allowing the noise signals to be described and plotted using units such as V^2/Hz. It is also common to use the root-mean-square value of a noise signal (the standard deviation of its probability distribution), which would have more intuitive units such as volts or V/Hz$^{1/2}$. Johnson noise is usually shown as a voltage source and labeled with the symbol e_n, but its meaning is really $e_{\text{RMS}} = \sqrt{\langle e_n^2 \rangle}$, the root-mean-square voltage.

Johnson noise has a flat frequency distribution over the attainable electronic spectrum and is therefore known as a "white" noise source. Note that even though the spectrum is flat, Johnson noise is more important at high frequencies, where the bandwidth B spans a broader range of frequencies.

Thus, to analyze the noise output of a circuit, we must associate a Johnson noise EMF with every resistance in the circuit, whether an actual resistor or the internal resistance of a signal generator or amplifier. Reactive circuit elements L and C do not contribute to the noise, except to the extent that actual reactive elements must be modeled with equivalent resistances. The two thermal noise models for a resistor are shown in Figure 13.6.

Figure 13.6 (a) The Johnson noise EMF associated with a resistance R is usually shown as a root-mean-square (RMS) voltage e_n, but in some cases it is useful to convert it to (b) a Norton equivalent current source.

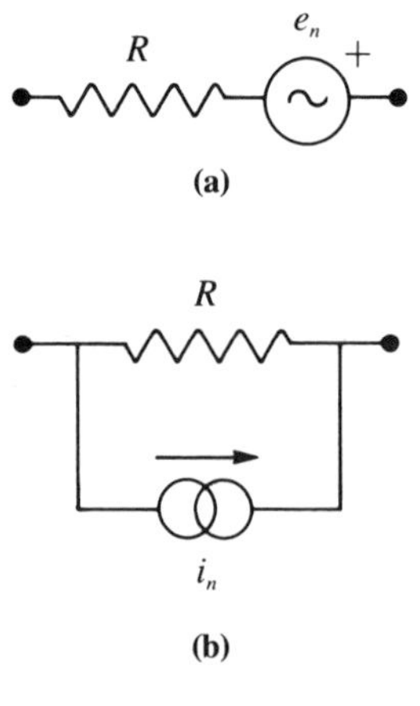

(a)

(b)

Example 13.1 Determine the RMS Johnson noise EMF associated with a 1-kΩ resistor over a 1-Hz bandwidth at room temperature.

Taking room temperature to be 293 K, Eq. (13.61) yields

$$\langle e_n^2 \rangle = 4(1.38 \times 10^{-23} \text{ joule/K})(293 \text{ K})(10^3 \text{ }\Omega)(1 \text{ s}^{-1})$$
$$= 1.6 \times 10^{-17} \text{ volts}^2$$

The root-mean-square (RMS) voltage for a 1-kΩ resistor is then

$$e_n = \sqrt{\langle e_n^2 \rangle} = 4 \times 10^{-9} \text{ volts} = 4 \text{ nV}$$

and from this result we can write the convenient room-temperature expression,

$$e_n = 4\sqrt{RB}$$

where R is in kΩ, B is in Hz, and the resulting RMS voltage is in nanovolts. Thus, a 1-MΩ resistor will contribute 0.13 mV of RMS noise over a 1-MHz bandwidth.

13.4.2 Counting Statistics or Shot Noise

A second fundamental noise source arises from the discrete nature of the charges that make up the current. This noise component is known as shot noise and was initially applied to the emission of electrons from the cathode of a vacuum tube. However, it is not limited to that application and really describes the Poisson statistics associated with counting charges as they cross some surface. It therefore applies to any current made up of a single charge component, but as shown in Example 13.2, the effect is generally significant only inside active devices such as transistors and tubes. The following plausibility argument shows how shot noise arises from statistical fluctuations.

In terms of the discrete charge q_e, a current I through a surface can be written as

$$I(t) = \frac{n q_e}{\Delta t} \tag{13.64}$$

where n is the number of charges and Δt is the observation time. If $I(t)$ has a DC component and we perform a series of experiments, each of duration Δt as shown in Figure 13.7, then we will find that the expectation value of the number of charges $\langle n \rangle$ is some number N, but that the number observed in any single experiment fluctuates according to Poisson statistics. The DC component of $I(t)$ is then

$$I = \frac{N q_e}{\Delta t} \tag{13.65}$$

Since $\langle n \rangle = N$ is large, we can approximate the Poisson distribution with a Gaussian of mean and variance N as indicated in Eq. (13.41) and use Eq. (13.65) to write the variance in n as

$$\langle \sigma_n^2 \rangle = N = I \frac{\Delta t}{q_e} \tag{13.66}$$

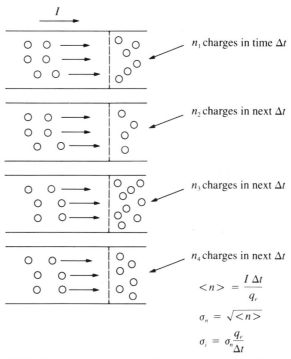

Figure 13.7 *Current composed of charges crossing a surface during the time interval Δt.*

Propagating the error (fluctuation) in the value of n through Eq. (13.64) gives the standard deviation of the current $I(t)$,

$$\sigma_I = \frac{\sigma_n q_e}{\Delta t} \tag{13.67}$$

and this is also the standard deviation σ_i of the AC noise component i. Taking the square root of Eq. (13.66) to get σ_n, we obtain

$$i_{RMS} = \sigma_i = \sqrt{\frac{q_e I}{\Delta t}} \tag{13.68}$$

If our signal detector has a maximum frequency of f_{max}, then its measurement duration Δt for a single fluctuation is half of the associated period or $1/(2f_{max})$. With this substitution, we have

$$i_{RMS} = \sqrt{2q_e I f_{max}} \tag{13.69}$$

or

$$\langle i^2 \rangle = 2q_e I f_{max} \tag{13.70}$$

By considering two different detection devices with different maximum frequencies f_1 and f_2 and using an argument similar to that in Section 13.3.3 for the variance of the difference of two signals, $\langle i^2 \rangle$ can be put into the standard form

$$\langle i^2 \rangle = 2q_e I B = 0.23 I B \ (nA)^2 \tag{13.71}$$

where $B = f_2 - f_1$ is the bandpass.

In common practice, shot noise is modeled as an AC current EMF i_n as shown in Figure 13.8, where once again the notation i_n means $i_{RMS} = \sqrt{\langle i_n^2 \rangle}$. Note that this noise contribution is a function only of the average current and the bandwidth of the detector. Although the absolute amount of shot noise increases with the average current I, the ratio i_{RMS}/I decreases with increasing average current. If the current I is really a signal current, then the "signal-to-noise ratio," I/i_{RMS}, will become worse (smaller) as the signal current decreases and fewer charges are counted.

Like Johnson noise, this shot noise contribution as a flat frequency distribution and is an AC white noise source added to deterministic signal current I.

Example 13.2 If the resistor in Figure 13.8 is at room temperature (293 K), determine the value of V above which the shot noise current exceeds the Johnson noise current.

Equating the equivalent Johnson noise current of Eq. (13.63) to the shot noise current of Eq. (13.71) gives

$$\frac{4kTB}{R} = \frac{2q_e BV}{R}$$

where the average current I has been replaced by V/R. Solving for V, we find

$$V = \frac{2kT}{q_e}$$

Plugging in the numbers at room temperature gives

$$V = \frac{2(1.38 \times 10^{-23} \text{ joule/K})(293 \text{ K})}{1.6 \times 10^{-19} \text{ coul}}$$

Figure 13.8 The current out of the resistor shown is composed of the average current I, a Johnson noise contribution e_n/R, and a shot noise component i_n.

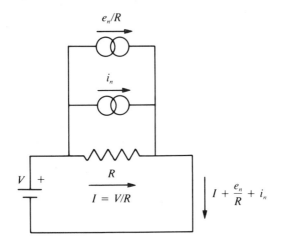

which reduces to

$$V = 50 \text{ mV}$$

Thus we see that a DC bias current from a power source may contribute shot noise that compares with Johnson noise, but the current from a typical low-voltage input signal will contribute an insignificant amount of shot noise.

This argument must be applied carefully inside active devices, where a large output current is proportional to a smaller controlling current. The charge flow in the photomultiplier tube of Figure 10.2e provides a clear example. The size of the counting fluctuation (noise/signal) $\sqrt{N}/N$ is determined by the small number of electrons N produced at the cathode, and this ratio persists after amplification by successive dynodes. The result is a noise/signal ratio that is much larger than would be generated by simply considering the number of charges in the current I at the anode.

13.4.3   Flicker or 1/f Noise

In addition to the noise sources described above, a typical electronic circuit will exhibit "excess" noise that cannot be explained by the sources just described. This excess noise is characterized by an increasing RMS amplitude at lower frequencies, and is variously described as flicker, $1/f$, or "pink" noise. The $1/f$ description is approximate only and refers to the power spectrum (Section 13.6) of the noise signal. Thus, the total $1/f$ noise in a bandwidth $B = f_2 - f_1$ can be estimated from

$$e_n{}^2 = V_e{}^2 \int_{f_1}^{f_2} df/f \tag{13.72}$$

where e_n is an RMS voltage and V_e is a proportionality constant. This expression can be integrated to give

$$e_n{}^2 = V_e{}^2 \ln\left(\frac{f_2}{f_1}\right) \tag{13.73}$$

If the noise in some particular bandwidth is given, we can use this equation to determine V_e and then find the noise in any other bandwidth. Note that if bandwidths are chosen such that f_2/f_1 is constant, then the integrated excess noise in each will be the same. Thus, the $1/f$ noise in the nine-cycle range 1 Hz to 10 Hz is the same as in the 900-kHz range 100 kHz to 1 MHz.

The amount of excess noise produced by an electronic component is very much dependent on fabrication details. For example, the resistance of a common carbon composition resistor is derived from the contact resistance between carbon granules, and normal changes at the contacts produces a relatively high flicker noise: typically 1 μV in a bandwidth $f_2/f_1 = 10$. By contrast, carbon film, metal film, and wire-wound resistors typically show 0.1 μV of flicker noise in the same bandwidth. Typically a low-noise operational amplifier with FET inputs also shows about 1 μV of flicker noise, whereas one with bipolar inputs is in the range of 0.1 μV. Using the method of Example 13.1, we see that a 1000-Ω resistor contributes about 1 μV of Johnson noise in a 10-Hz to 100-Hz bandwidth at room temperature. The relative importance of flicker noise decreases with increasing frequency. One manifestation of flicker noise is the slow drift in the offset voltage of an operational amplifier.

13.4.4 *Modeling the Noise in Real Amplifiers*

Figures 13.9a and 13.9b show a high-input-impedance, noninverting amplifier drawn first as a physical circuit and then using idealized noiseless components with the various noise contributions shown explicitly: e_s describes the Johnson noise of the source resistor, e_0 and i_0 describe the voltage and current noise

Figure 13.9 *(a) A physical noninverting amplifier circuit, and (b) its representation in terms of noiseless components. (c) A simplified model can be derived for the high-gain case where $(R_1 + R_2)/R_1 \gg 1$.*

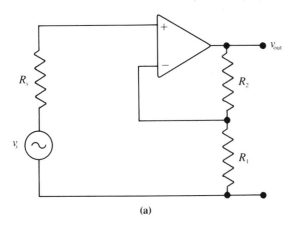

(a)

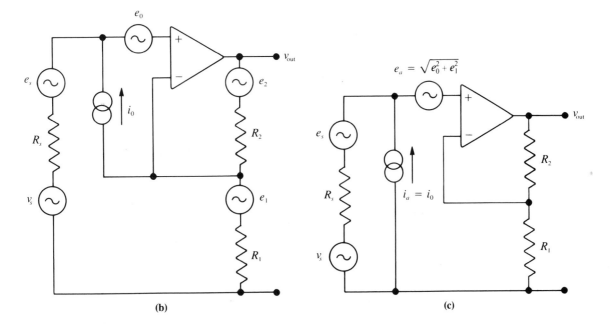

(b)   (c)

sources present in the amplifier, and e_1 and e_2 represent the Johnson noise from the gain-setting resistors. Specification sheets for low-noise operational amplifiers typically give the magnitudes of e_0 and i_0 in nV/$\sqrt{\text{Hz}}$ and pA/$\sqrt{\text{Hz}}$, often in the form of a graph showing the variation with frequency. Above 1 kHz, the best IC FET operational amplifiers attain values for e_0 and i_0 in the range of 10 nV/$\sqrt{\text{Hz}}$ and 1 pA/$\sqrt{\text{Hz}}$.

The procedure for calculating the contribution of multiple noise sources at the amplifier's output follows from the error propagation result of Eq. (13.57). If we treat the noise EMFs as ordinary signal generators, the amplifier's output signal can always be put into the form

$$v_{\text{out}} = \sum_i g_i e_i \tag{13.74}$$

where g_i is a function of the impedances and e_i represents one of the voltage or current noise sources on the schematic. Each term in this sum describes the contribution of one EMF and can be reinterpreted as the standard deviation of the output fluctuation caused by that particular EMF. Applying the arguments of Section 13.3.3 to these random fluctuations, the variance of the signal at the amplifier's output is seen to be

$$v_{\text{out}}^2 = \sum_i (g_i e_i)^2 \tag{13.75}$$

The summing process described by this equation is known as adding the noise contributions in quadrature. The root-mean-square output signal is simply the square root of this expression.

Applying this procedure to a simplified version of Figure 13.9b that retains noise contributions only from the gain resistors R_1 and R_2, it is possible to show that for a high-gain circuit the Johnson noise contribution from these gain resistors can be modeled by e_a as indicated in Figure 13.9c. Note that in this figure, all noise sources are described by input EMFs to the amplifier, a practice known as referring the noise to the input.

We are ultimately interested in the signal-to-noise ratio at the output of the amplifier, but by assuming an ideal, noiseless amplifier model, we can just as easily work with the signals present at the amplifier's input terminals. Assuming for the moment that the true, noiseless input signal v_s is zero, we can calculate the total noise voltage present at the amplifier's inputs. Using the model of Figure 13.9c, there are three sources of noise voltage: the Johnson noise from the source resistor, the voltage noise of the actual amplifier (including gain resistors), and the voltage produced by the actual amplifier's current noise flowing in the source resistor R_s (typically $R_s \gg R_1$ in this noninverting design).

We obtain the variance of the total noise signal by adding the squares of the signals as prescribed by Eq. (13.75):

$$v_n^2 = e_s^2 + e_a^2 + i_a^2 R_s^2 \tag{13.76}$$

Since e_s represents the Johnson noise from resistor R_s, we incorporate Eq. (13.61) and write the RMS output noise signal as

$$v_n = \sqrt{4kTR_s + e_a^2 + i_a^2 R_s^2} \qquad (13.77)$$

Clearly, the way to obtain the smallest noise contribution is to find an amplifier with the smallest values of e_a and i_a and use a signal source with the smallest possible output impedance. If the source impedance R_s is necessarily large, it is particularly important to choose an amplifier with a small current noise signal i_a; this normally means an amplifier with an FET input stage.

If some adjustable parameter of the amplifier allows e_a and i_a to be varied such that one increases while the other decreases, then the minimum noise contribution will occur when the parameter is chosen to make the last two terms in Eq. (13.77) equal. The most common example relates to the base and collector bias current of a bipolar transistor: The shot noise i_a associated with the base bias current clearly increases with I_B (or I_C), but the associated reduction in the collector to emitter resistance causes a decrease in e_a.

The condition for the last two terms of Eq. (13.77) being equal can be re-written as

$$R_s = \frac{e_a}{i_a} \qquad (13.78)$$

and gives rise to the notion of a "noise resistance" e_a/i_a which should ideally be equal to the source resistance. However, the idea only applies to situations where the noise EMFs can be varied to match R_s: Under no conditions should series resistance be added to make R_s equal e_a/i_a.

It may also be possible to make a noise improvement by inserting a transformer between signal source and amplifier. If the transformer is noiseless and has turns ratio n, then its effect is to convert the v_s, R_s signal source to an nv_s, $n^2 R_s$ source. Since it is the signal-to-noise ratio v_s/v_n that is truly important, we use Eq. (13.77) to write the "transformed" ratio

$$\text{S/N} = \frac{nv_s}{\sqrt{4kTn^2 R_s + e_a^2 + i_a^2 n^4 R_s^2}} \qquad (13.79)$$

If we divide the numerator and denominator of this expression by n, we see that the first term under the radical is constant as n increases, whereas the second decreases, and the third increases. The best signal-to-noise ratio occurs when the last two terms are equal, and can be obtained by choosing a turns ratio such that $n^2 R_s$ is equal to the noise resistance e_a/i_a.

13.4.5 Techniques Used to Extract Signal from Noise

Once the random noise in the signal source and amplifier combination has been reduced as much as possible, the only remaining recourse for white noise reduction is to average the signal, either over a long measurement time or over repeated measurements. The technique is known as signal averaging and amounts to interpreting the ensemble average of repeated measurements $\{v_k(t_1)\}$ as the underlying deterministic signal.

Signal averaging can also be used to remove some nonrandom periodic signals (such as the pickup of 60 Hz and its harmonics) if each experimental measurement is started at a random point in the 60-Hz cycle. Since the measurement time t starts at zero for each measurement, the 60-Hz signal will have a random phase with respect to t, and over many measurements its mean contribution at any particular t_1 will approach zero.

For DC or slowly varying input signals, the limiting noise signal is likely to be flicker or $1/f$ noise. Since this noise is largest at low frequencies, the general method here is to modulate the physical environment to change the slowly varying input signal to an amplitude-modulated AC signal, preferably with a frequency in the low-noise range of 1 kHz to 10 kHz. This is the basic principle behind the device known as a "lock-in amplifier."

13.5 Multivariable Analysis

In this section we introduce general techniques that can be used to analyze a pair of time-varying signals. The methods have a broad range of application, depending on the characteristics of the two signals used in the analysis. We can compare two measured signals when one signal is delayed with respect to the other (correlation), or we can compare one measured signal against a delayed version of itself (autocorrelation). In a variation of the correlation application, a measured signal can be compared against a known signal shape to determine the presence of the known signal and its time of occurrence. A nonstatistical but mathematically similar variation, known as convolution, provides a method by which a measured signal $x(t)$ can be combined with the measuring system's transfer function $h(t)$ to produce the input signal before it was corrupted by the measuring system.

13.5.1 Autocorrelation and Cross-correlation Functions

The autocorrelation and cross-correlation functions are analogous to the variance and covariance expressions already discussed, but are defined in terms of time-varying functions and admit the possibility of a variable time offset between the x and y data. The cross-correlation function r_{xy} is defined as the time average

$$r_{xy}(\tau) = \langle x(t)\, y(t - \tau) \rangle \qquad (13.80)$$

and is a function of the time offset parameter τ as indicated. In practice, one or both of the $x(t)$ and $y(t)$ functions represent measurements with both deterministic and random parts. Expressed in terms of an integral, this time average becomes

$$r_{xy}(\tau) = \lim_{T \to \infty} \frac{1}{2T} \int_{-T}^{T} x(t)\, y(t - \tau)\, dt \qquad (13.81)$$

and can be extended to

$$\mathbf{r}_{xy}(\tau) = \lim_{T \to \infty} \frac{1}{2T} \int_{-T}^{T} \mathbf{x}(t)\, \mathbf{y}^*(t - \tau)\, dt \qquad (13.82)$$

if complex functions $\mathbf{x}$ and $\mathbf{y}$ are to be used. We will need this latter form in the discussion of the power spectrum in Section 13.6, but we will continue to assume real signals for now. As in the case of single-variable statistics discussed earlier, these time-average definitions produce nonzero results only for signals of infinite extent (infinite energy but finite power); for physical signals of finite time duration and finite energy, the averaging process over infinite time would yield a zero result. Consequently, for finite time signals the definition can be modified to read

$$r_{xy}(\tau) = \frac{1}{T} \int_0^T x(t)\, y(t - \tau)\, dt \qquad (13.83)$$

where the signals $x(t)$ and $y(t)$ are understood to be nonzero only on the range $0 \le t \le T$.

To understand the significance of the cross-correlation function, consider the covariance σ_{xy} as defined by Eq. (13.53). Our data sequences are the values $\{x_k(t_1)\}$ and $\{y_k(t_1 - \tau)\}$ collected by repeated measurements at two different measurement times t_1 and $t_1 - \tau$. Following the earlier discussion but displaying the measurement time explicitly, we write this covariance as the ensemble average

$$\sigma_{xy}(t_1, \tau) = \int_{-\infty}^{\infty} \int_{-\infty}^{\infty} q(x, y)\, x(t_1)\, y(t_1 - \tau)\, dx\, dy \qquad (13.84)$$

where $q(x, y)$ represents an unknown and possibly correlated distribution function. Note that for $\tau = 0$, this expression reduces exactly to the covariance σ_{xy} evaluated at time t_1. Now, if x and y are replaced by independent random variables $\underline{x}$ and $\underline{y}$, this function should be zero for all τ according to the arguments of Section 13.3.3. If the random variables are stationary, σ_{xy} will not depend on t_1; and if the variables are ergodic, this ensemble average must yield the same expectation values as the time average of Eq. (13.83). Thus, the cross-correlation function will be zero for two independent random variables, and a nonzero experimental result would show the extent to which one measured variable depends on another.

The autocorrelation function $r_{xx}(\tau)$ is defined by

$$r_{xx}(\tau) = \langle x(t)x(t - \tau) \rangle \qquad (13.85)$$

which expands to

$$r_{xx}(\tau) = \lim_{T \to \infty} \frac{1}{2T} \int_{-T}^T x(t)\, x(t - \tau)\, dt \qquad (13.86)$$

If a random $\underline{x}$ has mean value zero, then by Eqs. (13.5) and (13.86),

$$r_{\underline{xx}}(0) = \sigma_{\underline{x}}^2 \qquad (13.87)$$

the variance of $\underline{x}_k$. If we think of a random process $\underline{x}(t)$ as being a sum of the impulse responses generated by a random sequence of impulses, then it is clear that $\langle \underline{x}(t)\underline{x}(t - \tau) \rangle$ must approach zero when τ becomes much larger than the decay time T_c of the individual response signals. Thus, it is possible that for τ greater than some T_c, the autocorrelation function $r_{xx}(\tau)$ will display the periodicity of the input signal $x(t)$ but with a greatly reduced noise content.

By analogy with the correlation coefficient defined by Eq. (13.58), we can define a correlation function coefficient ρ_{xy}

$$p_{xy}(\tau) = \frac{r_{xy}(\tau)}{\sqrt{r_{xx}(0)r_{yy}(0)}} \qquad (13.88)$$

The coefficient is a function of the delay time τ, and always lies in the range $-1 \leq p_{xy} \leq 1$. Because it is a unitless ratio, this definition avoids scale problems associated with signal amplitude and the integration interval of the time averages.

Example 13.3 Calculate the autocorrelation function for a sine wave of infinite time duration.

The autocorrelation function for the sine wave is

$$r_{xx}(\tau) = \lim_{T \to \infty} \frac{1}{2T} \int_{-T}^{T} \sin \omega t \, \sin \omega (t - \tau) \, dt$$

Expanding the second sine function, distributing the integration, and substituting $d(\omega t)/\omega$ for dt yields

$$r_{xx}(\tau) = \lim_{T \to \infty} \frac{\cos \omega \tau}{2\omega T} \int_{-T}^{T} \sin^2 \omega t \, d(\omega t) - \frac{\sin \omega \tau}{2\omega T} \int_{-T}^{T} \sin \omega t \, \cos \omega t \, d(\omega t)$$

The last integrand is an odd function and thus integrates to zero, leaving the first term, which integrates to

$$r_{xx}(\tau) = \lim_{T \to \infty} \frac{\cos \omega \tau}{2\omega T} \left(\frac{1}{2} \omega t - \frac{1}{4} \sin 2\omega t \right) \Bigg|_{-T}^{T}$$

$$= \lim_{T \to \infty} \frac{\cos \omega \tau}{2\omega T} \left(\omega T - \frac{1}{2} \sin \omega T \right)$$

For large T the first term will dominate, leaving as the final result:

$$r_{xx}(\tau) = \frac{1}{2} \cos \omega \tau$$

This example shows a useful characteristic of the autocorrelation function: A periodic component in the original signal shows up as a periodic component of the same frequency in the autocorrelation function. The use for this effect will become more obvious after we find the autocorrelation function for random noise in a later example.

13.5.2 *Comparison of Convolution and Correlation Integrals*

As used in this text, the convolution integral has nothing to do with statistics but is used to relate an output signal from a four-terminal network to the input signal and the transfer function $h(t)$ of the network. However, the convolution integral $z(\tau)$ is mathematically similar to the integral in the correlation function $r(\tau)$, as shown by the following comparison.

Assuming that the real functions $x(t)$ and $y(t)$ are nonzero on the range $0 \leq t \leq T$, we have

Convolution	*Correlation*

$$z_{xy}(\tau) = \int_{-\infty}^{\infty} x(t) \, y(\tau - t) \, dt \qquad r_{xy}(\tau) = \frac{1}{T} \int_{-\infty}^{\infty} x(t) \, y(t - \tau) \, dt \qquad (13.89)$$

or by a change of the dummy variable,

$$z_{xy}(\tau) = \int_{-\infty}^{\infty} x(\tau - t)\, y(t)\, dt \qquad r_{xy}(\tau) = \frac{1}{T} \int_{-\infty}^{\infty} x(t + \tau)\, y(t)\, dt \qquad (13.90)$$

By introducing new dummy variables within the integrals, it can be shown that

$$z_{xy}(\tau) = z_{yx}(\tau) \qquad\qquad r_{xy}(\tau) = r_{yx}(-\tau) \qquad (13.91)$$

Thus, the order of the constituent functions does not matter for the convolution integral, but a similar interchange in the correlation integral results in a time-

Figure 13.10 *(a) This sketch shows the difference between the convolution and correlation integrals for two real functions x(t) and y(t). Both integrals are zero when the functions are positioned as shown, but will be nonzero when τ is such that the functions overlap. (b) The integral values as a function of τ.*

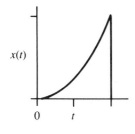
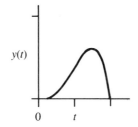

(a)

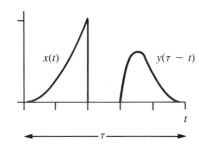
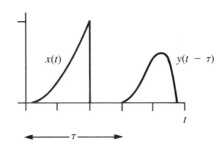

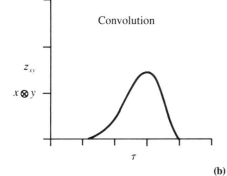
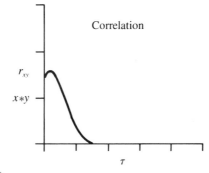

(b)

reversed correlation function. It is often convenient to use the compact notation $\otimes$ and $*$ to identify the convolution and correlation integrals, respectively:

$$z_{xy}(\tau) = x \otimes y \qquad\qquad r_{xy}(\tau) = \frac{1}{T} x * y \qquad (13.92)$$

A graphical interpretation of the convolution and correlation integrals is shown in Figure 13.10. Since both integrals involve a product of the two functions, contributions to the integral can come only from regions of τ where the two functions overlap. As τ changes, the y function slides along the abscissa, producing various amounts of overlap. For example, if $x(t)$ and $y(t)$ are identical functions, then the correlation integral will be a maximum at $\tau = 0$.

13.6   The Power Spectrum

In signal analysis applications it is common to refer to any squared signal as a power signal, but the Fourier transform of a signal such as $x^2(t)$ is not particularly useful, since $x^2(t)$ generally contains frequencies not present in $x(t)$. The power spectrum $R_{xx}(j\omega)$ is therefore defined in the frequency domain, and provides an alternative way of describing the frequency content of a signal.

If a signal $x(t)$ is nonzero only in the range $0 \leq t \leq T$ and has a Fourier transform $\mathbf{X}(j\omega) = A(j\omega) + jB(j\omega)$, then its power spectrum is defined as

$$R_{xx}(j\omega) = \frac{1}{T} \mathbf{X}(j\omega)\mathbf{X}^*(j\omega) \qquad (13.93)$$

where the product

$$\mathbf{X}(j\omega)\mathbf{X}^*(j\omega) = A^2(j\omega) + B^2(j\omega) \qquad (13.94)$$

will be real and even. The significance of this product is most easily seen by considering an $x(t)$ that is composed of a number of discrete frequency cosine and sine waves of varying amplitudes. When transformed, each of these components will contribute a delta function to either $A(j\omega)$ or $B(j\omega)$, respectively. Carrying these delta functions into the equations, it is clear that the power spectrum R_{xx} is composed of delta functions at these same frequencies, but with each having an area that is proportional to the amplitude squared of the original cosine or sine signal. Thus, the frequency content of $x(t)$ is reflected in the power spectrum.

Note that although the frequency content of $x(t)$ can be obtained from the power spectrum, we cannot reconstruct $x(t)$ from the power spectrum because the phase relationships have been lost: A delta function in the power spectrum cannot be identified as arising from either the A (cosine) term or the B (sine) term.

13.6.1   Fourier Transform of the Correlation Integral

Power spectra are related directly to the correlation integrals introduced above. In Chapter 12 we stated without proof that the Fourier transform of the convolution integral is the simple product of the transforms of the separate functions. Thus, if

$x(t)$ and $y(t)$ are measured signals with Fourier transforms $\mathbf{X}(j\omega) = \mathcal{F}[x(t)]$ and $\mathbf{Y}(j\omega) = \mathcal{F}[y(t)]$, then the Fourier transform of the convolution of $x(t)$ with $y(t)$ is

$$\mathcal{F}[x\otimes y] = \mathbf{X}(j\omega)\,\mathbf{Y}(j\omega) \tag{13.95}$$

Leaving the proof of this expression as a problem, we will now prove the similar result for the correlation integral. The Fourier transform of $x*y$ is

$$\mathcal{F}[x*y] = \int_{-\infty}^{\infty}\int_{-\infty}^{\infty} x(u+v)\,y(u)\,du\,e^{-j\omega v}\,dv \tag{13.96}$$

where we have used the Eq. (13.90) form of the correlation integral and have replaced t and τ with the timelike variables u and v; the outside integration over dv constitutes the Fourier transform of $x*y$. Reversing the order of integration, we have

$$\mathcal{F}[x*y] = \int_{-\infty}^{\infty} y(u)\left[\int_{-\infty}^{\infty} x(v+u)e^{-j\omega v}\,dv\right]du \tag{13.97}$$

The factor in brackets is the Fourier transform of the shifted $x(v)$ and can be rewritten using the shift theorem from Table 12.1. Substituting v for t, u for $-t_0$, and changing the name of the function to x, the shift theorem becomes

$$\mathcal{F}[x(v+u)] = X(j\omega)e^{j\omega u} \tag{13.98}$$

which is identical to the bracketed factor in Eq. (13.97). With this substitution, Eq. (13.96) becomes

$$\mathcal{F}[x*y] = X(j\omega)\int_{-\infty}^{\infty} y(u)e^{j\omega u}\,du \tag{13.99}$$

But, if $\mathbf{Y}(j\omega)$ is the Fourier transform of the real function $y(t)$, then this remaining integral must be $\mathbf{Y}^*(j\omega)$, leaving the result as

$$\mathcal{F}[x*y] = \mathbf{X}(j\omega)\,\mathbf{Y}^*(j\omega) \tag{13.100}$$

only slightly different than the convolution result of Eq. (13.95).

13.6.2 Power Signals in the Time and Frequency Domains

In Example 13.3, we showed that the autocorrelation integral r_{xx} of a sinusoidal function $x(t)$ is a sinusoid of the same frequency. The Fourier transform of this autocorrelation should then show a delta function at the frequency of $x(t)$. We can use Eq. (13.100) to extend this result to a more general $x(t)$ and show that the power spectrum is the Fourier transform of the autocorrelation integral.

If we form the autocorrelation function

$$r_{xx}(\tau) = \frac{1}{T}\,x*x \tag{13.101}$$

then Eq. (13.100) says that its Fourier transform will yield the power spectrum

$$[r_{xx}(\tau)] = \frac{1}{T}[X^*(j\omega)X(j\omega)] = R_{xx}(j\omega) \tag{13.102}$$

where $X(j\omega)$ is the transform of $x(t)$. Thus, the autocorrelation function contains the same frequencies as $x(t)$.

The inverse of Eq. (13.102) is also useful:

$$r_{xx}(\tau) = \mathscr{F}^{-1}[R_{xx}] = \frac{1}{T}\mathscr{F}^{-1}[X^2(j\omega)] \tag{13.103}$$

When working with time domain numerical data, this equation actually provides the most efficient way to evaluate the autocorrelation integral.

Note that in order for the transform pairs r_{xx} and R_{xx} to both be real, they must also both be even functions. Thus, when defining a power spectrum for discussion purposes, it is necessary to extend it as an even function into the nonphysical negative frequency region. Any power spectrum derived from the transform of real signals will automatically be a real and even function.

By analogy with the general correlation function r_{xy}, we can also define a cross power spectrum,

$$R_{xy}(j\omega) = \mathscr{F}[r_{xy}(\tau)] = \frac{1}{T}[X(j\omega)Y^*(j\omega)] \tag{13.104}$$

but its direct application is not as obvious as the power spectrum. Note that this cross-spectrum may be complex.

Example 13.4 Determine the autocorrelation function for a white noise signal.

Since white noise has a flat frequency distribution, its power spectrum is constant and can be written as

$$R_{xx}(j\omega) = A$$

The inverse Fourier transform of this is

$$r_{xx}(\tau) = \mathscr{F}^{-1}[R_{xx}(j\omega)] = \frac{1}{2\pi}\int_{-\infty}^{\infty} Ae^{j\omega\tau}\,d\omega$$

and, using Eq. (12.11), this integral evaluates to

$$r_{xx}(\tau) = A\,\delta(\tau)$$

a delta function of area A located at $\tau = 0$.

Two useful theorems further establish the connection between squared (power) signals in the time and frequency domains. The first of these, Parseval's theorem, applies to correlation functions in general and is easily derived. Let $r_{xy}(\tau)$ be given by

$$r_{xy}(\tau) = \frac{1}{T}\int_{-\infty}^{\infty} x(t+\tau)\,y(t)\,dt \tag{13.105}$$

From Eq. (13.100), its transform is then

$$\mathbf{R}_{xy}(j\omega) = \frac{1}{T} \mathbf{X}(j\omega)\mathbf{Y}^*(j\omega)$$

and the inverse transform of this must again give $r_{xy}(\tau)$:

$$r_{xy}(\tau) = \frac{1}{2\pi T} \int_{-\infty}^{\infty} \mathbf{X}(j\omega)\,\mathbf{Y}^*(j\omega)e^{j\omega\tau}\,d\omega \qquad (13.106)$$

At $\tau = 0$ this reduces to

$$r_{xy}(0) = \frac{1}{2\pi T} \int_{-\infty}^{\infty} \mathbf{X}(j\omega)\,\mathbf{Y}^*(j\omega)\,d\omega \qquad (13.107)$$

and replacing $r_{xy}(0)$ with Eq. (13.105) evaluated at $\tau = 0$ gives Parseval's theorem:

$$\frac{1}{T} \int_{-\infty}^{\infty} x(t)\,y(t)\,dt = \frac{1}{2\pi T} \int_{-\infty}^{\infty} \mathbf{X}(j\omega)\,\mathbf{Y}^*(j\omega)\,d\omega \qquad (13.108)$$

If we specialize to the autocorrelation integral with $y(t) = x(t)$, Parseval's theorem reduces to the power theorem,

$$\int_{-\infty}^{\infty} x^2(t)\,dt = \frac{1}{2\pi} \int_{-\infty}^{\infty} \mathbf{X}^2(j\omega)\,d\omega \qquad (13.109)$$

which provides the normalization between the area under the squared signal and the area under the power spectrum.

Example 13.5   Use the power theorem to show that the normalization of the Gaussian transform pair shown in Figure 12.4 is correct.

Using the notation shown in the figure, the power theorem becomes

$$\int_{-\infty}^{\infty} e^{-(t/\sigma)^2}\,dt = \sigma^2 \int_{-\infty}^{\infty} e^{-(\sigma\omega)^2}\,d\omega$$

where the factors of 2π have been cancelled. Moving one factor of σ to the left side, we can form the new variables $u = t/\sigma$ and $v = \sigma\omega$. Rewriting the left and right integrals in terms of u and v, respectively, produces two identical integrals.

13.6.3   The Causal Signal

A physical signal $x(t)$ derived from the output of an amplifier will necessarily be zero until after some turn-on time t_0. If we shift the time scale such that $t_0 = 0$, we can define a "causal" signal that is zero at $t \le 0$ but may be nonzero for $t > 0$. The impulse response functions used in the Laplace transforms of Chapter 12 are all causal signals. As with any other real signal, the causal signal can be written

$$x(t) = x_E(t) + x_O(t), \qquad (13.110)$$

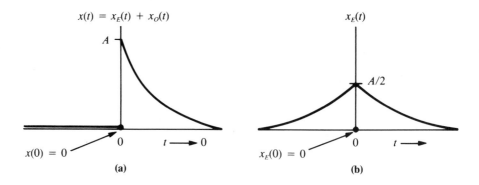

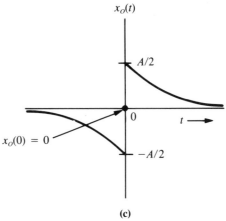

Figure 13.11 *(a) A causal signal showing its (b) even and (c) odd parts.*

where x_E and x_O are its even and odd parts as shown by the example in Figure 13.11. From the discussion in Section 12.2.4, the Fourier transform of this real signal must have the general form

$$\mathbf{X}(j\omega) = X_E(j\omega) + jX_O(j\omega) \qquad (13.111)$$

where X_E and X_O are real even and odd functions, and X_E and jX_O are the Fourier transforms of $x_E(t)$ and $x_O(t)$, respectively.

From Figure 13.11, it can be seen that the squared signals $x_E^2(t)$ and $x_O^2(t)$ are identical, and must therefore have the same frequency content. Thus we conclude that, except for phase factors (in this case $e^{j\pi} = -1$ on the negative frequency components of X_O), the frequency content of the even, odd, and causal signals are all identical. Stated another way, the inverse transforms of $X_E(j\omega)$ and $X_O(j\omega)$ produce functions that have identical shapes for $t > 0$, exactly half of the original $x(t)$. In many applications, this allows us to disregard the imaginary parts in the Fourier transform of a causal signal.

13.6.4 Autocorrelation Function of Band-Limited White Noise

The power spectrum shown in Figure 13.12a could be obtained by passing white noise through a low-pass, four-terminal network with a square transfer function extending from zero to ω_1; it thus serves as a simplified model to define the noise spectrum at the output of a signal amplifier. Recall that the power spectrum R_{xx} is always an even function, so it must extend to $-\omega_1$. Generally, the output of the amplifier will also contain a determinate signal component $u(t)$ that contains useful information. If this information is contained in the frequency spectrum of $u(t)$, we can extract it using the autocorrelation function, but since the autocorrelation func-

Figure 13.12 *(a) The power spectrum of white noise after passing through a low-pass filter with a idealized square bandpass region. (b) The autocorrelation function for this band-limited white noise.*

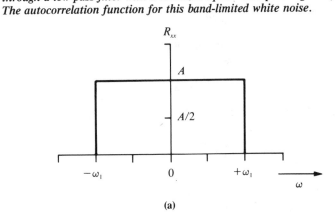

(a)

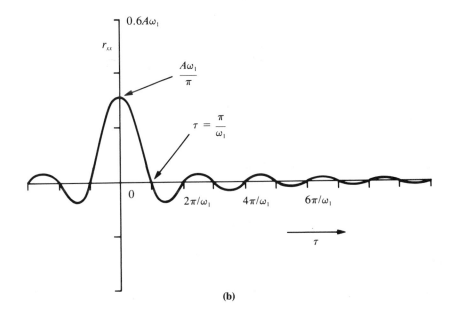

(b)

tion will also be influenced by the white noise, it is useful to know the autocorrelation function resulting from the noise alone.

The power spectrum of Figure 13.12a can be written as

$$R_{xx}(j\omega) = A \qquad -\omega_1 < \omega < \omega_1$$

and

$$R_{xx}(j\omega) = 0 \qquad \text{elsewhere} \tag{13.112}$$

From Eq. (13.103), the autocorrelation function is

$$r_{xx}(\tau) = \mathscr{F}^{-1}[R_{xx}(j\omega)] = \frac{A}{2\pi} \int_{-\omega_1}^{\omega_1} e^{j\omega\tau}\, d\omega \tag{13.113}$$

where the limits have been set by the nonzero region of R_{xx}. Expanding the exponent gives

$$r_{xx}(\tau) = \frac{A}{2\pi} \left[\int_{-\omega_1}^{\omega_1} \cos(\omega\tau)\, d\omega + j \int_{-\omega_1}^{\omega_1} \sin(\omega\tau)\, d\omega \right] \tag{13.114}$$

where the odd-function second term integrates to zero, and the first integrates to

$$r_{xx}(\tau) = \frac{A}{\pi\tau} \sin(\omega_1\tau) \tag{13.115}$$

Multiplying and dividing by ω_1 puts this into the form of a sinc function,

$$r_{xx}(\tau) = \frac{A\omega_1}{\pi} \text{sinc}(\omega_1\tau) \tag{13.116}$$

A sketch of this function is shown in Figure 13.12b. The first zero crossing of this function occurs when $\omega_1\tau = \pi$, showing that the central peak of this function becomes narrower as the bandwidth gets larger.

13.6.5 Autocorrelation Applications

In the preceding sections we determined that the autocorrelation function $r_{xx}(\tau)$ of a random noise signal will decrease with increasing τ, whereas the autocorrelation function of a sinusoidal oscillation of frequency ω_0 will be a sinusoid of the same frequency with constant amplitude. The power spectrum for such a combined signal is just

$$R_{xx}(j\omega) = A + B[\delta(\omega_0) + \delta(-\omega_0)] \tag{13.117}$$

where the negative-frequency delta function is introduced to keep the power spectrum even. From the Fourier transform addition theorem of Table 12.1, the autocorrelation is the sum of the individual autocorrelations, giving

$$r_{xx}(\tau) = A\,\delta(\tau) + 2B\cos(\omega_0\tau) \tag{13.118}$$

If we replace the noise component of R_{xx} with a more realistic band-limited form like Eq. (13.112), then the delta function in r_{xx} is replaced by a sinc function like Eq. (13.116). Even so, the noise component of $r_{xx}(\tau)$ will still decay with time τ. This situation is shown in Figure 13.13, where the $\cos(\omega_0 t)$ signal and white noise are passed through an idealized low-pass filter. The figure shows that

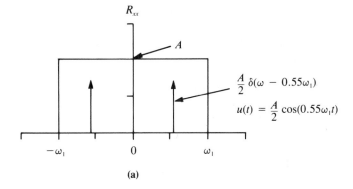

(a)

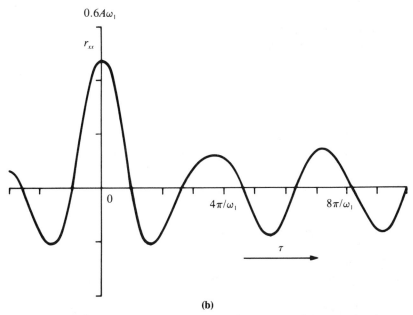

(b)

*Figure 13.13   (a) The power spectrum of a noisy cos(0.55 ω₁t) signal
after passing through an idealized low-pass filter with cutoff frequency
ω₁. (b) For large τ values, the autocorrelation function of this noisy
signal becomes a clean sinusoid with frequency 0.55 ω₁.*

at large τ, the dominate signal in $r_{xx}(\tau)$ is a sinusoid of frequency ω_0. Thus, the autocorrelation function can be expected to show a constant-amplitude sinusoid for each frequency component in the deterministic signal, whereas the contributions from random noise decrease as τ increases.

13.6.6   Coherence Time for Real Signal Generator

In practical applications the autocorrelation function cannot be expected to yield the perfect results indicated in the preceding argument. This is either because the signal to be measured does not persist for a sufficiently long time (it must obviously be larger than the maximum τ needed), or because it does not maintain a strict phase relationship with itself over the entire measurement time. This latter feature is related to the coherence time of the signal and will vary with the signal source. To be more specific, the time constant associated with the decay of the amplitude envelope of an autocorrelation function is a measure of the coherence time of its associated signal $x(t)$.

For any physical signal, the autocorrelation function will be a maximum at $\tau = 0$, and if the mean value of $x(t)$ is zero, the autocorrelation will always approach zero at sufficiently large τ.

13.7   Cross-correlation Applications

We have concentrated on the autocorrelation function because of its obvious uses in extracting signals from noise. The cross-correlation function is more useful in those cases where the shape of a short time duration signal is known but the time of its arrival is not. As an example, consider the distance measurement experiment shown in Figure 13.14a. If the speaker emits a burst of sound (any noise-like signal with a short coherence length will do), the microphone near the source will produce a signal $v_1(t)$, and the second microphone will generate a similar signal $v_2(t)$ delayed by a transit time t_0. The second signal may have a different amplitude from the first, but in the absence of reflections it will have the same shape and can be written as

$$v_2(t) = Av_1(t - t_0) \tag{13.119}$$

The cross-correlation of $v_1(t)$ with $v_2(t)$ will clearly have a peak at $\tau = t_0$ and can be used to determine the transit time.

When the input to a four-terminal network is white noise, the cross-correlation function taken between the input and output signals will be proportional to the impulse response $h(t)$ of the network. The experimental arrangement is shown in Figure 13.14b. This relationship can be derived by first forming the cross-correlation function r_{yx} between the input signal $x(t)$ and the output signal $y(t)$,

$$r_{yx} = \frac{1}{T} \int_{-\infty}^{\infty} y(t)\, x(t - \tau)\, dt \tag{13.120}$$

where the signals are nonzero only in the range $0 < t \le T$. From Eq. (13.100), the Fourier transform of this equation yields the cross-power spectrum

$$\mathbf{R}_{yx} = \frac{1}{T} \mathbf{Y}(j\omega)\, \mathbf{X}^*(j\omega) \tag{13.121}$$

and using $\mathbf{Y}(j\omega) = \mathbf{H}(j\omega)\, \mathbf{X}(j\omega)$, it reduces to

$$\mathbf{R}_{yx} = \frac{1}{T} \mathbf{H}\, \mathbf{X}\, \mathbf{X}^* = R_{xx}\, \mathbf{H} \tag{13.122}$$

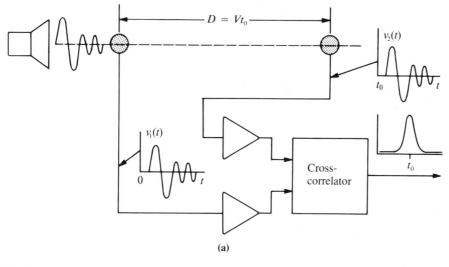

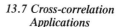

(a)

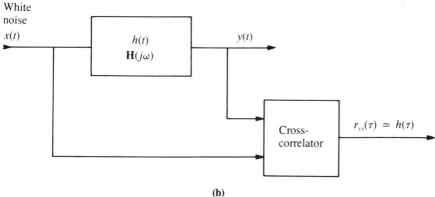

(b)

Figure 13.14   (a) An experimental arrangement to determine the transit time for a sonic impulse between two microphones. (b) An experimental arrangement to determine the impulse response of a four-terminal network.

Now, if $x(t)$ is white noise, its power spectrum R_{xx} (the Fourier transform of its autocorrelation function) will be a real constant, and the cross-power spectrum can be written as

$$\mathbf{R}_{yx}(j\omega) = A\,\mathbf{H}(j\omega) \tag{13.123}$$

Thus, the cross-power spectrum between a white noise input signal to a network and the resulting output signal will be proportional to the frequency-domain transfer function of the network. The inverse Fourier transform of Eq. (13.123) takes us back to the time domain,

$$r_{yx}(\tau) = \frac{A}{2\pi}h(\tau) \tag{13.124}$$

and we see that this cross-correlation function is proportional to $h(t)$, the impulse response of the network.

13.8 *Signal Analyzer Block Diagrams*

As with many things, the ideas just described are not as complicated as the mathematics might imply. Consider the block diagram of the cross-correlation device shown in Figure 13.15a. This device uses only three fundamental elements, signal delays, multipliers, and integrators, but repeats each element n times. One of the signals, in this case $y(t)$ for consistency with Eq. (13.89), is passed through a number of boxes that delay the signal by an amount τ but otherwise leave it unchanged. One crude way to accomplish this is to have the signal $y(t)$ on magnetic tape and pass it by a number of consecutive playback heads. The multiplication elements produce outputs of the form $x(t)y(t)$, $x(t)y(t - \tau)$, and so on, which are fed into an integrating circuit. After integrating for a measurement time T that is long compared to $n\tau$, the output of the integration circuits will correspond to points on a correlation plot. The sketch is easily converted to an autocorrelation device, yielding $r_{xx}(\tau)$ by feeding $x(t)$ into both inputs.

Figure 13.15 (a) A block diagram of an electronic device capable of producing outputs that are proportional to the cross-correlation function r_{xy}. If $x(t) = y(t)$, the outputs become the autocorrelation function r_{xx}.

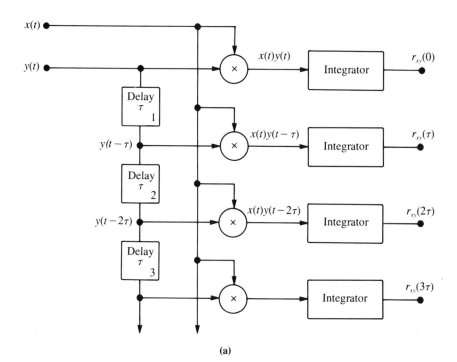

(a)

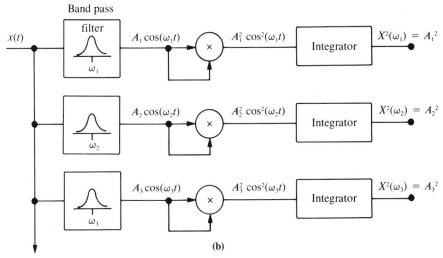

Figure 13.15 continued. *(b) A similar block diagram to produce the power spectrum R_{xx}.*

The block diagram of Figure 13.15b will produce an approximation to the power spectrum at several frequencies. The elements in this case are a number of band pass filters centered at different frequencies, multipliers for squaring the filter outputs, and integrators.

Although analog devices have been produced to generate the type of functions just described, they obviously require many components to produce a useful result. However, some of these operations become quite simple when performed by a digital computer on signals that have first been digitized. The autocorrelation function in particular can be quickly calculated from the list of numbers x_i obtained by repeatedly performing analog-to-digital conversions. If the delay between conversions is τ, then time-delayed products such as $x(t)x(t - n\tau)$ convert to $x_i x_{i-n}$. However, in making the analog-to-digital conversions at discrete time intervals, we have moved into the domain of discrete functions defined only at a finite set of points. Other problems and possibilities arise, and these are addressed in the next chapter.

Problems

1. Given the set of numbers $\{x\} = 2, 3, 3, 4, 6, 6, 6, 7, 8, 8$.
   (a) Determine the mean and standard deviation of $\{x\}$.
   (b) Form a new set of numbers $\{y\} = \{x - \bar{x}\}$, and use these numbers to demonstrate that $\sigma_y^2 = \overline{y^2}$.

2. Sketch a random signal $x(t)$ whose mean is zero but which is definitely not ergodic.

3. If $v = x + y$ and $\langle\ \rangle$ means time average, show that $\langle v \rangle = \langle x \rangle + \langle y \rangle$.

4. If $x(t) = 2t$ during the measurement interval $0 \le t \le 5$ and zero at all other times, determine $\langle x(t) \rangle$ using Eq. (13.8) and again using (13.16).

5. You have three different transistors in a bag. If you reach into the bag and remove two transistors, use a sketch to show the number of different combinations you could select. Verify this result using Eq. (13.26).

6. Verify Eq. (13.29) by expanding the factorials $K! = K(K - 1)(K - 2) \ldots$, then approximating each term with K as $K \to \infty$.

7. Verify Eq. (13.30) by expanding the left side using the binomial expansion $(1 + x)^m = 1 + mx + [m(m - 1)/2!]x^2 + [m(m - 1)(m - 2)/3!]x^3 + \ldots$ and the right side with $e^x = 1 + x + x^2/2! + x^3/3! + \ldots$.

8. The integrals of Eqs. (13.33) cannot be evaluated analytically. Using a calculator or computer, verify the first of these equations for the case where $\langle x \rangle = 0$ and $\sigma = 1$.

9. In Eq. (13.53) and similar equations for $\langle x^2 \rangle$ and $\langle y^2 \rangle$, replace $q(x,y)$ with the expression $\delta(x - y)p(x,y)$, then show that $\rho_{xy} = 1$.

10. *(a)* If two resistors are connected in parallel as shown in Figure P13.10, use Eq. (13.57) to show that the Johnson noise contribution from the equivalent resistance R_{Eq} is

$$\langle e_{Eq}^2 \rangle = \left(\frac{R_1 R_2}{R_1 + R_2} \right)^2 \left(\frac{\langle e_1^2 \rangle}{R_1^2} + \frac{\langle e_2^2 \rangle}{R_2^2} \right)$$

(b) If $R_1 = R_2 = R$, and the noise from each resistor is given by Eq. (13.61), show that evaluation of the equation from part (a) produces the same result as direct evaluation of Eq. (13.61) for the equivalent resistance.

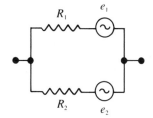

Figure P13.10

11. *(a)* Make a sketch showing the loop path followed by i_0 when the operational amplifier shown in Figure P13.11 is ideal. Is this the only current in R_1? *(b)* If G is the closed-loop gain of the amplifier $(R_1 + R_2)/R_1$ and $GR_s \gg R_2$, use the sketch to show that the output voltage resulting from the current noise is $v_{out} = GR_s i_0$. *(c)* With v_s, e_s, and e_a set to zero, show that the simplified circuit of Figure 13.9c yields the same result.

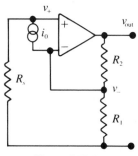

Figure P13.11

12. *(a)* Assuming that the operational amplifier in Figure P13.12 is ideal and that the three EMFs shown are locked in phase as indicated by the $+/-$ notation, show that $v_{out} = G(e_0 - e_1) + e_1 + e_2$, where G is the closed-loop gain of the amplifier. *(b)* Write an expression for $\langle v_{out}^2 \rangle$ if all three EMFs are random noise sources. *(c)* If $G \gg 1$ and under the assumption that e_1 and e_2 are generated by Johnson noise (last equation in Example 13.1), show that $\langle v_{out}^2 \rangle$ reduces to a two-term expression that can be represented as shown in Figure 13.9c.

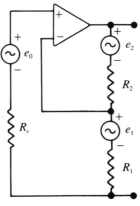

Figure P13.12

13. Using the techniques outlined in the previous problem, show that the three noise sources in the inverting amplifier of Figure P13.13a can be reduced to two as shown in Figure P13.13b if $G \gg 1$.

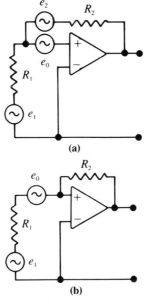

(a)

(b)

Figure P13.13

14. A noise source composed of a resistance R in parallel with capacitance C drives a noiseless amplifier that has infinite input impedance. The idealized amplifier has unity gain in its passband and zero gain outside the passband.

    (a) If the passband extends from zero to infinity, show that the output noise signal is
$\langle e^2 \rangle = kT/C$.

    (b) If the passband extends from $f_0/2$ to $2f_0$, evaluate Eq. (13.60) for the special case where $f_0 = 1/2\pi RC$. [Ans.: $\langle e^2 \rangle = 0.41kT/C$]

15. Evaluate numerically the autocorrelation integral of the function shown in Figure P13.15 at 10 points in the range $0 \le \tau \le 2T$ and graph the result. Note that the signal is zero for $t > T$.

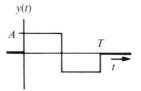

Figure P13.15

16. Evaluate numerically the cross-correlation integral for the two functions shown in Figure P13.16 at 10 points in the range $0 \le \tau \le 2T$ and graph the result. Note that both signals are zero for $t > T$.

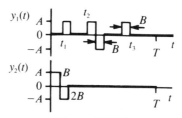

Figure P13.16

17. Sketch and label the amplitude of the odd and even parts of the causal signal shown in Figure P13.17.

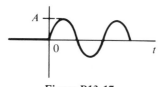

Figure P13.17

18. **(a)** If the filter bandwidth approximated by Figure 13.12a is 10 kHz, evaluate the ratio of the noise signal peak in the autocorrelation function at $t = 0$ to a maximum near $t = 1$ s. **(b)** At what time does the first negative peak occur? **(c)** What is the ratio at the location of the first negative peak?

19. **(a)** Use the power theorem of Eq. (13.109) to write an expression in terms of A and ω_1 for the area under an r_{xx}^2 curve derived from Figure 13.12. Remember that r_{xx} and R_{xx} are a Fourier transform pair. **(b)** Take $A = 1$ and $\omega_1 = \pi$, then count boxes on a graph to determine the fraction of the total area that occurs in the central peak and first lobe on each side. **(c)** If you now discard the central peak and first lobe on each side, and take the Fourier transform of the remaining r_{xx}^2 signal, approximately what value of A^2 do you now expect on the power spectrum within the original passband? **(d)** Explain the significance of this result with reference to Figure 13.13a.

20. Use the method outlined in Section 13.6.1 for the correlation integral to prove Eq. (13.95) for the convolution integral.

14

Discrete Signal Analysis

14.1   Introduction

Whenever a continuous, time-varying analog signal is transferred to a digital computer for analysis, the computer, by its very nature, can only know the value of the signal at a finite number of discrete times. As a result, the growing importance of the digital computer in data acquisition naturally coincides with an increasing interest in the techniques of discrete signal analysis. However, the application of these techniques is not limited to the general-purpose computer, for most of the algorithms can now be fabricated from special-purpose integrated circuits.

In this chapter we will treat discrete signal analysis as an approximation and extension to the analog methods developed earlier. As with the other signal analysis chapters in this text, this material is presented as an introduction to the more extensive treatments found in specialized engineering texts.

14.2   Sampled Signals

A typical sampled signal is shown in Figure 14.1a. When taken into the computer, the initial continuous wave $y(t)$ is replaced by a discrete signal defined only at the sample points y_n. The discrete signal is then given by the sequence $\{y_n\}$. In order to minimize the significance of any particular indexing symbol, we will sometimes use the notation $\{y\}$ to identify such a discrete sequence of points. In all applications, there will be some minimum time between sample points, determined in real time by the repetition rate of an ADC unit or in an off-line application by the

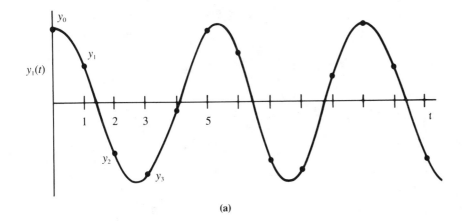

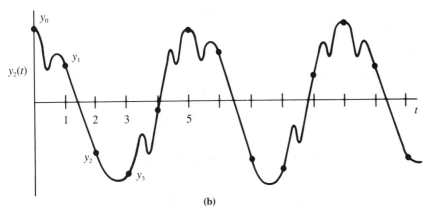

Figure 14.1   (a) A sampled signal $y_1(t)$ produces the discrete signal $\{y\}$ defined only at the solid dot sample points. (b) A second input signal $y_2(t)$ has higher-frequency components but produces exactly the same sample points $\{y\}$.

cost of entering more points into the computer. We will assume an equal time interval T between samples such that the discrete index n is proportional to t according to the relation $t = nT$. If the continuous signal was

$$y(t) = A \cos(\omega t) \qquad (14.1)$$

then by making the substitution $t = nT$, the discrete signal $\{y\}$ or $\{y_n\}$ of N points is found to be

$$y(nT) = y_n = A \cos(n\omega T) \qquad \text{for } n = 0, 1, 2, \ldots, N - 1 \qquad (14.2)$$

where $(N - 1)T$ represents the total measurement time. We assume N samples, spanning $N - 1$ intervals.

As in the case of continuous signals, we will assume causal signals such that $y_n = 0$ for any negative value of n.

The fundamental problem in discrete signal analysis is that the same set of sample points can be generated by an infinite number of continuous signals. Figure 14.1b shows how a second wave, with a higher-frequency content, can be drawn through the same sample points. By increasing the high-frequency content, an infinite number of other signals could also be drawn through these same sample points. The analysis programs in the computer know only about the discrete points, and therefore cannot distinguish between the many possibilities.

We will demonstrate that the sampling process effectively maps the frequency line into a circle where the frequency must be represented by a repeating angular variable. The inescapable consequence of sampling a signal at time intervals T is displayed most concisely by Figure 14.8, where it is seen that the $j\omega$ frequency axis of the s-plane is mapped into a unit circle on a new complex plane. The new scale is such that an angular frequency of $2\pi/T$ wraps completely around the circle and becomes indistinguishable from zero frequency.

14.2.1 Complex Signal Representation

Before developing this subject further, it will again be useful to extend the physical problem into the complex signal domain. The elements of the complex discrete signal {$\mathbf{y}$} are given by

$$\mathbf{y}(nT) = \mathbf{y}_n = Ae^{jn\omega T} \qquad \text{for } n = 0, 1, 2, \ldots, N - 1 \qquad (14.3)$$

which is equivalent to Eq. (14.2). We have chosen to describe a signal whose initial phase angle is zero; a more general expression would still require a multiplicative phase factor $e^{j\theta}$ or a complex amplitude $\mathbf{A} = Ae^{j\theta}$.

If $\omega T << \pi$, then the discrete phasors of Eq. (14.3) will map into the points at increasing angle as shown in Figure 14.2. Once the continuous signal of fixed frequency has been sampled at a constant rate, ωT just represents a constant angle, and the independent variable corresponding to the time is the index n.

From Eq. (14.3), it is clear that the point $n = 0$ lies on the real axis as shown in Figure 14.2a, but that there is no way to determine if the $n = 1$ point is at $\theta = \omega T$ or at ωT plus some integer multiple of 2π. With a continuous signal we could resolve this question by plotting the phasor at a time $T/2$, checking that it was really at $\omega T/2$; but with a discrete signal known only at intervals of T, this option is not available.

14.2.2 The Nyquist Frequency Limit

The frequency content of a sampled signal is further complicated by the fact that the complex phasor is really being used to represent a real signal. Consider the two cases where ωT is either slightly less or slightly greater than π, as indicated in Figure 14.3. If we introduce the angle variable x, then these two cases can be conveniently written as

$$\omega T = \pi - \frac{x}{2}$$

$$(14.4)$$

and

$$\omega T = \pi + \frac{x}{2}$$

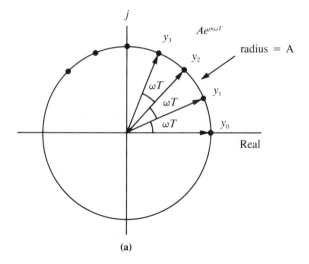

(a)

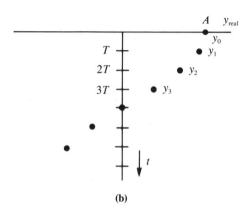

(b)

Figure 14.2 (a) The complex elements of a discrete signal $\{y\}$ with sample period T are $y_n = Ae^{jn\omega T}$. (b) The equivalent real signal elements are given by the real-axis projections $y_n = A\cos(n\omega T)$.

The condition $x \ll \pi$ shown in the figure is for visual illustration only; the arguments hold for any x. As can be seen from the figure, these two cases result in phasors that differ only in their imaginary parts and rotate in opposite directions with increasing sample index n. Since a real signal is represented by the projection of these phasors onto the real axis, the descriptions are equivalent.

To obtain the same result mathematically, we can expand the phasors of Figure 14.3a as

$$e^{j(\pi + x/2)n} = e^{-j2\pi n}e^{j(\pi + x/2)n} \tag{14.5}$$

where the extra factor on the right is just an alternative way of writing 1. These phasors have a frequency $n(\pi + x/2)$, and our goal is to show that the projection of these phasors onto the real axis cannot be distinguished from a similar projection of the phasors of Figure 14.3c whose frequency is $n(\pi - x/2)$. Equation (14.5) can be reduced to

$$e^{j(\pi + x/2)n} = e^{j(-\pi + x/2)n} \tag{14.6}$$

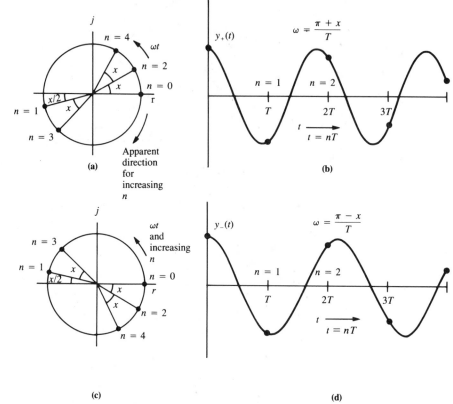

Figure 14.3 *(a) The phasors for the case when $\omega T = \pi + x/2$, and (b) the sample points on the continuous cosine signal. (c) An indistinguishable situation where $\omega T = \pi - x/2$, and (d) a slightly lower-frequency cosine signal drawn through the same sample points.*

and the real part, which we actually observe, can be written as

$$\cos n\left(\pi + \frac{x}{2}\right) = \cos n\left(-\pi + \frac{x}{2}\right) \tag{14.7}$$

But since $\cos(-\theta) = \cos(\theta)$, the right side can be rewritten to give

$$\cos n\left(\pi + \frac{x}{2}\right) = \cos n\left(\pi - \frac{x}{2}\right) \tag{14.8}$$

The parenthetical term in each of these expressions corresponds to ωT for some signal of angular frequency ω, and Eq. (14.8) shows clearly that the sample points cannot distinguish between a real signal with $\omega T = \pi + x/2$ and one with $\omega T = \pi - x/2$. Thus the frequency determined from sample points is restricted to an interval no larger than $\omega = \pi/T$. Note that other frequencies are not lost or absorbed as in a filter; they are simply mapped into the allowed interval.

If the sampling process is described by a sampling frequency $f_S = 1/T$ or an angular frequency $\omega_S = 2\pi/T$, then the maximum observable angular frequency ω_N in the discrete signal is given by

$$\omega_N = \frac{\pi}{T} = \frac{\omega_S}{2} \qquad (14.9)$$

The corresponding Nyquist frequency limit f_N is just

$$f_N = \frac{f_S}{2} \qquad (14.10)$$

or half the sampling frequency. An experimental device that is subject to the Nyquist frequency limit is shown in Figure 14.4a. If the original signal contains frequency components above ω_N, they will be mapped into the 0-to-ω_N range by repeatedly folding the continuous signal's frequency spectrum about the ω_N and 0 frequency points as shown in Figure 14.4b. Only in the lowest frequency interval from 0 to $f_S/2$ does the frequency of the output signal correspond to the frequency of the input signal. Low-frequency components in the output that arise from input components above $f_S/2$ are known as aliases.

The example of Figure 14.4 is in no way special; the same frequency aliasing occurs whenever a discrete signal is analyzed by a computer. Consider a square wave whose fundamental frequency is ω_0; the frequency content of this signal is shown in Figure 14.5a. If the square wave is sampled at a frequency $7.5\omega_0$, the

Figure 14.4 *(a) A simple scheme that uses an ADC to digitize the signal from a sine wave generator, then uses a DAC to regenerate an analog signal. (b) A plot of the frequency of the output signal as a function of the input signal frequency under an assumed sample frequency of f_S.*

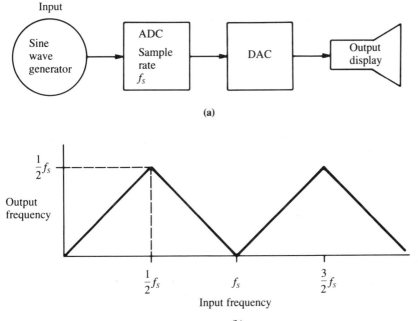

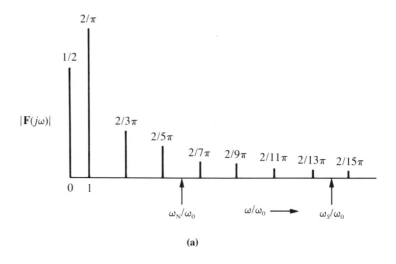

(a)

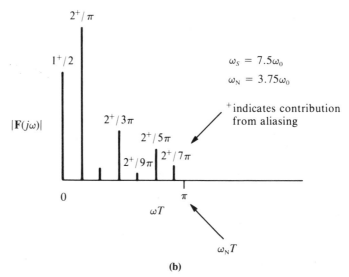

(b)

Figure 14.5 (a) The frequency content of the square wave signal of Figure 3.5 together with (b) the observed frequency content of the same signal after sampling at a frequency corresponding to 7.5 ω_0.

frequency spectrum will be folded about the Nyquist frequency $3.75\omega_0$, then again around zero, and so forth, as specified by Figure 14.4b. The results of the first fold about the Nyquist frequency ω_N are shown in Figure 14.5b. Because of the sampling frequency chosen in this example, later folds will overlap and add to the components shown. Note that the frequency axis on the spectrum of the sampled signal is expressed as an angle obtained by multiplying the angular frequency by the sample interval $T = \pi/\omega_N$.

Because of the aliasing problem, it is usually necessary to restrict the frequency content of a continuous signal before it is sampled. Thus, most ADC input devices are preceded by a low-pass analog filter that will eliminate or severely reduce any frequency content above the Nyquist limit. However, careful attention to Figure 14.4b will show that the Nyquist limit $f_S/2$ is really a bandwidth, and it is equally effective to employ an input bandpass filter with limits $nf_S/2$ and $(n + 1)f_S/2$. An actual frequency f in this range will be uniquely mapped into an f_{obs} in the 0-to-$f_S/2$ range according to the expressions

$$f_{obs} = f - \frac{nf_S}{2} \qquad \text{for even } n$$

$$(14.11)$$

and

$$f_{obs} = -f + \frac{(n + 1)f_S}{2} \qquad \text{for odd } n$$

Note that for a filter interval with odd n, the observed frequency decreases as the actual frequency increases.

The ability to digitally process a wide-frequency spectrum is thus limited by the rate at which the data acquisition system can sample the analog signal. For example, in order to digitally process an audio signal with frequency content spanning the range 0 kHz to 20 kHz, we must sample at a minimum rate of 40 kHz, corresponding to at most 25 μs between samples.

14.3   Filtering Sampled Data

Once a continuous signal has been digitized, it is subject to a wider range of analysis procedures than is practical in the analog domain. In particular, the time-domain convolution and correlation procedures mentioned in Chapters 12 and 13 are readily accomplished in the digital domain. Even the frequency-domain filtering process can be enhanced by digital techniques, either by the use of higher-order filters with many poles and zeros or by the even more powerful discrete Fourier transform (DFT) techniques.

To introduce the notation, we first consider a simple type of digital filter known as a moving-average filter. A general implementation of the moving-average filter is shown in Figure 14.6a. In this filter, an applied data sequence $\{x_m\}$ passes through a number of delay elements, and the output $\{y_n\}$ is formed as the sum of the delayed terms. The transfer function of this filter is determined by the amplifier gains a_i, which can be either positive or negative. Note that y_n will change only when a shift produces a new set of numbers at the amplifier inputs.

Figure 14.6b reminds us that the frequency response of an analog filter with transfer function $\mathbf{H(s)}$ is equally well defined by its impulse response $h(t)$ and displays the corresponding notation in the discrete domain. As shown in this figure, the transfer function in the discrete frequency domain is $\mathbf{G}(e^{sT})$, producing an impulse response $\{g\}$ when the sequence $\{p\} = 1/T, 0, 0, 0, 0, \ldots$ is applied as a discrete approximation to the delta function.

In terms of the N elements from a list of numbers $\{x\}$ representing the input signal and a set of K filter coefficients a_0 to a_{K-1}, we can write the general moving-average filter as

$$y_n = \sum_{k=0}^{K-1} a_k x_{n-k} \qquad \text{for } n = 0, 1, 2, \ldots, N-1 \qquad (14.12)$$

where the N elements of $\{y\}$ are the output from the filter. (Remember that $x_{n-k} = 0$ when $k > n$.) This expression is then the algorithm for implementing the filtering process on the input data $\{x\}$.

The insets in Figure 14.6a show the impulse sequence $\{p\}$ being applied as the input $\{x\}$ of the filter. With this input the output $\{y\}$ should be the filter's impulse response $\{g\}$. As seen in the figure, $\{p\}$ propagates to the left through

Figure 14.6   This moving-average filter can be implemented as either a hardware device with real delays and amplifiers or as a software procedure working on a data list of predigitized sample points $\{x_m\}$. In this example, a_2 would be negative, implying an inverting amplifier. (b) In the discrete domain, $\{p\}$, $G(e^{sT})$, and $\{g\}$ are the equivalents of the analog functions $\delta(t)$, $H(s)$, and $h(t)$.

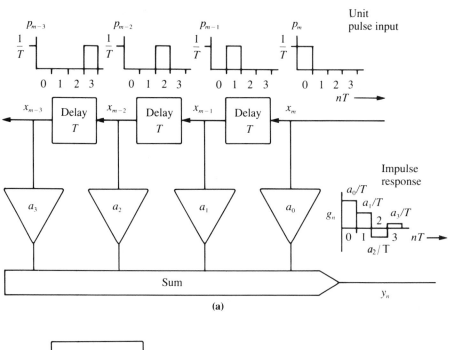

(a)

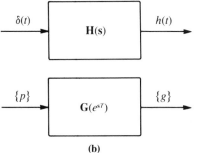

(b)

successive delays T, with its one nonzero term driving one amplifier channel after the other. At any specific time nT, only one amplifier a_n has a nonzero input, and only this amplifier contributes to the output, thereby making $\{g_n\}$ a sequence of pulses of height a_n/T. Likewise, if we use $\{p\}$ as the $\{x\}$ in Eq. (14.12), the $\{y\}$ output must be $\{g\}$. With the only nonzero term in $\{p\}$ being $p_0 = 1/T$, the result is clearly

$$g_n = \frac{a_n}{T} \qquad \text{for } n = 0, 1, 2, \ldots, K - 1 \qquad (14.13)$$

and zero for any index value greater than $K - 1$.

We now switch to the complex signal domain and let the input signal $\{\mathbf{x}\}$ be a single-frequency oscillation given by

$$\mathbf{x}_m = X(\omega)e^{j\omega mT} \qquad (14.14)$$

where T is the constant time between samples. Substitution of this $\{\mathbf{x}\}$ into Eq. (14.12) gives

$$\mathbf{y}_n = X(\omega) \sum_{k=0}^{K-1} a_k e^{j\omega(n-k)T} \qquad \text{for } n = 0, 1, 2, \ldots, N - 1 \qquad (14.15)$$

If we expand the exponent into a product, the $e^{j\omega nT}$ part can be factored out of the summation, leaving

$$\mathbf{y}_n = X(\omega)e^{j\omega nT} \sum_{k=0}^{K-1} a_k e^{-j\omega kT} \qquad (14.16)$$

$$\text{or} \quad \mathbf{y}_n = \left[\sum_{k=0}^{K-1} a_k e^{-j\omega kT} \right] \mathbf{x}_n \qquad (14.17)$$

$$\text{both for } n = 0, 1, 2, \ldots, N - 1$$

Thus, an oscillatory input signal $\{\mathbf{x}_n\}$ of frequency ω will produce an output $\{\mathbf{y}_n\}$ at the same frequency but with its amplitude modified by the complex factor in brackets. The bracketed term thus corresponds to the frequency-domain transfer function $\mathbf{H}(j\omega)$ of the previous analog chapters; we can therefore write the frequency response of this filter as

$$\mathbf{H}(j\omega) = \sum_{k=0}^{K-1} a_k e^{-j\omega kT} \qquad (14.18)$$

or with the standard substitution of $\mathbf{s}$ for $j\omega$,

$$\mathbf{H}(\mathbf{s}) = \sum_{k=0}^{K-1} a_k e^{-k\mathbf{s}T} \qquad (14.19)$$

This equation also defines the discrete transfer function $\mathbf{G}(e^{\mathbf{s}T})$,

$$\mathbf{G}(e^{\mathbf{s}T}) = \sum_{k=0}^{K-1} a_k (e^{\mathbf{s}T})^{-k} \qquad (14.20)$$

with the only difference being the functional form of **G.** We can use **H** and **G** interchangeably to describe the same filter, but some notational care is needed because we will soon introduce a new variable $\mathbf{z} = e^{sT}$; of course, at a given sT, both functions will evaluate to the same number.

We have been able to obtain this frequency-domain transfer function from the time-domain coefficients a_k using a prescription that at first glance may not seem to have much in common with the Laplace transform used to perform the same operation in the analog domain. However, as we show in Section 14.4, the process just discussed does amount to the digital equivalent of the Laplace transform; it is called the **z**-transform and can be effected by the simple operation of multiplying each a_k by an appropriate exponential factor and summing over k.

14.3.1 Moving-Average Filter Example

Given a sequence of sample points on a noisy signal, the most obvious way to obtain a smoother (less noisy) signal is to average adjacent pairs of points as shown in Figure 14.7a. Since this procedure tends to smooth out the most rapid fluctuations in the input signal, it is not surprising that the process can also be viewed as a moving-average low-pass filter.

The signal-averaging procedure just described uses the equation

$$y_n = \frac{x_{n-1} + x_n}{2} \tag{14.21}$$

to determine the averaged set of points $\{y\}$, and this equation can be identified as a special case of the general moving-average filter equation Eq. (14.12). Rearranging the averaging expression gives

$$y_n = 0.5x_n + 0.5x_{n-1} \tag{14.22}$$

and we see that this corresponds to a two-term (or two-point) moving-average filter with $a_0 = a_1 = 0.5$. The impulse response of this filter is obviously two adjacent pulses of equal height.

But if this averaging process is a filter with an impulse response, then it must also have a frequency response! Switching to complex signals and using Eq. (14.17), this two-element filter expression becomes

$$\mathbf{y}_n = 0.5(1 + e^{-j\omega T})\mathbf{x}_n \tag{14.23}$$

and from Eq. (14.18), its transfer function is

$$\mathbf{H}(j\omega) = \mathbf{G}(e^{j\omega T}) = 0.5(1 + e^{-j\omega T}) \tag{14.24}$$

As always, we can find $|\mathbf{H}|$ from the expression $|\mathbf{H}| = \mathbf{H}^*\mathbf{H}$, but this $|\mathbf{H}|$ is more easily found from its geometric representation as the sum of two phasors on the complex plane as shown in Figure 14.7b. Either way, $|\mathbf{H}(j\omega)|$ reduces to

$$|\mathbf{H}(j\omega)| = \left| \cos\left(\frac{\omega T}{2}\right) \right| \tag{14.25}$$

and is shown in Figure 14.7c. If the frequency scale is plotted in units of ωT as in the figure, then the Nyquist frequency limit is always π. In common with all discrete frequency plots, this function is symmetric around π on the interval 0 to 2π and is repeated on all following 2π intervals.

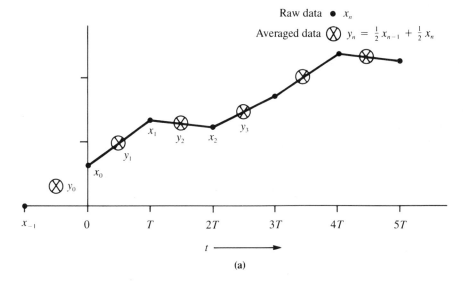

Raw data ● x_n

Averaged data ⊗ $y_n = \frac{1}{2} x_{n-1} + \frac{1}{2} x_n$

(a)

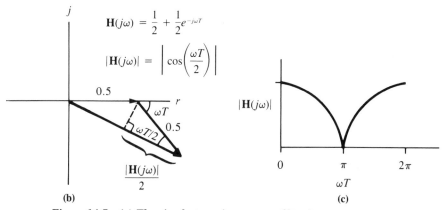

$$\mathbf{H}(j\omega) = \frac{1}{2} + \frac{1}{2} e^{-j\omega T}$$

$$|\mathbf{H}(j\omega)| = \left| \cos\left(\frac{\omega T}{2}\right) \right|$$

(b)

(c)

Figure 14.7 (a) The simplest moving-average filter just averages adjacent pairs of points. (b) The magnitude of $\mathbf{H}(j\omega)$ is most easily found from its phasor representation on the complex plane. (c) A plot of $|\mathbf{H}(j\omega)|$ over twice the Nyquist-limited frequency spectrum.

14.4 Z-Transform

Using an argument similar to the special case just described, we can obtain the **z**-transform by direct substitution into the Laplace transform. For a complex, time-varying signal $\mathbf{x}(t)$, the continuous Laplace transform definition of Eq. (12.25) is

$$\mathbf{X(s)} = \mathcal{L}[\mathbf{x}(t)] = \int_0^\infty \mathbf{x}(t) \, e^{-st} \, dt \qquad (14.26)$$

and if this is written as the limit of an infinite sum, we have

$$\mathbf{X(s)} = \lim T \to 0 \sum_{n=0}^{\infty} \mathbf{x}(nT) \, e^{-snT} \, T \qquad (14.27)$$

where T is the time interval between $\{\mathbf{x}\}$ points. By dropping the limiting process we obtain a discrete approximation of the Laplace transform:

$$\mathbf{X}(e^{sT}) = \sum_{n=0}^{\infty} \mathbf{x}_n \, T \, e^{-snT} \tag{14.28}$$

Defining a new variable $\mathbf{z} = e^{sT}$, this equation can be written

$$\mathbf{X}(\mathbf{z}) = \sum_{n=0}^{\infty} \mathbf{x}_n T \, \mathbf{z}^{-n} \tag{14.29}$$

The sampling interval T is a constant, normally set to 1 and dropped from this and following equations by the simple expedient of changing the unit of time from seconds to the sample interval T. With this change, the units of frequency are oscillations per sample interval rather than oscillations per second, a practical choice that is convenient for most applications (such as plotting frequency spectra). However, for clarity and as an aid in the dimensional analysis of various equations, we will retain T in the following expressions whenever it has significance.

The defining equation for the $\mathbf{z}$-transform of the infinite sequence $\{\mathbf{f}\}$ is

$$\mathscr{L}[\mathbf{f}_n] = \sum_{n=0}^{\infty} \mathbf{f}_n \, \mathbf{z}^{-n} \tag{14.30}$$

and with this definition, the discrete Laplace transform of Eq. (14.29) can be written as the $\mathbf{z}$-transform of the sequence $\{\mathbf{x}T\}$,

$$\mathbf{X}(\mathbf{z}) = T\mathscr{L}[\mathbf{x}_n] \tag{14.31}$$

where the constant T has been factored out of the summation. Note that whereas $\{\mathbf{x}\}$ is a sequence, $\mathbf{X}(\mathbf{z})$ is a polynomial with terms like $(1/\mathbf{z})^n$. The infinite sum used in the transform definition can be useful when a sum must be evaluated analytically, but in any real experiment the $\mathbf{f}_n$ will always vanish above some index $N - 1$, and the infinite sum can be replaced by a finite one for numerical calculation. A short table of $\mathbf{z}$-transform properties is given in Table 14.1.

As in Chapter 12, where the Laplace transform in the impulse response $h(t)$ was seen to give the frequency-domain transfer function $\mathbf{H}(j\omega)$, we now find that the $\mathbf{z}$-transform of a discrete system's impulse response sequence $\{g_nT\}$ will yield the frequency-domain transfer function $\mathbf{G}(\mathbf{z})$ of that system. For example, if we have a moving-average filter where the elements of $\{g\}$ are the coefficients a_k/T of the filter expression, Eq. (14.31) will reduce to Eq. (14.20).

The mathematical relationships between the variables $j\omega$, $\mathbf{s}$, and $\mathbf{z}$ show how the analog concepts developed earlier can be applied in the discrete domain. Consider the equivalent expressions

Table 14.1 z-Transform Theorems

Name	Signal	Transform
	$n = 0, 1, 2, \ldots$	
	$\{f_n\}$	$F(z)$
	$\{g_n\}$	$G(z)$
1 Linearity	$\{Af_n\} + \{Bg_n\}$	$AF(z) + BG(z)$
2 Delay (right shift)	$\{f_{n-N}\}$	$z^{-N}F(z)$
	$N \geq 0$	

$$\mathbf{z} = e^{\mathbf{s}T} \tag{14.32}$$

$$= e^{(\sigma + j\omega)T} \tag{14.33}$$

and
$$= e^{\sigma T} e^{j\omega T} \tag{14.34}$$

From these we see that the phasor $\mathbf{z}$ has magnitude $e^{\sigma T}$ and makes an angle ωT with the real axis. While the variable $\mathbf{z}$ can cover the entire complex plane, because $e^{j\omega T} = e^{(j2m\pi + \omega T)}$, an infinite sequence of $\mathbf{s}$ values will map into the same $\mathbf{z}$ point. Although not unique, an inverse transformation can also be defined. We will not present the general inverse transformation here, but some alternatives to the inversion operation will be developed in the recursive filter section below.

Figure 14.8   (a) Points on the s-plane map into new positions on the z-plane. The imaginary axis in the s-plane maps onto the unit circle in the z-plane, while the left and right halves of the s-plane correspond respectively to the inferior and exterior of the unit circle.

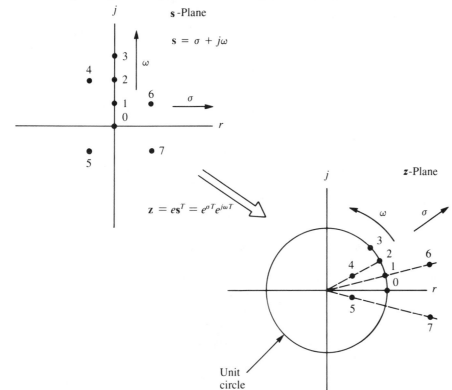

Figure 14.8 shows how points on the **s**-plane map into corresponding points on the **z**-plane. In particular, it is seen that points on the imaginary **s**-plane axis map into points on the unit circle in the **z**-plane.

With continuous signals, when we needed to determine the frequency response $\mathbf{H}(j\omega)$ from a filter's transfer function $\mathbf{H}(\mathbf{s})$, it was found by restricting **s** to the imaginary axis by requiring that σ be zero. The same procedure is appropriate for $\mathbf{G}(\mathbf{z})$, but now when we set $\sigma = 0$, **z** becomes a phasor of unit magnitude as shown in Figure 14.8, and $\mathbf{G}(\mathbf{z})$ becomes $\mathbf{G}(e^{j\omega T})$. Because of the cyclic nature of this complex exponential variable and its restriction to the unit circle on the **z**-plane, the frequency response $\mathbf{G}(e^{j\omega T})$ must repeat on 0 to 2π intervals in ωT and its real projection must be symmetric around π. As we have demonstrated, this behavior is expected in the discrete domain.

14.5   Digital Filters

In this section we will describe a more general and powerful filter algorithm, the recursive filter, and show how it can be used to simulate some basic analog filters.

14.5.1   The Recursive Filter

The moving-average filter defined by Eq. (14.12) and described in Figure 14.6a is only able to produce an impulse response $\{g\}$ that persists for a time KT, where K is the number of a_k coefficients. We can produce any impulse response desired, as long as we are willing to use enough a_k terms. However, for a given number of coefficients, we can greatly expand the persistence of the impulse response by using the recursive technique of feeding part of the output signal back into the summation as shown in Figure 14.9.

The algorithm for a general recursive filter is described by the formula

$$\mathbf{y}_n = \sum_{k=0}^{K-1} a_k \mathbf{x}_{n-k} - \sum_{m=1}^{M-1} b_m \mathbf{y}_{n-m} \qquad \text{for } n = 0, 1, 2, \ldots, N-1 \quad (14.35)$$

where the second summation starts from 1 in order to eliminate the uninteresting $b_0\mathbf{y}_n$ term from the right side of the equation. Unlike the moving-average filter (which is now seen to be a special case of the recursive filter with $b_m = 0$), this equation can produce a wide range of response functions using a reasonable number of terms. Since the terms in the $\{\mathbf{y}\}$ sequence may continue to be nonzero even after the $\{\mathbf{x}\}$ signal has vanished, the maximum index $N - 1$ is now limited only by the number of output points we wish to calculate.

> **Example 14.1**   Determine the coefficients of a moving-average filter that would produce the same impulse response as a recursive filter with $a_0 = 1$ and $b_1 = -1.1$.
>   Assuming real signals and the initial conditions $y_{-1} = 0$, and all $x_m = 0$ except for $x_0 = 1$, the first few terms of y_n evaluate to
>
> $$y_0 = 1 + 0$$
> $$y_1 = 0 + 1.1$$

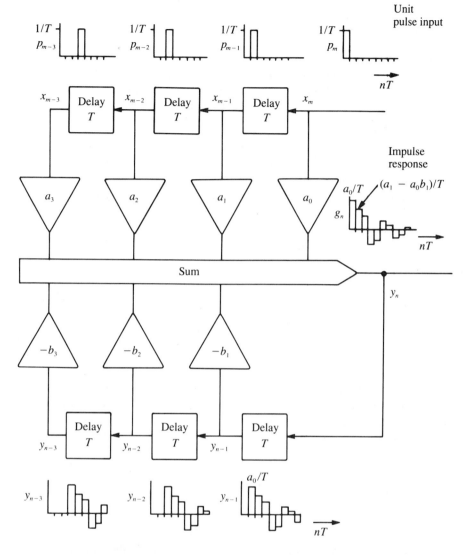

Figure 14.9 *For a given number of coefficients a_k and b_k, the recursive filter can generate a wider variety of impulse responses because the recycled output can drive the filter long after the input impulse {p} has moved through the x_m delays.*

$$y_2 = 0 + (1.1)^2$$
$$y_3 = 0 + (1.1)^3$$
$$\text{etc.}$$

These terms happen to be the Taylor expansion of $e^x = 1 + x + x^2 + x^3 + \ldots$ evaluated at $x = 1.1$, giving an unbounded exponential shape for the impulse response. Thus the filter is unstable, and the output continues to grow for as long as we care to calculate output points. To simulate this response with a moving-average filter would require a coefficient sequence

$$\{a\} = 1,\ 1.1,\ (1.1)^2,\ \ldots,\ (1.1)^{N-1},\ \ldots$$

with as many terms as we desire output points, hardly a reasonable alternative.

With the help of the linearity and shift theorems given in Table 14.1, we can write an expression for the frequency response of the general recursive filter in terms of its coefficients a_k and b_m. Taking the **z**-transform of both sides of Eq. (14.35) and applying the linearity theorem gives

$$\mathcal{Z}[\mathbf{y}_n] = \sum_{k=0}^{K-1} a_k \mathcal{Z}[\mathbf{x}_{n-k}] - \sum_{m=1}^{M-1} b_m \mathcal{Z}[\mathbf{y}_{n-m}] \tag{14.36}$$

Applying the shift theorem to the two transforms on the right side of this equation gives

$$\mathcal{Z}[\mathbf{y}_n] = \sum_{k=0}^{K-1} a_k \mathbf{z}^{-k} \mathcal{Z}[\mathbf{x}_n] - \sum_{m=1}^{M-1} b_m \mathbf{z}^{-m} \mathcal{Z}[\mathbf{y}_n] \tag{14.37}$$

and in this expression the **z**-transforms can be factored out of the sums. Via analogy with continuous systems, the frequency-domain transfer function for the discrete filter is defined by the equation

$$\mathbf{Y}(\mathbf{z}) = \mathbf{G}(\mathbf{z})\ \mathbf{X}(\mathbf{z}) \tag{14.38}$$

where **Y, G,** and **X** are just the **z**-transforms of $\{\mathbf{y}T\}$, $\{\mathbf{g}T\}$, and $\{\mathbf{x}T\}$, respectively. To obtain $\mathbf{G}(\mathbf{z})$ for the recursive filter defined by Eq. (14.35), we can move the last term in Eq. (14.37) to the left side of the equation, then form the ratio of **z**-transforms to give

$$\mathbf{G}(\mathbf{z}) = \frac{\mathbf{Y}(\mathbf{z})}{\mathbf{X}(\mathbf{z})} = \frac{\displaystyle\sum_{k=0}^{K-1} a_k \mathbf{z}^{-k}}{1 + \displaystyle\sum_{m=1}^{M-1} b_m \mathbf{z}^{-m}} \tag{14.39}$$

We now have $\mathbf{G}(\mathbf{z})$ in the form of a ratio of polynomials in the variable $\mathbf{z}^{-1}$, but this is an inconvenient form. Multiplying the numerator and denominator by $\mathbf{z}^K$ gives a more useful expression

$$\mathbf{G}(\mathbf{z}) = \frac{\displaystyle\sum_{k=0}^{K-1} a_k \mathbf{z}^{K-k}}{\mathbf{z}^K + \displaystyle\sum_{m=1}^{M-1} b_m \mathbf{z}^{K-m}} \tag{14.40}$$

which can be expanded to

$$\mathbf{G}(\mathbf{z}) = \frac{a_0 \mathbf{z}^K + a_1 \mathbf{z}^{K-1} + \cdots + a_{K-1}\mathbf{z}}{\mathbf{z}^K + b_1 \mathbf{z}^{K-1} + \cdots + b_{M-1}\mathbf{z}^{K-M+1}} \tag{14.41}$$

If we restrict our filter models so that $K - M + 1 \geq 0$ (requiring at least as many a terms as b terms), then there are no negative powers in this expression, and we have $\mathbf{G}$ in the form of a ratio of polynomials in $\mathbf{z}$. In this form we can associate the roots of the denominator and numerator with the poles and zeros of $\mathbf{G(z)}$, just as we did with $\mathbf{H(s)}$ in the continuous function case. Rewriting the polynomials in terms of their roots $\mathbf{z}_k$ and $\mathbf{z}_m$ gives a ratio of products,

$$\mathbf{G(z)} = C \frac{\displaystyle\prod_{k=0}^{K-1} (\mathbf{z} - \mathbf{z}_k)}{\displaystyle\prod_{m=0}^{M-1} (\mathbf{z} - \mathbf{z}_m)} \tag{14.42}$$

where C is a real constant. The points $\mathbf{z}_k$ locate transfer function zeros, and the $\mathbf{z}_m$ locate poles.

The pole-zero form of $\mathbf{G(z)}$ is most useful for determining the frequency response $|\mathbf{G}(e^{j\omega T})|$, since it can be evaluated graphically and qualitatively as shown in Figure 14.10. The procedure is the same as that described earlier for Figure 4.9, but the results can be very different here because of the circular geometry. The process can be described mathematically by rewriting Eq. (14.42) with $\mathbf{z}$ restricted to the unit circle ($\mathbf{z} = e^{j\omega T}$). In terms of phasors $\mathbf{R}_k$ and $\mathbf{P}_m$ from $e^{j\omega T}$ to the zeros and poles, respectively we have

$$\mathbf{G}(e^{j\omega T}) = C \frac{\displaystyle\prod_{k=0}^{K-1} \mathbf{R}_k}{\displaystyle\prod_{m=0}^{M-1} \mathbf{P}_m} = C \frac{\displaystyle\prod_{k=0}^{K-1} (e^{j\omega T} - \mathbf{z}_k)}{\displaystyle\prod_{m=0}^{M-1} (e^{j\omega T} - \mathbf{z}_m)}, \tag{14.43}$$

Figure 14.10 A graphical interpretation of $|\mathbf{G}(e^{j\omega T})|$ or $|\mathbf{H}(j\omega)|$ in terms of poles and zeros on the z-plane.

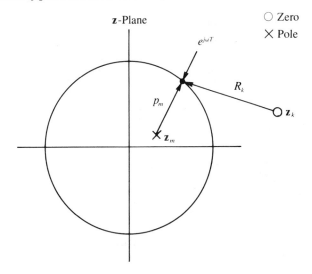

Via an argument similar to that used in the analog domain, each factor in these products is a phasor directed from a pole or zero to a pont $e^{j\omega T}$ on the unit circle. Each phasor is just a complex number and can be written in polar form as $A_n e^{j\theta_n}$, where each of the A_n and θ_n terms is a function of ω. The complex transfer function $\mathbf{G}$ can also be written as $|\mathbf{G}|e^{j\phi}$, where the frequency dependence of $|\mathbf{G}|$ and ϕ contains contributions from each pole and zero. The magnitude $|\mathbf{G}|$ at a particular frequency is simply C times the ratio of the products of distances (magnitudes), each distance being measured from the frequency point on the unit circle to a zero or pole on the $\mathbf{z}$-plane.

14.6   Simulating an Analog Filter

Because of the widespread use of analog filters, it is useful to establish a method for converting an analog filter design into an algorithm in the digital domain. It is important to realize that in making this conversion to the discrete domain, we are necessarily making some approximations. Three different methods for obtaining a digital filter algorithm from an analog transfer function $\mathbf{H}_a(\mathbf{s})$ are outlined below. Depending on the approximation, a given $\mathbf{H}_a$ can be converted to several different algorithms, each with its own transfer function $\mathbf{H}(\mathbf{s}) = \mathbf{G}(\mathbf{z})$ and impulse response $\{g\}$. The impulse method is the most accurate, but its application is also the most difficult, and the choice of a best method in a given application will obviously depend on many factors.

14.6.1   Matching the Impulse Response

If the impulse response of an analog filter is $h(t)$, we can sample it at time intervals T, and define the elements of an equivalent discrete transfer response $\{g\}$ as

$$g_n = h(nT) \tag{14.44}$$

The elements of a moving-average filter algorithm are now determined by simply making $a_n = g_n$ for as many terms as g_n remains significantly different from zero. A more practical recursive filter design is a bit more of a challenge and may require using the $\mathbf{z}$-transform of $\{gT\}$, which we write as

$$\mathbf{G}(\mathbf{z}) = \lim N \to \infty \sum_{n=0}^{N-1} g_n T \mathbf{z}^{-n} \tag{14.45}$$

Note that we have modified the form of Eq. (14.29) slightly to emphasize that N may be finite in some approximation.

There are two ways in which an equivalent recursive filter algorithm can be determined from $\{g\}$: (1) Take the $\mathbf{z}$-transform of $\{gT\}$ and try to put the resulting expression for $\mathbf{G}(\mathbf{z})$ into a form that can be matched to Eq. (14.41); or (2) try to find a recursion relation that generates the $\{gT\}$ sequence of points directly. Unfortunately, there is no guarantee that either method will yield an analytic expression for the filter coefficients. The application of these methods is demonstrated for a simple case in Example 14.2.

Example 14.2 Use the impulse response method to obtain a digital filter algorithm that matches the low-pass analog transfer function

$$H_a(s) = \frac{c}{s + c}$$

where $c = 1/RC$.

Using the Laplace transform from line 8 of Table 12.2, we can determine that the impulse response of this filter is $h(t) = ce^{-ct}$, and if sampled in intervals T, the equivalent discrete response is

$$\{g\} = ce^{-cnT} \qquad \text{for } n = 0, 1, 2, \ldots$$

The $\mathbf{z}$-transform of $\{gT\}$ is

$$G(\mathbf{z}) = cT \sum_{n=0}^{\infty} (\mathbf{z}e^{cT})^{-n}$$

where the infinite sum is of the form $\sum_{n=0}^{\infty} x^{-n}$ and is conveniently equivalent to the analytic expression $x/(x - 1)$. Using this identity, we get

$$G(\mathbf{z}) = \frac{cT\mathbf{z}e^{cT}}{\mathbf{z}e^{cT} - 1}$$

and by dividing numerator and denominator by e^{cT}, this expression can be put into the form of Eq. (14.41) with $K = 1$ and $M = 2$:

$$G(\mathbf{z}) = \frac{cT\mathbf{z}}{\mathbf{z} - e^{-cT}}$$

Comparing this equation with Eq. (14.41), we see that the only nonzero a_k and b_m coefficients are

$$a_0 = cT$$

and

$$b_1 = -e^{-cT}$$

Rewriting Eq. (14.35) with these coefficients gives the filter algorithm

$$y_n = cTx_n + e^{-cT}y_{n-1}$$

An alternative procedure is to write a few terms of the impulse response $\{g\}$ given above, and then observe that the same terms can be obtained with a recursion relation. With the input impulse being $x_0 = 1/T$ and other $x_n = 0$, we have

Desired Output	x Term + y Term
$g_0 = c$	$= c + 0$
$g_1 = ce^{-cT}$	$= 0 + e^{-cT}g_0$
$g_2 = ce^{-2cT}$	$= 0 + e^{-cT}g_1$
etc.	

Obviously, this sequence can be generated with the filter algorithm given above.

Both versions of this impulse response method require methematical insights that become less obvious as the transfer function becomes more complicated.

14.6.2   Mapping Poles and Zeros

The second method for obtaining a digital equivalent for an analog filter is easier and can always be accomplished, but it yields a poorer approximation to the desired response. The technique is to map the poles s_m and zeros s_z of the analog filter $\mathbf{H}_a(\mathbf{s})$ into corresponding poles $\mathbf{z}_m$ and zeros $\mathbf{z}_k$ on the $\mathbf{z}$-plane. The mapping of individual points is simply accomplished using $\mathbf{z} = e^{\mathbf{s}T}$ as shown in Figure 14.8, and the products of Eq. (14.42) can be expanded into polynomials like Eq. (14.41).

The catch is that the circular geometry of the $\mathbf{z}$-plane changes the phasor differences in Eq. (14.43) from the equivalent ones on the $\mathbf{s}$-plane, thereby allowing the new transfer function $\mathbf{G}(e^{j\omega T})$ to be significantly different from $\mathbf{H}_a(j\omega)$. This effect can be minimized only by choosing a sample rate that is large compared to the highest useful signal frequency, thus assuring that ωT is always stays near zero where the small portion of circular arc traversed by $e^{j\omega T}$ is well approximated by a straight line.

Example 14.3   Use the pole-zero mapping method to derive a digital filter algorithm for the low-pass filter of the previous example.

The analog transfer function from that example was

$$\mathbf{H}_a(\mathbf{s}) = \frac{c}{\mathbf{s} + c}$$

where $c = 1/RC$. This transfer function has no zeros and a single real pole at $\mathbf{s}_m = -c$, which maps into a $\mathbf{z}$-plane pole at $\mathbf{z}_m = e^{-cT}$. From Eq. (14.42), this pole corresponds to the discrete transfer function

$$\mathbf{G}(\mathbf{z}) = \frac{C}{\mathbf{z} - e^{-cT}}$$

where C is still an undefined real constant. We can pick a reasonable C by requiring that $|\mathbf{G}(\mathbf{z})| = |\mathbf{H}_a(\mathbf{s})|$ at the point $\mathbf{s} = 0$. Since $\mathbf{H}_a(0) = 1$ and $\mathbf{z} = 1$ when $\mathbf{s} = 0$, we find that

$$C = 1 - e^{-cT}$$

Multiplying the numerator and denominator of the resulting $\mathbf{G}$ expression by $\mathbf{z}$ puts it into the form of Eq. (14.41) and allows us to write the filter equation

$$y_n = (1 - e^{-cT})x_{n-1} + e^{-cT}y_{n-1}$$

If $cT \ll 1$, the exponential in the first term can be approximated by the first two terms of its Taylor series ($e^{-x} = 1 - x$), and this equation will reduce to an expression very similar to the previous example. Since $c = 1/RC$, we find that this method provides a good approximation only when the sample period T is much less than the natural time constant RC of the filter.

14.6.3   Bilinear Transformation

The even easier third method for obtaining the digital representation of an analog filter is to convert $\mathbf{s}$ to $\mathbf{z}$ by means of the bilinear transformation

$$s = \frac{2(z - 1)}{T(z + 1)} \qquad (14.46)$$

Starting with an $H_a(s)$, each occurrence of s is converted to z using this expression; the result is the desired digital transfer function $G(z)$.

The curious form of the bilinear transformation between s and z is easily derived. If we let $x = sT$ and expand $z = e^x$, we get

$$z = 1 + x + \frac{x^2}{2} + \frac{x^3}{3!} + \dots \qquad (14.47)$$

With the help of this expression, the polynomial ratio $(z - 1)/(z + 1)$ can be evaluated by long division to give $x/2$ plus a negative infinite series remainder, the largest term of which has magnitude $x^3/12$. If x is small enough that the remainder can be neglected, then Eq. (14.46) results. Like the pole-zero mapping method, this transformation method is accurate only when $|sT| \ll 1$. Since we will be evaluating G only on the unit circle, s is restricted to $j\omega$, making this requirement a more understandable $\omega T \ll 1$ or $f \ll 2\pi/T$. Thus, the signal frequencies must be small compared to the Nyquist limiting frequency π/T.

> **Example 14.4** Use the bilinear transformation to determine the digital filter approximation to the perfect integrator having transfer function $H_a = 1/s$.
>
> If the signal is sampled with period T, then when s is replaced with the bilinear transformation of Eq. (14.46), the discrete transform is just
>
> $$G(z) = \frac{T(z + 1)}{2(z - 1)}$$
>
> Multiplication by z/z puts this expression into the form of Eq. (14.41) with $K = 2$ and $M = 2$. The nonzero coefficients are
>
> $$a_0 = \frac{T}{2} \qquad a_1 = \frac{T}{2}$$
> $$b_1 = -1$$
>
> and the resulting filter expression is
>
> $$y_n = \frac{T}{2}(x_n + x_{n-1}) + y_{n-1}$$
>
> Investigation of Figure 14.11 will reveal that this algorithm is just the trapezoid rule for obtaining an approximate integral. The first term is just an area formed by the product of the average function value times the interval width, and the last term is just the summed area of all previous trapezoids.

14.7 Discrete Fourier Transform

In analogy with the discussion of continuous signals in Chapter 12, the discrete Fourier transform of a sequence $\{x\}$ can be defined as its z-transform evaluated on

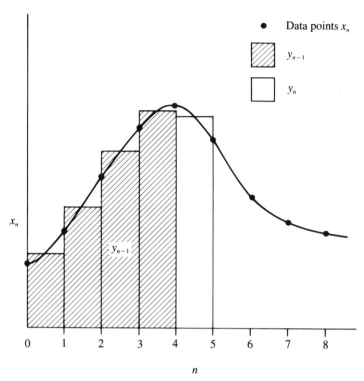

Figure 14.11   The bilinear transformation applied to the ideal analog integrator $\mathbf{H(s)} = 1/s$ produces a discrete filter that can be identified with the trapezoid rule for approximate integration.

the unit circuit where $\mathbf{s} = j\omega$. Applying this restriction to the **z**-transform definition of Eq. (14.30) gives

$$\mathbf{X}(e^{j\omega T}) = \mathscr{Z}[\{x_n\}] = \sum_{n=0}^{\infty} \mathbf{x}_n e^{-j\omega nT} \qquad (14.48)$$

where we have extended the definition to include complex $\mathbf{x}_n$. But we must make some changes, since in any practical problem the signal can be sampled at only a finite number of points. If we have a discrete signal known at N points,

$$\{\mathbf{x}\} = \mathbf{x}_n \qquad \text{for } n = 0, 1, 2, \ldots, N - 1 \qquad (14.49)$$

and take it to be zero outside this index range, then the transformed signal is exactly given by the finite sum

$$\mathbf{X}(e^{j\omega T}) = \sum_{n=0}^{N-1} \mathbf{x}_n e^{-j\omega nT} \qquad (14.50)$$

This $\mathbf{X}$ is a continuous function of the variable ω, but again, as a practical matter, we can calculate it at only a finite number of points. It will prove convenient to choose N points $\{\mathbf{X}(\omega_k)\}$ (same N as for the signal points) equally spaced between 0 and the sampling angular frequency ω_S. For real x_n, this choice wastes half of the frequency points (those beyond the Nyquist limit $\omega_S/2$), but it produces a symmetrical relationship between the transform and its inverse and is a feature of the fast Fourier transform implementation given below.

Following this prescription, we define N angular frequency points,

$$\omega_k = \frac{2\pi k}{NT} \qquad \text{for } k = 0, 1, 2, \ldots, N - 1 \qquad (14.51)$$

as shown in Figure 14.12a, and write the discrete Fourier transformed sequence $\{\mathbf{X}\}$ as

$$\mathbf{X}_k = \sum_{n=0}^{N-1} \mathbf{x}_n e^{-jkn2\pi/N} \qquad \text{for } k = 0, 1, 2, \ldots, N - 1 \qquad (14.52)$$

This equation defines the discrete Fourier transform (DFT), and as with continuous signals, the transformed sequence $\{\mathbf{X}\}$ describes the frequency content of the discrete signal $\{\mathbf{x}\}$. As expected for a discrete signal, the $\mathbf{x}_k$ values repeat for $k \geq N$ ($\omega T \geq 2\pi$). They also "reflect" around $k = N/2$ ($\omega T = \pi$) to the extent that $|\mathbf{X}_{N-k}| = |\mathbf{X}_k|$.

Note that the T factors appearing in the frequency and in the $\mathbf{z}$-transform have been canceled in the exponent of the DFT, and that we have chosen to deal with the sequence $\{\mathbf{x}\}$ rather than $\{\mathbf{x}T\}$, with the result that T does not appear in this definition; the independent variables are the sample indices n and k. The time interval in seconds will still be needed whenever it is necessary to convert from pure numbers back to time-dependent signals and frequencies described by time in seconds.

In analogy with the continuous transform, the inverse discrete Fourier transform (IDFT) is given by

$$\mathbf{x}_n = \frac{1}{N} \sum_{k=0}^{N-1} \mathbf{X}_k e^{jnk2\pi/N} \qquad \text{for } n = 0, 1, 2, \ldots, N - 1 \qquad (14.53)$$

This expression can be used to gain some additional insight into the meaning of $\mathbf{X}_k$. We first define a unit phasor whose phase angle is determined by the number of terms in the sequence

$$\mathbf{W} = e^{j2\pi/N} \qquad (14.54)$$

By raising this phasor to the various powers represented by the k index, we generate N unit phasors $\mathbf{W}^k = e^{jk2\pi/N}$, each of which makes an angle $\theta = 2\pi k/N$ with the positive real axis, an angle that is greater for larger k values. Raising one of these N unit phasors to an additional power n as in $\mathbf{W}^{kn}$, simply moves it to an angle $n\theta$. Since the index n is timelike ($t = nT$), each $\mathbf{W}^k$ can be interpreted as a phasor that rotates with angular frequency $2\pi k/NT$. In terms of these phasors, the inverse transform becomes

$$\mathbf{x}_n = \sum_{k=0}^{N-1} \frac{\mathbf{X}_k}{N} (\mathbf{W}^k)^n \qquad \text{for } n = 1, 2, \ldots, N - 1 \qquad (14.55)$$

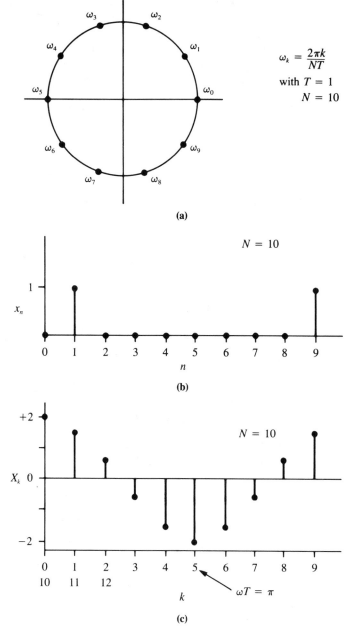

$$\omega_k = \frac{2\pi k}{NT}$$

with $T = 1$

$N = 10$

(a)

(b)

(c)

Figure 14.12 (a) The frequency points defined for the discrete Fourier transform span the entire unit circle. (b) The simplest "symmetric" input signal and (c) its discrete Fourier transform.

and can be written out in terms of $\mathbf{W}^k$ as

$$\mathbf{x}_n = \frac{\mathbf{X}_0}{N} + \frac{\mathbf{X}_1}{N}\,(\mathbf{W}^1)^n + \frac{\mathbf{X}_2}{N}\,(\mathbf{W}^2)^n + \;.\;.\;.$$

$$\text{for } n = 1, 2, \;.\;.\;., N - 1 \quad (14.56)$$

As the index n to the elements of the $\{x\}$ sequence increases, the first term is constant, the second rotates with frequency $2\pi/NT$, the third with frequency $4\pi/NT$, and so forth. Thus we see that each X_k/N defines the complex amplitude of a phasor rotating at angular frequency $2\pi k/NT$, in exact analogy with the continuous signal Fourier transform.

Example 14.5 (a) Determine the elements in the DFT of the 10-term sequence $\{x\} = 0, 1, 0, 0, 0, 0, 0, 0, 0, 1$ shown in Figure 14.12b. (b) If $\{x\}$ represents a signal sampled at intervals of 50 ms, determine the frequency scale in hertz.

(a) For the $\{x\}$ sequence given, the kth element in the transformed $\{X_k\}$ sequence is obtained by evaluating Eq. (14.52) using an $\{x_n\}$ with only two nonzero real terms, $x_1 = 1$ and $x_9 = 1$. This gives

$$X_k = e^{-jk0.2\pi} + e^{-jk1.8\pi}$$

and by considering the location of the second phasor on the complex plane, the expression can be rewritten as

$$X_k = e^{-jk0.2\pi} + e^{+jk0.2\pi}$$

The sum of these two phasors happens to be a real expression,

$$X_k = 2\cos(k0.2\pi)$$

and evaluates to the 10-term sequence plotted in Figure 14.12c:

$$\{X\} = 2, 1.6, 0.62, -0.62, -1.62, -2, -1.62, -0.62, 0.62, 1.6$$

Note that $X_{N-k} = X_k$ in this sequence because the x_n were real and symmetric, with symmetric defined as $X_{N-n} = X_n$. The more general rule for real signals is that $X_{N-k} = X_k^*$ but when $\{x\}$ is such that $\{X\}$ is not real, it is often useful to instead calculate the input function's power spectrum $\{X^*X\}$.

(b) To convert to hertz we must recognize that each X_k corresponds to the amplitude of a sinusoid at angular frequency $\omega_k = 2\pi k/NT$ as defined by Eq. (14.51). In this example, $N = 10$ and $T = 50 \times 10^{-3}$ s, making the frequency expression $f_k = k/0.5$ Hz. Thus, the interval between frequency samples in Figure 14.12c is 2 Hz.

14.7.1 Discrete Fourier Transform Theorems

Several of the Fourier transform theorems from continuous functions carry over into discrete function analysis with little change. Some of the more important theorems are given in Table 14.2.

Name	Signal terms	Transform terms
	$\mathbf{f}_n$	$\mathbf{F}_k$
	$\mathbf{g}_n$	$\mathbf{G}_k$
1 Linearity	$A\mathbf{f}_n + B\mathbf{g}_n$	$A\mathbf{F}_k + B\mathbf{G}_k$
2 Shift (delay)	$\mathbf{f}_{n-M}$	$\mathbf{F}_k e^{-j2\pi kM/N}$
3 Correlation	$\displaystyle\sum_{n=0}^{N-1} f_{n+m}g_n$	$\mathbf{F}_k\mathbf{G}_k^*$
4 Convolution	$\displaystyle\sum_{n=0}^{N-1} f_n g_{m-n}$	$\mathbf{F}_k\mathbf{G}_k$

The derivation of the shift theorem will serve to show the general character of discrete function operations. Using the notation of Eq. (14.54), we can write the transform of a signal $\{\mathbf{x}\}$ as

$$\mathbf{X}_k = \sum_{n=0}^{N-1} \mathbf{x}_n \mathbf{W}^{-kn} \tag{14.57}$$

its inverse as

$$\mathbf{x}_n = \frac{1}{N} \sum_{l=0}^{N-1} \mathbf{X}_l \mathbf{W}^{ln} \tag{14.58}$$

and the shifted signal's transform as

$$\mathbf{Y}_k = \sum_{n=0}^{N-1} \mathbf{x}_{n-M} \mathbf{W}^{-kn} \tag{14.59}$$

Then by using Eq. (14.58) to replace $\mathbf{x}_{n-M}$ in the shifted transform, we obtain

$$\mathbf{Y}_k = \sum_{n=0}^{N-1} \left[\frac{1}{N} \sum_{l=0}^{N-1} \mathbf{X}_l \mathbf{W}^{l(n-M)} \right] \mathbf{W}^{-kn} \tag{14.60}$$

By rearranging the exponents and reversing the order of the summations, this can be rewritten as

$$\mathbf{Y}_k = \sum_{l=0}^{N-1} \mathbf{X}_l \mathbf{W}^{-lM} \left[\frac{1}{N} \sum_{n=0}^{N-1} \mathbf{W}^{n(l-k)} \right] \tag{14.61}$$

Figure 14.13 The zero sum of the expression $\sum\limits_{n=0}^{N-1} \mathbf{W}^{(l-k)n}$ with $l \neq k$ is illustrated graphically for the special case of $l - k = 2$ and $N = 5$.

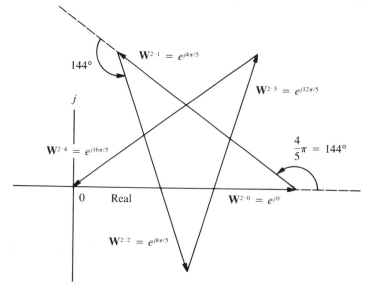

Now the factor in brackets amounts to a discrete delta function: if $l = k$, each term is 1 and the sum divided by N is also 1; for $l \neq k$, each term is a phasor and for any $l - k < N$, the sum of these phasors is always zero. This effect is illustrated for a specific case in Figure 14.13, which shows a sum of five phasors at angles 0, $4\pi/5$, $8\pi/5$, $12\pi/5$, and $16\pi/5$.

Replacing the bracketed factor in Eq. (14.61) with the discrete delta symbol δ_{lk}, the shifted transform can be written

$$Y_k = \sum_{l=0}^{N-1} X_l W^{-lM} \delta_{lk} \qquad (14.62)$$

and because of the discrete delta function this must be

$$Y_k = X_k W^{-kM} \qquad (14.63)$$

Expanding **W**, this becomes

$$Y_k = X_k e^{-j2\pi kM/N} \qquad (14.64)$$

proving the theorem. Using the shift theorem, the correlation and convolution theorems can be derived in a similar fashion.

14.7.2   The Ultimate Discrete Signal Filter

One important application of the DFT is to generate a signal filter that modifies the frequency content of a discrete input signal. The operation requires four steps: (1) Take the DFT of the input signal $\{x_n\}$ to get its frequency content $\{X_k/N\}$. (2) Form the sequence $\{H_k\}$ by evaluating the desired filter transfer function $H(j\omega)$ or $G(e^{j\omega T})$ at the frequency points $\omega_k = 2\pi k/NT$. (3) Effect the filter operation in the frequency domain by generating the sequence $\{H_k X_k/N\}$, and (4) transform this modified spectrum back to a discrete time signal $\{y_n\}$ using the inverse DFT.

This filter procedure involves many more numerical operations (multiplications and additions) than those discussed earlier, but it offers many more possibilities, and with the availability of the fast Fourier transform evaluation procedure developed in Section 14.8.1, the necessary transforms can be accomplished with reasonable speed, even in microcomputers.

14.8   Fast Fourier Transform

One of the reasons for hesitation in the use of the DFT is the effort or computer time needed to make the calculations. Before determining the number of operations required, we can make this and later manipulations easier by streamlining the form of the DFT. Using the complex notation

$$W = e^{j2\pi/N} \qquad (14.65)$$

the DFT defining Eq. (14.52) becomes

$$\mathbf{X}_k = \sum_{n=0}^{N-1} \mathbf{x}_n \mathbf{W}^{-kn} \qquad \text{for } k = 0, 1, 2, \ldots, N - 1 \qquad (14.66)$$

The quantity $\mathbf{W}^{-kn}$ can be written $(\mathbf{W}^n)^{-k}$, showing that the same set of N complex numbers $\mathbf{W}^n$ is used in each of the $\mathbf{X}_k$ calculations. Assuming that the $\mathbf{W}^n$ are already known, we see that each of the N $\mathbf{X}_k$ calculations requires N operations, where an operation is defined as a complex multiplication followed by a complex addition. To calculate the entire sequence $\{\mathbf{X}\}$ requires N^2 operations.

The value of the fast Fourier transform (FFT) algorithm developed by Cooley and Tukey in 1965 is that, for $N = 2^m$ points, the number of operations required to evaluate the DFT is reduced to $N \log_2(N)$ with no loss of information. The requirement that the number of points be some power of 2 (typically 64, 128, 256, 512, or 1024) is not a severe restriction, and it is even good practice to design experiments around these natural digital numbers. The reduction in calculational effort is quite significant: for $N = 512$ (2^9), evaluation of the DFT requires 262,144 operations, whereas the FFT requires only 4,608, a factor of 57 fewer! When programmed into a computer, this algorithm makes the FFT fast (or cheap) enough to be used as a general-purpose tool in signal analysis.

In most applications the FFT is implemented as a software subroutine: The user supplies a sequence of numbers and the FFT subroutine returns a second sequence that comprises the transform. The user does not really need to understand the very clever FFT algorithm, but for those who are still curious, the following description is provided.

14.8.1 *The FFT Algorithm*

The Cooley-Tukey DFT evaluation procedure outlined here is only one of several that exist, but it is the simplest and the fastest. Basically, the algorithm consists of an iterative procedure of breaking the N-point transform of $\{\mathbf{x}\}$ into two $N/2$-point transforms, then into four $N/4$-point transforms, and so on until finally we have N one-point transforms to perform. The sequence of N numbers $\{\mathbf{F}\}$ from these last transforms is obtained directly from $\{\mathbf{x}\}$ by a simple reordering process. The only calculational steps occur in the recombination of the numbers in $\{\mathbf{F}\}$ back into the desired sequence $\{\mathbf{X}\}$.

With the requirement that N be a power of 2, we start by dividing the sum of Eq. (14.66) into a sum over even terms plus a sum over odd terms:

$$\mathbf{X}_k = \sum_{n=0}^{(N/2)-1} \mathbf{x}_{2n} \mathbf{W}_N^{-2kn} + \mathbf{W}_N^{-k} \sum_{n=0}^{(N/2)-1} \mathbf{x}_{2n+1} \mathbf{W}_N^{-2kn}$$
$$\text{for } k = 0, 1, 2, \ldots, N - 1 \quad (14.67)$$

Note that we have added a subscript N to the $\mathbf{W}$ factors to remind us that $\mathbf{W}$ depends on the number of terms in the sequence. Now if we relabel the $N/2$ points in the even sequence as $\mathbf{e}_n$, those in the odd sequence as $\mathbf{o}_n$, and use the identity

$$\mathbf{W}_{N/2}^{-kn} = \mathbf{W}_N^{-2kn} \qquad (14.68)$$

these sums can be rewritten as

$$X_k = \sum_{n=0}^{(N/2)-1} e_n W_{N/2}^{-kn} + W_N^{-k} \sum_{n=0}^{(N/2)-1} o_n W_{N/2}^{-kn}$$

$$\text{for } k = 0, 1, 2, \ldots, N-1 \quad (14.69)$$

Comparison of the two sums in this expression with Eq. (14.66) shows that each will produce the elements of an $N/2$-point transform of the even and odd sequence, respectively. This observation is summarized by the expression

$$X_k = E_k + W_N^{-k} O_k \qquad \text{for } k = 0, 1, 2, \ldots, N/2 - 1 \quad (14.70)$$

where $\{E\}$ and $\{O\}$ are the $N/2$-term transforms of the even x and odd x sequences, respectively.

Now we seem to have a problem—for while Eq. (14.67) is defined for $k = 0$, $1, 2, \ldots, N - 1$, the even and odd transforms each contain terms only up to $N/2 - 1$. The solution comes with the observation that the $\{E\}$ and $\{O\}$ sequences obtained from the DFT are mapped onto the unit circle in the z-plane, and that our initial definition of the DFT frequency range (Eq. 14.51) spanned the entire unit circle. Thus, when $k \geq N/2$, the E_k and O_k frequency terms simply begin to repeat themselves according to the expressions $E_{k+N/2} = E_k$ and $O_{k+N/2} = O_k$. This observation allows us to write

$$X_k = E_k + W_N^{-k} O_k \qquad \text{for } k = 0, 1, 2, \ldots, N - 1 \quad (14.71)$$

and the only thing that distinguishes the second half of $\{X\}$ from the first is the W_N^{-k} factor. Using Eq. (14.65) to expand W, Eq. (14.71) becomes

$$X_k = E_k + e^{-jk2\pi/N} O_k \qquad \text{for } k = 0, 1, 2, \ldots, N - 1 \quad (14.72)$$

This one-step reduction has reduced the number of operations (we only count the complex multiplication and addition operation since it is by far the slowest) needed to generate the transform from N^2 to $2(N/2)^2 + N$. Having gone through this thought process once, we can imagine repeating it for each of the $N/2$-term $\{E\}$ and $\{O\}$ sequences, then for the four $N/4$-term sequences, and so on until we reach the point that $N = 1$ and the transform sum contains only the single $n = 0$ term. Looking back at Eq. (14.66), we see that since $W^0 = 1$, the one-term transform is just $X_0 = x_0$.

Even though the final N transforms are just terms in the original sequence $\{x\}$, the repeated division of $\{x\}$ into even and odd indexed lists has thoroughly scrambled the order of the terms by the time we reach the last step. Fortunately, the relationship of the original index n to the new index m is easily seen if the indices are written as binary numbers: The process is best described by an example. Consider the eight-term sequence $\{x\}$ given in Figure 14.14, which has its initial 0-to-7 index written in binary $n = 4a + 2b + c$, where a, b, and c are either 0 or 1. The initial division into odd and even is based on the value of c and reorders the terms as indicated by the lines to the second column. The second division is based on the value of b, and last on the value of a, which is seen to produce no further scrambling. Comparison of the binary indices in the last two columns will show that we have formed a new sequence by moving each term $x_n = x_{4a+2b+c}$ into a new location $m = 4c + 2b + a$. To minimize confusion, we will call this

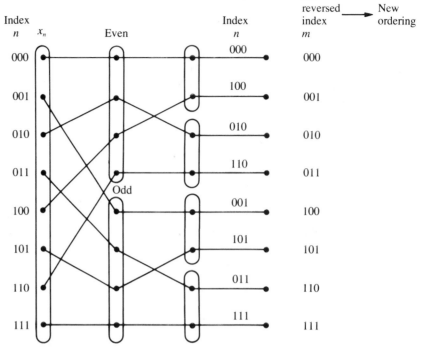

Figure 14.14   The reordering of an eight-term input data sequence $\{x\}$ as it is divided repeatedly into even and odd indexed halves. The indices are written as binary integers.

reordered sequence $\{\mathbf{g}_m\}$. The reordering process is accomplished for a general N by the FORTRAN subroutine INDEX, given in Section 14.8.2.

Using Eq. (14.72), we can now recombine these single-term transforms into two-term transforms. This first recombination has $N = 2$ and summation index $k = 0, 1$. Matching $\mathbf{E}_k$ and $\mathbf{O}_k$ to alternate even and odd terms in $\{\mathbf{g}_m\}$, using each term twice as discussed above, and evaluating $e^{-jk2\pi/N}$, we can form a new sequence $\{\mathbf{G}\}$ from the two-term results:

$$\begin{aligned}
\mathbf{G}_0 &= \mathbf{X}_0 = \mathbf{g}_0 + \mathbf{g}_1 \\
\mathbf{G}_1 &= \mathbf{X}_1 = \mathbf{g}_0 - \mathbf{g}_1 \\
\mathbf{G}_2 &= \mathbf{X}_0 = \mathbf{g}_2 + \mathbf{g}_3 \\
\mathbf{G}_3 &= \mathbf{X}_1 = \mathbf{g}_2 - \mathbf{g}_3 \\
\mathbf{G}_4 &= \mathbf{X}_0 = \mathbf{g}_4 + \mathbf{g}_5 \\
\mathbf{G}_5 &= \mathbf{X}_1 = \mathbf{g}_4 - \mathbf{g}_5 \\
\mathbf{G}_6 &= \mathbf{X}_0 = \mathbf{g}_6 + \mathbf{g}_7 \\
\mathbf{G}_7 &= \mathbf{X}_1 = \mathbf{g}_6 - \mathbf{g}_7
\end{aligned} \tag{14.73}$$

The two-term transforms that make up $\{\mathbf{G}\}$ are next combined into four-term transforms by two applications of Eq. (14.72) with $N = 4$ and the odd terms in $\{\mathbf{G}\}$ defined by a true index bit b. One more cycle with $N = 8$ and odd terms defined by a true c bit completes this 8-element FFT. The $\mathbf{g}_n$ coefficients, which happen to be just $+1$ or -1 in this example, are in general complex constants. The recombination process defined by Eq. 14.72 is accomplished more generally by the FORTRAN subroutine COMBIN listed below.

14.8.2 FORTRAN Implementation Suitable for Microcomputers

Since the FFT is a fundamental tool for all discrete signal analysis, we include a listing of an FFT subprogram that will accomplish this operation. The following version of the FFT program was derived from a listing given in *An Introduction to Discrete Systems,* by K. Steiglitz, but has been modified to run with improved performance on microcomputers that lack floating-point hardware and to return both the real and imaginary components of the DFT for instructional purposes. The code should be compatible with most FORTRAN compilers. A 4-MHz Z80 microcomputer takes about 17 s to obtain a complete 256-point transform.

The inverse FFT (IFFT) is not provided but is easily obtained using the FFT itself. The IFFT is given by

$$\mathbf{x}_k = \frac{1}{N} \sum_{n=0}^{N-1} \mathbf{X}_n \mathbf{W}^{kn} \qquad \text{for } k = 0, 1, 2, \ldots, N-1 \qquad (14.74)$$

and its complex conjugate is

$$\mathbf{x}_k^* = \frac{1}{N} \sum_{n=0}^{N-1} \mathbf{X}_n^* \mathbf{W}^{-kn} \qquad \text{for } k = 0, 1, 2, \ldots, N-1 \qquad (14.75)$$

Since the right side of this expression is like the FFT, the prescription for the IFFT is: (1) Take the complex conjugate of $\{\mathbf{X}\}$ and form the sequence $\{\mathbf{X}^*/N\}$; (2) apply the FFT algorithm to this sequence; and (3) take the complex conjugate of the resulting sequence to obtain $\{\mathbf{x}\}$.

FORTRAN Listing of FFT Subroutines

```
C
              SUBROUTINE FFT(S,FR,FI,NPTS)
C                     S( ) is the signal list of NPTS points
C                     FR( ) is the returned real part of the FFT
C                     FI( ) is the returned imaginary part of the FFT
              DIMENSION S(1),T(1),FR(1),FI(1)
              COMMON /COSIN/CSA(512),N
              DATA NFLAG/0/
C                     Check to see if a new cosine table is needed
              IF(NPTS.EQ.NFLAG) GO TO 80
C                     Point list different length from last CALL
              NFLAG=NPTS
              N=NFLAG
              IF(N.LE.512) GO TO 60
              WRITE(1,1000) N
1000          FORMAT(' YOU HAVE REQUESTED ',I5,' POINTS FOR FFT'
1             /' 512 POINTS IS THE LIMIT')
              RETURN
C                     Calculate a new cosine table
60            A=6.28319/FLOAT(NPTS)
              DO 70 I=1,NPTS
                 AL=A*FLOAT(I-1)
```

```
70                    CSA(I) = SIN(AL)
C                          Rearrange elements of S to give 1-point transforms
80                    CALL INDEX(S,FR,FI)
                      LENGTH = 2
C                          Combine 1-point transforms to give overall transform
100                   DO 120 J = 1,N,LENGTH
                      CALL COMBIN(FR,FI,J,LENGTH)
120                   CONTINUE
C

                      LENGTH = LENGTH + LENGTH
                      IF(LENGTH.LE.N) GO TO 100
                      RETURN
                      END
C
                      SUBROUTINE INDEX(S,FR,FI)
C                          This routine clears {FI} and fills {FR} with reordered {S}
                      DIMENSION S(1),FI(1),FR(1)
                      COMMON /COSIN/CSA(512),N
                      DO 50 IFORT = 1,N
                          FI(IFORT) = 0.0
                          I = IFORT − 1
                          J = 0
                          M2 = 1
10                        M1 = M2
                          M2 = M2 + M2
                          IF(MOD(I,M2).LT.M1) GO TO 20
                          J = J + N/M2
20                        IF(M2.LT.N) GO TO 10
                          JFORT = J + 1
                          FR(IFORT) = S(JFORT)
50                        CONTINUE
                      RETURN
                      END
C
                      SUBROUTINE COMBIN(FR,FI,J,N)
C                          This routine combines even and odd terms to give transform
                      DIMENSION FR(1),FI(1)
                      COMMON /COSIN/CSA(512),NPTS
                      ND4 = NPTS/4
                      NPN = NPTS/N
                      N2 = N/2
                      DO 10 L = 1,N2
                          LOC1 = L + J − 1
                          LOC2 = LOC1 + N2
                          NA = NPN*(L − 1) + 1
                          EMI = − CSA(NA)
                          NA = NA + ND4
                          EMR = CSA(NA)
                          ZR = EMR*FR(LOC2) − EMI*FI(LOC2)
                          ZI = EMR*FI(LOC2) + EMI*FR(LOC2)
                          FR(LOC2) = FR(LOC1) − ZR
                          FI(LOC2) = FI(LOC1) − ZI
                          FR(LOC1) = FR(LOC1) + ZR
                          FI(LOC1) = FI(LOC1) + ZI
10                        CONTINUE
                      RETURN
                      END
```

14.9 *Cyclic Convolution and Correlation*

The convolution and correlation integrals discussed in Chapter 13 can be evaluated numerically by considering discrete sampled signals rather than continuous ones. Since the techniques are very similar for both these integrals, we will discuss only one: the discrete signal correlation integral.

With T defined here as the interval between samples, the cross-correlation function of Eq. (13.90) can be approximated by the summation

$$r(\tau) = \frac{1}{NT} \sum_{k=0}^{N-1} x(kT + \tau)\, y(kT)\, T \tag{14.76}$$

where we have now limited the discussion to real signals as in the previous chapter. For convenience we choose to evaluate $r(\tau)$ only at times $\tau = nT$, giving the discrete result

$$r_n = r(nT) = \frac{1}{N} \sum_{k=0}^{N-1} x_{k+n}\, y_k \qquad \text{for } n = 0, 1, 2, \ldots, N-1 \tag{14.77}$$

Note that the factors of T have been canceled.

Given two signals $\{x\}$ and $\{y\}$, the correlation sequence $\{r\}$ can be directly evaluated in a computer. However, the evaluation of all N terms of $\{r\}$ requires N^2 operations much like the DFT. It is therefore reasonable to perform this evaluation using several FFT transforms rather than the direct sum indicated here. By analogy with the continuous signal theorem (the proof for discrete signals is left as a problem),

$$\text{DFT}[r] = \frac{1}{N} \text{DFT}[x]\, \text{DFT}[y]^* \tag{14.78}$$

and using capital letters to indicate the DFT of small-letter sequences, we can write

$$\mathbf{R}_m = \frac{1}{N} \mathbf{X}_m\, \mathbf{Y}_m^* \qquad \text{for } m = 0, 1, 2, \ldots, N-1 \tag{14.79}$$

The IDFT of $\{\mathbf{R}\}$ is then the desired correlation sequence $\{r\}$. Given an FFT subroutine, it can be applied three times (once as an inverse) to evaluate the correlation integral, saving both computer time and software programming effort.

Evaluation of the convolution integral differs from the correlation result of Eq. 14.79 only in that $\mathbf{Y}_m$ are used without taking their complex conjugates.

Evaluation of the correlation and convolution integrals using the DFT does have one side effect that changes the shape of the integral functions. As a result, when evaluated in this manner the processes are known as cyclic correlation and cyclic convolution.

As we have already seen, the DFT of an N-term time sequence $\{\mathbf{x}_n\}$ results in an N-term frequency sequence $\{\mathbf{X}_k\}$ that is cyclic for index values outside the range

$0 \leq k < N$. Likewise, the IDFT (which differs from the DFT only by complex conjugation) transforms $\{X_k\}$ back into an $\{x_n\}$ that is also cyclic outside the range $0 \leq n < N$. This cyclic character can dramatically distort the shape of the correlation and convolution functions as shown in Figures 14.15a and 14.15b. Fortunately, the problem is easily corrected by adding N zeros to each discrete sample sequence as shown in Figure 14.15c. The process of padding with zeros doubles the length of the data sequences and slows the DFT calculations, thereby eliminating some of the speed advantage of this method. Still, we note that generating

Figure 14.15   (a) Two typical continuous functions and their correlation function r(τ). (b) Discrete versions of the same signals, sampled 128 times, must be presumed to repeat outside the sample interval; this produces a cyclic correlation function, different from the continuous one. (c) The solution is to pad each signal with 128 leading zeros as shown. This yields a discrete correlation function whose first half matches the shape of the continuous one.

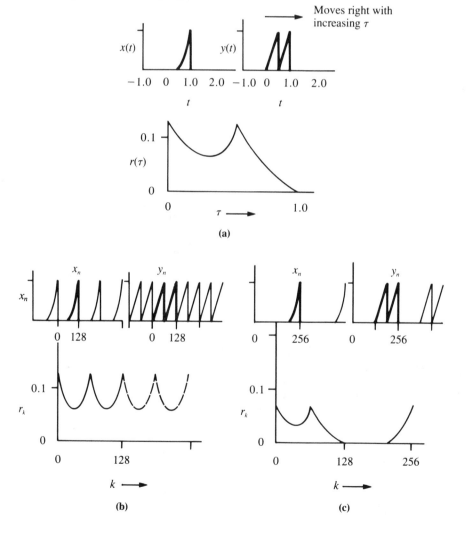

128 points on the correlation curve of Figure 14.15a by numerical integration took about 2.5 times longer than the equivalent FFT result of Figure 14.15c.

607

Problems

14.10   Applications

Whenever we deal with real laboratory signals (as opposed to mathemtaical ideal-izations), of necessity we must work with discrete signals. In many experiments the problems inherent to discrete data are minimized by the use of very small sampling intervals relative to the other times in the experiment, but in most cases the techniques of discrete signal analysis will not only speed the data processing but improve the understanding of the results. Using the FFT we can move freely between the time and frequency domains, producing impulse responses, frequency spectra, autocorrelation and cross-correlation functions, and convolution functions with ease. The range of applications is indicated by Table 14.3.

Table 14.3   Summary of Signal Analysis Applications

Time Domain	Frequency Domain
Impulse response $h(t)$	Transfer function $\mathbf{H}(j\omega)$
	Define transfer function of filter or amplifier.
Autocorrelation $r_{xx}(\tau)$	Power spectra $R_{xx}(j\omega)$
	Noise reduction in unknown periodic signals.
	Phase-independent frequency spectrum.
Cross-correlation $r_{xy}(\tau)$	Cross-power spectra $\mathbf{R}_{xy}(j\omega)$
	Signal detection in noisy data.
	Signal pulse shape known—find each occurrence.
	Periodicity known—find pulse shape.
	Extraction of $\mathbf{H}(j\omega)$ or $h(t)$ from cross-correlation of white noise signal in to signal out.
Convolution $c(\tau)$	$\mathbf{Y(s)} = \mathbf{H(s)X(s)}$
	Predict output $y(t)$ knowing $h(t)$ and input $x(t)$.
	Deconvolution to infer $x(t)$ from measurement of $y(t)$ and $h(t)$.

Problems

1. Sketch two cycles of the analog signal $x(t) = \cos(\omega t + \pi/4)$ and show the locations of samples taken at the Nyquist frequency. Take the first sample at $t = 0$.

2. The digital filter described by Figure 14.7 has no output at an angular frequency $\omega = \pi/T$. Sketch the sample points for an even sinusoidal oscillation at this frequency and explain in words why $|\mathbf{H}(j\omega)| = 0$.

3. An otherwise constant analog signal is contaminated by distorted 60-Hz pickup having a significant frequency content at 60 Hz and 120 Hz. If the signal is sampled at 110 Hz, what frequencies will be found in the sampled signal?

4. If the interesting frequency content of an analog signal lies in the range 1250 Hz $\leq f$ $\leq$ 1500 Hz: *(a)* What is the lowest sampling frequency that could be used to study this signal? *(b)* When sampled at this frequency, what will be the apparent frequency of a 1300-Hz component in the analog signal? [*Ans.:* 500 Hz, 200 Hz]

5. Determine the phase shift θ associated with the transfer function of the filter shown in Figure 14.7.

6. *(a)* Show that the frequency response of the small difference expression

$$y_n = x_n - x_{n-1}$$

is given by $|\mathbf{H}| = 2 \sin(\omega T/2)$.
(b) Show that the phase shift θ associated with this transfer function is $(\pi - \omega T)/2$.
(c) Plot both of these results on the range $0 \leq \omega T \leq 2\pi$.

7. Write the z-transform of the sequence $\{x\} = 0, 0, 0, 0, 0, 1, 1, 1, 1, 1$ and of the delayed sequence $\{x\} = 0, 0, 0, 0, 0, 0, 1, 1, 1, 1$.

8. If $x_n = 0$ when $n < 0$, derive the z-transform shift theorem as given in Table 14.1.

9. When the input is $\{x\} = 0, 0, 0, 0, 0, 1, 1, 1, 1, 1$, plot the output $\{y\}$ sequence for the filter algorithm of Example 14.2 when $cT = 1$ and again when $cT = 2$.

10. A two-pole, critically damped, low-pass analog filter has the transfer function $H_a(\mathbf{s}) = 1/(\mathbf{s} + c)^2$. Use Table 12.2 and the impulse response method to derive the recursive filter algorithm,

$$y_n = T^2 e^{-cT} x_{n-1} + 2e^{-cT} y_{n-1} - e^{-2cT} y_{n-2}$$

You will need the series sum result $\sum_{n=0}^{\infty} nx^{-n} = x/(x - 1)^2$.

11. *(a)* Find a recursive algorithm for the filter of the previous problem using the pole-mapping method. Make $|\mathbf{G}(1)| = |\mathbf{H}(0)|$. *(b)* If $cT << 1$, show that this algorithm reduces to the result of the previous problem. [*Hint:* To get a match, you will need to relabel the $\{x\}$ sequence.]

12. *(a)* Use the pole-zero mapping method to determine a discrete filter algorithm corresponding to the analog filter $\mathbf{H}_a = \mathbf{s}/(\mathbf{s} + c)$. Your algorithm should take account of the observation that $|\mathbf{H}_a| \to 1$ as $\mathbf{s} \to \infty$. *(b)* If $cT = 2$, evaluate the output sequence $\{y\}$ resulting from an eight-term unit step function $\{x\} = 0, 0, 0, 1, 1, 1, 1, 1$.

13. *(a)* Using the bilinear transformation technique, derive a filter algorithm corresponding to the analog transfer function $\mathbf{H}_a = c/(\mathbf{s} + c)$. Make $|\mathbf{G}(1)| = |\mathbf{H}(0)|$. *(b)* Reduce this expression for the special case of $cT = 0.1$ and compare with the results of Example 14.2 for the same case. *(c)* For both filters, calculate the $\{y\}$ sequence that results when $\{x\} = 0, 0, 0, 0, 0, 1, 1, 1, 1, 1$.

14. In general, the DFT of a sequence $\{x\}$ will be a complex sequence whose elements are $\mathbf{X}_k = A_k + jB_k$. Use Eq. (14.52) to evaluate and plot the sequences $\{A\}$ and $\{B\}$ as determined from the DFT of the 10-term sequence $\{x\} = 1, 0, 0, 0, 0, 0, 0, 0, 0, 0$.

15. Given an N-term $\{x\}$ sequence with all terms equal to zero except the first and last, which are 1:
(*a*) Expand the DFT summation and write an expression for $\mathbf{X}_k$ in terms of k and N.

(b) Write a similar expression for $X_k{}^2$.

(c) If $N = 10$, plot both x_k and $X_k{}^2$.

16. If $\{\mathbf{X}\} = \text{DFT}[\{\mathbf{x}\}]$, show that

$$\{\mathbf{x}\} = \frac{1}{N}\,(\text{DFT}[\{\mathbf{X}^*\}])^*$$

17. Evaluate graphically the bracketed term of Eq. (14.61) for the case $N = 4$ and $k = 3$.

18. *(a)* If the output $y(t)$ of a four-terminal network is measured and the transfer function $H(j\omega)$ of the network is known, write an expression for the original signal $x(t)$. Use a symbol like $\mathcal{F}$ to represent the Fourier transform. *(b)* Assuming that you have an FFT computer subroutine, rewrite this expression in terms of the DFT. (No, you do not have an IDFT routine!)

19. If $\mathbf{H}$ is the frequency-domain transfer function of a four-terminal network, R_{xx} is the power spectrum of the input signal $x(t)$, and R_{yy} is the power spectrum of the output signal $y(t)$, show that $\mathbf{H}^2 = R_{yy}/R_{xx}$.

20. Given a four-term sequence $\{x\} = 2, 3, 4, 5$, evaluate the four X_k using Eq. (14.52), then again from two two-term transforms following the prescription of Eq. (14.71).

21. Starting with the N-term cross-correlation sequence

$$r_n = \frac{1}{N}\sum_{l=0}^{N-1} x_{l+n}y_l \qquad \text{for } n = 0, 1, 2, \ldots, N - 1$$

show that the N-term cross-power sequence is given by

$$\mathbf{R}_m = \frac{1}{N}\,\mathbf{X}_m\,\mathbf{Y}_m{}^*$$

where $\mathbf{R}_m$, $\mathbf{X}_m$, and $\mathbf{Y}_m$ represent sequences obtained by applying the DFT to the corresponding small-letter sequences.

appendix a

Solution of Three Equations in Three Unknowns by the Method of Determinants

The method of determinants can be used to solve any set of n linear equations in n unknowns. As a practical matter, however, two equations in two unknowns can be easily solved by more primitive methods, and problems with more than three unknowns are prime candidates for the digital computer. Although a problem with three unknowns can also be relegated to the computer, an easily remembered algorithm for evaluating a 3×3 determinant simplifies the manual procedure.

Given a set of three equations in three unknowns x, y, and z, the first step is to put the equations in the standard form

$$A_{11}x + A_{12}y + A_{13}z = C_1$$
$$A_{21}x + A_{22}y + A_{23}z = C_2$$
$$A_{31}x + A_{32}y + A_{33}z = C_3$$

The solution for the variable x is then given by the ratio

$$x = \frac{|A1|}{|A|}$$

where $|A|$ is the 3×3 determinant of the matrix of coefficients A_{ij}, and $|A1|$ is the determinant of the same matrix with the first column replaced by the constant coefficients C_i. The solutions for y and z are similarly given by

$$y = \frac{|A2|}{|A|}$$

and

$$z = \frac{|A3|}{|A|}$$

The problem is thus reduced to one of evaluating four 3 × 3 determinants.

These 3 × 3 determinants are relatively easy to evaluate using the often ne-glected memory aid shown on Figure A.1. The method is an easily remembered extension of the 2 × 2 determinant evaluation rule shown on the same figure, but it is not generally taught, perhaps because the seemingly obvious extension to higher-order determinants does not work. As indicated in the figure, the three positive terms of the determinant sum are obtained from the top-left to bottom-right diagonals, two of which must be extended and looped back along a second diagonal to obtain a three-factor product. The three negative terms are similarly obtained from suitably extended bottom-left to top-right diagonals.

Figure A.1   A memory aid for evaluating a 2 × 2 and a 3 × 3 determinant. Positive terms in the determinant sum are linked by solid lines; negative terms are linked by dashed lines.

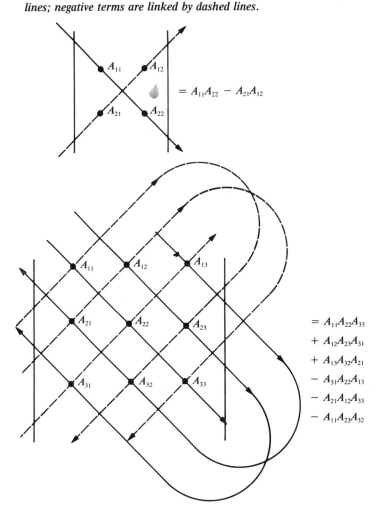

appendix b

The Miller Effect

The Miller effect provides a useful tool for calculating the input impedance of an inverting amplifier with feedback. The method can be applied to bipolar transistors, field effect transistors, or operational amplifiers; but we will derive the effect for an idealized, infinite input impedance, amplifier symbolized by the triangle in Figure B.1a. Since the Miller effect is relevant only to inverting amplifiers, we here define the amplifier's transfer function to be $-\mathbf{A}$, such that

$$\mathbf{v}_{\text{out}} = -\mathbf{A}\mathbf{v}_{\text{in}} \tag{B.1}$$

where the real part of $\mathbf{A}$ is positive. If the circuit is such that $\mathbf{A}$ changes when $\mathbf{Z}_F$ is connected (not uncommon for single transistor amplifiers), then $\mathbf{A}$ must be determined with $\mathbf{Z}_F$ in place.

With the amplifier connected as shown, the input is given by

$$\mathbf{i}_{\text{in}} = \frac{\mathbf{v}_{\text{in}} - \mathbf{v}_{\text{out}}}{\mathbf{Z}_F} \tag{B.2}$$

and substituting $\mathbf{v}_{\text{out}}$ from the previous equation gives

$$\mathbf{i}_{\text{in}} = \frac{\mathbf{v}_{\text{in}} + \mathbf{A}\mathbf{v}_{\text{in}}}{\mathbf{Z}_F} \tag{B.3}$$

Using this expression, the input impedance to the circuit $\mathbf{v}_{\text{in}}/\mathbf{i}_{\text{in}}$ (not just the amplifier) can be calculated. Because our assumed amplifier has an infinite input impedance, the input impedance looking into terminals ab is caused entirely by the feedback impedance $\mathbf{Z}_F$; we therefore label it $\mathbf{Z}_M$ for Miller:

$$\mathbf{Z}_M = \frac{\mathbf{v}_{\text{in}}}{\mathbf{i}_{\text{in}}} = \frac{\mathbf{Z}_F}{1 + \mathbf{A}} \tag{B.4}$$

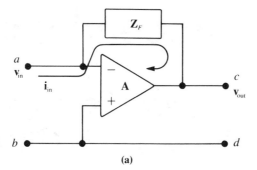

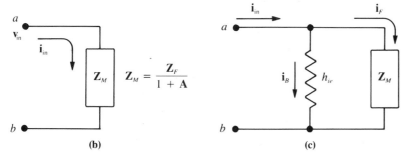

Figure B.1 (a) An ideal amplifier connected with a negative feedback impedance. (b) This circuit's input impedance Z_M is given by Miller's theorem. (c) The combined input impedances of a real amplifier.

As shown in Figure B.1b, the input looks like the feedback impedance reduced by the gain of the amplifier.

If the amplifier itself has a finite input impedance such as h_{ie} for a transistor, it would appear to be in parallel with $\mathbf{Z}_M$ as shown in Figure B.1c.

appendix c

Bipolar Transistor Amplifier with Collector-to-Base Feedback

Because of parasitic capacitance, collector-to-base feedback is present to some degree on every common emitter amplifier, usually with a deletory effect on the amplifier's high-frequency response. Collector-to-base feedback can also be implemented with an explicit element as shown in Figures 7.9b and 7.9c. To avoid complex variables, the following analysis is developed for resistive feedback, but the equations remain valid if resistances are replaced by complex impedances.

The Miller effect can be used to simplify the analysis. We first need to determine the voltage ratio $A = -v_C/v_B$ as defined by Eq. (B.1). The perfect transistor model in Figure C.1 is suitable for this purpose. A signal source v_S and its output resistor R_S have been explicitly included in this figure, since we will ultimately be interested in the overall transfer function v_C/v_S. From the figure we see that the base voltage is given by

$$v_B = h_{ie}i_B \qquad (C.1)$$

and that the collector voltage can be written

$$v_C = -R_C\left(h_{fe}i_B + \frac{v_C - v_B}{R_F}\right). \qquad (C.2)$$

Eliminating i_B between these two equations and simplifying yields the transfer function $-A$:

$$\frac{v_C}{v_B} = -A = -\frac{R_C(h_{fe}R_F - h_{ie})}{h_{ie}(R_C + R_F)} \qquad (C.3)$$

The input impedance to this circuit looking into the base of the transistor can now be found with the aid of Miller's theorem. This single-transistor amplifier

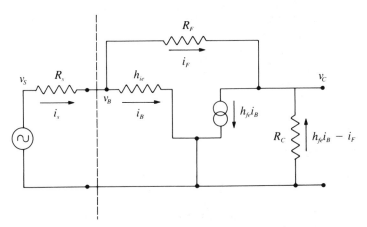

Figure C.1 *The AC equivalent circuit used to calculate the voltage gain for the CE circuit with resistive feedback between collector and base.*

differs from the ideal used to find Z_M in Figure B.1 because it draws some current i_B from i_S. However, one end of h_{ie} is connected to ground in Figure C.1, and its effect on the input impedance is easily included, as indicated on the equivalent sketch of Figure B.1c. Only the algebra remains: Find Z_M by substituting Eq. (C.3) into (B.4), then calculate the parallel combination of Z_M and h_{ie}. Assuming that $h_{fe} \gg 1$ and that $h_{fe}R_C \gg h_{ie}$, the result is

$$R_{\text{in}} = \frac{h_{ie}(R_C + R_F)}{h_{fe}R_C + R_F} \tag{C.4}$$

Further simplification of this equation can be made only after making specific choices for the circuit components.

Most circuits will satisfy the condition

$$R_F \ll h_{fe}R_C \tag{C.5}$$

allowing Eq. (C.4) to be reduced to

$$R_{\text{in}} = \frac{h_{ie}(R_C + R_F)}{h_{fe}R_C} \tag{C.6}$$

which is seen to be several times smaller than h_{ie}.

The collector-base feedback element is thus seen to have two major effects: It decreases the gain $-A$ given by Eq. (C.3); and it reduces the input impedance looking into the transistor base. Both of these effects are important to the overall voltage gain v_C/v_S. This transfer function can be written as the product of two terms,

$$\frac{v_C}{v_S} = \frac{v_C}{v_B}\frac{v_B}{v_S} \tag{C.7}$$

where the first term is given by Eq. (C.3) and the second is just

$$\frac{v_B}{v_S} = \frac{R_{\text{in}}}{R_S + R_{\text{in}}} \tag{C.8}$$

Substitution and algebraic reduction produces the overall transfer function

$$H = \frac{v_C}{v_S} = -\frac{R_C(h_{fe}R_F - h_{ie})}{h_{ie}(R_C + R_F) + R_S(R_F + h_{fe}R_C)} \tag{C.9}$$

If the terms involving h_{fe} dominate this expression, it will reduce to

$$H = \frac{v_C}{v_S} = -\frac{R_F}{R_S} \tag{C.10}$$

The output impedance of this circuit can be found from

$$R_{out} = \frac{v(\text{open})}{i(\text{short})} \tag{C.11}$$

Equation (C.10) provides us with an approximate expression for the open-circuit output voltage in terms of a fixed v_S, leaving only the task of finding $i(\text{short})$. When the output is shorted to ground, R_F and h_{ie} become parallel resistors between v_B and ground. If $R_F \gg h_{ie}$, we can neglect the current i_F, with the result that

$$i_B(\text{short}) = \frac{v_S}{R_S + h_{ie}} \tag{C.12}$$

Again neglecting the relatively small current i_F through the shorting wire, we have

$$i(\text{short}) = -\frac{h_{fe}v_S}{R_S + h_{ie}} \tag{C.13}$$

giving an output impedance of

$$R_{out} = \frac{R_F(R_S + h_{ie})}{h_{fe}R_S} \tag{C.14}$$

Assuming further that $h_{ie} \ll R_S$, this expression reduces to

$$R_{out} = \frac{R_F}{h_{fe}} \tag{C.15}$$

which can be much smaller than the normal common emitter output impedance of R_C.

Bibliography

Analog-Digital Conversion Notes, Analog Devices, Norwood, Mass., 1977.
  Covers a broad spectrum of data acquisition subsystems, mostly at the block-diagram level.

Bendat, J. S., and A. G. Piersol: *Measurement and Analysis of Random Data,* John Wiley & Sons, New York, 1966.
  Statistical analysis of continuous signals directed toward general experimental design.

Brigham, E. Oran: *The Fast Fourier Transform,* Prentice-Hall, Englewood Cliffs, N.J., 1974.
  Includes discussion of Fourier transforms, convolution, and correlation.

Brignell, J. E., and G. M. Rhodes: *Laboratory On-line Computing,* International Textbook Company, London, 1975.
  Practical discussion of on-line computing techniques with overview of mathematical techniques.

Champeney, D. C.: *Fourier Transforms and their Physical Applications,* Academic Press, New York, 1973.
  Contains a helpful pictorial table of Fourier transform pairs.

D.A.T.A. Book, D.A.T.A., Inc., San Diego, Calif., yearly.
  A set of several books giving one-line listings of diodes, transistors, opto-electronic devices, etc.

Fink, D. G., and D. Christiansen: *Electronics Engineers' Handbook,* McGraw-Hill Book Company, New York, 1982.
  Detailed information with references on many topics, including a chapter on transducers.

Graeme, J. G., G. E. Tobey, and L. P. Huelsman: *Operational Amplifiers, Design and Applications,* McGraw-Hill Book Company, New York, 1971.
  One of the more helpful books on operational amplifier applications.

Horowitz, P., and W. Hill: *The Art of Electronics,* Cambridge University Press, New York, 1980.

Extremely useful reference containing many circuits, ideas, and design hints presented with almost no mathematics.

Lindmayer, J., and C. Y. Wrigley: *Fundamentals of Semiconductor Devices,* D. Van Nostrand Company, Princeton, N.J., 1965.

Discusses semiconductors and device operation using both qualitative and mathematical arguments.

Motchenbacher, C. D., and F. C. Fitchen: *Low-Noise Electronic Design,* John Wiley & Sons, New York, 1973.

Detailed discussion of low-noise design techniques.

Parratt, L. G.: *Probability and Experimental Errors in Science,* John Wiley & Sons, New York, 1961.

Introductory text in probability and statistics.

Rabiner, L. R., and C. M. Rader: *Digital Signal Processing,* Institute of Electrical and Electronic Engineers Press, New York, 1972.

Useful collection of articles from the professional literature.

Reference Data for Radio Engineers, Howard W. Sams & Co., Indianapolis, Ind., 1979.

Engineering tables, graphs, and discussion biased toward radio frequency and transmission lines.

Robinson, F. N. H.: *Noise and Fluctuations,* Clarendon Press, Oxford, 1974.

The physical origins of noise and its effects on electronic systems.

Steiglitz, K.: *An Introduction to Discrete Systems,* John Wiley & Sons, New York, 1974.

Digital filters, z-transform, and the fast Fourier transform.

Williams, A. B.: *Electronic Filter Design Handbook, McGraw-Hill Book Company, New York, 1981.*

Tables, graphs, and circuit configurations needed to design multipole analog filters with specific characteristics.

Index

Operational Amplifier

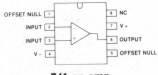

741 op amp

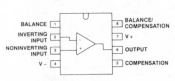

**NE5534 low noise,
10 MHz, bipolar op amp**

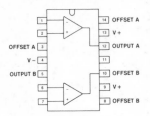

747 dual op amp

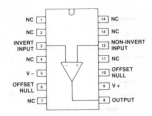

**LH0022 low noise
FET op amp**

π-network Attenuation

Attenuation	R_1/R_T	R_2/R_T
2X	3.000	0.750
4X	1.667	1.875
6X	1.400	2.917
8X	1.286	3.938
10X	1.222	4.950
20X	1.105	9.975
50X	1.041	24.99
100X	1.020	50.00
1000X	1.002	500.0

7400 Logic Families

Output Device	Gate Delay ns	Maximum FF clock MHz	Effective Input Pull-up Ohms	74	74LS	74ALS	74F
Family Characteristics				*Typical Fan-out* to Load Device of Type*			
74	10	35	4k	10	40	80	26
74LS	10	45	18k	5	20	40	13
74ALS	4	50	40k	5	20	40	13
74F	4	125	10k	12	50	100	33
74C	50	4	**	0	2	4	1
74HC	8	70	**	2	10	20	6

*Devices designated as buffers can typically drive 3 times the indicated fan-out.
**C and HC devices are CMOS with large input impedances.

Dual Inline Package (DIP) Outlines of 74LSxxx Logic
Notation: *Open collector outputs; ‡(double dagger);
Buffer output; †(dagger) Schmitt trigger inputs.

Combinational Logic

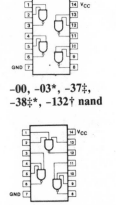

**−00, −03*, −37‡,
−38‡*, −132† nand**

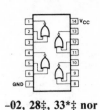

−02, 28‡, 33*‡ nor

**−04, −05*, −14†
inverter**

−08, −09* and

−10, −12* nand

−11, −15* and

**−13†, −20, −22*,
−40‡ nand**

−27 nor

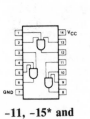

Fortney: Principles of Electronics   Harcourt Brace Jovanovich